POST-OIL ENERGY TECHNOLOGY

TECHNOLOGY

The World's First Solar–Hydrogen Demonstration Power Plant

POST-OIL ENERGY TECHNOLOGY

TECHNOLOGY

The World's First Solar–Hydrogen Demonstration Power Plant

BÉLA LIPTÁK

CRC Press
Taylor & Francis Group
Boca Raton   London   New York

CRC Press is an imprint of the
Taylor & Francis Group, an **informa** business

CRC Press
Taylor & Francis Group
6000 Broken Sound Parkway NW, Suite 300
Boca Raton, FL 33487-2742

© 2009 by Taylor & Francis Group, LLC
CRC Press is an imprint of Taylor & Francis Group, an Informa business

Library of Congress Cataloging-in-Publication Data

Liptak, Bela G.
   Post-oil energy technology : after the age of fossil fuels / Bela G. Liptak.
      p. cm.
   Includes bibliographical references and index.
   ISBN 978-1-4200-7025-5 (alk. paper)
   1. Power resources--Forecasting. 2. Energy industries--Technological
innovations. 3. Energy industries--Forecasting. I. Title.

TJ163.24.L56 2008
621.042--dc22   2008003669

Visit the Taylor & Francis Web site at
http://www.taylorandfrancis.com

and the CRC Press Web site at
http://www.crcpress.com

Dedicated to all parents and grandparents who want their children and grandchildren to live on a peaceful, happy, and healthy planet

Contents

The Author

Béla Lipták arrived in the United States at the age of 20 after the 1956 Hungarian Revolution. He was one of the students who drafted the "16 Points," the document which summarized the goals of that revolution. After arriving in the United States he became the first president of the Hungarian Student Association and continued to work for the liberation of Hungary and for the collective human rights of Hungarian minorities in the Danubean region.

He has published and/or edited 27 books, including the *Environmental Engineer's Handbook,* and prepared the "Compromise Plan" for the Danube River, which was diverted by Slovakia into a concrete-lined bypass canal. The "Compromise Plan" resolves the conflict between the shipping and power generation considerations on the one hand and environmental concerns on the other.

He was an engineering department head at Crawford & Russel, an adjunct professor at Yale University, and a course director at the Center for Professional Advancement. In 1995 he was invited to teach as a Fulbright Scholar. He was a book reviewer for *American Scientist,* received the Control Hall of Fame Award in 2001 and the Life Time Achievement Award of ISA in 2005. Today, he is the president of the consulting firm Lipták Associates P.C. His work includes the design and optimization of industrial processes, energy conservation, pollution prevention, and safety-related projects.

Protecting the future of his grandchildren—Ivan, Katie, Ava, Emmi, and Adam—is his main reason for writing this book.

Introduction

In the beginning, human evolution was directed by Nature. During the Industrial Age, however, we humans started to rule Nature, change the climate and influence our own evolution by changing the ecosystems. Our individual and national selfishness created a polluted world whose future is at risk. The Earth should not become a product of human design; it is time to make our peace with the planet. Lately, Nature has issued warning signals and we should listen! We should listen and understand that in this age of "self-guided evolution," the survival of human civilization is up to us. The planet is not at risk; we are!

Today we live far beyond our means and the damage we inflict on the planet could become irreversible. Understanding the price that future generations could pay for the indoor ski slopes in the Persian Gulf and for the executive jets filling the sky is just as essential as the knowledge that we still have time to solve the problems that we have created. We should not give reason for our grandchildren to ask: "Why didn't you act?" We should find the moral courage to resolve this crisis.

During the last century, the world population quadrupled, and during the last 20 years it has increased by another third. During the last 50 years, global energy consumption also quadrupled and the carbon dioxide concentration of the atmosphere has reached the highest level in 650,000 years. Together with overpopulation and nuclear proliferation, global warming is one of the most serious problems we face. Climate change has no military solution! Nuclear warheads will not reduce carbon emissions or increase the size of our fossil or nuclear fuel deposits. They will only increase the probability of energy wars.

Overpopulation and unsustainable consumption are stressing the planet and if the "tipping point" is reached, natural disasters can occur. Our children have the right to hold us responsible if we neglect to do what needs to be done. It is essential that during the next 50 to 100 years, we convert our exhaustible fossil fuel and nuclear-based economy to one that is inexhaustible, safe, and clean. This book describes such an economy based on solar–hydrogen. We have to do this because even if global warming was not taking place, our fossil and nuclear fuel deposits will run out.

We have to understand that global warming is not like a pendulum. Global temperature by itself will not return to where it was. The thermal processes of our planet have such immense inertia that they cannot be stopped overnight. On the other hand, a well-planned, orderly transformation can gradually slow the warming trend and stabilize our climate.

We should also understand that the transition to a clean and renewable economy need not disrupt, but in fact can improve, the world economy. The building of new technologies and new infrastructure can create new jobs, new markets, and a global economic boom that will advance the Third World.

The U.N. Human Development report of 2007 concluded that if the global temperature was to rise by only 2°C, 600 million people would go hungry, 200 million would be displaced by floods, and 400 million would be exposed to malaria, dengue, and other diseases—not only in the Third World. It has, for example, been reported that tropical viruses have moved into Italy. The U.N. also indicated that in case of inaction the planet would warm 6°C by 2030 and the consequences would be catastrophic. There is no question that our goal must be to stabilize the climate. On the other hand, during the required transition to a sustainable lifestyle, we can live with a few degrees of rise in global temperature and with a few inches of increase in the level of the oceans. What is needed is a gradual transition that, on the one hand, accepts that the conversion cannot occur overnight, that we might have to raise some levees and embankments but, on the other hand, aims at the longer-range goal of stabilizing carbon emissions.

Therefore, once the demonstration plant described in this volume is built and we have proved that renewable energy power plants do not cost more than fossil or nuclear ones, from that point on, all new power plants should be renewable. This book also emphasizes that the transition can be speeded by energy conservation. While the ultimate solution is a solar–hydrogen economy, the immediate tasks include the use of the existing technologies serving energy conservation.

Our planet receives more solar energy in about a half an hour than humankind uses in a year. The solar energy that can be collected on 1% of the Sahara is sufficient to meet our global electricity needs. Solar collectors covering a small part of the arid Southwest of the United States are all that are needed to supply all the electricity needed by the nation. If, in addition to the gradual conversion to a solar–hydrogen economy, all new homes are weatherized and covered with solar roofs, if we start building the infrastructure for millions of plug-in electric cars, the transition will not only be manageable but will also create the economic boom of the century.

This book is organized into five chapters. The first chapter describes the causes of global warming, its consequences, the steps various governments have taken to slow it, and the impacts of these steps. This chapter also describes the existing conservation technologies and the existing renewable energy processes, including the ones needed to build a solar–hydrogen demonstration power plant. In addition to the present technologies, new ones are also evolving, such as ultrathin ink-based semiconductor collectors or the reversible fuel cells, which during the day generate hydrogen and at night use it to make electricity. If this trend is combined with drastic cost reduction from mass production, the transition can be cost effective.

The next two chapters give specific technical information, describing energy optimization techniques that could reduce the energy consumption of the industrialized countries by 25% and that of the Third World by even more. For example, the energy consumption of the Chinese and that of former

Soviet-era buildings can be cut in half, and the energy use of boilers, chillers, compressors, refineries, and conventional power plants can be optimized.

The second goal of Chapters 2 and 3 is to show that we already have the know-how to build a full-sized, renewable energy demonstration plant. We know how to optimize large solar power plants, which might consist of some 25,000 disk collectors. We also know how to herd the collectors to track the Sun and how to provide solar energy storage in hydrogen and other forms. We know how to operate reversible electrolyzers that can convert solar energy to hydrogen and can also act as fuel cells, converting hydrogen back into electricity when needed.

We also have the technology to build millions of "energy-free" homes, and the electric grid can provide for their energy storage. Lastly, we also know how to burn hydrogen in modern fuel cells, which operate at twice the efficiency of gasoline engines and emit no pollutant, only distilled water.

The fourth chapter of this book provides the detailed design of what I hope will be the world's first full-sized solar–hydrogen demonstration power plant. In this chapter and elsewhere in this volume, I provide estimates of the present efficiencies and costs. However, the main goal of this book is to move the whole topic of the renewable energy economy from estimates to proven facts. Once this demonstration plant is built, we will have these facts.

The concluding fifth chapter compares the energy options available to mankind. It provides quantitative data on the present trends of CO_2 emissions, energy consumption and population growth and on the consequences of continued reliance on exhaustible (fossil and nuclear) energy sources. I also explain why dependence on thermal nuclear energy is likely to lead to dependence on plutonium-fueled breeder reactors. The chapter calculates the costs and time needed to convert to a totally renewable energy economy and also discusses the consequences of inaction.

After reading this book, it is hoped that the reader will not only realize the moral imperative for action, but will also understand that the conversion to the solar–hydrogen economy is technically feasible and economical, and can be done in a calm, orderly fashion.

To achieve these goals—to start this, the third Industrial Revolution—will require vision and commitment, but so did the landing on the moon. The scale of the effort could exceed that of the Marshall Plan. It is debatable how much time we have or how much climate change we can live with. It is also debatable how much of our economic resources should be devoted to stabilizing or reversing humankind's growing carbon footprint. What is not debatable is that we have to do it if humankind's future is to be different from that of the dinosaurs!

1

The Case for Renewable Energy Processes

1.1 Global Trends

During the last 100 years, we have created a global warming time bomb while nearly exhausting our energy reserves. The population of the planet and the global energy consumption both quadrupled, and the global gross world product (GWP) increased sixfold. We treat the atmosphere, the rivers, and the ocean as "open sewers" while consuming 5.5 billion tons of coal, 33 billion barrels of oil, and 100 trillion cubic foot of natural gas every year. The total yearly electricity consumption is 15 trillion kWh and that is also estimated to triple during the next 50 years. While nuclear warheads will not increase our energy reserves, there are enough in storage to destroy the planet 25 times over, and now the arms race is also expanding into space. In 2008 the total global GWP was about $68 billion, the GDP (gross domestic product) of the United States was about $15 trillion and her military budget for 2009 exceeds that for Social Security. The public believes that the main problems we face are the economy and the wars without realizing that both are the result of the energy crisis.

This trend of increasing population and per capita consumption of energy and other resources is approaching the "tipping point." The climate-driven disruption of food, water, and land resources is threatening stability by causing mass migration, disease, and malnutrition. Islands and coastal regions face flooding and hurricane damage. Changing rainfall patterns are turning the semiarid regions of the planet into arid ones, the melting of glaciers threatens drastic reductions in the water supplies of vast areas, including India and China, and even the fishing rate of the oceans has exceeded the sustainable limit. We are racing toward self-destruction like a car with its gas pedal floored and stuck.

Today the per capita yearly energy use on the planet ranges from 1,000 kWh/yr in Africa to 16,000 kWh/yr in North America. These numbers alone should convince us that the advanced industrial nations have not only the need, but also the duty to correct the damage they have done and the only way to do that is to implement clean and inexhaustible energy technologies.

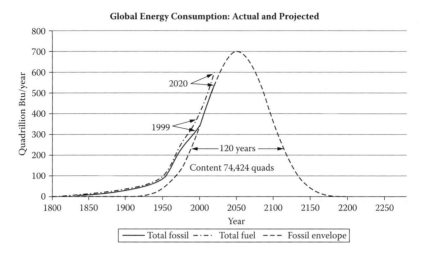

FIGURE 1.1
Availability of the known fossil fuel reserves. (From NASA, 1999.)

My reason for writing this book is to show that the future does not need to be bleak, that we can overcome this planetary emergency. We do have both a free energy source and the know-how to convert today's oil-based economy into an economical, clean, and inexhaustible one in the future.

1.1.1 Global Energy Reserves and Trends

The time for holding conferences and writing articles is over. The Secretary General of the United Nations, on September 24, 2007, put it this way: "The time for doubt has passed." It is time to build those demonstration plants that will clearly establish the feasibility and costs of the various alternative energy systems. It is time to start to replace fossil fuels with clean and renewable energy sources such as solar–hydrogen.

Figure 1.1 shows the global fossil energy resources and the rate at which they are being consumed.

When discussing global energy consumption, the unit of energy most often used is the quad (Q). One Q equals 1 quadrillion Btu (10^{15} Btu), 1.055 exajoules (EJ), 172 million barrels of oil equivalent (boe), or 0.293 petawatt hour (pWh = 10^{12} kWh)* of electricity. One Q is also equivalent to the yearly energy produced by over two dozen nuclear power plants, the energy content of 10,000 supertankers of oil, 400,000 railcars of coal, or 28 billion cubic meters of natural gas.

As shown in Figure 1.1, the total fossil fuel reserves of the globe are estimated to be about 75,000 Q, and the total energy consumption is rising at

---

* kilo (k) = 10^3, mega (m) = 10^6, giga (g) = 10^9, tera (t) = 10^{12}, peta (p) = 10^{15}, exa (e) = 10^{18}, googol = 10^{100}.

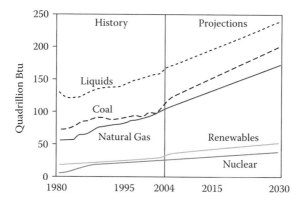

FIGURE 1.2
World marketed energy use by fuel type, 1980–2030. (From Energy Information Administration (EIA), *International Energy Annual 2004* (May–July 2006), http://www.eia.doe.gov/iea. Projections: EIA, System for the Analysis of Global Energy Markets (2007).)

a rate higher* than the supply of fossil fuels. The difference between the curves is being met from nuclear and renewable energy sources. The fossil envelope (dotted line) describes the likely future consumption of fossil energy (coal, oil, and natural gas). The area under this curve is the total of the known fossil reserves on the planet.

The curve projects a maximum yearly fossil fuel production capability of about 700 Q, which could be reached by 2050. It also projects the total exhaustion of the fossil energy supply by the year 2200. Besides being exhaustible, the burning of fossil fuels also releases carbon dioxide (CO_2) into the atmosphere. In spite of these facts, the global dependence on fossil fuels is projected to increase (Figure 1.2).

The global energy consumption during the half-century 1950–2000 increased from 100 to 400 Q. Today, it is about 450 Q and is rising at a rate of about 15 Q/yr. It is expected to reach 600 Q by the year 2020. The present energy consumption of the United States is about 100 Q. The global and domestic energy needs are being met mostly by fossil and nuclear sources (Table 1.3).

1.1.2 Traditional Energy Sources and Costs

As shown in Table 1.3, nearly 90% of the global energy demand is met by fossil and nuclear sources. Of these sources, nuclear is unsafe, fossil is polluting, and both are exhaustible. The costs of all forms of fossil fuels increased drastically during the last 5 years. In early 2008 the wholesale price of a million Btus of energy in the form of coal was about $6, in natural gas about $10,

---

* 2007 Solar Energy—Complete Guide to Solar Power and Photovoltaics, Practical Information on Heating, Lighting, and Concentrating—U.S. Department of Energy.

TABLE 1.3

Global and Domestic Energy Sources

Energy Sources	Global (%)	United States (%)
Oil	35–37	39–40
Coal	25–26	23–24
Natural gas	20–25	21–24
Wood and biomass	9–10	*
Nuclear	7.5	8
Hydroelectric	2.4	7
Solar	0.6	a
Geothermal	0.4	a
Wind	0.05	a

* The combined total of these three sources is about 2.5% and is included in the 7% listed for hydroelectric.

and in the form of oil about $30. On the average, power plants pay about 80% of the commercial prices while residents pay from 150% to over 200%.

Table 1.4 lists the types of electric power plants and their shares in the total production of electricity in the United States. The total energy consumption of the United States is about 100 Q, of which industry consumes 32.5%, transportation 28%, residents 21%, and commerce 18%. Of the total energy consumption, about 15 Q is used in the form of electricity.

One of the reasons for the present energy crisis is that powerful business interests are trying to extend the use of fossil and nuclear power. Figure 1.5 describes (according to the Nuclear Energy Institute) the cost of electric power generation in the United States during the decade 1995–2005. One might observe that nuclear and coal costs are shown as being the lowest, and hydroelectric and wind power costs are not listed at all. It should also be noted that although the Nuclear Energy Institute shows the coal and nuclear electricity generation cost at about 2.0¢/kWh. This has changed drastically in the last couple of years, because the cost of both uranium and of all fossil fuels at least doubled. For example, in the spring of 2008 the wholesale price of a metric ton of coal was around $140, while in 2003 it was $25. Similarly, the cost of a kilogram of uranium increased from $10 to $75 during the last decade. Therefore, such data as provided by the Nuclear Energy Institute in Figure 1.5 is totally misleading. The reality is that by the time electricity reaches a household, it is 10 times that shown in Figure 1.5 for nuclear (in 2007 in Connecticut, we paid about 18¢ for a kilowatt hour of electricity).

Another major problem with the global electricity system is the lack of continent-wide electric grids that could serve both to distribute and to store electricity. This "net metering" capability (Section 1.4.3.1) is essential to make

TABLE 1.4

Total Electricity Generation in the United States, by Energy Source and Type of Producer, 1995–2006 (1000 mWh)

Period	Coal	Petroleum	Natural Gas	Other Gases	Nuclear	Hydroelectric Conventional	Other Renewables	Hydroelectric Pumped Storage	Other	Total
1995	1,709,426	74,554	496,058	13,870	673,402	310,833	73,965	−2,725	4,104	3,353,487
1996	1,795,196	81,411	455,056	14,356	674,729	347,162	75,796	−3,088	3,571	3,444,188
1997	1,845,016	92,555	479,399	13,351	628,644	356,453	77,183	−4,040	3,612	3,492,172
1998	1,873,516	128,800	531,257	13,492	673,702	323,336	77,088	−4,467	3,571	3,620,295
1999	1,881,087	118,061	556,396	14,126	728,254	319,536	79,423	−6,097	4,024	3,694,810
2000	1,966,265	111,221	601,038	13,955	753,893	275,573	80,906	−5,539	4,794	3,802,105
2001	1,903,956	124,880	639,129	9,039	768,826	216,961	70,769	−8,823	11,906	3,736,644
2002	1,933,130	94,567	691,006	11,463	780,064	264,329	79,109	−8,743	13,527	3,858,452
2003	1,973,737	119,406	649,908	15,600	763,733	275,806	79,487	−8,535	14,045	3,883,185
2004	1,978,620	120,771	708,854	16,766	788,528	268,417	82,604	−8,488	14,483	3,970,555
2005	2,013,179	122,522	757,974	16,317	781,986	270,321	87,213	−6,558	12,468	4,055,423
2006	1,990,926	64,364	813,044	16,060	787,219	289,246	96,423	−6,558	13,977	4,064,702

Source: Energy Information Administration (EIA).

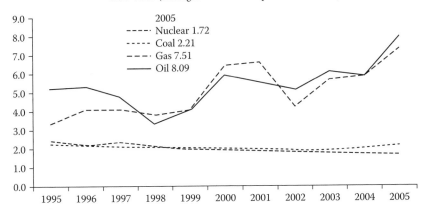

FIGURE 1.5
Electricity production costs during the decade of 1995–2005. (From the Nuclear Energy Institute.)

the energy of intermittent energy sources (solar, wind, tide, etc.) continuously available.

The actual average wholesale cost of 1 kWh of electricity in 2007 in the United States (in cents), as a function of fuel used to generate it (according to the *New York Times,* July 11, 2007) was as follows:

Pulverized coal	5.7
Nuclear	6.4
Coal gas	6.6
Natural gas	7.3
Wind	9.6
Biomass	10.7
Solar thermal	12.0

If a $50/ton carbon emission charge is mandated, renewable energy cost would become more competitive as follows:

Nuclear	6.4
Natural gas	9.2
Wind	9.6
Pulverized coal	10.1
Coal gas	10.8
Biomass	11.2
Solar thermal	12.0

Therefore there are generation costs (Figure 1.5), wholesale costs (above), and transportation and distribution costs. The total for us, the end user, June, 2007 came to 18.9¢/kWh. The national average cost of electricity generation in the United States in 2007 was about 4–5¢/kWh, but it varies from region to region. In 2007 the national average "retail" prices by end-user sector were as follows:

Residential 10.7¢/kWh

Commercial 9.9¢/kWh

Industrial 6.4¢/kWh

Transportation 10.4¢/kWh

All Sectors 9.2¢/kWh

The cost of electricity is also a function of the time of its use, because the summer "peak price" can be two or even three times the "night time" electricity cost. Because solar electricity is usually produced during "peak" periods (when the air-conditioning loads are high), the value of solar electricity is usually greater than fossil or nuclear.

1.1.2.1 Oil and Natural Gas

At the beginning of the industrial revolution, the ultimately recoverable global oil supplies were estimated as 3.3 to 4.8 trillion (10^{12}) barrels. Of this, about 1.2 trillion has already been consumed and the U.S. Department of Energy (DOE) estimates that 1.32 trillion barrels of oil (including Canadian oil sand) remain recoverable (Table 1.6). The U.S. oil reserves on accessible land are estimated to be 21 billion barrels, offshore as 18 billion, in the Arctic National Wildlife Reserve as 12 billion, and in oil shale in the Green River basin as 800 billion barrels. The United States consumes about 8 billion barrels of oil a year of which about 5 billion are imported. It would take 10 years to start oil production from deep water wells and it would cost three times as much as from accessible land. In 2007, production cost on land was $20 per gallon, $40 in shallow waters, $50 from oil shale, $60 from deep waters or from oil sands. In contrast, solar power plants for hydrogen production can be built in 4 years and can produce a gallon gasoline equivalent (gge) quantity of hydrogen for $5.00/gge.

The recoverable global natural gas reserve is slightly over 6,000 quadrillion (10^{15}) cubic feet (Table 1.7). The shortage of natural gas in Europe and elsewhere is also resulting in wasteful and expensive projects. One such project is the 2,000 mile pipeline, which by 2017 would bring natural gas from the North Slope of Alaska to the lower United States at a cost of $30 to $60 billion. (This investment would be enough to build ten to twenty 1,000 mW solar–hydrogen power plants, for which the fuel is free and inexhaustible.) Russia and the European Union (EU) are also building gigantic gas pipelines as the supply of energy is also being exploited for political purposes.

TABLE 1.6

Proved Global Oil, Natural Gas, and Coal Reserves* as of January 2007

By Region and/or by Nation	Oil Reserves in Billion (10⁹) Barrels	Percent of Global Oil Reserves	Natural Gas Reserves in Quadrillion (10¹⁵) ft³	Percent of Natural Gas Reserves	Coal Reserves in Billion (10⁹) Short Tons	Percent of Global Coal Reserves
GLOBAL TOTAL**	1,317	100	6,182	100	997.7	100
North America	**213**	**16.1**	**276**	**4.5**	**276**	**27.6**
United States**	21	1.6	204	3.3	267	26.7
Canada***	179***	13.7	58	0.9	7	0.7
Mexico	12	0.9	15	<0.5	negligible	<0.5
Latin America	**102**	**7.8**	**240**	**3.9**	**22**	**2**
Venezuela	80	6.1	152	2.5	negligible	<0.5
Brazil	12	0.9	11	<0.5	11	1.1
Equador	4.5	<0.5	3	<0.5	negligible	<0.5
Argentina	2.5	<0.5	16	<0.5	negligible	<0.5
Peru	1	<0.5	9	<0.5	negligible	<0.5
Trinidad	1	<0.5	19	<0.5	negligible	<0.5
Colombia	negligible	<0.5	negligible	<0.5	7	0.7
Europe	**16**	**1.2**	**180**	**2.9**	**65.7**	**6.6**
Norway	8	0.6	82	1.3	negligible	<0.5
Netherlands	0.1	<0.5	50	0.8	negligible	<0.5
UK	4	<0.5	17	<0.5	negligible	<0.5
Denmark	1.3	<0.5	2.5	<0.5	negligible	<0.5
Italy	0.6	<0.5	5.8	<0.5	negligible	<0.5
Poland	0.2	<0.5	5.8	<0.5	15.4	1.5

Romania	0.6	<0.5	2.2	<0.5	negligible	<0.5
Czech Rep.	negligible	<0.5	negligible	<0.5	6.1	0.6
Serbia	negligible	<0.5	negligible	<0.5	18.2	1.8
Germany	negligible	<0.5	negligible	<0.5	7.4	0.7
Turkey	negligible	<0.5	negligible	<0.5	4.6	<0.5
Greece	negligible	<0.5	negligible	<0.5	4.3	<0.5
Hungary	negligible	<0.5	1.5	<0.5	3.7	<0.5
Eurasia	**99**	**7.6**	**2,015**	**32.6**	**250.5**	**25**
Russia	60	4.6	1,680	27.2	173	17.3
Kazakhstan	30	2.3	100	16.2	34.5	3.4
Turkmenistan	0.6	<0.5	100	16.2	negligible	<0.5
Azerbaijan	7	0.5	30	0.5	negligible	<0.5
Uzbekistan	0.6	<0.5	65	1	negligible	<0.5
Ukraine	0.4	<0.5	39	0.6	37.6	3.8
Middle East	**739**	**56.4**	**2,566**	**41.5**	**negligible**	**<0.5**
Saudi Arabia	262	20	240	3.9	negligible	<0.5
Iran	136	10.4	974	15.8	negligible	<0.5
Iraq	115	8.8	112	1.8	negligible	<0.5
Kuwait	102	7.8	55	0.9	negligible	<0.5
Un. Arab Em.	98	7.5	214	3.5	negligible	<0.5
Qatar	15	1.1	910	14.7	negligible	<0.5
Oman	5.5	<0.5	30	<0.5	negligible	<0.5
Yemen	3	<0.5	17	<0.5	negligible	<0.5
Africa	**114**	**8.7**	**484**	**7.8**	**55.5**	**5.5**
Libya	41	3.2	53	0.9	negligible	<0.5

Continued

TABLE 1.6 (Continued)

Proved Global Oil, Natural Gas, and Coal Reserves* as of January 2007

By Region and/or by Nation	Oil Reserves in Billion (10^9) Barrels	Percent of Global Oil Reserves	Natural Gas Reserves in Quadrillion (10^{15}) ft^3	Percent of Natural Gas Reserves	Coal Reserves in Billion (10^9) Short Tons	Percent of Global Coal Reserves
Nigeria	36	2.7	182	2.9	negligible	<0.5
Algeria	12	0.9	162	2.6	negligible	<0.5
Egypt	4	<0.5	58	0.9	negligible	<0.5
South Africa	negligible	<0.5	negligible	<0.5	53.7	5.4
Asia, Oceania	**33**	**2.5**	**419**	**6.8**	**327.3**	**32.7**
China	16	1.2	80	1.3	126.2	12.6
Indonesia	4	<0.5	98	1.6	negligible	<0.5
India	6	<0.5	38	0.6	negligible	<0.5
Malaysia	3	<0.5	75	1.2	negligible	<0.5
Pakistan	0.3	<0.5	28	<0.5	negligible	<0.5
Thailand	0.3	<0.5	15	<0.5	negligible	<0.5
Australia	negligible	<0.5	negligible	<0.5	86.5	8.7
Hong Kong	negligible	<0.5	negligible	<0.5	101.9	10.1

* The data is obtained from the Energy Information Agency (EIA) of the U.S. Department of Energy (DOE). For detailed, nation by nation, continuously updated data on oil and natural gas see: http://www.eia.doe.gov/emeu/international/reserves.html. For coal see: http://www.eia.doe.gov/pub/international/iea2005/table82.xls

** Yearly oil production in the United States from 2001 to 2006 dropped from 112 to 98 million barrels. Natural gas (NG) production remained constant at around 50 billion cubic foot per day (bcfd). In 2007, conventional NG sources produced 28 bcfd, tight sand 15 bcfd, coal methane 5 bcfd, and shale about 5 bcfd for a total of 53 bcfd.

*** DOE obtained the data for oil and natural gas from the *Oil and Gas Journal*. This data, for Canada, includes the reserves contained in their oil sand.

TABLE 1.7

Global Fossil and Nuclear Fuel Reserves, Consumptions and Costs

Type of Fuel	Oil	Natural Gas	Coal	Nuclear
Global Reserves	1,310 billion (10^9) barrels*	6,200 tcf**	998 billion (10^9) tons	5 million tons***
Global Yearly Consumption—2007	~32 billion (10^9) barrels	~120 tcf	~7 billion (10^9) tons	~77 thousand tons
Reserve/Production (R/P) Ratio****	~40 years	~50 years	100–150 years	~65 years
Approximate Market Values	1998—$15/barrel 2008—$140/barrel	1998—$2.5/1000 cf 2008—$10/1000 cf	1998—$25/ton 2008—$140/ton	1998—$10/pound 2008—$75/pound

* Including the oil recoverable from Canadian oil sand.

** Trillion (1,012) cubic feet = tcf.

*** The uranium reserves are estimated as a function of the market value. This 5 million ton figure is based on a market value of $130/pound. If the market value is $60/pound, the profitably recoverable reserve is only 2 million tons. The absolute total uranium reserves are about 15 million tons.

**** As the global consumption of any of the fuels rises, the R/P ratio naturally drops.

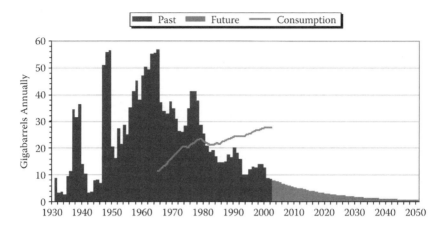

FIGURE 1.8
Oil discovery (3-year average—past and projected) 1930–2050. (*ASPO Ireland Newsletter,* http://www.aspo-ireland.org/index.cfm/page/newsletter).

With the global yearly oil consumption given in Table 1.7 as 32 billion barrels (U.S. 7.5, China 2.5), the 1.32 trillion barrel oil reserve (including Canadian oil sand), the total global oil reserve corresponds to about 40 years of consumption. This is called the reserve/production (R/P) ratio of oil reserves. Since 1990 the American demand for oil has grown by 20%, the Chinese by 200% and it is estimated that by 2030, the global demand for oil will increase by another 40%. Since the last oil embargo, the dependence of the United States on imported oil has increased from 35% to 60% and costs $700 billion yearly. At the same time the discovery rate of new oil reserves has dropped below the rate of rise in consumption (Figure 1.8).

In June 2008 the wholesale price of a barrel of oil was close to $140 and the wholesale price of natural gas was about $10 per 1,000 ft^3. Both of these prices reflect increases of about 400% during the last 5 years.

What makes the situation worse from an American perspective is that the remaining oil reserves are mostly in unfriendly or unstable regions of the world (Table 1.9). The logical response to such a situation would be to make the United States energy independent by leading a global effort to develop a renewable energy economy. Yet, this transition is very slow to start because of the powerful interest groups that favor the continued use of fossil fuels and the military means to "protect" the oil supply.

The oil lobby wants the government to continue providing oil- and gas-drilling tax incentives and subsidizing domestic exploration, and wants to drill in federal waters without paying royalties. The opponents of the oil lobby would prefer to transfer government support and tax breaks from the oil industry to the renewable energy industries. Similar to oil, the coal, natural gas, and the uranium reserves are also limited (Table 1.7).

U.S. oil imports in 2006 came from Canada (1.799 million barrels per day), Mexico (1.734 million barrels per day), Saudi Arabia (1.549 million barrels per day), Venezuela (1.008 million barrels per day), Nigeria (0.996 million barrels

TABLE 1.9

Proved Crude Oil Reserves of the Planet by Regions*

Region	Percent of Global Total	Proved Reserves (billion barrels)	Share of OPEC Members	OPEC Members in the Region
Global total*	100	1,195*	75%	**
Middle East	62.5	743	99%	**
Latin America	10.5	123	62%	Venezuela
Africa	10	118	74%	Libya, Nigeria
E. Europe, Russia	11	129	0%	—
Asia/Pacific	3	39	10%	—
N. America	2 (3***)	27 (45***)	0%	—
W. Europe	1	15	0%	—

* Data is for the year 2006 from OPEC and does not include the oil content of Canadian sand. For up-to-date information see http://www.opec.org/library/Annual%20 Statistical%20Bulletin/interactive/FileZ/XL/T10.XLS

** Saudi Arabia (38%), Iraq (16%), Kuwait (14%), Iran (14%), UAE (14%), Qatar (2%), Oman (1%), Syria (0.4%), others (0.6%).

*** Including artic and offshore reserves.

per day), Iraq (0.617 million barrels per day), Angola (0.525 million barrels per day), Algeria (0.474 million barrels per day), Ecuador (0.282 million barrels per day), and Russia (0.216 million barrels per day). The total crude oil imports averaged 12.2 million barrels per day (or about 4.5 billion barrels per year) in 2006. The United States consumes 20.6 million barrels of oil per day, which is nearly 25% of the world's total. Half of the planet's oil supply has already been used up and its recovery has already peaked. In 2007 the oil companies spent $100 billion on oil exploration and found less oil than was pumped from the ground. Fourteen of the twenty largest oil companies are state-owned giants, and Exxon, Mobil, BP, etc., control only 10% of the oil supplies. Since 1980 oil consumption of all industrial nations has dropped (Germany 20%, Sweden 33%, etc.) except the United States—which increased by 21%.

1.1.2.2 Coal

Five years ago the wholesale price of a metric ton of coal was about $25; in early 2008 it was up to $140 and rising. In contrast to oil and natural gas, the United States has very substantial coal reserves amounting to 27% of the global total (Table 1.6). Yearly coal consumption by the United States is 1.1 billion tons. Today in the United States, about 2 trillion kWh of electricity (about 55% of the total) are produced from coal (Table 1.4), and by 2030, that number is projected to rise to 3.3 trillion kWh (or 62%). The carbon emission from electric power generation is about 2.3 billion metric tons (90% from coal), and that emission is also projected to rise to 3.3 billion metric tons by 2030. Some projections in the past suggested that at this rate American coal reserves would last 250 years, but in 2007 the National Academy of Sciences reduced that projection to about 100 years (Table 1.7).

The coal industry is asking for billions in government aid to develop the technology to make gasoline and diesel fuels from coal and to develop technology for power plants to capture and store (sequester) their carbon emissions, but the feasibility and cost of such projects are unknown. Some suggest that these techniques would double power plant costs. It is also feared that if mining is expanded into the layers above and below the seams that have already been excavated, safety risks will increase, and underground water flows will be disrupted.

The generating capacity of the average fossil power plant in the United States is 500–600 mW, and on the average these plants operate at 70% loading. The variables that determine their construction cost include the process used, the building enclosure, the fuel, and the pollution control or carbon-sequestering process used. An outdoor coal burning power plant with only the minimum required pollution controls costs about $1 billion ($1,500/kW). Such plants are being built in China and in other less-developed parts of the world. Enclosed power plants provided with carbon-sequestering or carbon-recapturing features cost over $2 billion ($3,000/kW), and a coal-to-synthetic diesel fuel plant (which does not exist today) might cost $3–$5 billion ($4,500/kW–$7,500/kW).

The U.S. Energy Department projected in 2006 that 151 coal-fired power plants could be built by 2030 to meet the projected 24% increase in demand. On the other hand, resistance to building new coal-fired power plants is also growing. For example, the regulatory agency in Kansas (Kansas State Department of Health and Environment) rejected a permit request for a 700 mW unit near Holcomb, and the same happened in Jerome County, Idaho. More than 60 proposed coal-fired power plants were delayed or cancelled in 2007. For example, citizen opposition caused the cancellation of the building of a coal-fired power plant near the Florida Everglades. One reason for this type of resistance is that the United States has no national regulation for carbon or mercury emissions. (Mercury can damage the nervous system.)

In the United States, the coal lobby (coal mining interests joined by Peabody Energy Corp. and the Air Force) is lobbying the government for loan guarantees to build six to ten coal-to-liquid plants at over $3 billion each. In addition, the lobby is asking for a tax credit of 51¢/gal of fuel produced, arguing that coal-to-fuel technology would reduce dependence on foreign oil. The lobby is also asking for a 25-year purchase contract for coal-based fuel from the Air Force.

Coal-to-liquid technology—developed by Lurgi in Germany during World War II and later improved upon in South Africa—produces about twice as much greenhouse gases as does the burning of regular gasoline or diesel. The technologies for capturing and storing greenhouse gases are unproven and very expensive.

The much-discussed carbon storage technology (also called *sequestering*), besides being unproven and limited by underground space availability for storage, also adds about $1 billion to the cost of each coal-to-fuel power plant.

1.1.2.3 Nuclear

The size of the exploitable proven uranium reserves is a function of the market value of uranium (because as it rises the mining of lower concentration ores became profitable). At $125/kg it is estimated between 4 and 5 million tons. As the present yearly demand for uranium is about 77,000 tons, this reserve is sufficient for about 65 years (R/P ratio = 65). Other estimates suggest that the known reserves will last about 80 to 100 years. On the other hand, if more nuclear power plants are built, the consumption will also rise and therefore the P/R ratio could drop.

Because it has no carbon emission, nuclear power is preferred to fossil fuels by some and it competes with renewable energy. The attitude toward nuclear power also seems to be changing: for example, in Italy, an earlier public referendum banned the use of all nuclear power; today new plants are contemplated. The attitudes of Holland, Belgium, Sweden, and Germany also seem to be changing despite the fact that uranium supplies are limited. Building nuclear power plants takes up to 20 years, and it costs significantly more than the cost of building solar powered plants. In the United States nuclear power supplies nearly 20% of the nation's electricity and new installations are contemplated.

The arguments favoring nuclear energy include low operating cost (not low capital cost) and lack of carbon emissions. The main arguments against it include its high capital and decommissioning costs and concerns about its safety. The two most well-known nuclear accidents were Three Mile Island (1979) and Chernobyl (1986). Chernobyl was the worst nuclear accident in history. The 1979 accident at Unit #2 of the Three Mile Island power plant was the worst civilian nuclear accident outside the Soviet Union, as the reactor experienced a partial core meltdown. According to the claims of Greenpeace, a total of nearly 200 nuclear meltdown "near misses" have occurred in the United States, but most plant operators do not disclose events like radioactive leaks. This was reported to be the case with the Exelon Corporation.

Besides leaks that occur during routine operation, damage to nuclear power plants due to earthquakes, aging, and terrorist attacks (including Internet virus attacks) is also of concern. For example, in August 2007 the cooling tower of the Vermont Yankee plant collapsed. Also, in May 2008, Cameco, the world's largest uranium producer's Port Hope plant was shut down in Ottawa, Canada, because of concerns about radioactive leakage into Lake Ontario.

Nearly 443 nuclear fission power plants are in operation around the world, and of these, 103 are located in the United States (Figure 1.10). The American plants were built at a total investment of about $0.5 trillion. Plant construction takes over 10 years, and no new orders have been issued for nuclear power plants for decades. Between 1970 and 1980, some 100 applications were submitted, but all were turned down. During the last 50 years, 253 nuclear

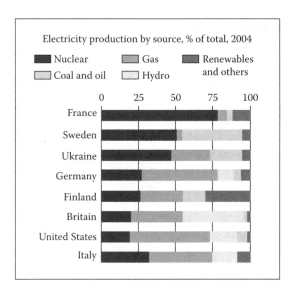

FIGURE 1.10

The dependence on nuclear power as a share of total electric power generation. (From Energy Information Administration, Official Energy Statistics, 2004.)

power plants were ordered in the United States. Of these, 71 were canceled before construction started, 50 were canceled after construction began, and 28 were shut down after they started running. It is projected that by 2050 nearly all currently operating nuclear power plants will have exceeded their useful life. On the other hand, the Nuclear Energy Institute estimates that 8 new plants will be in operation by 2016, and by 2030 an additional 16.4 gW capacity will be provided by nuclear plants.

In the United States, the last plant start up was in 1996, but in 2006–2007 some 20 new permit applications were submitted, including one at Calvert Cliffs, Maryland, by UniStar and others by Constellation Energy, Areva (French–German firm), and NRG. In addition, the Tennessee Valley Authority applied to build in Alabama, and Dominion applied in Virginia. The first application to go before the Nuclear Regulating Commission is the NRG plant to be built in Texas.

In many parts of the world, nuclear energy is considered as an alternative to fossil fuels, because it does not generate greenhouse gases. Naturally, the nuclear industry emphasizes this view and minimizes nuclear power's non-renewable and unsafe nature. It also claims that the spread of nuclear technology will not lead to the proliferation of nuclear weapons. This debate has not been resolved yet. Others argue that generation III, IV and breeder reactors will provide easy access for terrorists to obtain plutonium from which they can fabricate primitive nuclear weapons. The world's 443 nuclear plants generate 375 gW of electricity and save 2.2–2.6 billion tons in CO_2 emissions yearly. Nuclear opponents argue that this technology is too expensive and draws funds away from the renewable energy sector. A study by Princeton scientists suggests that to provide a 15% carbon emission cut by 2050 would require building 1,070 new nuclear power plants.

France remains committed to nuclear power (Figure 1.10). India is currently building six plants, Russia five, China four, and the United States and Canada one each. In the future China, Russia, South Africa, India, and other countries are planning many more, while Sweden voted to phase them out by 2010, and Germany is committed to phasing them out by 2020.

One of the main arguments against the use of nuclear energy is the unresolved problem of nuclear waste storage. Most nuclear power plants are provided with only a 30- to 40-year temporary waste storage capacity, and in most cases that capacity is filling up. In addition, some of these storage facilities are above ground, which is considered unsafe. Although as of today no reliable solution has been found for nuclear waste storage, there are several related projects in progress. One such plan is that of the Finnish waste disposal company Posiva, which is digging a tunnel at Olkiluoto, an island west of Finland, in the hopes that it will receive approval to store nuclear waste half a kilometer underground.

In the United States the government promised to build a nuclear waste repository in the Yucca Mountains and intended to start accepting waste from the nuclear industry by 1998. Because of the resistance of Nevada and because of technical problems, the repository is not expected to open until 2020. It is for that reason that some 60 nuclear plant owners are suing the government to recover their costs resulting from continued local storage. In the suit filed by Progress Energy, the U.S. Court of Federal Claims awarded $82 million to this plaintiff. At present each reactor produces 20 tons of waste per year, and this waste is stored in steel casks at 122 temporary waste sites in the country. These casks can be penetrated by regular weapons and can release radioactive cesium gas. An additional concern is the waste produced by nuclear weapons production, which can be concentrated enough to build "dirty" nuclear bombs.

Nuclear power raises safety issues related to radioactive waste disposal as well as the largely unsolved problem of decommissioning (after 50 years, which is nearing for many plants). In many cases evacuation plans are unrealistic, untested, or do not even exist. For example, the license of the Indian Point plant in New York State expires in 2013. This plant leaked in the past and is vulnerable to terrorist attack in a highly populated area on Long Island. There is a black market in nuclear weapons technology (such as the smuggling by Abdul Qadeer Khan to Libya and Iran), as well as trafficking in highly enriched uranium for building dirty bombs (arrests have been reported in Slovakis, Russia, Kazahstan, China, and Lybia), and also software attacks on nuclear power plants (such as the January 25, 2003 Slammer worm attack on Ohio's Davis-Besse nuclear power plant).

The advantages of nuclear power plants include the fact that they operate at a 90% capacity factor (loading). Also, 1 kg of natural uranium generates about as much electricity as 20,000 kg of coal. In contrast to fossil fuels, nuclear power does not contribute to global warming. In the past, the cost of nuclear fuel was low, about 0.5¢/kWh, and that price was relatively stable, well below the costs of the major competing fossil fuels.

The size (generating capacity) of light-water nuclear power plants is usually twice that of their fossil counterparts or about 1.2 gW (1,200 mW). The plants' life spans are in the range of 40 to 60 years, and their total cost includes not only construction and operation, but also waste disposal and decommissioning. Insurance costs can also be high because there have been cases when plants were not allowed to operate at all, for example, the ill-fated $5 billion Shoreham facility which was never allowed to operate.

The initial cost of construction of a nuclear power plant is at least $5 billion, and its decommissioning can also be in the billions. Therefore, the cost of a second-generation light-water nuclear plant in the United States can easily reach $5,000/kW.

During the several decades of the freeze on nuclear construction, some of the design problems have been solved (for example, the new General Electric design places the cooling water circulating pumps inside the reactor), but the freeze created other problems. One such problem is that most existing nuclear plants are controlled by traditional controls, using digital systems only as monitoring devices. Therefore, when new, fully computerized plants are built, they might experience "growing pains."

In breeder reactors, the most common isotope of uranium, U-238, can be converted by neutron bombardment into fissionable Pu-239 (plutonium-239). The excess nuclear fuel can be used in other reactors or to build nuclear weapons.

Nuclear fusion does not require uranium fuel and does not produce radioactive waste, and has no risk of explosive radiation-releasing accidents, but it takes place at a temperature of several million degrees. Nuclear fusion occurs in the sun, its fuel is hydrogen and, as such, it is an inexhaustible and a clean energy source. The problem with this technology is that, because it operates at several million degrees of temperature, its development is extremely expensive, and it will take at least until 2050 before the first fusion power plant can be built (Tokomak fusion test reactors). It is estimated that it will be 50 times more expensive than a regular power plant, and its safety is unpredictable. In short, the only safe and inexpensive fusion reactor is the Sun!

A comparison between the main features of nuclear and solar power plants is given below:

- Initial cost of a 1,000 mW plant: $5 billion (solar about $3 billion).
- Construction time is 10 to 20 years (solar 3 to 4 years).
- Fuel cost increased from $10/kg to $75/kg in 10 years (solar is free).
- Uranium-235 deposits will be exhausted in 60–80 years (solar is inexhaustible).
- Risk of accidents due to human error or earthquakes (none with solar).
- Nuclear proliferation can contribute to terrorism (no such risk with solar).

- Decommissioning after a 40- to 60-year life span is very expensive (no such expense for solar).

- No permanent solution exists for waste storage anywhere on the globe (none required for solar).

1.2 Global Warming

There are a few people who still argue about the reasons for global warming, but practically no one questions its existence. A few argue that because of the natural variability of the Earth's climate, or because of the effects of volcanic and solar activity, it is not possible to say that the change in the climate is caused mostly by humans. Some people take this position under the influence of business or political interests, and others base it on past experience, such as when the climate did change owing to the increase in solar activity in 1934. Climate skeptics refer to sunspot cycles or to shifts in ocean and atmospheric patterns and ignore the fact that these effects are only superimposed on the main cause of global warming: human activity.

Based on the heat balance of the planet, the temperature rise caused by the atmospheric increase of the CO_2 concentration should have been larger and should have occurred faster than it did. The reason for this is the tremendous thermal capacity of the oceans and the great amount of heat being absorbed by the melting of polar ice. Once this ice has melted, warming will drastically accelerate. This will result in a "new planet" that is warmer and has a reduced land area because of the flooding of coastal regions. I would prefer that carbon emission taxes did not just raise the overall tax burden, but were used to lower the cost of conversion to renewable energy systems. It is also essential that the number of jobs created by the renewable energy industry exceed the loss of jobs in the fossil-power-related industries. This can be guaranteed, for example, by supporting the use of solar shingles on the roofs of new buildings.

The rising concentrations of greenhouse gases increase the temperature of the Earth, and human activity is an important contributor to that trend. In May 2001, a National Research Council study on global warming commissioned by the White House stated, "Greenhouse gases are accumulating in Earth's atmosphere as a result of human activities, causing surface air temperatures and subsurface ocean temperatures to rise. The changes observed over the last several decades are mostly due to human activities, but we cannot rule out that some significant part of these changes is also a reflection of natural variability."

In 2002, President Bush created the Climate Change Science Program, which, after spending $1.7 billion on several years of study, reported that the climate is in fact changing. Yet global warming has become a political issue.

This is unfortunate, because such debates slow down reaching an agreement on the corrective steps that have to be taken. I hope that this book will show that the issue of global warming is not a claim of the left or the right, because both sides of the globe are warming!

In February 2007, the United Nation's Intergovernmental Panel on Climate Change reported, with 90% confidence, that human activity is the main cause of global warming. According to the 2007 final report of this panel, greenhouse gases have risen 70% since 1970 and could rise an additional 90% by 2030. At the UN meeting in Bali in December 2007, the World Meteorological Organization reported that the last decade was the warmest ever on record. Therefore, it is not "alarmist fear-mongering," but simply a fact that global warming is changing our climate and that the concentration of greenhouse gases is growing at an alarming rate in the atmosphere. On the other hand, it is also logical to ask these questions: How much climate change can we live with? How fast should we restabilize the climate in order not to disrupt the global economy and, particularly, the development of the third world?

To date, global temperature has increased by 0.74°C (1.33°F), and in 2007 the UN Panel on Climate Change predicted a rise of 2–6°C (3.5–11°F) in global temperature by 2100 if no corrective action is taken. However, even if the global temperature was to rise by only 2°C, the consequences would include wildfires, droughts, rising sea levels, decline in the oceans' oxygen content causing the collapse of marine ecosystems, more intense hurricanes, and more outbreaks of insect-borne diseases. Likewise, some 600 million people would go hungry, 200 million people would be displaced by floods, and 400 million people would be exposed to malaria, dengue, and other diseases. Some climate models predict snow-free Arctic summers by 2030 and suggest that the polar bears of Alaska will be wiped out by 2050.

Seeing the unexpectedly rapid increase of emissions in Asia, in November 2007, the UN panel revised its prediction saying that a rise of 6°C (11°F) could be reached by as early as 2030 and, as a result, one quarter of the world's species could be eliminated. Yet, as of today, we are still using the atmosphere as a free garbage dump for our emissions.

I hope that this book will convince the reader that the orderly transition to a clean and inexhaustible solar–hydrogen economy has to start now and that it must be completed by the end of this century.

1.2.1 Greenhouse Gases

"Greenhouse gases" allow ultraviolet sunlight to strike the Earth's surface, but when some of that energy is reflected back into space, the greenhouse gases absorb this infrared radiation and trap the heat in the atmosphere. This heat accumulation, which started at the beginning of the industrial age, upsets the balance between the solar energy received and the heat radiated back into space.

Water vapor, carbon dioxide (82% of our emissions), methane (9%), nitrous oxide (5%), and aerosols (2%) are all greenhouse gases. Also, ozone blocks the entering ultraviolet radiation from the Sun, and when the ozone layer is destroyed, the resulting ozone hole allows more energy to enter, which contributes to the melting of the ice at the poles and to global warming. Aerosols, air conditioning fluids, and other organic chemicals (CFCs [chlorofluorocarbons]) in the atmosphere also react with ozone (O_3) and are responsible for increasing the "polar ozone hole."

Anthropogenic (human-made) gases are released as byproducts of industrial processes. One of the major sources of greenhouse gases is the pulp and paper industry, because 20% of all greenhouse gas emission is caused by deforestation as the carbon stored in the trees is released into the atmosphere. In addition, the trees that are cut down not only stop absorbing carbon dioxide, but also eliminate biodiversity. (An acre in Borneo contains more tree species than all of North America.) Indonesia today is losing tropical forests the size of Maryland every year, and deforestation accounts for the emission of more greenhouse gases than all the cars in the world.

In addition to CO_2, methane comes from landfills, coal mines, oil and gas operations, and agriculture. Nitrous oxide (NO) is emitted from burning fossil fuels and through the use of certain fertilizers or industrial processes. In terms of its effect on the climate, the most important greenhouse gas is CO_2.

Our planet was created in such a way that plants will consume CO_2, while animals generate it. The concentration of CO_2 in the atmosphere reflects the balance between plant and animal life on the planet. Prior to the industrial age, the movement ("flux") of carbon between the atmosphere, the land, and the oceans were kept in balance by nature's photosynthesis. In the last centuries, this balance has been upset not only by overpopulation and deforestation, but also by lifestyle changes—resulting in increased per capita energy consumption.

1.2.2 The Carbon Dioxide Cycle

Figure 1.11, prepared by the U.S. Energy Information Administration, describes the global carbon cycle. It provides data that was collected in 2001. Since that date, the yearly anthropogenic carbon emissions (measured in carbon equivalent terms) increased from 6.3 to about 9 billion metric tons (over 1 ton per capita in the world). In November 2007, the National Academy of Science reported actual emissions for 2006 as 8.4 billion tons. *Carbon equivalent* means that the emission of 3.7 tons of CO_2 is counted as the emission of 1 ton of carbon, so the 8.4 billion tons per year of carbon that enters the atmosphere owing to fossil fuel combustion corresponds to 33 billion tons per year of CO_2 because of the molecular weight ratio of CO_2 to carbon (44/12).

The total quantity of carbon on Earth is about 41,000 billion metric tons (92% in the oceans, 6% on land, and 2% in the atmosphere). Prior to the Industrial Age, the concentration of CO_2 in the atmosphere was stable and balanced. Two hundred and ten billion tons of carbon dioxide entered the atmosphere and approximately the same amount was taken from the atmosphere by the photosynthesis of plants. That balance has been upset by fuel combustion, deforestation, and changing land use as the population increased.

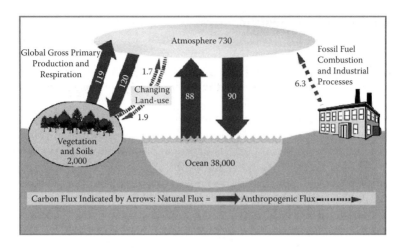

FIGURE 1.11
Global carbon cycle in billion metric tons of carbon per year. (From Intergovernmental Panel on Climate Change, 2001, U.S. Energy Information Administration.)

Since the beginning of the Industrial Age, the atmospheric concentration of CO_2 has increased from 280 to 380 parts per million (ppm). Global temperature is already the highest since the Middle Ages; the last decade witnessed the highest temperatures ever recorded. The atmospheric concentration of CO_2 is the highest in 650,000 years.* If this trend continues, this concentration will rise to 550 ppm by the end of this century. (According to some projections, this level could be reached by 2030 if the emissions of Asia continue to grow at the present rate.)

World CO_2 emissions are expected to increase by 1.9% annually from 2001 to 2025. Much of the increase in these emissions is expected to occur in the developing world, where emerging economies such as China and India fuel their economic development mostly with fossil energy. China's CO_2 emissions in 2007 exceeded those of the United States, and emissions of the developing countries are expected to surpass those of the industrialized countries about the year 2018. In terms of per capita carbon dioxide emissions, the United States is still the "leader" with a 21 tons/per capita/per year emission (Russia 11.8, EU 8.6, China 5.1, India 1.8), and these emissions continued to increase in all parts of the world except the EU, where it has been reduced by 2%.

Today, the average American generates 21 tons of CO_2 a year (some of the "green" cities in California and elsewhere reduced this to 9–13 tons/per capita), whereas the global per capita average is only 4 tons a year. The generation of each kilowatt-hour of electricity from fossil fuels releases from 270 to 1,050 g (0.27–1.05 kg) of CO_2 into the atmosphere, in addition to other

---

* According to the 2007 UN Intergovernmental Panel on Climate Change, the CO_2 content of the planetís atmosphere in the past oscillated between 200 and 300 ppm and has NEVER been as high as it is today. For details refer to: http://en.wikipedia.org/wiki/Carbon_dioxide_in_the_Earth's_atmosphere.

pollutants such as NOx, SO_2, and particulates. On average, an automobile generates about 160 g/km of CO_2 (250 g/mi). Even a hybrid automobile generates 3 tons of CO_2 each year.

In comparison, the average tree consumes only 1 ton of CO_2 in a lifetime, and an acre of rainforest consumes about 500 tons yearly. When agribusiness, the ethylene industry, or pulp and paper corporations turn forests or rainforests into farmland, they also destroy an effective consumer of CO_2. The world's fastest-disappearing forests are in Indonesia, where they are cut down either to make paper pulp or to be replaced by palm oil plantations. Palm oil is mostly used to make biodiesel fuels. Tropical deforestation not only results in CO_2 emissions (20% of the global total), but it also poisons the rivers.

In the United States, greenhouse gas emissions come from the following sources: power plants (33%), transportation (28%), industry (20%), agriculture (7%), commerce (7%), and households (5%). A 500 mW coal-fired power plant releases 100 tons of CO_2 every hour. The electric power generation capacity of the United States is around 1,000 gW (the equivalent of 2,000 such power plants). Counting all power plants, including the smaller ones, the total is some 16,000.

1.2.3 Consequences of Carbon Emissions

According to the 2007 final report of the Intergovernmental Panel on Climate Change, the consequences of global warming could include the mass extinction of species and a worldwide deterioration of political and security conditions. Rising sea levels will threaten people on Asia's coastlines, and the heat in Africa will cause water and food shortages, the shrinking of lakes, large-scale migrations, and the collapse of weak states such as Burundi, Congo, and Rwanda. The nations that will be harmed the most are the ones least responsible for climate change. The 840 million people in Africa generate only 3% of the global emission of CO_2, yet they face the greatest risk of drought and disruption of their water supplies.

The United Nation's report—prepared by over 1,000 scientists—predicts cultural and social disruptions, loss of wetlands, flooding of river deltas, bleaching of coral reefs, permafrost thawing, acidification of oceans, drop in crop output, widespread water shortage, and even starvation in parts of southern Europe, the Middle East, Africa, Mexico, Southern Asia, and the American Southwest. Deforestation, soil erosion, storms, droughts, and devastation of agriculture are likely to result as temperatures exceed the heat tolerance of crops. These trends can combine to cause migration, ethnic strife, social destabilization, and wars.

Figure 1.12 shows the global trend in carbon emissions. As a result of these emissions, the global temperature has already risen by 0.74°C (1.3°F). If this trend continues, by 2030 the global temperature can rise by 6°C (11°F). Just because of the thermal expansion of the waters, this will result in the rise of ocean levels by about 60 cm (23 in.). According to the UN panel report in November 2007, if the ice sheets over Western Antarctica and Greenland continue to melt, the sea level rise can reach 40 ft in a few centuries.

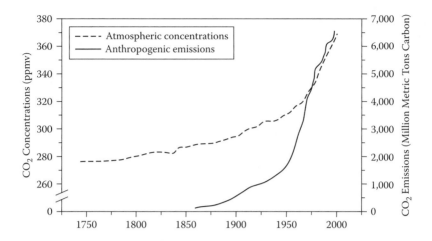

FIGURE 1.12

Trends in atmospheric concentrations and anthropogenic emissions of carbon dioxide. (From Oak Ridge National Laboratory, Carbon Dioxide Information Analysis Center, http://cdiac. esd.omi.gov/.)

In addition to these direct effects, there are also indirect consequences. One consequence is that, as the oceans warm, they absorb less CO_2; the other is that, as the Canadian, Siberian, and Alaskan permafrost melts, the rotting organic matter will release vast amounts of CO_2 and methane. The rising CO_2 concentration of the atmosphere also reduces the pH of the oceans. In addition, about 10 teratons of carbon (tera = 10^{12}) are stored in the frozen methane hydrates of the Arctic regions, which will also be released if the ice melts.

The corrective steps needed must not be allowed to disrupt the world economy. Therefore, a reasoned balance is needed between the costs of being able to live with the consequences, such as raising the dams, and the costs of eliminating their cause, i.e., stopping carbon emissions. There is no question that the goal is to stabilize the climate, but it is also true that we can live with a few degrees of temperature rise and a few inches of sea level increase while we convert our economy to a renewable one. For example, once it has been proved that renewable energy power plants do not cost more than fossil or nuclear ones, it would be reasonable to decide that from now on all new power plants will be of the renewable energy type.

The environmental cost of climate change will be paid by all of humankind. Sir Nicholas Stern, former chief economist at the World Bank, estimated that if no changes occur, this cost by 2020 will amount to 20% of the GWP. The Cato Institute estimates that as of this writing 5 to 10% of the GDP is already being lost due to global warming (http://www.cato-at-liberty.org/2006/11/03/global-warming-costs-benefits/. The global GWP in 2007 was about $65 trillion and the GDP of the United States was about $15 trillion).

In February 2007, the UN Intergovernmental Panel on Climate recommended banning the construction of all conventional coal-fired power plants.

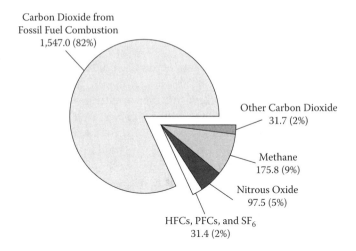

FIGURE 1.13

The anthropogenic greenhouse gas emissions in 2001 in the United States in million metric tons of carbon equivalent. (From Energy Information Administration.)

Yet, in China, every week another coal-fired plant is being built, and in other countries such as Slovakia, they are under construction.

Figure 1.13 describes the quantities and percentages of greenhouse gas emissions in the United States in 2001. Today, of the global emissions of about 8.4 billion tons, the emissions of the United States amount to over 2 billion.

1.2.4 Ice Caps and Glaciers

The cap of floating ice on the Arctic Ocean covers 6.5 million square kilometers (2.5 million square miles). The National Snow and Ice Center in Colorado reported in 2007 that over 40% of that ice (six times the area of California) has melted. This has opened a ring of open water in the Arctic (the Northwest Passage) over North America and Eurasia. According to some estimates, the total ice cover of the Arctic could melt by 2040; this melting trend is continuing at a rate of 7.8% per decade and some believe that it is now irreversible. This degree of melting is unmatched in the last 100 years and is even higher than the temporary melting experienced in the 1930s, when solar flares caused a temporary spell of global warming.

It seems that the poles are warming faster than the rest of the planet. It was believed that because of the tilt of the Earth's axis, the North Pole was melting faster. This is disputed by the March 2008 report of the British Antarctic Survey that reported the fracture of the Wilkins floating ice shelf at the South Pole, which is the size of the state of Connecticut.

The World Glacier Monitoring Service at the University of Zurich reviewed 30 glaciers in 9 mountain regions, including the moving frozen rivers of the Andes, Himalayas, and elsewhere. Its members concluded that the loss of glaciers will reduce summertime sources of river water and therefore will reduce and eventually eliminate sources of both drinking water and hydroelectric power in locations such as South Asia and on the western slopes of the Andes.

If Greenland's 2-mile-thick ice sheet melts, coastal cities and villages could be flooded. Greenland lost 50 cubic miles of "nonfloating" ice (ice on solid ground), and the glaciers are also melting. During the last 150 years, the world's glaciers lost 50% of their surface area—half of that in the last 30 years. The dwindling of ice that accumulated on solid ground will raise the level of the oceans. The melting of the floating ice in the Arctic also contributes to global warming, because water absorbs 80% more solar energy than ice, which reflects most of it. Therefore, if the surface area of water increases, heat absorption also rises.

Measurements in Bermuda and the Bahamas indicate that before the last ice age (130,000 years ago), corals flourished 20 feet above today's sea level. Such high sea levels could have been caused by the melting of the West Arctic Ice Sheet and the ice on Greenland.

All this does *not* mean that the combination of human and natural influences on ice cap formation is easy to understand; after all, satellite observation of polar ice started only in 1979, and the ice cap area around Antarctica is at record highs. Yet, for the first time, scientists have confirmed that the ice on Earth is melting at both ends. How is that possible? How can the southern ice cap grow and the ice be melting at both ends? It seems that the area of the southern ice cap is growing because of increased humidity in the air, and its thickness is dropping because of the warming of the waters below.

Two new studies find that despite the increasing snowfall that results from the increased moisture content of the air, Antarctica's ice sheet is losing more than the snow that is being added. According to the National Academy of Sciences, the Earth's surface temperature has risen by about 1°F in the last century, and according to NASA glaciologist Jay Zwally, "The warming ocean comes underneath the ice shelves and melts them from the bottom, and warmer air from the top melts them from the top, so they're thinning and eventually they get to a point where they go poof!" If the melting speeds up to a rapid runaway process called a *collapse,* coastal cities and villages could be in danger.

James Hansen, director of NASA's Earth Science Research, said, "Based on the history of the Earth, if we can keep the warming to less than 2°F, I think we can avoid the disastrous ice sheet collapse." Hansen and other scientists point out that a rise of at least 1°F—and another few feet of sea level—seems virtually certain to happen, even if corrective actions are initiated now.

In February 2007, the United Nation's Intergovernmental Panel on Climate Change (IPCC) recommended banning all construction on land that is less than 39 inches above sea level. In July 2007, the Union of Concerned Scientists predicted that if inaction continues, the Financial District and the subways in New York City will probably be flooded.

1.2.5 Ocean Currents and Hurricanes

Global warming and melting ice caps affect the formation of hurricanes and also change ocean currents. These currents are great heat conveyors and cause interactions between oceanic and atmospheric processes. Once these aerodynamic and hydrodynamic processes have started, their flywheel effect (inertia) makes it very hard to stop or reverse them.

The "Great Heat Conveyor Belt" (GHCB) transports thermal energy from the equator to the north. As this gigantic conveyor moves heat north (see Figure 1.14), a smaller heat pump (the Gulf Stream) does the same on the East Coast of the United States.

Three forces combine to drive the GHCB. The first is the Coriolis effect, which is caused by the rotation of the globe and results in a faster movement of the ocean's surface near the equator. The second is the level of the Pacific

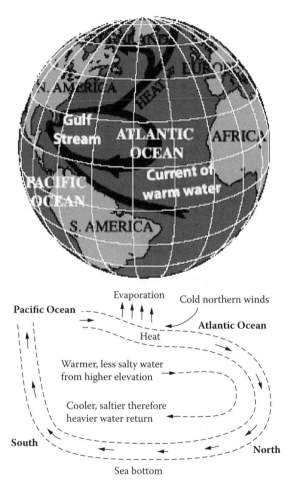

FIGURE 1.14
Ocean currents: The "Great Heat Conveyor Belt."

Ocean, which is slightly higher than that of the Atlantic Ocean. The third driving force is the difference between the salinity and the temperature of the waters near the equator and near Greenland in the north.

The oceans' currents develop as follows: As the warm and relatively low-salinity (and therefore lighter) water from the equator travels north on the surface of the Atlantic, some of it evaporates. This increases its salinity and density. As this water continues north, cold winds cool it, making it even heavier. Finally, near the coast of Greenland, it sinks down to the bottom of the Atlantic. After sinking, this gigantic underwater river (far larger than all of our planet's rivers combined) then travels south, gradually warms up, and becomes diluted. As its salinity and density drop, it rises to the ocean's surface in the Pacific, where the water of the GHCB completes the loop.

Global warming affects ocean currents. Fresh water from the melting ice at Greenland lowers the density of the current, and therefore slows the sinking of this gigantic river. As the GHCB slows, it brings less heat to Europe and the East Coast of the United States, and therefore these regions eventually (in several decades) will cool. The slowing of this heat conveyor and the reduction in the heat it transports to the north also increase the intensity of hurricanes.

Tropical storms form over warm tropical oceans when local sea surface temperatures are above 26.5°C (80°F), causing evaporation and the rising of the high-humidity air up in the atmosphere ("the chimney effect"). The globe's rotation (Coriolis effect) causes this rising air column to start rotating and to form a vortex (tropical depression). Heat from the ocean surface is drawn up through the center of the vortex and is released to the atmosphere as the water vapor around the perimeter of the vortex condenses into rain (Figure 1.15). The more heat is available at the surface of the ocean, the more potential there is to generate heavy rain and high wind.

A hurricane operates as a heat pump. Its evaporator is at its core where the vacuum sucks in the moist and warm air from the ocean's surface. Its condenser is on the perimeter, where the moisture condenses into rain and where its energy is released. The driving force for this circulation is the chimney effect created by the pressure and temperature differences between the core and the perimeter.

According to the 2007 Fourth Assessment Report of the Intergovernmental Panel on Climate Change (IPCC-AR4), it is "more likely than not" (better than even odds) that there is a human contribution to the observed trends of hurricane intensification since the 1970s. In the future, "it is likely (better than 2-to-1 odds) that future tropical cyclones (typhoons and hurricanes) will become more intense, with larger peak wind speeds and more heavy precipitation associated with ongoing increases of tropical sea surface temperatures." According to the IPCC-AR4, "increases in the amount of precipitation are very likely (better than 9-to-1 odds) in high latitudes, while decreases are likely (better than 2-to-1 odds) in most subtropical land regions."

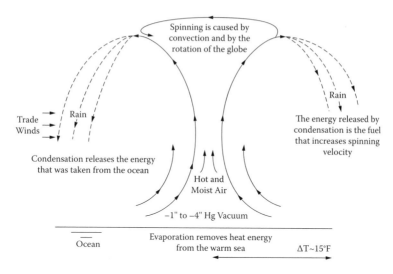

FIGURE 1.15
Hurricanes are formed by the development of hot areas in the ocean. This high temperature zone is the eye of the hurricane, where the oceanís heat vaporizes the water and the moist air is pulled up by the chimney effect, resulting in low pressures at the core. The hot air moves to the perimeter where it releases its energy content as its moisture condenses.

1.2.6 Advanced Strategies to Reduce Carbon Emissions

The price of fossil energy does not reflect either the cost of the damage inflicted by the production of electricity and transporting the fuel, nor does it reflect the damage caused by the emission of the combustion products. This "underpricing" causes overconsumption without providing incentives to invest in alternatives. The atmospheric concentration of CO_2 can be reduced by (a) not generating it in the first place, (b) reducing its rate of generation by conservation, (c) capturing it after being generated but before it enters the atmosphere, (d) removing it after it has entered the atmosphere by planting trees, (e) "fertilizing" vegetation in the oceans (dumping iron dust, which causes phytoplankton to bloom), etc., and other so-called "geoengineering" remedies such as injecting sulfate particles into the stratosphere to screen out the sun's radiation to cool the poles, or blocking sunlight by making clouds more reflective, or launching sun-reflective mirrors into stationary orbit.

Governments can help reduce carbon emissions by (1) taxing on the basis of emissions or on the basis of consumption, (2) supporting/subsidizing alternative/renewable energy technologies, (3) imposing energy and/or fuel efficiency standards, (4) imposing emission standards, (5) charging a fee for greenhouse gas emission, (6) charging a tariff on imported products on the basis of the consumption of carbon during their production, and (7) by "cap-and-trade" arrangements. In addition to encouraging their governments to take the above steps, the average citizen can also take such steps as the ones discussed in Section 1.3.6. For example, if everybody planted just one tree each year, that alone would make a great difference in a few decades.

Probably the most effective method of reducing carbon emissions is direct taxation on the basis of carbon emissions and consumption or use of fossil fuels. If these taxes are used to support the development of renewable energy technologies through tax credits, the total tax burden would not change, and the revenues could be used, for example, to support increased use of public transportation. Similarly, fuel taxes or the taxation of energy guzzlers is preferred, because this allows companies to incorporate that tax into their investment plans and let market forces guide the transition. The other, more bureaucratic, methods are complicated, hard to administer, and easy to evade.

It is debatable what the emission rate is that would stabilize the atmospheric CO_2 concentration at an acceptable level. Similarly, it is not known how much the emission charge should be to reach that target rate. According to an MIT study released in March 2007, if a global carbon emission charge of $25/ton was applied, this would result in stabilizing the global CO_2 emission at today's rate of 28 billion tons per year by 2050.

According to the 2,500 scientists of the UN's Intergovernmental Panel on Climate Change, it is essential to stabilize greenhouse gas emissions by 2015. They also estimated that a carbon emission charge of $50/ton would result in stabilizing the CO_2 concentration in the atmosphere at around 550 ppm (since the beginning of the Industrial Age this concentration has increased from 280 to 380 ppm). This carbon emission charge would increase gasoline prices by 15% and electricity prices by 35%, but would reduce the global economic growth by 10%.

Other estimates suggest that in order to lower CO_2 emission by 2030 to the level it was in 2000, the carbon emission charge should be $50–$100. Robert Stavins of Harvard University estimates that a $100/ton fee would increase the cost of coal-based electricity by 400% and natural gas–based electricity by 100%. Such increases would make the cost of solar, wind, or other alternative sources of electricity more than competitive.

Yet another role governments could play is to provide markets for renewable energy systems by installing solar collectors and wind turbines on government and military buildings. Similarly, governments could help expand the electric power grid into an integrated and nationwide grid so that all green energy generators could use the grid for storage.

1.2.6.1 Taxation or Cap-and-Trade

Both the *cap-and-trade* and the carbon *tax* approaches have advantages and disadvantages:

- *Cap-and-Trade* has the advantage of penalizing the worst polluters. Its disadvantages include that no polluter is forced to cut its excessive emissions, because even the worst polluters are allowed to continue what they have been doing for a price which they can pass onto the public by raising prices. Another disadvantage is that carbon penalties do not directly benefit the transformation to a renewable energy economy because the penalty money stays in the hands of (more efficient) carbon emitters.

- *Taxing Carbon Emissions* has the disadvantage (if applied to all emitted carbon) of equally penalizing all carbon emitters without considering if that ton of carbon is emitted after the plant has applied its best effort of minimizing it or not. The advantages include its ease of enforcement and the fact that the collected money is not returned to carbon-emitting industries, but is available to be used to finance transformation to an inexhaustible and clean energy economy.

"Cap-and-trade" is a way to "commodify" climate change and allow corporations to deal with it by buying their way out of making reductions in their own emissions. It gives the right to pollute, as one needs only to write a check and somebody else will reduce the emission. In spite of all these imperfections, cap-and-trade is still better than doing nothing, because it does give a financial incentive to reduce emissions.

The task of setting the "cap"—the carbon allocation (or pollution permit) for each industry—requires first the establishment of a national total target quantity (moving or stationary) of carbon allocation and next, it requires the distribution of that total quantity among the various emitters. In this sense carbon allocation is just like taxation, except that the emitter has the option of investing in better technology to lower its emissions or paying for the excess emission by purchasing an equal amount of "carbon credit" from some other plant that emits less than its allocation. The lower the allocation of an industry, the more expense that represents, but instead of paying it in taxes to the government, it is spent to buy credits from other emitters.

If allocations were set by industry, such as a uniform cap for the whole power industry (which represents some 40% of the American carbon emission), it would probably double the cost of electricity in locations where electricity is made from coal while representing a windfall for the nuclear power industry. If caps were set as a function of carbon emission, it is not clear how "indirect" emissions would be handled. For example, the oil and refinery industry is directly responsible for only 4% of the total carbon emission, but it fuels a transportation industry, which is responsible for 35%.

If the carbon price on the global "carbon market" settles at $41 per ton,* the value of the global carbon credits market would be over $200 billion, which is more than that of most commodities. It is argued that this trading will drive jobs to countries that do not limit carbon. The more industrialized nations emit more carbon and therefore the cap-and-trade approach would penalize them more. It is for this reason that the opponents of this approach call it "economic disarmament." The EPA in the United States estimated that a 2008 bill on cap-and-trade would have reduced the nation's GDP by about 2%.

Instead of waiting for market forces to take effect, large-scale public funding should be provided for renewable energy development and demonstration

---

* One ton of carbon is contained in 3.7 tons of carbon dioxide.

projects. Yet, where there are no fixed limits or taxes on carbon emissions, governments should at least cap emissions on a tradable permits basis (cap-and-trade) so that if a plant reduces its emissions below the limit set by the permit, it can sell the unused portion to less efficient plants for a profit.

Under the Clean Development Mechanism plan of the UN, $4.8 billion was paid in 2006 to developing countries for reducing their CO_2 emissions. These credits have been sold to richer nations at an average price of $10.70/ton. Of the total of $4.8 billion, China received $3 billion, and all of Africa, $0.15 billion. In comparison to the $10.70/ton rate, in the EU market carbon credits are traded at $28/ton. The Dell Company, for example, has bought carbon offsets in Hungary that will be backed by the planting of trees.

California is already capping emissions and is allowing trading of emission allotments, and a dozen northeastern states are copying these cap-and-trade strategies in order to reduce the emissions of their power and transportation industries.

The problems with cap-and-trade can be illustrated by what is happening to the rainforests. Although 1 acre of rainforest consumes 500 tons of CO_2 yearly, rainforests are being cut down and converted to farmland. The cap-and-trade value of not cutting them down, at the UN rate of $10.70/ton is $5,350 per acre, and at the EU market rate, $14,000. Yet, because "not doing something" cannot be traded, rainforests are being cut down to gain more farmland and to grow ethanol crops that will produce less than $500 a year.

1.2.6.2 Sequestering, Carbon Capture and Conversion into Methanol

It has also been proposed to store the CO_2 emissions underground (sequestering) in the limestone and sandstone cavities from which natural gas or oil has been extracted. This "creation of carbon landfills" is being tested in Thornton, California. The U.S. Department of Energy (DOE) is funding seven testing projects of sequestration technologies. One advocate of this technology is H. J. Herzog of MIT. He estimates that it would take $1 billion a year for 8–10 years to build a test project to evaluate the feasibility of this technology.

Post-combustion carbon capture equipment can be added to existing power plants, but this is very expensive. In the United States several such projects have been abandoned (in Illinois, Florida, West Virginia, Ohio, Minnesota, and Washington State). Storing the carbon dioxide underground requires very unique soil and rock formations to prevent leakage or groundwater pollution.

Gasifying coal so that mercury, sulfur, and soot can be removed before burning eliminates only these pollutants, but not CO_2 emissions and even so the process is very expensive. For this reason the Federal Government "FutureGen" in Mattoon, Illinois pulled out of its showcase project after spending some $40 million on it, when the projected costs doubled to nearly $2 billion. Other gasification proposals (by Duke Energy in the United States and others in Australia, Denmark, and Sweden) are still being evaluated, but completion of the tests is not expected until 2020.

In my view "clean coal" is a pipe dream, but still a lot of research money is being spent to figure out a way of capturing carbon after it has been generated by combustion.

Europe's first land-based carbon dioxide sequestration pilot projects are planned in Ketzin, and at Schwarze Pumpe, southeast of Berlin, Germany. The Ketzin plants will store 60,000 tons of CO_2 at a depth of 850 m (3,000 ft) in a salty water aquifer (60,000 tons is an insignificant capacity as a coal-fired power plant generates millions of tons a year). The 30 mW Schwartze Pumpe plant will store the carbon dioxide in porous stone at 500 atmospheres. Internationally, commercial-scale geologic sequestration (greater than 1 million metric tons of CO_2 per year) projects are underway at Weyburn (Canada), Sleipner (Norway), and in Salah (Algeria).

In the North Sea near Norway a million tons of CO_2 is injected into geological formations under the sea. In the United States bills have been offered in the Congress to require that all new fossil power plants be provided with state-of-the-art carbon capture and storage (CCS) technology, but even if passed, this conversion will take decades and will be very expensive. In addition, CCS reduces power plant efficiency by as much as 30%. Sequestering would add about $1 billion to the capital cost of each power plant. In addition, fossil power plants are not always close to underground storage caverns. A listing of all the carbon dioxide storage projects around the world (as of late 2008) can be found on the Web: http://sequestration.mit.edu/tools/projects/storage_only.html. The fossil power plants are not always close to underground storage caverns. So, this strategy does not seem practical, nor is it economical, even if it was possible to seal the wells so that the high-pressure CO_2 would not escape. In short, sequestering does not appear to be a realistic solution.

In addition to sequestering, the fossil-power industry is developing a variety of CO_2 recapturing and storage methods. One process uses chilled ammonia to remove the CO_2 from the flue gases and, after removal, compresses or condenses it, before injecting the compressed gas or liquid into the ground at about 9,000 ft below the surface.

Another carbon-capturing invention is to convert the captured CO_2 into methanol. If this process matures by the time the solar–hydrogen demonstration power plant described in this book is built, and if there is a CO_2 source near the plant, I will incorporate it as a subsection of the plant that is described in Chapter 4 of this book.

1.2.7 Energy Politics, Economics, and the Lobbies

Several authoritative assessments including that of the Intergovernmental Panel on Climate Change have stated that man-made emissions are largely responsible for global warming and that swift action is needed to avoid potentially tragic consequences. Yet, the debate on global warming is still in progress. Powerful interest groups (http://www.globalwarmingheartland.org) are sponsoring conferences and publications to cast doubt on the very existence of global warming or argue that it requires no urgent action. These interest groups usually profit from the energy status quo and therefore cannot be convinced by arguments. The only way to close this debate is to build the solar–hydrogen demonstration power plant described in this book and *prove* that it is both feasible and economical.

The steps governments can take to slow global warming include the removal of fossil fuel subsidies, which amount to up to $200 billion a year. This compares to the present support for low-carbon technologies amounting to an estimated $33 billion annually. Governments can also boost the support of "green" R&D. The International Energy Agency estimated that the support for renewable energy R&D declined by 50% between 1980 and 2004. It is estimated that in order to achieve a CO_2 stabilization target of 550 ppm (today's concentration is 380 ppm) support for innovation needs to rise from just over $30 billion to $90 billion by 2015 and to $160 billion by 2025.

Other steps that can be taken are increasing global targets for energy efficiency by 2.5%/yr; strengthening building codes for new and existing structures; giving penalties or disincentives for builders who choose the cheapest, least energy efficient designs; policies that promote mass transit, especially rail and international performance standards for industrial and household appliances. Other measures include the promotion of utility pricing that favors energy efficiency and improves energy savings in existing power plants and in the electricity transmission infrastructure.

The conversion from the inexhaustible to a renewable energy economy is controversial, because the interests of developed and underdeveloped nations are in conflict, and there is a segment of the population that is in favor of only "energy independence" (reducing oil imports) but is otherwise not concerned about global warming. This segment of the population considers global warming as either a hoax or a minor problem.

Conversion is also slowed by lack of public education. It is widely believed, for example, that our lifestyle would have to change, comfort would have to be sacrificed, and mobility reduced, if we converted to a renewable energy economy. This is not true. The solar–hydrogen economy of the future will not differ from the present one. Our lifestyle will be supported by the same quantity of energy; it will differ only in the source of that energy.

Still, education can result in voluntary lifestyle changes that will serve energy conservation. Some of these changes, such as the use of bicycles or public transportation, are obvious, but few realize, for example, the consequences of humankind's eating more meat and less vegetables than in the preindustrial age. This trend affects the plant-to-animal ratio on the planet. The meat industry not only produces gigantic quantities of CO_2—as both animals and animal wastes release large quantities of CO_2—but its demand for animal feed also increases the demand for agricultural products, which results in the destruction of more forests.

Progress has been slowed by some governments' refusal to provide leadership. It is unfortunate that conversion to renewable energy became a right–left issue, as if only the left of the planet is warming. On the issues of tax breaks for the oil industry, coal-to-liquid subsidies, oil shale exploration or

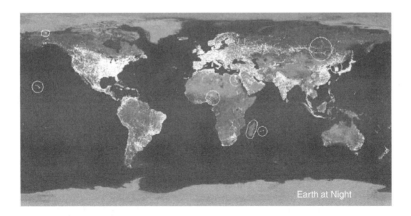

FIGURE 1.16
Illustration of the distribution of energy use on the planet. (Courtesy of C. Mayhew and R. Simmon and NASA/GSFC archive.)

expanding nuclear power, the left and the right are taking diametrically opposite positions. In the United States the areas where the political right and the left are close to agreement are the drilling for arctic and offshore oil on the one hand and on the use of energy efficient light bulbs, subsidizing ethanol, and increasing automobile mileage on the other.

Reducing carbon emissions costs money, and that money has to come from an increase in the price of manufactured goods. This price must reflect the harm that production does to the planet. The "carbon price" can be calculated on the basis of carbon emission or carbon consumption, but as long as some nations have and others do not have carbon caps, the second group will have an economic advantage in the marketplace ("leakage"). The introduction of "tradable permit fees" that treat domestic and imported products equally could be a good first step in balancing the economic playing field. Thus, the price of all emission-intensive products would rise, no matter where they were made.

The economics of reducing carbon emissions would be simple if all countries accepted binding emission caps, but at the 2007 UN conference in Bali, the United States and China refused to make a firm commitment to reduce emissions because they do not want to raise the prices of their products. The United States has agreed only to further negotiations.

Most greenhouse gases are emitted by the industrialized nations (Figure 1.16). These nations are also the ones that will be least affected by and are best able to adjust to climate change, whereas poor nations near the tropics (Africa, South America, and South Asia) will suffer the most from it. Also, the 2.5 billion people worldwide who live on less than $2 a day and go to bed hungry every day do not have the means to contribute to the solution.

1.2.7.1 The United Nations

In 1992, almost all nations signed a treaty that set voluntary goals for curbing the emission of greenhouse gases, but it achieved little. Five years later, the

Kyoto Protocol of 1997, which was an addendum to the original pact, set CO_2 emission caps for the industrial nations but not for the developing ones such as China and India. Three dozen industrial nations ratified this pact, hoping to limit the ecological, financial, and human costs of global warming. This protocol will expire in 2012. For the industrialized nations, the Kyoto Protocol sets the target of reducing greenhouse gas emissions to 5% below the 1990 level by 2012. This treaty was not signed by China or the United States, because these nations did not accept limits on their domestic emissions. With 2008–2012 being designated as the compliance period for the Kyoto Protocol and with many countries and organizations contemplating future commitments, a worldwide effort is underway to lay out a road map for a "post-Kyoto" agreement.

The Kyoto agreement did have some useful effects, however. Between 2004 and 2006, the global investment in renewable and low-carbon technologies increased from $28 billion to $72 billion. On September 24, 2007, Secretary-General Ban Ki-moon announced the start of negotiations on a new agreement to succeed the Kyoto Protocol. In December 2007, 15 years after the original 1992 treaty, an international meeting was called in Bali, Indonesia, to revive the climate treaty. At this meeting of 187 nations, it was proposed to cut the emissions of industrial nations from 25 to 40% below the 1990 level. This was not adopted because of the refusal of the United States and China. When the European Union threatened to boycott the separate meeting of industrialized nations sponsored by the United States and planned in Hawaii, the United States agreed to a 2-year deadline for formulating an addendum to the 1992 treaty. This treaty will replace the Kyoto agreement, but it will not be a global pact that binds the United States or the developing countries to accept mandatory caps on their emissions.

At the meeting in Bali, the focus was not on the nations that are emitting but on the ones that are suffering from these emissions and, therefore, the emphasis shifted from mitigating climate change to adapting to it. The financing of this adaptation process is proposed to come from the rich nations that will be receiving carbon credits for investing in renewable projects in developing countries. In particular, developing countries such as Indonesia or Brazil would be compensated for protecting their rainforests. Carbon trading is expected to become an $86 billion industry in 2008 (much more later), and 2% of that trade is proposed to be channeled to an adaptation fund that would help poor nations to cope.

Less-developed nations such as China believe that the industrial nations polluted their way into prosperity and became concerned about environmental damage only after their economies had already matured. Therefore, they feel it is unfair that the industrial nations—which did and are still emitting most of the greenhouse gases—should expect the less-developed ones to curtail their economic development to cut global carbon emissions. At the 2007 UN conference in Bali, China and India continued to refuse limiting their emissions.

In 2007, the UN's Intergovernmental Panel on Climate Change concluded (http://hdr.undp.org/en/) that water shortages, droughts, flooding, and severe storms are the results of unsustainable energy consumption, and a fundamental shift is needed both in regulations and market incentives. It also concluded that even after emissions are cut back, temperatures and sea levels will continue to rise for many decades, because these gases are long-lived and the inertia of these global thermal processes is gigantic. As long as more CO_2 is sent to the atmosphere than is being taken from it, the buildup will continue. The greenhouse gas concentration that accumulated during the last centuries will start to be reduced only when that material balance is reversed, which will only occur when carbon emissions are cut to preindustrial levels.

In 2008 a panel of United Nations experts expressed the view that some energy-related developments such as the manufacturing of ethanol are creating hunger and should be abandoned. In 2008 alone wheat prices increased by 130% and soy prices have risen by 87%. The panel also indicated that, while food represents 60% to 80% of the spending in the developing regions, overall food prices increased by 83% during the last 3 years.

1.2.7.2 United States

The United States is at the top of the list of leading nations, yet, in the area of global warming, instead of acting as a leader, to date, it has obstructed progress, thereby contributing to tension between rich and poor nations. To be consistent with her past, America should be a role model for the world. It should show that, while being innovative and becoming the most energy-efficient nation on the planet, it can also stay prosperous and secure. America should lead by example through initiating an effort matching the scale of the war on terror.

Instead, today, the United States is the world's largest emitter of CO_2. It emits about 2 of the over 7 to 9 billion tons of yearly carbon emission on the planet, and that number is projected to rise to 3.3 billion tons by 2030. The carbon dioxide emissions of American power plants increased in 2007 by 2.7%, the largest jump in a decade according to the EIP (Environmental Integrity Project). In spite of emitting more greenhouse gases than any other nation, in 2001 the United States withdrew from the Kyoto agreement and as of this writing is still refusing to accept any carbon emission targets or limits, and the American carbon emitters are still free to dump their waste into the atmosphere and do not pay anything for it. Likewise, as of this writing, the government is still in favor of "voluntary, self-imposed limits." This is similar to setting your own speed limit on the highway.

Yet there is hope that this will soon change, because the presidential candidates of both parties announced that they are in favor of both increasing fleetwide mileage on automobiles and of mandating cap-and-trade policies to lower greenhouse gas emissions. Therefore, it is likely that by the end of 2009, polluters will have to buy permits to emit carbon dioxide and the government will spend billions in subsidies on renewable energy. It is not clear if the pending bills will permit American companies to buy emission offsets overseas or to allow tariff-free imports from nations that have not imposed carbon emission limits on their industries.

Part of the reason for this change in attitude is the government climate report issued in May 2008 (www.climatescience.gov) indicating that without drastic reductions in emissions, disruption of water supplies, agriculture, forestry, and overall deterioration of the ecosystem will occur. Rainfall in the West is projected to drop by 20% while in the Midwest and East it will increase, resulting in the spreading of nonnative and invasive weeds, insects, and pests. In addition the spikes in summer temperatures will contribute to crop failures.

The United States does not have a bipartisan energy policy. It is still being debated if the tax incentives should be spent on finding more oil, building more nuclear plants, or supporting renewable energy technologies. Similarly, there is no agreement on the increased use of renewable energy in power plants or on limiting carbon emissions. Part of the reason for this is political, but regional differences between energy sources also play a role. Natural gas is an important energy source in the South and West, while nuclear energy is mostly used in the East. Coal is important throughout the country except the West, while hydroelectric power is significant in the Northwest of the country. This could be part of the reason why most of the states that seek to cap carbon emissions are western states.

On January 23, 2008 the World Economic Forum announced the "pollution ranking" of 149 countries and the United States was listed in the bottom tier when greenhouse gas emissions were considered and also at the bottom on regional smog.

On the international front, the United States argues that no meaningful progress can be made without the support of China and India, and is trying to organize the industrialized nations into a separate interest block. The past actions of the United States can be characterized as a bipartisan failure to act and as deliberate blocking of opportunities. In the summer of 2008 the G-8 industrialized nations (United States, Japan, Russia, Germany, France, England, Canada, Italy) in a meeting in Hokkaido, Japan agreed to reduce greenhouse gas emissions by 50% during the next 42 years (by 2050). This is totally inadequate. According to the International Energy Agency, meeting the G-8 target will require yearly building of 32 nuclear and 50 carbon capturing fossil power plants at an annual cost of 1.1% of the gross world product (GWP). Since by the end of the century the nuclear and fossil fuel reserves will be mostly exhausted, meeting the G-8 plan by nuclear and fossil means makes little sense. Yet the G-8 meeting was useful, because it brought attention to not only the inadequacy of the plan, but also to facts such as the per unit of production energy used in Japan is half that of the EU or the United States, and is 1/8th that of China. The 2007 energy bill provided $25 billion in loan guarantees for building new nuclear plants but only $10 billion for all the renewable energy plants. To date, it seems that on energy conservation only "carrots" such as tax incentives have been used, but no "sticks" such as emission limits or taxes. The 2007 energy bill did not include a cap-and-trade system for CO_2 and it did not require that utilities generate 15% of their energy from renewable sources.

The United States consumes one fifth of the world's oil and possesses only 3% of its oil reserves. She also has the world's lowest gasoline prices, the least developed public transportation system, and the longest daily commutes. Yet the oil industry suggests that more drilling could lead to energy independence. The most unfortunate aspect of the 2007 energy bill is that the oil industry was successful to keep $12 billion in tax breaks and to obtain federal support of oil exploration and for permits to drill in the Arctic National Wildlife Refuge. The 2007 energy bill also fails to extend vital tax credits, which expire in 2008 and would have been used to subsidize wind, solar, geothermal, and other renewable energy industries. In his budget request for 2008, the president was asking for $1.24 billion for renewable energy (for the EERE) and $309 million for his Hydrogen Fuel Initiative for the Energy Department. In February 2008 the House approved a bill to extend the $17 billion in tax credits to promote renewable energy production and to end the incentives for the oil and natural gas producers, but as of this writing the President has not signed it.

A "fuel or food" debate is also in progress, because biofuels are made from agricultural products and therefore they drive up food prices. Today 33 nations are at risk of social unrest because of the rise in food prices (most of their families spend 75%–80% of their income on food). In the United States, a fifth of the corn crop is used to brew ethanol and as more corn is planted shortages develop in other produce, such as soybeans.

The agricultural lobby favors biofuels, whereas the food industry is against subsidizing the production of corn-based ethanol because it increases the cost of corn (the diet of hogs is 80% corn) and, thereby, the cost of food. Yet since 1999, $50 billion in subsidies and tax breaks have been spent on supporting corn ethanol. (Corn is a major product of states such as Iowa). The 2007 energy bill requires that 36 billion gallons of renewable fuels be produced and blended into gasoline by 2020. This is a fivefold increase from the current production levels. If this goal is reached, one sixth of the projected gasoline consumption will be replaced by renewable fuels by 2020. Of the authorized $36 billion, $21 is for "advanced" biofuels (cellulosic ethanol from crop and wood wastes or from perennial grasses), the rest is for corn-based ethanol. At the end of 2008 it seems that the ban on offshore drilling will unfortunately be lifted. The 11 billion barrel offshore reserves and the 7 billion reserves in the Arctic National Wildlife Reserve are sufficient to meet the total American oil consumption for about 2.5 years and the oil production will probably not start until 2020. It is estimated (*New York Times*, 9/15/2008) that the total benefit of obtaining these 18 billion barrels of oil is $1.7 trillion and the investment required to do so is $0.4 trillion.

The same reduction in oil imports could be achieved right now (not by 2020), if automobile mileage was increased by only 5 mpg. Also, if the same $0.4 trillion was invested to build solarñhydrogen power plants in the Southwest, the amount of hydrogen produced would completely eliminate the need for all oil imports, not for 2.5 years, but forever. On top of all that, the jobs created by building the hydrogen infrastructure would result in the greatest economic boom of the century.

Until 2007 the fleet mileage average of American cars stood at 25 mpg. The 2007 energy bill requires this average to rise to 35 mpg by 2020. This requirement of the 2007 energy bill resulted from a compromise that stripped out some of the tax support for renewable energy and eliminated new incentives for plug-in hybrid vehicles from the bill.

The auto industry also pressured the EPA to reject the 43 mpg target of 17 states, forcing them to sue the EPA. The difference between the two targets (35 mpg vs. 43 mpg) is more than the total oil imported from Saudi Arabia and Iraq.

In 2007 the powerful lobby of the electric utilities successfully fought against requiring them to use 15% renewable energy in their energy mix by 2020. This requirement was included in the 2007 energy bill of the House but was dropped during the negotiations before the president signed the bill because of the pressure by the "polluter lobby" led by the Edison Electric Institute and supported by the National Association of Manufacturers and the Chamber of Commerce. This lobby also successfully fought against legislation that would have taxed or placed a cap on carbon emissions. The 2007 bill also provided $10 billion to turn coal into fuel and $2 billion for coal gasification. The coal industry and the refineries are also lobbying for federal support for carbon-capturing technology research and are seeking approval to "improve mining efficiency" by open pit mining (blowing the tops off mountains), thereby transforming woodlands into toxic rubble.

Bills have been offered in the Congress (Markey–Waxman in 2008) to stop the building of low-tech coal plants and to require that all new plants be provided with state-of-the-art CCS. Yet, the proportion of CCS plants will remain small. By 2030 it is estimated that the proportion of CCS plants in operation will be between 9% (EIA) and 15% (EPRI) of the total.

As the value of the dollar is dropping and as the domestic coal demand is constrained due to the tightening of environmental regulations, coal exports are increasing and reached about 8% of the total coal production in 2008.

The nuclear lobbies were also successful in obtaining $25 billion in loans, which are fully guaranteed by the government for building new nuclear power plants, and $2 billion for a uranium enrichment plant. On the other hand, the problems of nuclear waste storage, security, and decommissioning have not been solved. The nuclear industry is planning to build 28 new reactors at about $5 billion each.

In contrast to the industrial lobbies, the courts seem to be "catching up" with the interests of the general public and have already played a positive role. In April 2007, the Supreme Court in a 5:4 decision rejected the administration's contention that CO_2 is not a pollutant and decided that under the Clean Air Act, the EPA has the power to regulate greenhouse emissions. It took eight years for the lawsuit filed by the EPA and eight states against the American Electric Power Company to force a reduction in its carbon emissions.

The penalty for inaction was $15 million in civil penalty and $60 million for environmental mitigation. Later, based on this ruling, Georgia denied the permit for a coal power plant (2008). Lately the industrial lobbies have been trying to take away the authority of the EPA to enforce the mileage rules and give it to the more industry-friendly Transportation Department's Highway Safety Administration. In 2008 a three-judge federal appeals panel in Washington ruled that the EPA limits on mercury emissions from power plants are insufficient. In 2008 Kivalina, a coastal village in Alaska, sued 5 oil companies, 14 electric utilities, and a large coal company, arguing that they indirectly caused the flooding of the village through global warming which necessitated the relocation of the village.

In another case, the federal appeals court in San Francisco ruled that raising the fuel economy of light trucks from 22.5 to 23.5 mpg as proposed by the Transportation Department was not enough. This appeals court also held that exempting larger SUVs (such as the Hummer H_2) from fuel economy standards is unacceptable. Also, in some cases the American courts have agreed that the states have the authority to formulate their own regulations.

In addition to the 2007 energy bill, the federal government has taken some small steps, such as providing a production tax credit for wind-generated electricity. This federal subsidy amounted to 2.3¢/kWh produced in 2007. The tax credit in 2006–2007 cost the treasury $2.75 billion. Being the largest single user of energy in the country, the government could do much more. It could, for example, advance energy conservation in all government buildings and require its appliance suppliers to meet high-efficiency standards. The federal government could also advance the use of renewable energy by such steps as standardizing the use of solar shingles on the roofs on all government and military buildings.

In contrast to the inaction at the federal level, 20 American states and the District of Columbia require the utilities within their states to use some renewable energy. In addition, 17 states are working on tailpipe emission standards that would force car manufacturers to increase the fuel efficiency of their cars. California and 16 other states want to raise the average mileage in their states to 43 mpg, which new standard would affect half of the automobiles in America. Currently 25 states support mandates for renewable energy and 18 states want to cap industrial carbon emissions.

Probably the most meaningful efforts by American States are the Regional Greenhouse Gas Initiative (RGGI) of the 10 northeastern states and the Western Climate Initiative organized by California. They limit the maximum emissions of plants. If emissions are below that limit, they allow these plants to sell the difference (the margin) to plants that emit more than the allowed limit. This way the cleaner plants can sell their "margin" at auctions. The first such auction was held in September, 2008. Because the emission limits were set high, the supply of carbon allowances was greater than the demand for them and the value of a ton of carbon allowance was only about $5, while in Europe where the maximum emission limits are set lower, it is about $38 per ton.

An unexpected new development is that the EPA refused to give permission to the states to set their standards on CO_2 emission higher than that of the federal government. This decision was praised by the car manufacturers, but the 17 states have sued the EPA to overturn the decision in the courts.

The outcome will set a precedent by deciding that in case of the failure of the federal government to sufficiently regulate emissions, the states do or do not have the right to act on their own.

The states leading in the passage of renewable energy legislation are Florida, New Jersey, Massachusetts, and California. California has set the ambitious goal of creating 3,000 megawatts of solar power by 2017 and has earmarked $3.2 billion for subsidizing solar installations in order to place solar collectors on one million rooftops. The city of Berkeley pays the upfront cost of residential solar installations recouping that investment over 20 years in increased property taxes. California also has a vision of a hydrogen highway with 100 fuel stations running the length of the state by 2010. In San Francisco, the Bay Area Air Quality Management District is recommending a direct charge of 4.4¢ per ton of carbon dioxide emitted (16.1¢ per ton of carbon).

Other states are not far behind. Hawaii plans to obtain 70% of her energy needs from renewable sources by 2030. Pennsylvania in 2008 passed an energy bill that provides $850 million for investments in renewable energy, including $200 million in solar energy installations. California and Florida are planning to reduce their greenhouse gas emissions 25% by 2020. Other states have also set global warming–related goals. For example, Maine will cap its emission at 5.9 million tons and will reduce it 10% by 2019. In addition, a number of states, including California, New Jersey, New York, Connecticut, and Florida are supporting solar, wind, and other renewable energy installations by rebates and tax incentives. Massachusetts is planning to become the first state that requires its heating oil to contain 2% biofuels and in waiving its $23 gasoline tax on cellulosic ethanol.

As we have indicated, progress is slow not only because of the influence of powerful lobbies, but also because it is difficult to make lawmakers think beyond the next election in the interests of future generations.

1.2.7.3 Europe

The 15 original European Union countries agreed to implement the Kyoto Protocol and to curb global warming by establishing carbon emission quotas for thousands of factories. (Since that time the EU has grown to 27 states.) By 2005, the Kyoto Protocol resulted in a 2% reduction in carbon emissions relative to the 1990 levels. After 2012, when the Kyoto Treaty runs out, the EU is considering placing overall emission caps on national governments.

The EU aims for an 8% reduction in CO_2 emissions by 2012 and 20% by 2020, and recommends a global target of cutting emissions to 50% of the 1990 level by 2050. The carbon tax planned for 2010 is €16 (about $20) per ton of CO_2 released. European utilities, if they stay below their emission quotas, can sell that margin on the auctions of the open market at about $25 to $40 per ton of carbon. The EU requires that by 2010 its members increase the proportion of renewable energy use in their overall energy consumption from 5.3% to 12%, and generate 22% of their electric energy from renewable sources.

The EU also requires that fuels used for transportation contain 5.75% bio-fuels by 2010, and 10% by 2020. The Union also provides a guarantee to purchase biofuel crops at prices that exceed the market value of wheat. This is likely to result in an increase in the cost of food items. At the end of 2007, deep cuts in automobile emissions and a penalty for not meeting them were proposed. The cut proposed is from 146 to 120 g/km of CO_2, and the penalty, beginning in 2012, is 20 €/g, rising to 95 €/g by 2015.

Another European initiative is to regulate airline emissions by charging a carbon tax for emissions that exceed the allowed limit. The cap-and-trade approach would allow airlines to buy or sell the difference between their actual emissions and the limit. Because of the resistance of the airlines and the lack of international standards, it will probably take years before such a system is implemented.

Denmark, Finland, Norway, and Sweden have been taxing carbon emissions since the 1990s. The results have varied with the tax collection methods. In Finland, Norway, and Sweden the carbon taxes had little impact on emissions, as industry just included them in its cost of operation while the governments treated these taxes as general revenue. In Denmark, the collected carbon taxes were invested in subsidizing the development of alternative energy technologies, and as a consequence the per capita emission by 2005 dropped below the 1990 level.

On the other hand, because of record high oil and natural gas prices and because of the aversion to nuclear energy, some European power plants (Enel) are converting from natural gas or oil to coal. In addition, some 50 coal-fired power plants are planned to be built in Europe during the next 5 years.

Among the individual European states, Germany is the leader in supporting the renewable energy technology, which already generates 14.2% of German electricity. Germany aims at reducing greenhouse gas emissions to 40% below the level of 1990 by the year 2020. In Germany, the power companies "buy back" the solar-generated electric power at a rate that is secured for 20 years. This is called a Feed-in Tariff (FiT) program. Because of the reliable income guaranteed by the FiT program, many homeowners installed solar collectors on their roofs as did farmers in their fields.

Other European nations have also taken some concrete steps. For example, England is planning to reduce greenhouse gas emissions 26% by 2020. Britain's Minister for the Environment Hilary Benn said that the United States must end its opposition to legally binding target limits on carbon emissions. French President Nicolas Sarkozy said, "Collective action is imperative."

1.2.7.4 Asia

In this section I am focusing on the position of China, because almost the same could be said for India and other Asian nations. Naturally, there are exceptions, because Japan, for example, with the world's second highest GDP is also one of the world's most energy-efficient nations, using only half the energy per unit of production as the EU or the United States and 1/8th that of China.

In a couple of decades, China's oil consumption will double and the number of her cars will grow sevenfold. Today, 68% of China's energy comes from burning coal, and if this trend continues, in 25 years it will emit as much carbon as all the industrial nations combined.

In China, the energy consumed to produce $1 GDP ranges from the equivalent of 1.0 to 2.5 tons of coal. In the United States, $1 GDP is generated by 0.4 tons. In other words, China uses 250–600% more energy per unit of production. The Chinese government knows how disastrously low their energy efficiency is and aims at reducing their energy use by 20% in the next 5 years, but they appear to be failing even at meeting that modest target.

China argues that the industrial nations damaged the planet (Figure 1.16) and therefore they should be the first to pay for the consequences. China also argues that the international emission quotas for nations should be set on the basis of "per capita" and not "total" CO_2 emission. Yet it is planning to increase the share of its renewable energy use to 10% by 2010. Although this goal is less than that of the EU (22%), it is more than that of the United States, which, even in its 2007 energy bill, did not set any fixed target at all. Also, China proposed that the industrial nations should disseminate nonpolluting energy technologies and invest 0.5% of their GDPs to do so. The United States rejected that proposal.

China's oil demand is expected to double by 2020, the number of cars on her roads increased sevenfold since 1990, and she will have 300 million cars by 2030.

Last year China added 96 gW of new coal-burning power plants—without any CO_2 cleanup—to its electricity-generating capacity. Today, China is building almost two fossil-fired power plants a week. In 2006, she added 114 gW to her fossil-generating capacity. Last year she burned 2.7 billion tons of coal. This is 75% of the quantity that was projected for 2020. Since 1990, American emissions rose by 18%, whereas Chinese emissions increased by 77%. Because the Chinese economy is three times as "carbon intensive" as the American, dollar for dollar more reduction could be achieved if investments were made to reduce Chinese, instead of American, power plant emissions.

In China and other less industrialized nations, industrial energy efficiency could be substantially improved. Chinese buildings require twice the energy as do buildings in similar climates in Europe or the United States. The efficiency of their power plants is 40% below the efficiency of combined-cycle plants in the United States. Their steel industry requires 20%, their cement 45%, and ethylene producers require 60% more energy than the international average. For these reasons, state-of-the-art optimization and energy conservation techniques (the topic of Chapter 2) alone can make a big difference.

1.2.7.5 OPEC Countries

In the Persian Gulf, Abu Dhabi has started to build Masdar City, a mini-municipality of 50,000, which will be car-free and will be shaded by solar panels to provide its own electricity. Its water supply will come from solar-powered desalination plants and all its wastes will be composted or recycled. Masdar City is part of a $15 billion investment in renewable energy technology by Abu Dhabi.

While this is the largest renewable energy project in the world, it is small compared to the $150 billion investment which the 13 OPEC countries will make in new oil projects through 2012.

1.2.8 Transportation Trends

By 2020, the number of cars on the planet will reach 2 billion. Today 70 million automobiles are produced yearly, and it is projected that this number will double in 5 to 6 years. In this same time period, 10 million of the cars built will be electric cars. The transition from today's oil-based economy is just beginning, but some trends are already observable. These include the pressure on manufacturers to (1) improve their fuel economy, (2) develop gasoline engines that can also burn ethanol–gasoline mixtures and diesel engines that can use biodiesel fuels, (3) introduce multifuel vehicles, (4) introduce regular and plug-in hybrid cars, (5) develop electric cars using new high-density batteries, (6) develop electric cars with hydrogen fuel cells, and (7) develop hydrogen combustion engines.

The car manufacturers have not yet decided whether the future belongs to the all-electric car, the hydrogen fuel cell, the hydrogen internal combustion engine, or some multifuel vehicle. The efficiencies of the various engines are as follows: gasoline, 25%; diesel, 35%; hydrogen internal combustion, 38%; and hydrogen fuel cell, 45 to 60%. Car manufacturers do not want to improve the mileage of their fleets and are delaying the conversion to high-efficiency vehicles, because their most profitable products are cars such as the large and inefficient SUVs. The fleetwide mileage of American cars is 27.5, while in the EU it is 43, and in Japan it is 46. The average American drives about 12,000 miles a year, while Europeans drive 8,000, and Japanese 7,000.

Cars with better mileage on today's market are listed below. Their mileage is given for both city and highway driving (city/highway). They are mostly hybrids—2-Seater: Smart Car for Two (33/41), Mini-compact: Mini Cooper (28/37), Sub-compact: Toyota Yaris (29/36), Compact: Honda Civic Hybrid (40/45), Mid-size: Toyota Prius (48/45), Large: Honda Accord (22/31), Small wagon: Honda Fit (28/34), Mid-size wagon: VW Passat (21/29).

The energy cost of worldwide transportation of food and other goods is also rising, because goods are no longer used or consumed locally, but are brought to the most profitable, sometimes distant markets. This creates not only pollution but also food shortages in the less prosperous parts of the world. On a per mile basis, the most energy-intensive method of transportation is air freight and the least expensive methods are railroad and ship transports. On the other hand, in terms of total energy consumed and the size of the carbon footprint created, ship transport is twice as damaging as is aviation.

1.2.8.1 Biofuels and Multifuel Vehicles

The 2007 energy bill of the United States sets the target to increase ethanol production, which should increase from the present 6.5 billion gallons a year to 36 billion by 2020. Biofuels are proposed to be made from soy oil, animal fat, agricultural waste, municipal garbage, wood chips, sugarcane, and switchgrass, but their largest source remains corn. Ethanol advocates are pushing for high (20–30%) ethanol blends, but in the United States, out of 200 million motor vehicles, only 6 million are certified to run on blends higher than 10%. Also, some ethanol opponents claim that ethanol's mileage is worse than gasoline's, and increasing ethanol concentration not only lowers mileage but also increases pollutant emissions.

In 2008 a panel of UN experts expressed the view that the production of biofuels, and particularly ethanol, is reducing the food supply of the planet and creating hunger. The panel proposed the abandonment of these processes and indicated that global overall food prices have increased by 83% during the last 3 years. This has tragic consequences as food represents 60% to 80% of the spending of a family living in the developing regions of the planet. Jean Ziegler of the UN called biofuels "a crime against humanity."

In addition to the biofuel sources that come from food products, non-food grasses and reeds can also be harmful. Being an alien, powerful and invasive species, they can overrun adjacent farms or natural lands. Some of these species, such as *Jatropha*, are also poisonous and if they invade pastures, they can harm the grazing herds. Other species, such as giant reeds can drain wetlands and clog drainage systems. Therefore, not all the sources of cellulosic biofuels are useful or desirable.

By 2020, the EU wants to have 10% ethanol in its transportation fuel, and China is aiming for 15%. The United States Senate is proposing a biofuel production target of 36 billion gallons by 2020, of which 21 billion would come from corn-based ethanol. As a result, corn prices doubled (from $1.65 to $3/bushel), and the number of ethanol plants also nearly doubled (from 81 to 129).

It is estimated that food production would be severely disrupted if more than 15 billion gallons of corn ethanol were produced. At the same time, the OECD (Organisation for Economic Co-operation and Development) estimated that replacement of 10% of America's motor fuels with biofuels would require about one third of all the cropland that is devoted to the production of cereals, oilseeds, and sugar crops.

Brazil is the leader in biofuel technology. "Flex-fuel cars" that can also run on biofuels comprise 70% of the new cars in Brazil. In other regions of the world, such as in Hawaii, large (28 million gal/yr) biofuel plants are being built.

The ethanol concentration in gasoline can vary from 10% (called E10) to 70% and up to 85% (called E85). All automobiles can use E10 fuel, but only the flex-fuel vehicles (6 million cars in the United States, which is 2.4% of the automobile fleet) can run on E85. The flex-fuel car gas tanks are distinguished by yellow caps. During the last 3 years, the number of gas stations selling ethanol blends rose from 285 to 1413, but that number is still only 1% of the gas stations in the United States. Ten times that number is needed if most American motorists are to have access to a pump within 5 miles of their homes.

Lately, natural gas is also being used because of its lower cost, but the conversion is expensive ($10,000).

The three main American car manufacturers are planning to have half of their fleets run on ethanol blends, and the government is subsidizing ethanol production. On the other hand, little has been done about the distribution of ethanol. Ethanol is corrosive, there are no ethanol pipelines, and ethanol trucks and railcars are few; 40,000 ethanol railcars are on backorder. Because of all these problems, in 2007, the price of corn ethanol collapsed from $3.60/gal to $1.80/gal.

1.2.8.2 Hybrid Cars

Today, hybrid automobiles are used more to make political or cultural statements than because their mileage is much superior to that of conventional automobiles. Almost all manufacturers are in the process of developing hybrids. They include Cadillac's Provoq, Chevrolet's Volt; Chyrsler's EcoVoyager; Dodge's ZEO; Fisker Automotive's Karma; Ford Escape; GMC Yukon; Honda Civic; Jeep's Renegade; Mazda Tribute; Mercedes-Benz's S 300 BlueTec; Mercury Mariner; Nissan Altima; Saturn's Flextreme; Tesla's Roadster; and Toyota's Camry, Highlander and Prius.

Toyota's Prius uses the heavy, range-limited nickel-metal hydride battery, basically because it is safe. The Prius recaptures the braking energy (instead of wasting it in friction) and runs on electric power in stop-and-start traffic, but its all-electric mode of operation is limited. Toyota's Prius is also available in a version that has been converted to hydrogen by Quantum Fuel Systems. Toyota is planning to have a hybrid version of all its 2010 models.

GM plans to increase the electric mode of operation of the Chevrolet Volt, which is designed as a "plug-in hybrid" that can be recharged overnight. This manufacturer plans to use high-energy-density lithium-ion batteries to obtain an all-electric range of 40 mi, and hopes that these batteries will be safe and reliable by 2010. This car is claimed to provide 150 mi/gal and is expected to be available by 2010.

Honda has dropped the hybrid versions of its Insight and Accord, but (in addition to its hybrid Civic) is working on a completely new hybrid car. Conversion kits are also available* to add extra batteries and convert the Prius and Ford's Escape into plug-in hybrids that provide up to 75 mi/gal.

In order to reduce peak demand on power plants, it has been suggested that stored electricity in the batteries of plug-in hybrid cars could be used during peak periods. The contract between the owner of the car and the

---

* From A123 Systems of Watertown, Massachusetts.

utility that operates the grid would stipulate that at night, when electricity is inexpensive, the owner would charge the batteries, and during the peak periods of the day, if the car is not in use, it would be left plugged in and the charge in the batteries would be available to the utility. Naturally, controls would be provided to always leave enough power to start the gasoline engine until recharged.

1.2.8.3 Electric Cars

In the past, electric vehicles (EVs) were thought of as glorified golf carts. That is no longer the case. Today, about ten electric car designs are in production, including the Tesla Roadster, GM's Sequel, Chevrolet Volt, Electrum Spyder, Phoenix SUT, Tango, Think City, Venturi Fetish, Wrightspeed X1, and Zap-X, and more from Nissan, Mitsubishi Motors, Fuji Heavy Industries, Renault, Project Better Place, and Bajaj Auto are on the way.

Their advantages and limitations are multiple. From the American perspective, energy independence and the EV's operating cost are the main advantages of replacing the internal combustion (IC) engines with batteries: energy independence, because electricity is made mostly from American coal, and operating cost, because the price of a gallon of gasoline can pay for the electric energy used to drive 100 mi. When compared with hydrogen fuel cells, another advantage is that the electric "fuel" distribution infrastructure already exists, as only an electric plug is needed to refill the batteries.

The main disadvantage is cost: the purchase price of an EV today is still around $100,000. This will drop drastically when Nissan American comes out with its electric vehicle in 2010 and Bajaj Auto starts producing its $2,500 car in 2011. Other disadvantages include the limited driving range of the electric cars, long charging time, short battery life, and the small cargo space because of the weight and size of the batteries. To overcome the charging time, there are some high-voltage charger designs (Altair Nanotechnologies) that claim to reduce the normally required time from several hours to about 10 minutes.

Ideas to overcome the charging-time problem include the introduction of new types of "filling stations" where all the batteries as a single pack would be replaced in a couple of minutes with already charged ones. These electric filling stations could also offer multiple fuels. Yet another design variation being experimented with is to add solar collectors to the roof of the car and use the electricity generated to recharge the batteries.*

As to past history, during the last 25 years the cost of batteries has been reduced by a factor of 12, and according to some estimates (California Air Resources Board), if lithium–ion packs were mass produced, their unit cost would be between $3,000 and $4,000.

A few years ago Ford Motor Company leased some 300 electric cars in California. Later, although their owners were satisfied, the cars were recalled and sold in Norway, without a clear explanation of the reason for this recall.

---

* From solar electric vehicles in California.

1.2.8.4 Fuel Cells vs. Batteries

Until new batteries that can provide much higher energy densities without compromising safety are discovered, fuel cells will continue to outperform today's heavy and large storage batteries. On the other hand, it is less expensive to build electric cars with batteries than with fuel cells. Today's batteries are less expensive than fuel cells, but their energy density is insufficient, and their weight and size are too high to provide the required driving range. The final outcome of the battery-versus-fuel cell race cannot be predicted. All that is obvious right now is that there are substantial developments in both fields.

In the area of fuel cells, reliability and availability have much improved. Recent U.S. military experience with phosphoric acid fuel cells found that the mean time between failure (MTBF) was almost 1,800 h and the availability was 67%. This is comparable with the MTBF service intervals for diesel generators. These fuel cells also favorably compare with the service interval needed for a typical gas turbine generation set. Still, much more development is required to obtain a commercially viable product. Today, the typical fuel cell system still requires servicing every 3–4 days to replace its scrubber packs.

The early electric cars used the old lead–acid batteries. Today's hybrids are provided with more robust nickel–metal units. The EVs of the future are likely to be provided with lithium–iron batteries, found in today's laptops and cell phones. Much work remains to be done in this area to increase safety and life span (to 100,000 mi of driving), while reducing their cost. Nissan and Mitsubishi are both making major investments in building lithium-ion battery mass production plants.

New battery developments include the ultracapacitor hybrid barium titanate powder design (EEStors). These devices can absorb and release charges much faster than electrochemical batteries. They weigh less, and some projections suggest that in electric cars they might provide 500 mi of travel at a cost of $9 in electricity. But these are only the projections of researchers.

Another direction of battery development involves high temperature and larger units. NGK Insulators, Ltd., in Japan uses sodium–sulfur batteries operating at 427°C (800°F) that are able to deliver 1 mW for 7 hours from a battery unit. The size of these units is about the size of a bus. Such units could be used at electric filling stations that are not connected to the grid.

1.2.8.5 Hydrogen Fuel Cell Cars

Hydrogen fuel cells are widely used in forklift trucks and for power backup in data centers. The reason for their popularity, in contrast to lead–acid batteries is that they last longer (8 hours instead of 2 hours), require no recharging, can discharge energy faster, and their price is becoming competitive.

In June 2008, Honda started the production of its FCX Clarity (a sleeker version of its Accord) which runs on hydrogen and provides a driving range of 280 miles, a mileage of 78 mpg of gasoline-equivalent hydrogen, and an

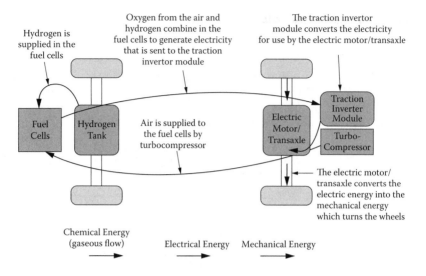

FIGURE 1.17
The hydrogen fuel cell car causes zero emissions of greenhouse gases or pollutants. (Courtesy of H₂Gen Innovations, Inc.)

ability to accelerate from 0 to 60 mph in 9 seconds. Honda is going to lease the vehicle for $600/month; although the car itself cost several hundred thousand dollars to produce in 2008, the price is predicted to drop drastically as mass production starts. The FCX Clarity is driven by a 100 kW fuel cell, which is the size of a desktop PC.

In the cars with high-efficiency hydrogen fuel cells, the motor is electric. Fuel cell efficiency is about 60%, whereas the efficiency of gasoline IC engines is only 25%. It is this high efficiency of the fuel cells that makes them prime candidates to provide electricity for the electric cars of the future (Figure 1.17). As of today, the fuel cell–based electric vehicles (FCEVs) are the best long-term options for the zero emission vehicles (ZEVs) of the future. Assuming that the fuel cell efficiency is 2.5 times that of a gasoline engine, and assuming that gasoline costs $3.50/gallon and hydrogen is $5/kg, their fuel cost will be about half the standard rate, or about 10¢/mile.

The newer fuel cell–based automobile designs include Honda's "FCX Clarity," Volkswagen's "space up! Blue" and "Touran"; BMW's "Hydrogen 7"; Toyota's "FCHV SUV and Highlander"; Nissan's "X Trail"; Hyundai's "i/Blue"; GM's "Volt, E-Flex, Hydrogen3, and Sequel"; Mercedes/Benz's "F/Cell"; Ford's "Edge, Daygo, and Focus"; Cadillac's "Provoc"; and Chevrolet's "Equinox SUV and ecoVoyager." At the end of 2007, the only hydrogen fuel cell vehicle that is certified by the American IRS, EPA, and CARB as a zero emission vehicle (ZEV) is Honda's "FCX" which qualifies for a $12,000 federal tax credit.

Their fuel tanks contain either high pressure gas or liquid hydrogen. High pressure gas units operate at pressures from 150 to 700 bar (2,200 to 10,000 psig). GM's 100 HP Hydrogen3 model wagons can store hydrogen in either gas or liquid form. One version of the hydrogen fuel tanks, Quantum Technologies'

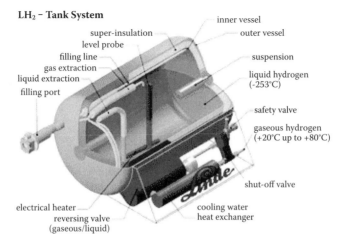

FIGURE 1.18

Hydrogen fuel tanks hold 3 kg or more of hydrogen, which, due to the high efficiency of a fuel cell car provides a driving range of 150 miles or more. (Courtesy of EERE/DOE.) (Top) High pressure hydrogen fuel cylinders. (Courtesy of Quantum Technologies.) (Bottom) Liquid hydrogen fuel tank. (Courtesy of Linde.)

TriShield composite cylinders, are shown in Figure 1.18. They can hold up to 3 kg of hydrogen at 350 bar (5,000 psig), which is sufficient for a 200-km journey in a standard sedan. The Lawrence Livermore National Laboratory designed a tank for BMW to hold liquid hydrogen and found that it held the cryogenic liquid for 6 days without venting. Their goal is extending the "no-vent storage period" to 15 days. For more details on hydrogen storage refer to Section 1.5.8.

These cars provide a range of up to about 300 miles and an acceleration of 8 to 10 seconds from 0 to 60 mph. Their mileage is from 50 to 60 miles per kilogram of hydrogen. The fuel cells usually are PEM units and range from 40 to 80 kW in size. The lithium–ion battery sizes range from 8 to 20 kWh. BMW's Hydrogen 7 model burns hydrogen in its fuel cell up to a range of 200 kilometers (125 miles) and when that range is exceeded it can be switched to burn gasoline. GM's "Volt" and Ford's "Dayglo" and "Edge" have fuel cell hybrids that operate with lithium–ion batteries. Mazda is working on a dual mode, gasoline/hydrogen rotary engine to be used in its Premacy Hydrogen RE Hybrid with 346 volt lithium batteries.

In the United Kingdom, Lotus Engineering Ltd. will be converting the black cabs of London by installing 25 kW hydrogen PEM fuel cells made by Intelligent Energy for the 2012 Olympics. In the United States, the largest fuel cell order for public transport to date (mid-2008) has been the eight 120 kW PureMotion fuel cells ordered by California's AC Transit from UTC Power. In Germany, the aerospace agency DLR and Lange Aviation are planning to convert the Antares 20E glider to hydrogen fuel cells as well.

1.2.8.6 Hydrogen IC Engine

The volumetric energy density of H_2 is less than that of gasoline. Therefore, to provide the same driving range, the hydrogen fuel tank needs to be three times the size of a gasoline tank. Today, a typical passenger car has a range of 575 miles and is provided with an 18-gallon tank, whereas an 18-wheeled semitruck has a 750 miles driving range and requires two 90-gallon tanks. Actually, the volume of the hydrogen tanks can be somewhat smaller than three times because the efficiencies of hydrogen IC and fuel cell engines are better than the efficiency of gasoline engines (gasoline, 25%; hydrogen IC, 38%; and hydrogen fuel cell, 45–60%).

BMW, DaimlerChrysler, GM, Honda, and Toyota are in the process of placing both IC and fuel cell units into the hands of ordinary drivers to gain experience and to collect data. Their prototype units cost about $1 million each. The manufacturers aim for a "pilot commercialization phase" by 2010–2012 at a unit cost of $250,000. They expect full production by 2013 at a unit cost of $50,000, and this cost will drop as the volume of production increases.

The list of vehicles that can run on H_2 is constantly growing. Quantum Fuel Technologies Worldwide converted Toyota Priuses to hydrogen fuel. BMW is marketing its 7 Series, 12-cylinder, 260-horsepower car with an IC engine that can burn liquid hydrogen or run on gasoline, whereas the BMW 750hL is designed to burn liquid hydrogen. The IC engine of the Ford E-450 shuttle bus burns 5,000 psig hydrogen gas.

In connection with using H_2 as a fuel for transportation, there is a lot of activity, but no firm direction or conclusion yet. In Iceland, one can rent a hydrogen-fueled car from Hertz. In Japan, as part of its national hydrogen program, a 200,000 m^3 tanker ship has been designed for transporting H_2. Also in Japan, an H_2-fueled commuter train is in operation, using H_2 at 35 mPa (5,000 psig or 350 bar) to fuel a 125 kW "Forza" proton exchange membrane (PEM) fuel cell by Nuvera (http://www.rtri.or.jp).

Hydrogen buses operate in Montreal and Bavaria, an H_2-powered passenger ship sails in Italy, and the 2008 Olympics in Beijing featured hydrogen vehicles. Russia has flown a jet, fueled partly by hydrogen. In the United States, the Defense Advanced Research Project Agency (DARPA), NASA, and the Air Force are jointly developing an Earth-orbit airplane fueled by

H_2. Two teams (in Turin and Madrid) are converting two light planes so that they can use hybrid fuel cell–battery electric engines.

1.2.8.7 Hydrogen Filling Stations

As of this writing, there are some 160 hydrogen fuel stations worldwide, for a list of hydrogen fueling stations, see http://www.fu elcells.org/info/charts/h2fuelingstations.pdf or http://www.naftc.wvu.edu/naftc/data/refueling/h2fuelingstations.pdf.

In the United States, there are 170,000 gas stations. In terms of infrastructure during the transition from oil to hydrogen, 12,000 filling stations would be needed to provide access to 70% of the population.

Hydrogen tanks for high-pressure gas are made of carbon fiber. Cryogenic (liquid) hydrogen tanks are double-walled with the space between the walls evacuated to provide good thermal insulation. The Lawrence Livermore National Laboratory also designed small fuel tanks for cars holding liquid hydrogen and found that they held the cryogenic liquid for 6 days without venting. Their goal is extending "no-vent storage" to 15 days.

Hydrogen filling stations are already in operation in Japan and Germany, and in the United States in Vermont, Florida, and California. Some of these stations dispense both gas and liquid hydrogen such as the one designed by Air Products at the University of California in Irvine. In 2008, Air Products also started up a new solar-powered hydrogen fueling station at the Sacramento Municipal Utility District. In Burlington, Vermont, the Department of Public Works' hydrogen fuel station uses wind energy to produce 12 kg/d of H_2. Air Products and Chemicals participated in the design of this wind-to-hydrogen generator. Figure 1.19 illustrates some of the hydrogen handling facilities that are already in operation.

In Orlando, Florida, Ford airport buses are served at a Chevron energy station, where 115 kg/d of H_2 is generated by H_2Gen Innovation units. In Munich, a fuel station designed by Linde can dispense both liquid and gas. At that fuel station, H_2 is stored above ground in a 17,600 liter tank and is

FIGURE 1.19
The technology for hydrogen storage, transportation, and dispensing at regular filling stations already exists. (Courtesy of the US Department of Energy's "Hydrogen Fuel Initiative.")

dispensed at a rate of 50 L/min. GH_2 is produced from LH_2 by evaporation followed by two steps of compression reaching 350 bar (5,000 psig) at 15°C.

1.3 Non-Solar Renewable Technologies

The energy consulting firm Cambridge Energy Research Associates estimates that the combined investment to date in renewable energy, and carbon capture technologies has been $125 billion and that it could surpass $7 trillion by 2030. The renewable energy sources include hydroelectric dams; wind turbines; solar cells; geothermal, wave, or tidal power; biofuel; and methane obtained from rotting trash, manure, or landfills. The term *renewable* is used in the sense of "not exhaustible," because when biofuels, methanol, or methane are burned, they do emit CO_2. Nuclear power is not renewable, and it is estimated that the uranium-235 deposits, if used at the present rate, will be exhausted in about 60 to 70 years.

Some of the available renewable energy sources are also limited. Much of the available hydropower has already been utilized during the last century, the availability of new sites is limited, and resistance against flooding large areas is increasing. The availability of geothermal energy is also limited to certain locations on the planet. Solar energy is intermittent, and the generated electricity is still somewhat expensive; in most areas it is competitive only with electricity generated during peak periods. Solar storage (in areas where the grid is unavailable) is also expensive. It should be noted that new technologies are available to solve both the cost (mass production, thin film, etc.) and the storage (grid, hydrogen, and thermal) problems. Ocean power (thermal, wave, and tide) availability is limited to certain areas and requires further development. Yet technological advances and the increase in the cost of fossil and nuclear energy are making these forms of energy (particularly wind, geothermal, and solar–hydrogen) more competitive.

1.3.1 Ethanol, Biofuels, Biodiesel, and Bioplastics

Biofuels have triggered a "food versus fuel debate." A few years ago turning farms into fuel factories appeared to be a good idea and both Europe and the United States supported it. For example, the American Congress mandated a fivefold increase in the use of biofuels. Today these views and policies are being reconsidered because they drive up food prices, which in turn contribute to starvation.

Biofuels and bioplastics are renewable; they reduce the need for oil imports, but some of them cause more greenhouse gas emissions than do regular fossil fuels and others such as biodiesel plants also cause water pollution. When forests are cut, they not only stop absorbing CO_2, but the destroyed vegetation will release staggering amounts of greenhouse gases as they are burned. Therefore the production of biofuels from most sources can actually exacerbate global warming.

On the other hand, if these fuels are made from sources that do not cause deforestation and do not compete with food supplies, such as agricultural waste, algae, coconut and babassu nut, and if their refining meets pollution standards, they can be useful. One of the most promising biofuel source is algae because it can be produced in large quantities. Cellulose- or sugarcane-based fuels are less desirable because the clearing of land for their production can cause deforestation, but they are superior to corn-based ethanol. Similarly, non-food grasses and reeds can also be harmful because they can overrun adjacent farms or natural lands, or drain wetlands and clog drainage systems. Therefore, the sources of cellulosic biofuels should be used only selectively and after careful analysis.

The Agriculture Secretary of the United States, Edward T. Schafer, said in May 2008 that biofuel production was responsible for only 2% to 3% of the global increase in food prices while it reduced crude oil consumption by 1 million barrels per day. Others suggest a much higher impact on food prices, but nobody disputes that biofuel production, particularly if speeded by government subsidies, does increase food prices. The UN Food and Agricultural Organization reported in 2008 that wheat prices increased by 80% in 2007, and the cost of corn doubled in 2 years, both occurred because a rising proportion of the crop is used for biofuel production. Today, a Nigerian family spends 75% of its income on food and 33 nations are at risk for social unrest because of rising food prices.

Biofuels are being used both directly by burning and indirectly by first being converted to liquid fuels. This conversion is an energy-intensive process. In many parts of the world, wood, sugarcane, animal wastes, etc., are still being burned as fuel. New biomass facilities, such as the world's largest biomass plant (350 mW costing $830 million) in England, are also being built. This plant will burn some 3 million tons of wood per year.

Elsewhere, wood chips, sugarcane, switchgrass (also called *Miscanthus)*, corn husks, prairie grass, or soybean and corn are being converted into liquid biofuels. Various waste materials are also used to make ethanol, butanol, biodiesel, and other substitutes for gasoline.

The DOE also supports research to develop genetically engineered trees. The goal of this effort is to reduce the amount of lignin in the wood, making it easier to break down its cellulose into sugar, which then can be converted into ethanol. This approach does not seem to be a good idea because, although saving on acid and steam in the processing of the wood, the reduction in lignin will make the trees structurally weaker and less resistant to pests. Therefore, the use of switchgrass for cellulose and the direct use of sugar are better options.

By 2020 the EU wants to replace 10% of its transportation fuel with biofuels, while China is aiming for at least 15%. In the United States, 6 billion gallons of ethanol were produced in 2006, 15 billion is projected by 2012, and 36 billion by 2020. In 2006, in the United States 250 million gallons of biodiesel, were also produced, and in 2008 the production is expected to reach 2 billion gallons. In Europe, in 2005, biofuels accounted for about 1% of the fuel used.

In 2008 Virgin Atlantic Airlines successfully tested a 25% biofuel mixture in a flight without passengers between London and Amsterdam. The goal was to show that the coconut- and babassu nut-based fuel would not freeze at 30,000 feet (10,000 meters) altitude.

Bioplastics can be made by fermentating corn sugar (propaneidol-based Cerenol by DuPont), starch and cellulose can be used to make nylon (Michigan State University), or propaneidol can be used to produce stain-resistant textiles (DuPont).

1.3.1.1 Ethanol

It seems that the production of ethanol drives up food prices and contributes to starvation in poorer nations. Others argue that the main cause of food shortages are droughts, overpopulation, improved living standards, and the soaring demand for meat in China and India. This is because it takes seven calories of corn to produce one calorie of meat. When the ethanol is made from agricultural waste or algae, the side effects are more favorable than if it is made from corn, but the refining process can still be a source of pollution. For quantitative data the reader can access http://www.iol.co. za/index. php?set_id=1&click_id=143&art_id=vn20080210085730876C308900.

The U.S. bioethanol industry is growing rapidly. Production in 2007 was 6.5 billion gallons from 139 bioethanol refineries. A further 4 billion gallons of capacity are expected to come online by the end of 2008. In 2006, 14% of the corn crop in the United States was used to produce ethanol and probably as a result, corn prices increased by 25% in 2007. In the United States 90 plants operated in 2006 and 160 in 2007. Just in Iowa, 42 ethanol and biodiesel plants are in operation and an additional 18 are under construction. A study by the Organization for Economic Cooperation and Development calculated that in order to meet 10% of the fuel requirements of the United States, Canada, and the EU, 30% to 70% of their crop area would have to be devoted to biofuels.

The use of ethanol reduces the dependence on oil, but it does not reduce CO_2 emissions; it also increases the health risks due to ozone emissions. Ethanol from corn also increases the use of nitrogen-based fertilizers, which cause water pollution. Ethanol production from corn is inefficient: 80% of the ethanol energy content is used up in making it (its "energy balance" is 1.3:1). Often, the energy source for making the ethanol is imported oil. In addition, ethanol cannot be transported by pipeline, because it absorbs water and other impurities; it is also corrosive and cannot be distributed through gasoline pipelines, trucks, railcars, or barges. In addition the overwhelming majority of cars cannot use gasoline blends in which ethanol exceeds 15%.

In contrast to the 1.3:1 energy balance of corn ethanol, sugarcane-based ethanol has an energy balance of 8:1. Cellulosic ethanol made from prairie and switchgrass has an energy balance of 6:1, and even gasoline has an energy balance ratio of 5:1. Therefore, the overall environmental impact of corn ethanol can be worse than that of gasoline. One advantage of cellulosic ethanol is that it can come from perennial plants and also from whole plants, not only from the seeds. Algae or switchgrass does not require large amounts of energy to grow or to convert to ethanol and therefore its net energy gain is better. On the other hand, non-food grasses and reeds can also be harmful, because they can overrun adjacent farms or natural lands, can be toxic or drain wetlands, and clog drainage systems. The United States places a tariff of 54¢ on each gallon of Brazilian ethanol, while American ethanol producers receive a tax break of 51¢/gal.

The farm industry supports the production of corn-based ethanol, although the food industry opposes it, because the increased demand for corn is raising the cost of animal feed and, therefore, the cost of dairy, poultry, and other products.

Sugarcane- or cellulose-based ethanol has less of an effect on food prices, but its expanded production can end up destroying wildlife habitat and forests, threatening the survival of the rainforests, and polluting water supplies. The DOE is establishing three bioenergy research centers in order to evaluate the various processes of turning cellulose into fuel.

Ethanol can also be produced from "non-food" materials, such as garbage or wastewater sludge, which are "negative-cost" feedstocks. If all American wastes (industrial and municipal) were converted to biofuels, not only would some 50 to 100 million gallons of fuel be obtained, but the emission of methane from landfills and other wastes would also be eliminated. Plasma gasification, a commercially available process, can also simultaneously increase the fuel supply and reduce greenhouse gas emissions.

1.3.1.2 Biodiesel

Biodiesel fuels are made from vegetable and seed oils, woodchips, soybeans, animal fats, and recycled restaurant grease. A typical fast food restaurant produces about 250 pounds of grease a week. In New York State, a 20¢/gal tax credit supports the use of biodiesel. In Greenpoint (Brooklyn, New York), a 110 million gallon biodiesel refinery is awaiting approval. Federal policy is still tilting to ethanol in spite of the fact that turning biomass into gasoline is simpler because it does not require changes in pipelines and car designs.

A number of companies are working on biomass projects. BlueFire Ethanol's process converts organic material to fuel, Dynamotive Energy Systems is making oil from biomass, and Virent Energy Systems is working on turning biomass into sugar and then sugar into gasoline.

Byproducts of biofuel production are glycerin and lignin. The production of each gallon of biodiesel also produces a pound of glycerin. These materials can be used to replace oil-based products with bio-based ones. It is expected that in the decades to come, the development of the biofuel industry will result in the building of multiproduct biorefineries.

In the past, plastics have been made from hydrocarbons. Now, with the development of new catalysts, they can also be made from agricultural products. The organic alternatives to petrochemical polymers cannot only

FIGURE 1.20
In many locations, the electricity generated by wind farms is already cost competitive.

contribute to lowering dependence on oil, but can also replace nondegradable plastic products with degradable ones.

1.3.2 Wind Turbines

In the coming years the global wind turbine market is projected to be $65 billion per year. The worldwide leader in wind power is Germany, having invested $9 billion in it. In 2007, her 20,000 wind turbines generated 5% of electricity consumption, but Germany is running out of places where new turbines can be located. The second largest user of wind power is Spain. The DOE estimates that by 2030 it could reach 20% of American electricity capacity. Electricity generation capacity is growing at an annual rate of 30% and in 2007 reached 17 gW. In the United States, wind-based total energy capacity is around 1% of the national electricity generation capacity and is nearing 15 gW (Texas: 4.3; California: 2.4; Minnesota, Iowa, Washington and Colorado: about 1.2 ea.; Oregon, Illinois and Oklahoma about 0.8; and New Mexico and New York: about 0.5 gW). The largest wind turbine suppliers include General Electric Energy, Siemens, Power Generation, Vestar Wind Systems, Aero Vironment, Clipper Turbine Works, Sualon Energy, and Gamesa Corp.

In Texas, 3% of the electricity is already being generated by wind turbines and a fivefold increase is projected by 2012. The highest wind potentials are in Texas, Montana, and the Dakotas. These states are sparsely populated and do not have good electricity transmission. Therefore, an important aspect of converting to a renewable energy economy is to develop a nationwide electric grid.

The size of wind turbines has also increased as new and lighter construction materials have become available. Today, thousands of 75 m (250 ft) tall wind turbines with 45 m (140 ft) wings are in operation in Texas and in many other locations, each generating about 1 mW.

In May 2007, the National Academy of Sciences released a study indicating that by 2025 the use of wind turbines could reduce the nation's CO_2 generation by 4.5%. In the United States, between 2000 and 2006, the quantity of electricity generated by wind farms quadrupled.

Figure 1.20 illustrates a typical wind farm. The DOE estimates that in the next 15 years wind generator capacity in the United States will reach between

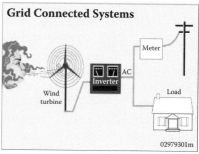

FIGURE 1.21

Grid connected solar and wind energy installations. (Left) Grid connected home in Colorado meets nearly 100% of the homeowner's annual electric needs. (Right) Grid connected wind turbine installation. (Courtesy of EERE/DOE.)

20 and 70 gW, which is 2 to 7% of the national electricity generating capacity. Their share of electricity production will naturally be much less because wind turbines do not operate continuously. In Europe, some estimates suggest that if fully exploited, they could meet 30% of their electricity needs by wind power.

The wind turbine–generated electricity is already cost competitive (5–9¢/kwh) with natural gas-based electricity and in some locations also with nuclear- or coal-based electricity. During the last few years, the cost of wind turbines in the United States has been increasing because the price of commodities and the value of the Euro have been rising (most turbine components are imported from Europe).

In many parts of the world, government support is available to finance wind turbine installations. In the United States, the production tax credit for wind-generated electricity is 1.9¢/kWh. This, in 2006–2007, cost the treasury $2.75 billion. In California, for example, a $32,000 wind turbine (about 12 ft in diameter and 30–100 ft tall) is supported by a $16,000 state rebate.

The average generating capacity of a "home turbine" in the United States is 2–10 kW, and the federal government is considering a one-time credit of $3,000/kW. For household applications, the cost of a 3 kW wind turbine is about $6,000 (Hgenerators™). The daily production of a 3 kW generator is about 20 kWh/d if the wind is blowing for 210 h/month and the average wind velocity is 12 m/s. If the generated electricity exceeds the needs of the home, it can be stored by sending it to the grid (Figure 1.21). Naturally, it can also be stored in batteries or in the chemical energy of H_2. If H_2 is generated, it can also be used as transportation fuel.

It is estimated that to generate and liquefy 10 tons of LH_2 per day would require a wind farm with a 100 mW rated (a 30 mW average) capacity. This installation would call for about 200 wind turbines with 40 m spans at an installed cost of about $75 million. Calculating at a 20-year payback, the cost of electricity would be about 6¢/kWh, and the cost of 1 kg of LH_2, about $7.*

---

* http://www.andrew.cmu.edu/user/dk3p/papers/53.DeCarolis.2002.IsTheAnswerBlowingInTheWind.f.pdf.

In many locations, excess electricity is accepted by power companies on a "net metering" basis (meters "run backward" when renewable energy is being sent into the grid). The accumulation of this excess can compensate for (or exceed) the amount of electricity needed when there is no wind. The Dutch utility Nuon reports that in 2005 the electricity cost on the spot market was 5.6¢/kWh when there was no wind, and it dropped to 3¢/kWh when the wind speed reached 13 m/s, because the fuel (wind) is free.

A small company, General Compression, is experimenting with the direct storage of wind energy in compressed air. The air-compressing windmills eliminate the use of electrically driven compressors and thereby increase the efficiency of the operation. In another new development the Alameda, California–based Makani Power Inc. is developing high-altitude wind technologies.

In order to meet wilderness preservation considerations, it has been suggested that birds, bats, and other animals will be better protected if slowly rotating, larger blades are utilized. As to the aesthetic aspects, some artists are working on coming up with more aesthetic shapes and colors for these structures.

1.3.2.1 Wind Turbine Installations

In the United States in 2007 the generating capacity of the wind industry expanded by 30% and the consulting firm, Emerging Energy Associates, estimates that between 2007 and 2015 some $65 billion will be invested in wind power. Shekk and TXU Corporation are planning to build the world's largest (3 gW) wind farm in the Texas panhandle. An even larger (4 gW) farm costing about $10 billion is planned on a 150,000 acre area, also in the Texas panhandle. This project is called the "Pama Wind Project" and will be built by Mesa Power LLP. The first phase is a 1 gW unit consisting of 667 turbines, and the total project is expected to be completed by 2014. Another large project is the 2 gW $8 billion wind farm being built by Spain's Iberdrola. This project will double the capacity of the company's operating plants in the United States. The world's first floating offshore wind turbine with a peak production of 2.3 mW is being built in Karmay, Norway. One of the world's largest wind parks was built by the Vattenfall Power Company between Sweden and Denmark.

The two largest wind farm operators in the world are Iberdrola of Spain and FPL Energy in the United States. FPL operates 6,400 wind turbines on 50 wind farms in 15 states and generates electricity for 1.2 million homes. There is a lot of activity as corporations are discovering the financial potentials of wind technology. Energias of Portugal has recently purchased Horizon Wind Energy in the United States, Acciona acquired EcoEnergy, Iberdrola bought CPV Wind Ventures, and the list goes on. A new 75 mW wind farm is planned in North Dakota, windmills are proposed to be installed on the bridges and skyscrapers of New York City, and a 450 mW wind power plant is planned for Clyde in Scotland.

American Electric Power is installing a 6 mW wind farm with battery storage for $27 million, or at a unit cost of $4,500/kW, using NGK Insulator's sodium–sulfur batteries made in Japan. The rationale for the installation is that although the wind turbines operate mostly at night when the value of electricity is low, by storing the electricity generated until the next peak period, its value is much increased.

In Burlington, Vermont, at the Department of Public Works' hydrogen fuel station, wind turbines and an H-series PEM electrolyzer from Proton Energy Systems are used to produce 12 kg/d of H_2. Air Products and Chemicals participated in the design of this wind-to-hydrogen installation.

Wind-based hydrogen generators are in operation in Patagonia, Argentina, Spain, Norway at Utsira Island, at Hovik, Minnesota, and in Canada at Prince Edward Island. The "Wind$_2$H$_2$ Project" in Boulder, Colorado, links 5 kW wind turbines to water splitters generating 20 kg of H_2 a day. This H_2 is stored as a gas at 240 bar (3,500 psig). PEM water splitter units of 5 kW size are available from Proton Energy, and units with 50 kW capacity alkaline electrolyzers are available from Teledyne.

1.3.3 Ocean Energy

The rotation of the Earth and the circling of the Moon cause the ocean tides, which represent enormous renewable energy potentials. Tidal generators (Figure 1.22, top right) operate like wind turbines, but at much lower velocities, because water is 832 times heavier than air. Another renewable energy source is the thermal energy of the oceans. This source of energy has been studied for nearly four decades, but its exploitation is just beginning. The highest-energy waves can be found off the western coasts in the 40–60° latitude range north and south.* In these regions, the power of the wave fronts varies between 30 and 70 kW/m, with peaks up to 100 kW/m in the Atlantic southwest of Ireland, the Southern Ocean, and off Cape Horn. If this energy resource could be harnessed, it could meet 10% of global electricity demand.

Some of the designs to collect the energy of ocean waves are shown in Figure 1.22. On the top left of Figure 1.22, is the tapered channel (tapchan) design, in which the waves enter a raised lagoon through a converging inclined channel, from which the water returns back into the ocean through a turbine generator. The tapchan feeds into a reservoir constructed on cliffs above sea level. The narrowing of the channel causes the waves to increase in height as they move toward the cliff face. The waves spill over the walls of the channel into the reservoir and the stored water is then fed through a turbine.

Gates and turbines are installed along the dam. When the tides produce an adequate difference in the level of the water on opposite sides of the dam, the water flows through the turbines that turn an electric generator to produce electricity. Tidal fences can also harness the energy of tides. They can be used in areas such as channels between two landmasses. Tidal fences

---

* Quoted from Dr. Tom Thorpe of ETSU (author of the 1998 Wave Energy Commentary).

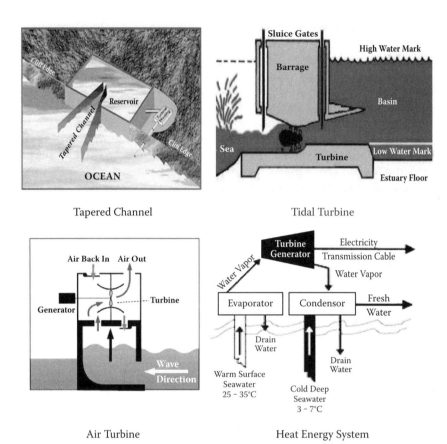

Tapered Channel   Tidal Turbine

Air Turbine   Heat Energy System

FIGURE 1.22
Ocean energy collection designs. (Courtesy of EIA/DOE.)

have less impact on the environment than tidal barrages although they can disrupt the movement of large marine animals.

The top right of Figure 1.22 shows the tidal turbine that can be used in many tidal areas. They are basically wind turbines that can be located anywhere there is strong tidal flow. They are arrayed underwater in rows, as in some wind farms. The turbines function best where coastal currents run between 3.6 and 4.9 knots (4 and 5.5 mph). In currents of that speed, a 15-meter (49.2-feet) diameter tidal turbine can generate as much energy as a 60-meter (197-feet) diameter wind turbine. Ideal locations for tidal turbine farms are close to shore in water depths of 20–30 meters (65.5–98.5 feet).

The Severn Barrage is a proposed tidal power station to be built across the Bristol Channel where the River Severn has a tidal range of 14 meters, producing in excess of 17 billion kWh of electricity annually. Along the length of the Severn Barrage, open sluice gates would allow the tide to flow in. These gates would then be closed at high tide trapping enormous quantities of water behind the barrage.

In the bottom left of Figure 1.22, an oscillating water column is shown, which consists of a partially submerged concrete or steel structure that has an opening to the sea below the waterline. It encloses a column of air above a column of water. As waves enter the air column, they cause the water column to rise and fall. This alternately compresses and depressurizes the air column. As the wave retreats, the air is drawn back through the turbine as a result of the reduced air pressure on the ocean side of the turbine.

The temperature difference in the ocean can also produce electricity as shown at the bottom right of Figure 1.22. This thermal gradient operated heat pump, which converts the temperature difference in the ocean into a pressure difference of the working fluid, utilizes this pressure difference to generate electricity. It exploits the 40°F temperature difference between the deep and surface waters of tropical seas. This method of electricity generation can be viewed as a heat pump in reverse. Here, the heat content of the warm surface waters is used to convert (evaporate and compress) the working fluid into a high pressure vapor. This vapor stream passes through a pressure let-down turbine generator and the exiting low pressure vapor is condensed at the low temperature of the deep ocean.

This heat pump moves the heat from the surface of the ocean into its lower depths and in the process recovers some of the energy obtained by the working fluid as its pressure is reduced through a turbine generator. Sea Solar Power Inc. estimates that a 100 mW power plant could be built for $336 million.

The pendulor wave-power device consists of a rectangular box that is open to the sea at one end. A flap is hinged over the opening and the action of the waves causes the flap to swing back and forth. The motion powers a hydraulic pump and a generator.

A more recent design is the Pelamis Wave Energy Converter (Ocean Power Delivery, Ltd.). The Pelamis (named after a sea snake) is under development in Scotland. It is a series of cylindrical segments connected by hinged joints (Figure 1.23). As waves run down the length of the device, they move the joints, and the hydraulic cylinders incorporated in the joints pump the oil, which drives a hydraulic motor via an energy-smoothing system. The electricity generated in each joint is transmitted to shore by a common undersea cable. The slack-moored device is planned to be around 130 m long and 3.5 m in diameter. The full-scale version is planned to have a continuously rated power output of 0.75 mW.

In the United States in 2007, the Federal Energy Regulatory Commission issued four permits to test various wave-energy designs. One of the wave-energy test sites will be operated by Oregon State University at Reedsport, Oregon. It is expected to generate 50 mW of electric power over an area of several square miles. Finavera Renewables of Vancouver is developing a 2 mW commercial plant in Northern California and estimates that tidal power could generate 10% of the global electricity consumption.

Other installations in Canada include a pilot project in Victoria, designed by Clean Current Power Systems and three in the Bay of Fundy. Other projects

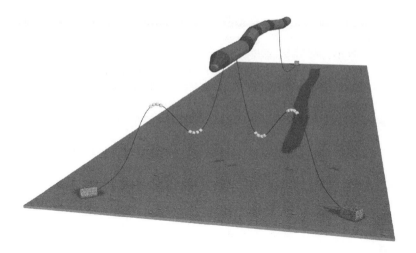

FIGURE 1.23
The Pelamis Wave Energy Converter. (Ocean Power Delivery Ltd.)

are in progress in Pembrokeshire, Wales, where a deep-sea tidal energy farm is being built by Lunar Energy, a 1.2 mW installation by Marine Current Turbines off the coast of Ireland and a 3 mW plant in Scotland. A 10-mile river barrage system across the Severn River is also under development in the U.K. This 8.6 gW plant, when completed, should produce 17 tWhrs/yr of electricity, meeting 5% of the nation's electricity needs. A tidal turbine system is also planned at Paimpol, France.

1.3.4 Geothermal Energy

Geothermal energy is just as inexhaustible and renewable as solar energy, but in contrast with solar energy, it is continuously available. Its production costs are low, but the initial costs can be high if the production wells are deep and the ground is rocky. On the other hand, if natural hot springs exist in the area or if the wells have already been drilled for oil or natural gas, initial investment can also be low. The total global potential of theoretically exploitable geothermal energy is in millions of quads, whereas the global yearly energy consumption today is under 500 Q.

The temperature in the core of the Earth—due to the decay of radioactive isotopes—is on the order of 4,000°C, the temperature of the lava of volcanoes is about 1,200°C, and the temperature of thermal springs can reach 350°C. If the groundwater temperature exceeds 150°C, flash steam power plants can be built, and if it is between 100 and 150°C, binary cycle power plants can be operated.

Worldwide, less than 10 gW of electricity is produced by geothermal plants, mostly in Iceland, Australia, El Salvador, and more recently, in Hungary. A 2–5 mW binary plant is being built in Hungary, near Szentlőrinc as a joint venture of Enex (Iceland), GreenRock Energy (Australia), and MOL Group

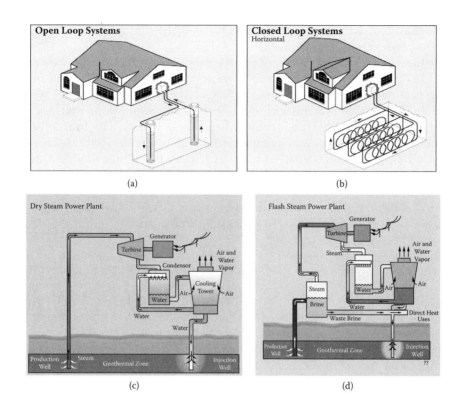

FIGURE 1.24
Geothermal energy can be used for heating, cooling, and electricity generation. (a) Open loop geothermal heat pump: GHP. (From U.S. DOE/EERE.) (b) Closed loop geothermal heat pump: GHP. (From U.S. DOE/EERE.) (c) Dry Steam Power Plant. (From Idaho National Laboratory operated for U.S. DOE by Battelle Energy Alliance.) (d) Flash steam power plant. (From Idaho National Laboratory operated for U.S. DOE by Battelle Energy Alliance.)

(Hungary). The cost of the project is estimated at $3.7 million, and the depth of the wells is from 2,900 to 3,200 m. In Connecticut, UTC Power signed agreements with Raser Technology, UT for 30 mW Rankine cycle-based geothermal power plants.

When the groundwater temperature is less than 100°C, geothermal heat pumps (GHPs) can be used to heat buildings in the winter and cool them in the summer (see the open (a) and closed (b) versions in Figure 1.24). Some buildings at Bard College in New York State, for example, are heated by GHPs that obtain the heat from the groundwater, which is 15°C (60°F). In the heating mode, GHPs operate in the same way as a domestic refrigerator, but instead of removing the heat from the inside the refrigerator and pumping it to the warmer outside, it takes the heat from the colder groundwater and moves it into the hot water (or hot air) of the building's heating system.

In dry steam power plants, the steam is generated directly by the production wells (Figure 1.24c), while in the case of the flash steam power plants (Figure 1.24d), the wells produce hot water from which steam can be separated to drive conventional steam turbine generators. The size of these plants ranges from 100 kW to 150 mW.

In binary cycle power plants, where the water in the ground is between 100 and 150°C, a heat pump is used, and a secondary working fluid operates turbine generators. In this case, the 100–150°C water heats (vaporizes) the working fluid in an evaporator heat exchanger, and this vapor in turn drives the turbine generator, while the turbine exhaust is recondensed by conventional cooling. These plants range in size from 100 kW to 40 mW.

In all three cases (flash steam, binary cycle, and GHP installations), after the heat content of the geothermal water is utilized, the spent fluid is reinjected into the underground water reservoir.

In the case of enhanced geothermal systems (EGS), thermal energy is obtained by pumping water onto hot rocks in the ground rather than harvesting hot water already there.

1.3.5 Fuel Cells

Fuel cells generate electricity, heat, and distilled water by reacting H_2 with oxygen (air). This process emits no CO_2, no carbon monoxide, no sulfur dioxide, no volatile organic compounds, and no fine particles. The only byproduct of the oxidation of H_2 is distilled water. There are some 1,035 firms that are active in the field of fuel cell development and manufacturing (http://www.fuelcells.org/directory). This field of technology is fast advancing, the delivery of new units increased by 75% to approximately 100,000 units in 2007.

For closed-cycle applications, such as for spacecraft, submarines, or transportation vehicles, the combinations of lightweight, reasonable power density, and compact size are favorable features in comparison with equivalent-capacity battery-based systems. In the International Space Station, for example, both electricity and water are provided by fuel cells. Fuel cells have not only been used in space exploration, but also in submarines (because they generate no noise or vibration). They have also been used to recover the energy from methane that is generated by wastewater, by garbage dumps, and more recently in automobiles as an alternative to the IC engine.

Single-fuel cells are rarely able to produce enough power for commercial applications. The cells are therefore connected into stacks. Commercial stacks frequently have more than 100 and sometimes as many as 400 cells. Today's fuel cells in the 2 to 4 mW size range cost from $3,000/kW to $4000/kW. According to the U.S. Department of Energy, the most widely deployed fuel cells today cost about $4,000 per kilowatt. According to the U.S. Department of Energy, the most widely deployed fuel cells today cost about $4,000 per kilowatt. The Solid State Energy Conversion Alliance (SECA) Cost Reduction Program is cutting costs to as low as $400 per kilowatt. The DOE Hydrogen Program of the United States made significant progress since its initiation is 2003. By 2007 it has reduced the high-volume (0.5 million/year) cost of fuel cells from $275/kW in 2002 to $107/kW and its goal is to reach $30/kW by

2015. This year, in Folsom, California, Altergy Systems started up the world's first automated assembly line for manufacturing fuel cells. By contrast, a diesel generator costs $800 to $1,500 per kilowatt, and a natural gas turbine can be $400 per kilowatt or even less. This year, in Folsom, California, Altergy Systems started up the world's first automated assembly line for manufacturing fuel cells.

Fuel cell suppliers include Ballard, FuelCellEnergy, GE, Nuvera, UTC, and many others. Once mass-produced, lightweight, and low-cost fuel cells are available, they will be able to power electric cars, which cause no pollution. They may also generate electricity from H_2 that is produced by solar energy. At Kingston Point, New York, an 800 ft^2 solar–hydrogen home has been built by the New York Institute of Technology (NYIT) that is being powered by a 5 kW PlugPowerGenCore fuel cell.

Single-cycle fuel cell efficiencies range from 47 to 50%. The efficiency of combined-cycle fuel cells is about 60%, and if the generated heat is also recovered (in the form of hot water), the total efficiency can be around 80%. In comparison, the efficiency of gasoline engines is around 25%, of nuclear power plants about 35%, and of subcritical fossil fuel power plants, 37%.

The combustion of a mol of hydrogen produces 286 kJ (1.0 kJ = 0.948 Btu = 0.278 Wh) of energy. If the combustion of hydrogen takes place in a fuel cell, only 237.3 kJ of this energy can be recovered in the form of electric energy, whereas 48.7 kJ will be lost in the form of heat to the environment. Therefore, in an ideal (theoretically perfect) fuel cell, efficiency could reach 237/286 = 83%.

Viewing electrolysis (discussed later) and fuel cell reactions as a pair (see Figure 1.48 top left in Section 1.5.4), the enthalpy change is 286 kJ in both reactions; but in the case of electrolysis the environment contributes 48.7 kJ of heat energy, whereas in the fuel cell reaction 48.7 kJ is dumped into the environment in the form of heat.

1.3.5.1 Fuel Cell Designs

A fuel cell consists of an electrolyte that is sandwiched between two electrodes (Figure 1.25). A large variety of fuel cell designs exist, which include the following:

Phosphoric acid fuel cell (PAFC)—Phosphoric acid electrolyte with platinum catalyst. It can use hydrocarbon fuel and is suited for stationary applications. It can generate both electricity and steam. As many as 200 units in sizes ranging from 200 kW to 1 mW are in operation.

Proton exchange membrane (PEM)—Solid organic polymer electrolyte with platinum catalyst. It requires hydrogen fuel and is suited for automobiles.

Molten carbonate fuel cell (MCFC)—Carbonate electrolyte with conventional metal catalyst. It can use coal gas and natural gas fuel, and is suited for 10 kW to 2 mW power plants.

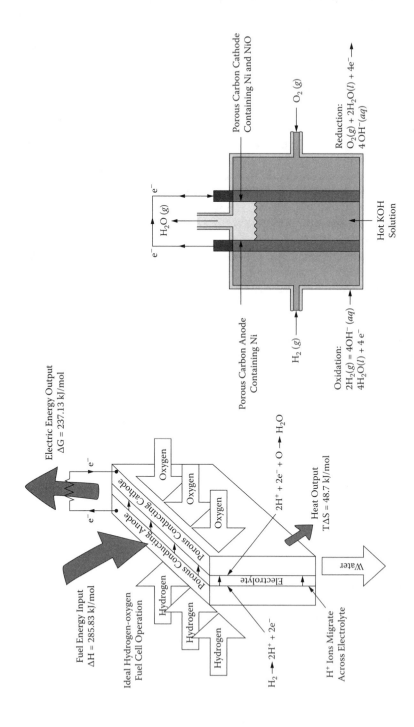

FIGURE 1.25
The operation and energy balance of a fuel cell.

Solid oxide fuel cell (SOFC)—Solid zirconium oxide electrolyte; it is suited for large-scale central electric power plants.

Alkaline—The electrolyte is an aqueous solution of alkaline potassium hydroxide soaked in a matrix. It is used by NASA on space missions to generate both electricity and water.

Direct methanol fuel cell (DMFC)—Polymer membrane electrolyte; no fuel reformer is needed, because the catalyst draws the H_2 directly from the liquid methanol. This design can be used for cellular phones and laptops.

Regenerative fuel cell (RFC)—Solar-powered electrolyzer separates water into H_2 and oxygen (O_2) to produce heat, electricity, and water. The operation can be reversed, so the H_2 is used as a fuel to generate more heat and electricity.

Zinc-air fuel cell (ZAFC)—Electricity is produced as zinc and oxygen are mixed in the presence of an electrolyte, producing zinc oxide. This design can replace batteries.

Protonic ceramic fuel cell (PCFC)—The ceramic electrolyte can electrochemically oxidize fossil fuels, eliminating the need for fuel reformers.

The various fuel cell designs (Figure 1.25) can also be categorized according to their operating temperatures, electrode designs, types of electrolytes, and the type of fuel used (Table 1.26). Electrolytes can be either acidic or alkaline liquids, solids, or solid–liquid composites. The basic core of the fuel cell—called the stack—consists of the manifolds, anode, cathode, and electrolyte.

For costs and other information on the various designs, refer to the Web pages of the various suppliers:

Ballard Power Systems, Inc.	http://www.ballard.com
CFCL Ltd.	http://www.cfcl.com.au
E-TEK	http://www.etek-inc.com, http://www.denora.it
Fuel Cell Today	http://www.fuelcelltoday.com
Fuel Cells 2000	http://www.fuelcells.org
FuelCell Info.Com	http://www.fuelcell-info.com
FuelCell Energy	http://www.ercc.com
Forschungszentrum Julich	http://www.fuelcells.de
Honeywell	http://www.honeywell.com
Plug Power	http://www.plugpower.com
U.S. Fuel Cell Council	http://www.usfcc.com
UTC Fuel Cells	http://www.utcfuelcells.com
Westinghouse Electric Co.	http://www.westinghouse.com

TABLE 1.26

Features and Characteristics of a Number of Fuel Cell Designs

Fuel Cell Type	PAFC	PEMFC	MBFC	DMFC	AFC	MCFC	SOFC	ZAFC
Feed Component								
H_2	Fuel	Fuel	Diluent	Fuel	Fuel	Fuel	Fuel	Fuel
CO	Poison (0.5–1.5%)	Poison	Poison	Poison	Poison	Fuel[a,d]	Poison (@ 0.5 ppm)	Poison (@ 1 ppm)
CH_4	Diluent	Diluent	Fuel	Diluent	Poison	Diluent[b]	Fuel[a]	Fuel
CH_3OH	Diluent	Diluent	Fuel	Fuel	Poison	Fuel	Fuel	Fuel
C_xH_x[c]	Diluent	Diluent	Fuel	Diluent	Poison	Diluent	Fuel[a]	Fuel
CO_2	Diluent	Diluent		Diluent	Poison	Reagent	Diluent	Diluent
H_2O	Diluent	Reagent (membrane needs water)	Reagent (membrane needs water)	Reagent	Reagent	Reagent	Reagent	Diluent
Sulfur (H_2S, COS)	Poison (@ 50 ppm)	No information	Fuel*	No information	Poison (@ 50 ppm)	Poison (@ 0.5 ppm)	Poison (@ 0.5 ppm)	Poison (@ 1 ppm)
Charge Carrier	H^+	H^+	H^+	H^+	OH^-	CO_3^-	O^{2-}	OH^-
Efficiency	40–50%	40–50%	1–2%	20–40%	70%	60%	60–85%	
Operating Temp.	150–220°C	<100°C	15–40°C	65–220°C	50–120°C	650°C	650–1000°C	700°C

	H$_3$PO$_4$	Teflon (inert)	Teflon (inert)	Teflon (inert)	Potassium hydroxide (KOH)	Lithium carbonate/ potassium carbonate or sodium carbonate	Zirconia/ yttria	Zinc oxide
Electrolyte								
Electrodes Anode	Platinum	Platinum/ carbon	Platinum/ carbon	Platinum/ ruthenium	Platinum	Nickel	Nickel/ nickel oxide	Zinc anode
Electrodes Cathode	Platinum	Platinum/ carbon	Platinum/ carbon	Platinum	Platinum	Nickel	Manganese oxide/ vanadium oxide Lanthanum/ strontium/ tin doped	

Note: PAFC: phosphoric acid fuel cell; PEMFC: proton exchange membrane fuel cell/polymer electrolyte membrane fuel cell; DMFC: direct methanol conversion fuel cell; AFC: alkaline fuel cell; MCFC: molten carbonate fuel cell; SOFC: solid oxide fuel cell; ZAFC: zinc air fuel cell.

a In reality, CO, with H$_2$O, shifts to H$_2$ and CO$_2$, and CH$_4$, with H$_2$O, reforms to H$_2$ and CO faster than reacting as a fuel at the electrode.

b A fuel in the internal reforming MCFC.

c Molten carbonate fuel cells, solid oxide fuel cells, and zinc air fuel cells all can consume complex hydrocarbons (C$_x$H$_x$).

d CO can strip nickel at high temperatures by forming Ni(CO)$_4$ gas.

e Microbiological fuel cells use bacteria to catalyze the conversion of carbohydrates to fuel, low efficiencies can be balanced against low fuel cost, including domestic sewage.

* The use of sulfur compounds was subject to further testing. Results are yet to be published.

I am proposing to develop a fuel cell that can also operate in the "reverse" state as an electrolyzer (see Figure 4.2 reproduced below). Once available, these units will be able to store solar energy for the night in the form of hydrogen which can later be used when solar radiation is not available. Households having fuel-cell engines in their cars will also be able to use it to generate hydrogen at night (when electricity is cheap) and use that hydrogen in the fuel cells of the family's electric car during the day. These reversible fuel cells (RFC) will be both less expensive and much lighter than the combined weight of regular electrolyzers and fuel cells.

Another new development is the fuel cell that can vary its power generation capacity as a function of the hydrogen flow supplied to the cell by the generated water being pushed out or drawn back into the fuel cell.

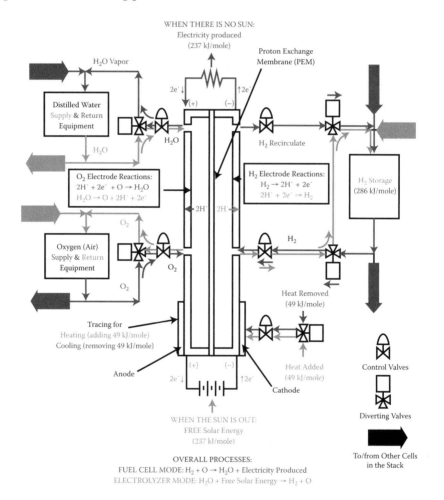

FIGURE 4.2 (Discussed in Chapter 4)
(See color insert following page 140.) A single cell of my reversible fuel cell (RFC) design, using the basic proton electrolyte membrane (PEM) type fuel cell.

1.3.5.2 *Alkaline and Phosphoric Designs*

One of the first fuel cell designs was low-temperature alkaline fuel cells (AFCs) used in the U.S. space program. They served to produce both water and electricity on the spacecraft. Some of their disadvantages are that they are subject to carbon monoxide poisoning, are expensive, and have short operating lives. The AFC electrodes are made of porous carbon plates laced with a catalyst. The electrolyte is potassium hydroxide. At the cathode, oxygen forms hydroxide ions, which are recycled back to the anode. At the anode, hydrogen gas combines with the hydroxide ions to produce water vapor and electrons that are forced out of the anode to produce electric current.

PAFCs are considered to be first-generation mature designs that are often used in larger vehicles and buses. These are medium-temperature (200°C or about 400°F) units, generally available in the 60–200 kW size range. They can be up to 85% efficient when used to generate both electricity and heat, but are about 60% efficient when generating electricity only. They are large, heavy, and cost around $4,000/kW.

1.3.5.3 *Proton Electrolyte Membrane*

In the proton-emitting membrane or proton electrolyte membrane (PEM) design, the membrane electrode assembly consists of the anode and cathode, which are provided with a very thin layer of catalyst, bonded to either side of the proton exchange membrane. With the help of the catalyst, the H_2 at the anode splits into a proton and an electron, while O_2 enters at the cathode. On the inside of the porous anode is a thin platinum catalyst layer. When H_2 reaches this layer, it separates into protons (H_2 ions) and electrons. One of the reasons why the cost of fuel cells is still high is because the cost of the platinum catalyst is rising. One ounce of platinum cost $361 in 1999 and increased to $1,521 in 2007.

The protons pass through the electrolyte, while the free electrons are conducted—in the form of a usable electric current—through an external circuit. At the cathode, the electrons combine with the O_2 in the air and with the H_2 protons that migrate through the PEM to produce water and heat. Air flows through the channels to the cathode. One PEM fuel cell design suited for use in passenger vehicles is shown in Figure 1.27. They operate at low temperatures (88°C or 190°F), provide high power density, and are low in weight. Their byproduct is hot water, and their electric power–generating capacity ranges from 60 to 200 kW. They use a solid polymer electrolyte, porous carbon electrodes, and a platinum catalyst. Developers are currently exploring other catalysts that are more resistant to carbon monoxide poisoning.

The core of the Ballard fuel cell consists of a membrane electrode assembly (MEA) that is placed between two flow-field plates. The flow-field plates direct H_2 to the anode and O_2 (from air) to the cathode. To obtain the desired amount of electric power, individual fuel cells are combined to form fuel cell stacks. Increasing the number of cells in a stack increases the voltage, and

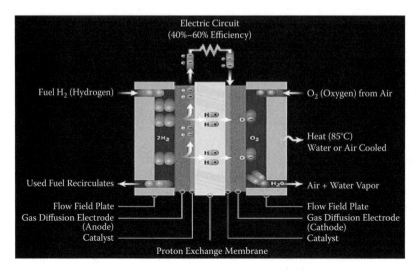

FIGURE 1.27
(See color insert following page 140.) PEM fuel cell. (Courtesy of Ballard Power Systems Inc.)

increasing the surface area of the cells increases the current. It takes only a few seconds for cold fuel cells to start producing electricity.

In high-temperature and larger-size (250 kW–3 mW) fuel cells, carbonate or solid oxide materials are used. Their operating temperatures range from 650 to 980°C (1,200–1,800°F), and their byproduct is high-pressure steam.

1.3.5.4 Reversible Fuel Cells (RFCs)

In the solar–hydrogen demonstration power plant I have designed, the functions of the electrolyzers and of the fuel cells are combined into reversible units (RFCs). These dual-state fuel cells are expected to be much lighter than the combined weights of separate electrolyzers and fuel cells. The RFC during the day will operate in the electrolyzer mode converting solar energy into the chemical form (hydrogen) while at night it will switch to its fuel cell mode and will convert the chemical energy stored in hydrogen back into electricity.

It takes the same amount of energy to split water into hydrogen and oxygen as is obtained when oxidizing hydrogen into water. The only difference between the two operations is that electrolysis increases the entropy and therefore not all the energy needs to be supplied in the form of solar electricity, as the environment contributes 48.7 kJ/mol of thermal energy. Inversely, when the RFC is operated in the fuel cell mode, part of the energy in the fuel, which is released as the hydrogen is oxidized, will be released as heat. Therefore, the electrolysis mode of operation can require heating and fuel cell mode cooling (or heat recovery).

In a solar–hydrogen power plant, when *excess* solar energy is available, the RFC is switched into the electrolyzer mode to split water into hydrogen and oxygen. The hydrogen is collected at about 3 bar (45 psig) of pressure and is either liquefied or is compressed to some high pressure (about 1,000 bar = 15,000 psig) and sent to storage.

On the other hand, when solar electricity is *insufficient* and needs to be supplemented, the RFC is switched into the fuel cell mode of operation where the oxidation of one mol of hydrogen will generate 237.1 kJ/mol of electrical energy plus 48.7 kJ/mol of thermal energy. This waste heat can also be used for heating buildings or for preheating boiler feedwater.

1.3.6 Energy Conservation

Global conversion to a solar–hydrogen economy will take a generation or two. In the meanwhile, we should utilize the "low-hanging fruits" of the transition period, which are the various forms of energy conservation. Conservation is inexpensive, and it brings immediate results without increasing emissions. Today the waste of energy in transportation is about 70%, in electric power generation it is about 65%, in building operations about 35%, in the general industry about 30% and in electricity transmission about 10%.

Governments could take steps to increase global targets for energy efficiency by about 2.5%/yr—by strengthening building codes for new and existing structures; setting penalties or disincentives for builders who choose the cheapest, least energy efficient designs; devising policies that promote mass transit especially rail; and by setting minimum performance standards for industrial and household appliances. Other possible measures include motivation for the power industry for energy savings in existing power plants and in the electricity transmission infrastructure.

Besides conservation by industry, individual families can also consider conservation by changing their life style. Public education is needed to make people realize for example that the energy in the same gallon of gasoline that we use to drive 25 or 35 miles, can take a railroad car 400 miles, or that walking or using a bicycle provides needed exercise while saving energy. Similarly, we do not realize that to produce one calorie in pork takes seven calories of grain. Even such small steps as using reusable shopping bags, turning off lights, computers, etc., when not needed, letting the hot water in the bathtub cool before draining it or using self-powered solar street lights can add up to substantial savings if billions of people use them.

In the United States, energy conservation, if implemented just in the commercial and industrial sectors, could eliminate the need for all oil imports from the Middle East. The total energy consumption of the United States is about 100 Q, and commerce and industry consume 51%, or approximately 51 Q, of all energy used. As detailed in my book *Optimization of Unit Operations*, energy consumption of commercial buildings, power plants, refineries, and the chemical industry can be reduced by an average of 15–25%, or by nearly 10 Q, while the total oil import from the Middle East is about 6 Q.

Conservation in the transportation sector holds similar potential. The U.S. transportation sector consumes 28% of the total energy use, or approximately 28 Q. Today, in the United States it requires 35% more gasoline to drive a mile than it does in Europe. If U.S. transportation efficiency was increased to the European level, the savings would amount to 9 Q, which alone exceeds the total oil imports from the Middle East.

Residential use of energy in the United States is 21%, or approximately 21 Q. If the homes used more efficient appliances, low-energy light bulbs, better boilers and heaters, more efficient heating and cooling, double windows, better insulation, and the like, residential energy consumption would drop by at least 25%. Naturally, if the owners of some U.S. homes installed grid-connected solar roofs and utilized plug-in hybrid cars, millions of the residences in the United States would become "energy free."

Today, in the average American home 40% of the energy used is spent to operate appliances, 35% for heating/cooling, 15% for making hot water, and 10% on lighting. Considering that of the total American energy consumption 51% is used in commerce and industry, 28% for transportation, and 21% by residences and because conservation technologies can reduce the energy consumption of transportation by 35%, industrial use by 20%, and residential use by 25%, oil imports could be completely eliminated if the already understood conventional conservation strategies were implemented.

Improved energy efficiency is even more important in underdeveloped nations. In China, for example, the energy use of building conditioning could be cut in half. Similarly, the energy use of China's power plants, refineries, and cement and steel industries could all be reduced by 25–50%. Chapter 2 of this volume describes specific optimization and energy conservation techniques that can be adopted immediately to eliminate such energy waste in the various industries.

Automation and process control can also contribute to energy conservation through the timing of energy use. Because electricity costs much more at periods of high demand, it makes good sense to use "demand response" in the use of electricity. Such automation systems can be installed by plants and residences, or can be obtained from the utilities. For example, some of the refrigeration, air-conditioning, and lighting loads in stores, offices, or homes can be automatically cut back or turned off during peak periods by computers located hundreds of miles away.

In evaluating the values of new processes or products, one should consider both their energy payback ratio (EPR) and their carbon emission payback period (CEPP). The EPR is the ratio of energy obtained to energy invested in manufacturing and assembling the product. The EPR of photovoltaic (PV)-type solar collectors is about 10:1. In contrast, the EPR of oil shale production is only 4:1.

The CEPP is the time required to generate enough clean electricity to compensate for the CO_2 emission that occurs during the manufacture and installation of the product. For example, the manufacture of today's solar collectors (using fossil energy) produces about 400 kg of CO_2/m^2 of collector, and by the time they are installed, that number rises tenfold. In an average power plant, 1 kg of CO_2 is emitted per kWh of electricity produced. Naturally, if renewable energy is used in manufacturing, no CO_2 emission occurs. On the other hand, if fossil energy was used and that collector installed in an area where it generates 2,000 kWh/year of clean electricity, the time required to compensate for the carbon emissions accompanying its manufacturing CEPP is about 2 years.

1.4 Solar Energy

Our planet receives as much solar energy in 30–40 minutes as humankind uses in a year. Solar energy is the most abundant energy source on the planet. It is already being used as the energy source of space vehicles and space stations. (The energy source on the International Space Station is an acre-size solar collector receiving an insolation of 1.37 kW/m².) The capacity of a collector is expressed in terms of its peak power production (wp).*

Small solar power plants can be built on the outskirts of towns, while large ones can be built in the less inhabited high-insolation areas of the planet. Solar panels can be mounted above the ground so that grass and flowers can continue to flourish between and below the rows of panels. Care should be taken that sufficient amounts of rainwater can drop through and between adjoining panels so that vegetation can survive. The first costs of solar power plants are similar to those of carbon capturing advanced fossil or nuclear power plants (Table 5.1), while the time for their construction is much less (4 instead of 10 years). As the cost of natural gas and oil increases, solar electricity is already competitive at 12¢ per kWh and during peak periods it is less expensive than fossil- or nuclear-generated electricity.

As of this writing, only 0.6% of the energy used in the world is being generated by solar energy; but in the last decades that number has been doubling every year or two. The nations with the largest installed capacities of solar collectors are China (36 gW), Japan (9 gW), Turkey (7 gW), Germany (3.5 gW), Greece (3.5 gW), and the United States (1.5 gW). On a per capita basis, the leading users of solar energy are Cyprus, Israel, and Greece.

The global total of installed solar collector capacity today (2008) is about 60 GW and according to Emerging Energy Research and the Prometheus Institute, it will reach about 300 GW by 2020. As to their size and design distribution, the small (10–100 kW) photovoltaic (PV) units will total 170 gW, the medium-sized (1–10 mW) concentrating PV (CPV) units about 6 gW, the

---

* This is the amount of electric power that a PV module is able to generate when it receives 1,000 W/m² of solar irradiation at 25°C. Therefore, the actual rate of power generation can be much less when the insolation is low, such as at night.

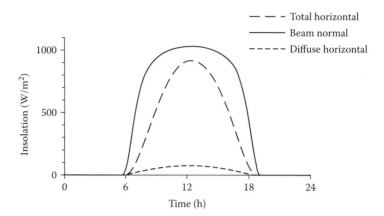

FIGURE 1.28
Insolation on a clear March day in Daggett, California.

medium-sized tracking PV (TPV) systems 100 gW and the large 10–>100 mW concentrating solar thermal (CST) plants about 12 gW.

1.4.1 Insolation: Global and the United States

Insolation (the solar energy received by an area of 1 m²) varies with the geographic location, weather, orientation of the collectors, and with diurnal and seasonal weather variations. Solar power plants have been built on all five continents, mostly in the high-insolation areas near the equator. There are also plans to collect solar energy on artificial islands in the oceans and in other locations including space stations and solar satellites.

The insolation at the top of the atmosphere is 1,368 W/m². An important factor in the total solar energy available in a particular area is the number of hours during which the sun shines per year. In Germany or London it is around 1,500 hrs/yr, in Florence, Italy it is 2,000 hrs/yr, and in San Diego it is 3,000 hrs/yr. Figure 1.28 shows the solar insolation on a clear March day in Draggett County, California.* This radiation is practically zero from 6 p.m. to 6 a.m., whereas during the daylight hours between 6 a.m. and 6 p.m., it averages about 600 W/m². This corresponds to a daily average of about 0.3 kWh/m²/hr. Therefore, in March each square meter of collector area (if continuously pointed toward the sun) receives 7.2 kWh of solar energy a day.

If the March insolation in Figure 1.28 also represents the yearly average, then each square meter of collector area receives 2,628 kWh/yr. If these numbers are correct, and if the collector efficiency is 20% (solar collector efficiencies can range from 5 to over 30%), then each square meter of collector area will produce 526 kWh/yr. Based on such calculations, one can calculate the area requirements of solar–hydrogen power plants in any location (see Section 4.2).

---

* Landolt-Börnstein—Group VIII, Volume 3C, *Renewable Energy*, published by Springer, Berlin, 2006.

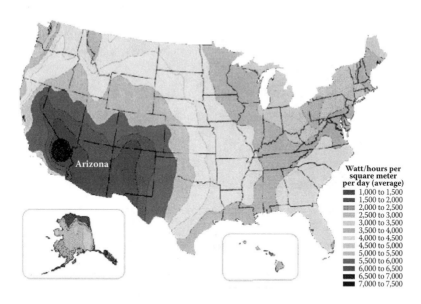

FIGURE 1.29
(See color insert following page 140.) Annular average of daily solar energy received (insolation) in different locations in the United States. (Courtesy of the National Renewable Energy Laboratory (NREL)/U.S. Department of Energy.)

Similarly, the total yearly insolation can be calculated for any other location on the planet. Figure 1.29 provides data on the yearly average insolation for the United States in units of kWh/m²/d, and Figure 1.30 provides the same information for the entire planet. Therefore, the yearly potential for solar power generation can be calculated for any location on the planet.

For example, if the yearly average insolation in the New York area is 3.5 kWh/m²/d (1,277 kWh/m²/yr), and if the efficiency of the collector is 15%, then each square meter of collector area will produce yearly power of 192 kWh of electricity.

Today, the per capita energy use on the planet ranges from 1,000 kWh/yr in Africa to 16,000 kWh/yr in Canada. Therefore, if the efficiency of a solar energy collection system is assumed to be 15%, and if that solar energy is collected in a high-insolation region such as South California (500 kWh/m²/yr), the per capita collector area required in Africa would be 2.0 m², and in Canada 32 m². If this same electric energy was generated by collectors in the New York area, the aforementioned per capita collector areas would have to be doubled.

1.4.2 Area Required to Meet Global Needs

Here I discuss the collection area requirements to meet global energy needs, but before that, smaller municipal applications will be discussed. Municipal solar power plants of the size range of 2 to 10 mW require 10 to 50 acres of

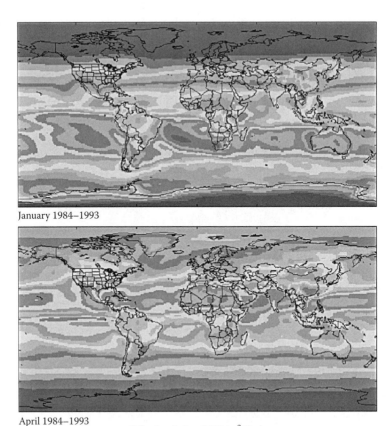

January 1984–1993

April 1984–1993

Solar Insolation (kWh/m²/day)

0   >8.5

FIGURE 1.30
(See color insert following page 140.) Global map of insolation in units of $kWh/m^2/d$. (Source: Earth Observatory, NASA.)

collector area and can supply 1,000 to 5,000 homes. These plants can consist of rows of solar panels mounted at the outskirts of municipalities and towns.

Assuming that solar collector efficiency is 20% and assuming that the yearly solar energy received by each square meter of collector area is 3,000 $kWh/yr/m^2$, each square kilometer will generate 6×10^8 kWh/yr. Therefore, the collector area required to generate one Q (Q = 10^{15} Btu = 2.93×10^{11} kWh) of electric energy is $[(2.93 \times 10^{11})/(6 \times 10^8)] = 488$ km². (Today it takes about 33 nuclear or about 66 fossil fuel power plants to generate 1 Q/yr.)

Therefore, to collect the 750 Q that is expected to be the global energy consumption by 2050 would require $488 \times 750 = 366,000$ km², which is 4% of the area of the Sahara (9 million km²). Therefore, the solar radiation reaching less than 4% of the Sahara would suffice to provide all the energy required by mankind in 2050. Another way to express the abundance of solar energy is to note that the total area of high insolation on the planet is more than 5,000 times the area needed to meet mankind's energy needs. (The land area of the planet is 150 million km^2; the oceans occupy 361 million km^2.)

Considering that the electricity used by the United States is about 4.5 × 10^{12} kWh/yr (15.35Q), if this amount of energy was to be produced by solar farms in the arid southwestern regions (the area of the Mojave Desert alone is 15,000 mi^2), the area required to generate 1 Q at a yearly insolation is 2,250 kWh/m^2/yr and with a collector efficiency of 20% is 2,250/3,000 × 488 × 15.35 = 5,618 km^2 or 2,194 mi^2. Therefore, 14.6% of the Mojave Desert would need to be covered to meet the total electric energy requirement of the United States.

If, instead of the Mojave Desert, one attempted to meet the national electricity demand by covering the roofs of buildings with solar shingles, the following would be required. Assuming that the average insolation in the United States is about 1,500 kWh/m^2/yr (Figure 1.29), roof shingle efficiency is 10% and the average industrial and private building has a southern facing roof area of 200 m^2, such an "average" building could collect about 1,500 × 0.1 × 200 = 30,00 kWh yearly. Therefore, to meet the total U.S. electricity demand (4.5 × 10^{12})/(3 × 10^4 = 1.5 × 10^8) 150 million buildings would have to be so covered. Obviously, if the collector efficiencies increase, this number would drop. Still, currently both the Mojave Desert and the roof options seem unreachable, but reaching that goal by the end of the century using a mix of both is not unrealistic.

In the Southwest region of the United States, the area requirements of smaller photovoltaic systems can be estimated at about 6 acres per mW or about 300 acres for a smaller city requiring 50 mW. In the same geographic location, if part of the roof area is covered with collectors and is connected to the grid, the system usually can make the home energy free, including transportation by charging the batteries of an electric car with the excess electricity.

1.4.3 Solar Energy Storage

The availability of solar energy is subject to diurnal-, seasonal-, and weather-related variations. Therefore, if solar energy is to meet continuous energy demands, it must be stored. On small installations, such storage can be provided by high-density batteries. (NGK Insulators Ltd. of Japan, for example, manufactures sodium–sulfur batteries that can store 7 mWh.) On midsized installations, pumped hydrostorage can be considered.

For larger installations, the compressing of air into underground caverns has been suggested. This method is inefficient, and to my knowledge only two compressed air energy storage systems are in operation (neither serving the storage of solar energy).

In even larger installations, one solution is the storage of heat in high-temperature (400°C or higher) oil or in molten salt. Hot oil storage is more common, but in other cases such as the 64 mW solar power plant called Solar One, in Boulder City, Nevada, excess solar energy is planned to be stored in molten salt. These plants store the molten salt at high temperature and when electricity is needed, use that heat to boil water and use the high pressure steam to spin turbines that generate electricity. By providing such storage, the plant can sell the electricity at peak periods when the price of electricity is high.

1.4.3.1 Storage by "Net Metering"

The most economical and practical storage methods are to store the solar energy in thermal or chemical form (hydrogen), but the electricity grid can also provide storage indirectly. In parts of the world where an electric grid does exist (about 20%), a good solution is to send the solar electricity to the grid when in excess and take it from the grid when solar generation is insufficient. In the United States, many states require utilities to buy the excess solar electricity at retail and not wholesale prices. This drastically improves the economics of the installation as the difference between wholesale and retail prices is substantial (see Section 1.7). Another major problem with the global electricity system is the lack of continent-wide electric grids that could serve both to distribute and to store electricity. This "net metering" capability is essential to make the energy of intermittent energy sources (solar, wind, tide, etc.) continuously available.

Many power companies are willing to provide this net-metering service, because excess solar energy is usually available at times when the air conditioning loads are also high and, therefore, by receiving the excess solar electricity during such peak periods, the utility can lower its operating costs by "shaving" its peak demand. When renewable energy is being sent to the grid, the electric meters of the solar generator "run backward," creating a credit that can be used at night or to supplement solar generation at other times.

This "net metering" is the least expensive method for small solar generators to store energy and for utilities to meet and reduce their peak demands. When an integrated grid operates in a net-metering mode, the loading of the individual power plants "floats" (modifies the rate of generation) as a function of the total amount of electricity received from the various renewable energy sources.

Solar power plants can use power inverters with intelligence to improve grid power quality at the outer branches of the electric grid where power quality is hard to maintain. Municipal solar power plants can also integrate into the existing electricity grid by feeding power into the grid voltage levels of 20 kV, which eliminates the expense of a substation for down-transforming the power from high (multi-100 kV) transmission voltages.

1.4.3.2 Storing as Chemical Energy

In the case of large solar power plants that are located in remote areas, such as the Sahara or on floating islands where no electric grid exists, solar energy can be stored by converting it into chemical energy and storing it in that form. The carriers of this chemical energy can be gases, liquids, or solids. A variety of processes are available for chemical storage.

In one process, mirrors concentrate the Sun's rays on zinc oxide and vaporize it at a temperature of 1,200°C. The vaporized zinc is later recondensed into a powder. This zinc can then be transported, and when reacted with water vapor, produce hydrogen fuel while using the O_2 from the water to recombine the zinc back into zinc oxide.

A pilot plant jointly funded by the EU, the Paul Scherrer Institute (PSI), and the Swiss Federal Institute of Technology of Zürich (ETHZ) has successfully demonstrated this process in a 300 kW pilot plant at the Weizmann Institute of Science (WIS) in Rehovot, near Tel Aviv, Israel. The plant has produced 45 kg of zinc per hour. This method of solar energy storage is not yet commercially available.

The best-developed method of solar electricity storage is to send it to an electrolyzer that splits water into O_2 and H_2. In this process, the O_2 is released (used or sold), and the H_2 is stored either as high-pressure gas or as a cryogenic liquid. This process will be described in the discussion of hydrogen processes after the forthcoming description of various solar collector designs in Section 1.5.

1.4.4 Solar Collector Designs

Solar collectors are either concentrating or flat, their operation is either stationary or they can track the Sun, and they can convert the solar energy into either thermal energy (heat) or electricity by the photovoltaic (PV) process. (The idea of tracking the sun is not new; sunflowers and other plants have been doing that.) There are also solar collector designs that can generate both heat and electricity. The concentrator designs include parabolic troughs, parabolic dishes, central receivers, and Fresnel lenses. This last design uses refraction instead of reflection.

In addition to the existing technologies, research is continuing on designs for new and more efficient converters. For example, researchers at Penn State University are working on new ways to harness the power of the sun by using highly ordered arrays of titania nanotubes for H_2 production and increased solar cell efficiency.

Another area of development is in lower-cost thin- and ultrathin-film designs. One such product is made by Nanosolar of copper indium gallium selenide (CIGS), which is claimed to achieve up to 19.5% efficiency and is as thin as a newspaper. This claim is yet to be proved. The collector cost is also reduced, because the substrate material on which the ink is printed is much less expensive than the stainless steel substrates that are often used in thin-film solar panels. The manufacturer claims a five- to tenfold reduction in the collector cost (about $1/W) compared to conventional PV cells.

1.4.4.1 Thermal Solar Collector Designs (SEGS and DSG)

Thermal collectors on the roofs of private homes are usually flat and stationary units that serve to provide residences with heat and hot water. The yearly energy needed for heat and hot water of a well-insulated home in the Northeast United States is about 150 million Btus (44,000 kWh), which is about the same as the energy content of 1,000 gal of oil. According to a 2007 study by the McKinsey Global Institute, the residential energy consumption in the United States is 21.3 Q, or 21% of the total energy used. Of the 21.3 Q total, 8.5 Q is used to operate appliances, 7.5 Q for heating/cooling, 3.2 Q for the generation of hot water, and 2.1 Q for lighting.

Today thermal–solar plants are the most popular. Besides their lower cost and fairly good efficiency (compared to photovoltaic designs), their main advantage is the ability to store energy for continued nighttime operation. One such 354 mW plant in the Mojave Desert has been in operation for several decades. Ten new plants are planned for Arizona, California, and Nevada, with an electricity generating capacity equaling that of three nuclear plants. Other thermal–solar plants are under construction in Spain, Algeria, and Morocco and another nine are planned in Israel, Mexico, China, South Africa, and Egypt.

In the United States solar–thermal–electric generating system power plants have been in operation since 1985. In the Southwest they generate up to 1,000 kWh/yr/m^2 of electricity. They include SEGS I (1985, Daggett, CA, 14 mWe, 5,000 m^2/mWe); SEGS II (1986, Daggett, CA, 30 mWe, 6,000 m^2/mWe); SEGS III, IV, V, VI, VII (1987–1989, Kramer Junction, CA, 30 mWe, 6,500 to 8,500 m^2/mWe); SEGS VIII and IX (1990–1991, Harper Lake, CA, 80 mWe, 6,000 m^2/mWe); and Solar One (2007, Boulder City, NV, 64 mWe (5,500 m^2/mWe).

The first costs of the plants are competitive with conventional fossil or nuclear plants while the time needed for construction is one fifth (3 to 4 instead of 10 years). One reason why solar power is competitive is the increase in the costs of oil and natural gas, which makes solar electricity competitive (12¢/kWh) and actually less expensive than fossil or nuclear during peak periods.

In solar power plant applications, the largest solar energy generating system (SEGS) in the world is operated by Southern California Edison. The photograph in Figure 1.31 shows part of that solar power plant. It has been in operation at Kramer Junction in the Harper Valley of the Mojave Desert in California since 1985. This facility consists of nine solar electric plants, with a combined capacity of 354 mW. The facility has 400,000 mirrors, which are distributed over 1,000 acres (4 km^2). It was built by Luz International and is owned and operated by FPL Energy, a subsidiary of Florida Power and Light.

In this design, parabolic mirror reflectors (troughs) are used to track the trajectory of the Sun and to concentrate the sunlight onto absorber tubes that are located at the focal line of the parabolic mirrors. Inside the absorber tubes, heat-resistant oil is circulated. This heat-transfer fluid serves to transport the collected heat into boilers that generate steam to drive the turbine generators.

In this particular installation, the hot oil circulates at 400°C (752°F), but at other solar power plants (for example, Boulder City, Nevada, by Solargenix Energy, using molten salt), the operating temperature is even higher. One advantage of the SEGS design is that the solar energy can be conveniently

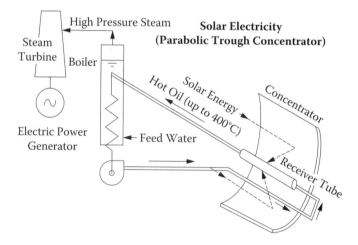

FIGURE 1.31

A 354 mW thermal solar plant at Kramer Junction in the Harper Valley in California that has been in operation since 1985 (courtesy of NREL/DOE). (Top) This is one of nine solar electric energy–generating plants at Kramer Junction, California and it uses parabolic troughs to collect the Sun's energy. (Courtesy of National Recoverable Energy Laboratory–NREL/DOE.) (Bottom) In the receiver tubes, hot oil transports the concentrated solar heat to steam boilers, which drive the turbine generators.

FIGURE 1.32

A 10 mWe central receiver concentrator plant with molten salt–based thermal storage, located at Daggett, California. (Courtesy of Sandia National Laboratories.)

stored (as thermal energy) and used later as needed to compensate for diurnal-, seasonal-, and weather-related variations in the availability of solar energy (insolation).

Thermal solar collectors are also available in so-called central concentrator or solar concentrator designs. In these configurations, a large number of independently movable flat mirrors (heliostats) are used to reflect the solar radiation onto a central receiver on the top of a tower. Each heliostat moves about two axes. The receiver typically is a vertical bundle of tubes in which the heat-transfer fluid (water or oil or molten salt) is heated by the reflected and concentrated insolation. The molten salt technology also provides thermal energy storage.

Designed by the DOE, such a design was implemented at Daggett, California, and named *Solar One*. In 1995 this plant was converted and expanded to a total area of 82,750 m², and is referred to as *Solar Two*. A 10-mW central receiver-type generating plant located at Daggett, California is shown in Figure 1.32.

In some of the more recent SEGS designs, the circulating fluid temperature has been more than doubled, whereas other designs generate steam directly. These improvements have increased the amount of power produced by about 15%. The main limitation of direct steam generation (DSG) is the difficulty of providing energy storage, which is not a limitation in SEGS.

The main advantage of DSG is its improved efficiency in comparison with the SEGS design. This is because the two-circuit system (hot oil collection and water/steam Rankine power circuit) of the SEGS design is replaced by a single circuit in which the collector circuit is directly coupled to the

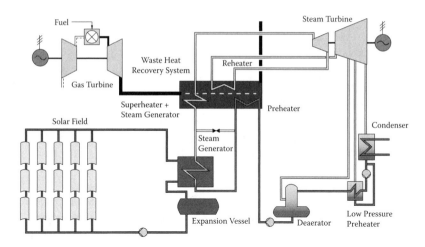

FIGURE 1.33
(See color insert following page 140.) The integrated solar/combined cycle system (ISCCS) design, a thermal solar collector field is integrated with a steam- and gas turbine-based electric power generation. (Courtesy of FLAGSOL, Cologne.)

power circuit. This eliminates the oil-to-water heat transfer equipment and the associated loss in efficiency. With DSG systems, superheated steam at 100 bar pressure and 400°C temperature can be generated.

1.4.4.2 The Combined-Cycle Plant (ISCCS)

A combined-cycle plant utilizing a solar parabolic trough collector system is shown in Figure 1.33. This design is referred to as the ISCCS (Integrated Solar and Combined Cycle System). SEGS steam cycle plants have been constructed with capacities ranging from 15 to 80 mWe; ISCCS plants can be larger and the electricity they generate is usually less expensive, because as the size of the plant increases, the unit cost of electricity drops. When solar energy collection is coupled with a combined-cycle power plant (ISCCS configuration), the overall generating capacity of combined-cycle power plants (ISCCS) can exceed 200 mW.

1.4.4.3 Photovoltaic (PV) Collectors

Sunlight is composed of photons with energy corresponding to the range of wavelengths within the solar spectrum. When photons strike the collector cell, they may be reflected, pass through, or be absorbed, but only the absorbed photons generate electricity. This is because the construction material (the silicon atoms in the crystal) has to receive 1.1 eV to cause its valence electron (electron in the outermost shell) to move into the conduction zone.

A typical silicon PV cell is composed of a wafer consisting of an ultra-thin layer of phosphorus-doped silicon (N-layer with a negative character).

This layer is placed on top of a thicker layer of boron-doped silicon (P-layer with positive character). These layers are connected by the P-N junction. When sunlight strikes the surface of the PV cell, an electric field is generated, which provides momentum and direction to the light-stimulated electrons, resulting in a flow of current when the solar cell is connected to an electric load.

Photovoltaic (PV) cells are almost always arranged on a panel to form a solar module. Modules are then linked in series to form what is known as a solar array. The size of a solar module or array is given in terms of its peak power production (Wp) under standard condition (STC). This condition is also referred to as "one sun," and this STC condition is defined as a solar irradiation of 1.0 kW/m^2 on a clear summer day with the sun approximately overhead, the cells facing directly toward the sun and the cell being at 25°C. Therefore, a 100 Wp module will generate 0.1 kW/hr/m^2 under STC conditions so it is 10% efficient, and a 150 Wp module under STC conditions will generate 0.150 kW/hr/m^2 and therefore it is 15% efficient. The cost of installed PV systems is in the range of $3 to $7 per Wp with $5/Wp or $4,000/kWp being typical for larger systems. For the same system capacity, the required collector area drops as the collector efficiency rises but the system cost in units of $/Wp remains approximately the same.

Presently, the availability of solar-grade silicon is limited, and its cost is over $30/kg and rising. The largest supplier of polycrystalline silicon (Hemlock Semiconductor Corp.) is building new production facilities, and users such as Sharp Corporation and BP Solar are working on designing thinner panels requiring less silicon.

In the PV cells, germanium or gallium compounds provide electricity when exposed to solar radiation. The absorbance spectrum of the PV cell material can go from UV wavelength (~250 nm) to IR radiation (~1,500 nm). The conversion efficiency of standard PV cells is between 10 and 15%.

In newer PV designs, the range of spectral wavelengths has been extended from the visible interval (380–780 nm) toward the UV wavelengths (250–380 nm) and the IR range (780–1,500 nm). Because of the stratification of different materials, each of which is adapted to a portion of the working spectral range, these novel devices achieve high performance over a wavelength range from 200 to 1,800 nm.

These units are claimed to provide conversion efficiencies of up to 25%, or even 30% if the irradiation level is high enough. High irradiation is defined as 200–300 suns (1 sun = 0.1 W/cm^2), which correspond to 20–30 W/cm^2 of solar irradiation. To obtain such levels of solar irradiation, the solar light has to be focused 200–300 times. For a light concentration ratio of 100×, the high-performance PV cells are claimed to have a conversion efficiency of 28%.

According to the International Energy Agency (IEA), most of the existing PV installations are in Germany (40%) and the United States (13%). In Germany, the power companies "buy back" the solar-generated electric power at a rate that is secured for 20 years. This is called a Feed-in Tariff

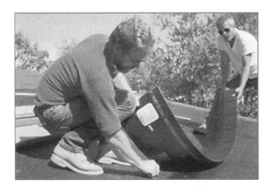

FIGURE 1.34
Solar shingles can be installed on the rooftops of homes. (Courtesy of NREL/DOE.)

(FiT) program. By 2020, the total solar electricity generation commitment* in the United States is expected to reach 7.3 gW. Of this total, California is committed to 3 gW and New Jersey to 1.5 gW.

Pacific Gas and Electric is building a 550 mW thin film (OptiSolar) and a 250 mW regular PV (SunPower) plant on a combined area of 12.5 square miles.

In the last few years thin film designs have been marketed to cover roofs of buildings. For example, Uni-Solar® laminates were used by GM in a 1 mW installation in Rancho Cucamonga, California.

1.4.4.4 Flat Collectors and Roof Shingles

Flat-plate PV collectors contain an array of individual cells connected in a series/parallel circuit and encapsulated within a sandwich structure, the front of which is glass or plastic. Unlike thermal collectors, the back of these collectors is not insulated, because for best performance, they need to be cooled by the atmosphere. If this energy could be used and thereby this loss could be eliminated in new designs, the conversion efficiency could be much improved.

Flexible thin-film solar cell strips and collectors are also available. Figure 1.34 illustrates the rooftop installation of solar shingles developed by the National Renewable Energy Laboratory (NREL). U.S. Department of Energy's NREL reported in March 2008 that their researchers created a world record in efficiency for thin-film solar cells (19.9%), nearly tieing the record for traditional photovoltaics (20.3%). Thin cells require less expensive materials to build and use less of them. This results in a substantial cost reduction.

Shingles can be made from a thin film of amorphous silicon on plastic. They can generate solar electricity for private homes or commercial buildings. Flexible roofing laminates can be designed to be bonded on 0.4 m (16-inch) wide, flat Galvalume® pans. A 2.75 m (9 ft) PV roof pan is rated to generate a maximum of 64 watts.

---

* *Solar Generation* magazine, Sept. 2006 p. 24.

Other PV shingles are 0.3 m (12 in.) wide and 2.2 m (7.2 ft) long, and are nailed in place on roof decks on top of 30 lb felt sheeting (Figure 1.34). Electric lead wires, #18AWG, extend from under each shingle and pass through the roof deck, allowing the wire connections to be made in the roof space connections. After the solar shingles are installed to match the course of the conventional shingles, an electrical system installer wires them together under the roof, and then wires the array of shingles to the combiner box. The electrical contractor takes the wires from the combiner box and runs them down to the power-conditioning equipment (charge controller, meters, system disconnect switches, fuses, inverters, and batteries). For a detailed design of PV collector controls, see Section 2.18.3.

1.4.4.5 Ultrathin-Film Nanowire Designs

Nanocrystals and nanowires are utilized in a new generation of solar collectors (a nanometer is one billionth of a meter). In conventional solar cells, at the P-N junction one photon splits one electron from its "hole companion" as it travels to the electron-capturing electrode. If solar collectors are made of semiconducting nanocrystals that disperse the light, according to TU Delft's professor Laurens Siebbeles, an avalanche effect results and one photon can release two or three electrons, because this effect maximizes photon absorption while minimizing electron-hole recombination. This effect of the photon-scattering nanoparticles substantially increases cell efficiency.

Nanowires increase efficiency because they *directly* deliver the electrons from the interface of the nanowire with the polymer to their electrode, while the electrode holes travel in the opposite direction to the tip of the wire and pass through the very thin polymer layer before reaching their electrode. Earlier carbon nanotube and nanowire designs were not directly connected to their electrodes and therefore did not provide the electrons with a direct path.

Conventional photovoltaic cells are made of slabs of crystalline material. The light absorption of these slabs increases with thickness, but so does the difficulty of the electrons to escape. It is this trade-off that is overcome by nanowires that are "grown like hair" on the fabric of the electrode surface in lengths of about 5 microns and with 10 to 100 nanometers in diameter. The great promise of nanowire technology is the reduction in cost, because so little material is needed of the exotic Group III-V materials (gallium arsenide, indium gallium phosphide, etc.) and because of the increase of cell efficiency (Figure 1.35). Today's thin-film collectors are 5–8% efficient, regular PV collectors are about 15% efficient, while nanowire collector efficiencies can theoretically reach 40%.

Another advantage of the nanowire design is that the nanowires can be grown on cheap metal substrates that can conform to irregular and curved electrode surfaces such as the roofs of homes or cars.

Nanosolar's ultrathin-film design is made of copper indium gallium selenide (CIGS) and is claimed to achieve up to 19.5% efficiency. Although such

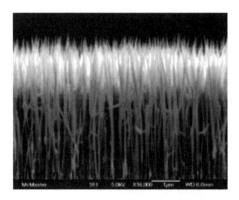

FIGURE 1.35
A side profile of gallium arsenide nanowires growing on a silicon substrate. The nanowires grow upward from the substrate, creating a surface that's able to absorb more sunlight than a flat surface is. (Credit: McMaster University.)

efficiency is unlikely, the design, using a semiconductor ink, is original. These solar cells can be produced by depositing the ink on a flexible substrate (the "paper") in which the nanocomponents in the ink align themselves via molecular self-assembly. This approach improves production speed and eases the deposition of a uniform layer that provides the correct ratio of elements everywhere on the substrate. The cost of these collectors is expected to be low because the substrate material on which the ink is printed is much less expensive than the stainless steel substrates that are often used in thin-film solar panels. The manufacturer projects that once mass production starts, these cells will be available at about $1/W (http://www.nanosolar.com).

1.4.4.6 Large-Scale PV Installations

In large-scale applications (Figure 1.36 top left), the solar power generators are often connected to the electric grid in the area. The shown PV power plant capacity is 2 mW, and the plant is located near a defunct nuclear power plant.

The Fresnel lens concentrator, shown in Figure 1.36 top right, uses refraction rather than reflection to collect the solar energy. These units are molded out of inexpensive plastic and provide higher efficiency than the standard PV collectors. Point-focusing Fresnel lenses are also available. In Figure 1.36 bottom left, parabolic dish collectors are shown. Both the Fresnel lenses and these collectors can be rotated around two axes to continuously track the Sun.

Parabolic PV collectors can combine the steam generation capability of the thermal collectors with direct electricity generation by PV cells, as shown in Figure 1.36 bottom right. In this design, silicon solar cells are bonded to the circulating hot oil pipe and serve to generate electricity, while the high-temperature oil is used to produce steam.

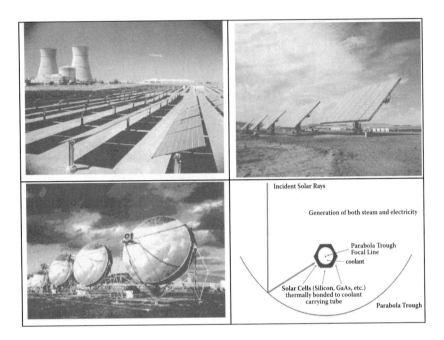

FIGURE 1.36

Flat and concentrating photovoltaic solar collector designs. (Top left) Photovoltaic power plant near Sacramento, California. (Courtesy of the National Renewable Energy Laboratory/U.S. Department of Energy.) (Top right) Photovoltaic collectors using Fresnel lenses. (Courtesy of Amonix Inc.) (Bottom left) Two-axis tracking parabolic dish collectors. (Courtesy of Schlaich, Bergermann und Partner.) (Bottom right) Parabola trough collector designs are available to simultaneously generate both steam and electricity.

Today, the energy payback period of PV collectors for thin-film PV systems is about 3–5 years and for multicrystalline silicon PV systems about 6–8 years. As manufacturing techniques improve and the use of preinstalled roof PV collectors spread,* these payback periods are likely to drop. With a minimum life span of 25 years, the ratio of energy obtained to energy invested in manufacturing and assembling the PV collectors is about 10:1. This ratio compares favorably, for example, with the energy payback ratio of oil shale, which is only 4:1.

The CO_2 emission payback period of manufacturing PV collectors is estimated to be 2–3 years. This is the time required to generate enough clean electricity to compensate for the CO_2 emission that occurs during the manufacturing of the collectors. If the PV collector is manufactured by using fossil fuel–generated energy, about 400 kg of CO_2 will be generated per square meter of collector. By the time it is installed, this number can increase manyfold. Naturally, if renewable energy is used in the manufacturing, no CO_2 emission occurs.

---

* A prefab home builder in Japan, Sikisui Chemical, provides preinstalled PV systems in half of the homes they build.

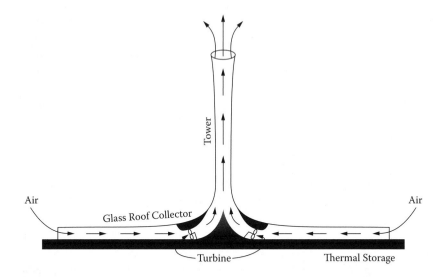

FIGURE 1.37
The solar updraft tower converts solar-based thermal energy into concentrated aerodynamic energy that drives turbine generators.

1.4.4.7 Solar Updraft Towers

The chimney effect of tall towers generates an updraft because the atmospheric pressure is higher at the bottom of the chimney and lower at the top. Therefore, the heavier colder air from outside of the chimney pushes the lighter warm air upward inside the chimney. Static home cooling systems utilize this effect to pull the cold air into homes from underground ducts during the summer. On the other hand, in the winter, this chimney effect can be harmful because it sucks cold air into the building and thereby increases the heating load on high rise buildings. (The method of eliminating such chimney effects is described in Chapter 2, Section 2.3.2.)

In solar towers, thermal storage can be provided by covering the ground with heat-absorbing surfaces so that power generation continues at night. In the case of this design, both the investment cost (about $30/m²) and the solar energy collection efficiency (about 5%) are low, while the energy payback period is expected to be 3–5 years.

The solar updraft tower is illustrated in Figure 1.37. It is a solar energy converter that converts solar-based thermal energy into aerodynamic energy (wind). In this system, air is heated under a circular greenhouse-like canopy. The roof of this canopy slopes upward from the perimeter toward the center, where the tower stands. Under the canopy, the Sun heats the air, which rises up the tower and generates electricity by driving an array of turbine generators.

This low-tech solar energy collector concept is over 100 years old. The Spanish Colonel Isidoro Cabanyes first proposed it in 1903, but the first patent for this design was applied for by Robert E. Lucier in 1975. The first 50 kW working model was built in 1982 in Ciudad Real, south of Madrid, Spain.

This project was funded by the German government. The chimney of this model had a 10 m (33 ft) diameter and was 195 m (640 ft) tall, whereas the diameter of the canopy was 244 m (800 ft) in diameter (about 11 acres, or 46,000 m², in area). This prototype operated for 9 yr and reached a maximum production of 50 kW.

Today, much larger installations (50–200 mW) are planned in Australia, China, and the American Southwest. This year, construction will start on the world's first commercial plant in Australia. The 2 mi diameter canopy of this 50 mW power station at Tapio will be 2.4 m (8 ft) tall at the perimeter and will rise to 15 m (50 ft) at the center near the tower. The tower will be 79 m (260 ft) in diameter and 488 m (1,600 ft) in height. It is anticipated that this power plant will meet the electricity needs of 100,000 homes. The government-financed plant is expected to cost $75 million.

Another plant in China, designed by EnviroMission, will consist of a 38 km² canopy and a 1 km (0.62 mi) tall tower to provide 200,000 homes with electricity. The 200 mW power plant will abate about 1 million tons of greenhouse gases and will cost some $800 million.

1.4.4.8 Stirling Solar Power Generators

In addition to SEGS, ISCCS, PV, and updraft tower methods for collecting solar energy, Stirling solar dish systems are also used (Figure 1.38). These units focus the Sun's energy onto H_2 in sealed Stirling engines. As the H_2 is heated to 732°C (1,350°F), it expands and drives the pistons of the engine. Stirling engines are used on submarines because they are quiet (no combustion takes place). In solar applications, their main advantage is their high efficiency (30%), which is nearly double that of the best PV collectors. These closed systems do not need to be refilled, only their mirrors need washing every couple of weeks. The operation can be fully automated including start-up in the morning, shutdown in the evening, tracking the Sun, and remote monitoring over the Internet.

The present cost of a 25 kW, 944 ft² stand-alone prototype unit is about $150,000. This price is likely to drop as production volume rises. Stirling Energy Systems, Inc., is operating a six-dish test unit at Sandia National Laboratories and has contracts for full-size power plants. Their construction started in 2008 and will be completed in 2012. A 500 mW plant will be built at Victorville, California, in the Mojave Desert for Southern California Edison. This construction is estimated to take 3–4 years. The plant will consist of 20,000 dishes over a 4,500 acre area. Another 300 mW plant will also be built at Imperial Valley, Calexico, California, for San Diego Gas and Electric on an area of 2,000 acres.

The Stirling engine contains a fixed quantity of gas (H_2) that is moved back and forth between the hot and the cold ends of the engine. As the gas is moved, it expands and contracts, and this change in volume is used to drive

FIGURE 1.38
California Edison's 25 kW dish/Stirling system, McDonnell Douglas/Southern California. The 944 square foot concentrator consists of 82 spherically curved glass mirrors, each 3 ft by 4 ft. The United Stirling 4-95 Mark II engine (four cylinders of 95 cc displacement) uses hydrogen at a maximum gas pressure of 2,900 psi. This engine delivered 25 kW output at 1,000 W/m2 insolation. (Courtesy of Stirling Energy Systems—SES.)

the engine. In Figure 1.39, the hot end (which is heated by the Sun) is on the right, and the cold end (cooled by the ambient air) is on the left.

In phase #1 of the cycle, when most of the gas is in the "power piston" (the hot end on the right), the gas is heated by the Sun and it therefore expands, forcing the power piston to the left. This is when the engine does the work,

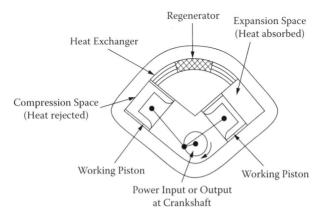

FIGURE 1.39
One version of the Stirling engine, with the "power piston" on the hot (right) side of the engine and the "displacer piston" on the cold (left) side.

FIGURE 1.40
Solar satellite collector planned by Japan for 2040.

and the higher is the hot-end temperature, the more work it does. In phase #2 of the cycle, as the gas expands, it forces more gas into the "displacer piston" (the cold end on the left).

In phase #3 of the cycle, the gas is cooled by the ambient air in the cold end, while the displacer piston compresses the gas. Cooling contracts the gas (lowers its pressure), which makes it easier to compress it. In this phase of the cycle, energy is consumed by the engine (but less than the energy that was generated in phase #1). The lower the temperature, the less energy is consumed in compressing the gas.

The displacer piston on the cooling end on the left controls the direction in which the pistons move. It determines if most of the gas is in the piston on the right or left. In phase #4 of the cycle, the displacer piston on the left moves the compressed gas back into the power piston on the right. As the gas is returned into the power piston, it is heated at this hot end by the concentrated solar energy, and the cycle repeats.

1.4.4.9 Solar Satellites

Japan plans to launch giant solar satellites by 2040. The satellites are planned for a geostationary orbit at 22,000 mi above the Earth's surface. They would generate about the same amount of power as a nuclear power plant (1 gW). The collected solar energy would be beamed back to Earth by microwaves onto several-kilometer-diameter antennas. Figure 1.40 shows the gigantic power-generating wing panels of the satellite. The weight of the structure is estimated at about 20,000 tons, and the cost is estimated at $17 billion. (The cost of a nuclear power plant is about $5 billion.)

1.4.4.10 Floating Solar Islands

A Swiss-designed artificial island is planned near the equator (Figure 1.41) in the Gulf state of Ras Al-Khaimah. A 100-m diameter prototype is under construction. It is one tenth of the size of the planned solar island. The solar concentrators will convert circulating water into steam, which will produce both electricity and H_2. The electricity will be used for plant operation, and the H_2 will be sold as fuel. The designers feel that the plant will be cost competitive

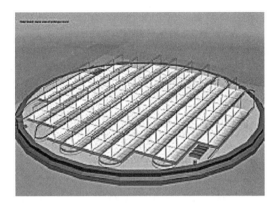

FIGURE 1.41
Solar island floating on an inflated ring structure. (Courtesy of Neuchâtel's Swiss Center for Electronics and Microtechnology—CSEM.)

because the "land cost" is minimal, no support structure is needed, and the island need not be connected to land as the H_2 can be transported by ships.

1.4.5 Operating and New Solar Power Plants

The first concentrating trough-type solar power plant in the United States was built in 1988. It is the 1 mW Saguaro plant located north of Tucson, Arizona, and was built for Arizona Public Service (APS). It covers 1 km² and has parabolic trough-shaped mirrors.

Today, the largest solar power plant in the United States is the 22-year-old thermal plant in California's Mojave Desert, which has a combined total capacity of 354 mW. At Kramer Junction, California, nine solar power plants, each 30 mW or larger, have been in operation for two decades. The yearly insolation in the area is 2,940 kWh/m². Plant efficiencies range from 10 to 17%, and their capital costs range from \$2,500 to \$3,500 per kWp.* The cost of generated electricity from these plants drops as their size increases, and ranges from 10 to 17¢/kWh.

Other installations include one by a Spanish company named Acciona Energy, which recently started up a 64-mW solar thermal power plant near Boulder City in Nevada. This company is also planning two thermal solar plants in southern Spain (50 mW each) in Palma del Rio for an investment of 0.5 billion euros.

---

* The capacity of a collector is expressed in terms of its peak power production (Wp). This is the amount of electric power that a PV module is able to generate when it receives 1,000 watts per square meter of vertical solar irradiation at 25°C cell temperature. This is also called one sun. If this level of insolation existed for 24 h every day of the year, each m² of collector area would receive 8,760 kWh/yr. The *actual* rate of power generation is naturally less. The value of Wp/m² of a module is also called its power density or efficiency. Therefore a 10% efficient module when receiving one sun will generate 0.1 kWp/m² and a 15% will generate 0.15 kWp/m².

A number of new solar power plants are under construction. In 2007, First Solar signed a contract to produce 685 mW of solar collectors over 5 years for $1.28 billion, or at a unit cost of $1.87/W. This might correspond to a $3/W installed cost. Southern California Edison is erecting a 500 mW plant designed by Solel Solar Systems of Beit Shemesh, Israel. It is scheduled to start up in 2009.

Pacific Gas and Electric (PG&E) is building the largest solar thermal power plant in the world (560 mW) in the Mojave Desert, using troughlike arrays of mirrors that concentrate the sunlight onto steam-generating oil pipes over a 9 mi^2 area. This power plant is expected to produce electricity at slightly more than 10¢/kWh. Seventeen percent of the electricity produced by PG&E comes from renewable energy sources, and the company is aiming at reaching a 20% target in the near future. Southern California Edison (SCE) has also signed a similar contract for a solar thermal power plant.

Both Southern California Edison and San Diego Gas & Electric have signed contracts with Stirling Energy Systems of Phoenix for two power plants of higher efficiency (30%) than the solar thermal ones. A 500 mW plant will be built at Victorville, California, in the Mojave Desert for Southern California Edison. This construction is estimated to take 3–4 years. The plant will consist of 20,000 dishes over a 4,500 acre (18 km^2 = 7 mi^2) area. A 300 mW plant will also be built at Imperial Valley, Calexico, California, for San Diego Gas and Electric over 2,000 acres (8 km^2 = 3 mi^2). In 2008, Southern California Edison also started an $875 million project that will cover some 2 square miles of unused rooftops and generate 250 mW of electricity.

In Boulder City, Nevada, in 2006 the construction of a $250 million, 64 mW solar thermal power plant named *Solar One* started. It uses Schott's new PTR 70 solar receivers and is built by Solargenix Energy.

Both Israel and China are building large solar power plants at investments of about $2,500/m^2 of collector area. The initial capacity of the solar thermal power plant in the Negev Desert in Israel is planned to be 150 mW, and it is estimated to cost $350 million. The plant is planned to expand to 500 mW at a cost of $1 billion.

In Dunhuang City, Gansu Province in China, the construction of a 100 mW $765 million solar power plant has been approved. The plant will have 3.1 km^2 of solar collectors at an installed cost of $2,450/m^2. Because the Sun shines for 3,362 h/yr in the area, assuming that the plant will run fully loaded when solar energy can be collected and that the electricity will be sold at 12¢/kWh, the value of the produced electricity will be $40.3 million per year.

In Portugal, an 11 mW solar power plant has been designed by SunPower and Catavento. It consists of 52,000 collectors over a 62 acre area (0.25 km^2 = 0.1 mi^2). At an investment of $75 million, it is expected to generate 20 gWh/yr.

A 40 mW photovoltaic installation started up in Eastern Germany near Leipzig in 2007 using thin-film technology. The total area of the plant is 4.5 km^2 and the electricity generated is sent to the grid. The insolation in the area is about 1,000 kWh/yr and the efficiency of the thin-film collectors is estimated to be 5 to 7%.

Another concentrating solar power (CSP) plant is planned in Madinat Zayed in Abu Dhabi.

1.4.6 Solar Collector Costs, Efficiencies, and Suppliers

Swimming pool and rooftop water heaters can convert solar energy into heat with an efficiency of 50–70%. The efficiencies of electricity-generating solar collector systems are much lower. They operate at efficiencies from 5% to about 30%.

Thin-film solar shingles used on residential roofs today are only about 5 to 8% efficient. The efficiency of solar towers is also about 5%. Photovoltaic (PV) cell efficiencies range from about 5% for amorphous silicon (A-Si) designs, 9% to 10% for CdTe modules, and 13% to 16% for crystalline silicon modules. SEGS efficiencies are between 10 and 25%.

A recently introduced ultra-thin semiconductor ink–based design made of copper indium gallium diselenide by Nanosolar is claimed to have 7% to 19% efficiency and a cost of $1/Wp, or about one seventh of the cost of regular PV cells, but these numbers have not yet been substantiated, and the first units were only shipped in late 2007. These copper indium gallium selenide units use semiconductor ink and a printing process in depositing the thin film. The United States DOE's National Renewable Energy Laboratory reported in March 2008 that its researchers had created a world record in efficiency for thin-film solar cells (19.9%), nearly tieing the record for traditional photovoltaics (20.3%).

Fresnel lens–type PV concentrators have operated at 26% efficiency (Amonix, Inc.). The efficiency of concentrating PV designs can reach 25–30%, and DSG thermal systems can also reach 30%. Similarly, the efficiency of dish concentrators using Stirling heat engines is also about 30%. Table 1.42 provides a summary of solar collector costs, efficiencies, and suppliers.

A new, experimental PV design by Spectrolab using concentrated sunlight has reached an efficiency of 40%. This milestone achievement was verified by the DOE's National Renewable Energy Laboratory (NREL) in Golden, Colorado.

The average cost of solar electricity in the United States is about 12¢/kWh. The wholesale costs of 1 kWh of electricity in 2007 in the United States (according to the *New York Times*, July 11, 2007) are as follows: pulverized coal—5.7¢, nuclear—6.4¢, coal gas—6.6¢, natural gas—7.3¢, wind—9.6¢, and biomass—10.7¢.

The wholesale costs do not include transportation and distribution costs and therefore the actual cost the user pays can be twice as much. For example, in September, 2007, in our household we paid ¢18.9/kWh. In 2007 the American national average ìretail" prices paid by the various enduser sectors

TABLE 1.42

Solar Collector Efficiencies, Costs, and Suppliers

Collector Types	Efficiency	Costs ($/Wp,* ¢/kWh or $/m²)	Power Density (Wp*/m²)	Suppliers
Solar Wind Tower	Under 5%	$3/Wp $30/m²	10	EnviroMission
Thin Film** PV, Roof Shingles	5%–8%	Installed: $3/Wp $200/m², 10¢/kWh Ave. home $60,000	60	EDF Energies Nouvelles, PlugPowerGenCore, Ovonic, UniSolar, Konarka, Miasolé, Nanosolar, HelioVolt, Opt Solar
PV, Non-Concentrating CdTe: CrystallineSi	8%–17% 9%–12% 3%–16%	Module: $4/Wp Installed: $5–$7/Wp $400/m², 12¢/kWh	120	Aleo Solar, BP Solar, EPV, eSolar, First Solar, General Electric, Isofoton, Kyocera, Photowatt Int., Q-Cell, Sanyo, SCHOTT Solar, Sharp, Shenzhen Xinhonghua Solar-Energy Co., Signet Solar, SkyPower, Solarwatt, Solarworld, Solan, SunEdison, SunPower, Suntech, Webel Solar Energy, TATA BP Solar
Concentrating Thermal SEGS (Oil)	17%–25%	Installed: $3–$4/Wp $500–$750/m² 10¢–17¢/kWh	150–250	eSolar Inc., FPL Energy, SCHOTT Solar, Skyfuels, Solel Co., Suntech
Concentrating Thermal DSG (Steam)	20%–30%	Installed: $4/Wp $800–$1000/m² 12¢–20¢/kWh	150–250	Ausra, eSolar Inc., SCHOTT Solar, Luz II (BrightSource), Solel Co.

Concentrating Fresnel, Reflected PV	25%–30%	Installed: $6/Wp $1800–$2400/m² 20¢/kWh	300–400	Spectrolab, Amonix, Suntech, Solarex, SunPower, H2GO/Solfocus, Energy Innovations, BP Solar, Evergreen Solar, First Solar, Kyocera Corp., Q-Cells AG, Sekisui Chemical, Sharp Corp., Solar Systems, Solar World AG, SunPower, Tokuyama Corp.
Stirling Solar Dish	30%	$6/Wp $2,400/m²	400	Infinia, Stirling Energy Systems, Inc.
Nanocrystal and Nanowire	Up to 40%	Under development	~300–400 (projected)	Nanosolar, Researchers: Prof. Siebbeles (TU Delft), Prof. LaPierre (McMasters U.), Prof. Yu (U of California, San Diego)
Hot Water Collectors	50%–70%	$250/m² ($25/ft²) about $25,000/home	—	Many

* The power output of solar modules is expressed in terms of their "peak" power output (Wp). This is the amount of electric power that a module generates when it receives 1.0 kW/m² of vertical solar irradiation at 25°C module temperature (also called one sun). Therefore, a 10% efficient module receiving one sun, will produce 100 Wp/m². If this solar irradiation exists for 8 hours per day, this module will generate 0.8 kWh/m²/day. Naturally both the number of sunny hours of the day and the intensity of insolation vary (see Figures 1.29 and 1.30 and www.bpsolar.com). In addition the angle of irradiation and/or cell temperature can also be less than optimum.

** United States Department of Energy's National Renewable Energy Laboratory reported in March 2008 that their researchers created a world record in efficiency for thin-film solar cells (19.9%), nearly tieing the record for traditional photovoltaics (20.3%).

were as follows: residential, 10.7 ¢/kWh; commercial, 9.9 ¢/kWh; industrial, 6.4 ¢/kWh; transportation, 10.4 ¢/kWh; all sectors, 9.2 ¢/kWh.

The cost of electricity during summer peak periods can be two or even three times the night time electricity cost. Because solar electricity is usually produced during peak periods, the value of solar electricity is usually greater than fossil or nuclear.

Today, when fossil energy is used in the process of manufacturing solar modules the mining of the raw materials and the manufacturing of the cells still result in the emission of greenhouse gases. According to the Brookhaven National Laboratory, this emission is about one tenth of fossil power plants or other nonrenewable energy sources. Naturally, when the global economy is fully converted to solar energy, the energy used in manufacturing the modules will also come from solar energy and therefore this carbon footprint of solar energy will also disappear.

1.4.7 Estimated Solar System Costs

In determining the costs of solar power plants, one must consider the following factors and components: (1) the value of the land used, (2) the cost of the collectors, (3) installation costs, (4) operating costs, (5) the cost of money, and (6) government support. The purpose of this book is to show that hard and reliable cost and performance data are not yet available and will be obtained only when full-size demonstration plants are built.

If the solar power plants are built in locations such as the Mojave Desert or Sahara, the land cost might not be excessive. Some argue that there could even be benefits in having these plants in such areas because of the shade, which the collectors provide, and the jobs, which their installation and maintenance create.

Installed collector costs range from about \$250/m² for roof-mounted solar hot water heaters to about \$4,000/m² for concentrating, electricity-generating, high-efficiency collectors. In units of peak watt (Wp) generation, the present unit cost for large installations is in the range of \$2 to \$5/Wp. (Nanosolar Inc. claims that its thin-film designs will cost \$1/Wp or less, but I know of no installation yet.) Of the total installed cost, about 75% is the first cost, 10% is the engineering design, and 15% is the installation cost. For individual home installations in the United States, contractors can be found on http://findsolar.com/index.php?page=findacontractor. To estimate the cost of a specific solar installation, including state and federal support in the United States, go to: http://findsolar.com/index.php?page=rightforme.

As an example of estimating payback periods, consider the Mojave Desert where the yearly average solar energy reaching a square meter solar collector is over 3,000 kWh/m²/yr. If the collector is 20% efficient and the generated electricity is sold for 12¢/kWh, the yearly income is \$72/m² (3,000 × 0.2 × 012 = 72). If an oil-based concentrating thermal system (SEGS) is used to collect the solar energy and its installed cost is \$750/m², the payback period is (750/72) or about 10.5 years. If the interest on the investment and operating costs are included, the payback period rises to about 12ñ13 yrs. If the use of renewable energy is supported by the government by cap-and-trade arrangements, these numbers are reduced.

The payback period on smaller installations is similar. For example, the cost of a 50 kWp PV installation in Washington Heights, Manhattan on the roof of a one-story garage was $7,400/kWp. The total installed cost of the system was $370,000, but the state grants and tax credits covered $265,000 of that. Therefore the cost to the owner of the garage was $105,000, which the owner financed with a 10-year loan. During that 10-year period, the payment on the loan is the same as the monthly bill for commercial electricity would have been. Once the loan is paid, the solar collectors will continue to generate electricity, plus the value of the garage will also increase.

Most solar power plants require little maintenance. This maintenance includes the (approximately) weekly cleaning of the reflecting mirrors. In case of thermal solar power plants, the boiler–turbine–generator segment of the plant requires the same amount of maintenance as their conventional counterparts. For financing of the power plants, an estimated 25-year life and 5% interest can be assumed. The solar energy fuel cost as of now is zero, although in the future, nations owning the land on which the energy is collected might charge a fee.

Although the building of the first solar demonstration plants will require the support of national governments or the United Nations, such support may not continue for long. After the price of carbon emissions is established, the market forces alone are likely to support the building of new renewable energy power plants.

It is estimated that the installed cost of a 1 gW thermal solar power plant is about $3 billion. The mass production of solar collectors is just beginning, and it is probable that with it will come a substantial drop in collector prices. The cost of a new nuclear power plant, if one includes the waste disposal and decommissioning costs, is about $5 to $6 billion. On average, nuclear plants generate 1 gW of electricity, which is about twice the electricity production of typical fossil power plants. The cost of a 1 gW fossil fuel power plant (two 0.5 gW plants), if carbon-capturing technology is included and if carbon emission charges are also considered, is the same as nuclear plants.

Therefore, even if an installed solar–hydrogen power plant costs $1 billion more than a regular solar plant, this is still under the costs of state-of-the-art nuclear or fossil power plants of the same size range. If the solar collector cost is estimated at $3,000/kW, the life expectancy of the equipment at 25 years, and the interest on investment at 5%, the unit cost of electricity generated will be about 12¢/kWh. This cost is already competitive with fossil-generated peak electricity costs and even with nonpeak electricity prices in some areas. (In June 2007, in Connecticut in my household, we paid 18.9¢/kWh for our electricity.)

To meet the electricity requirement of an average home (depending on the geographic location), 5–10 kWp solar collector capacity needs to be installed. The initial cost of such an installation today ranges from $40,000 to $70,000.

The payback period at today's electricity costs is between 10 and 20 years (average about 14), but this number will be drastically reduced as the cost of fossil electricity increases caused by factors such as the coming carbon-cap legislation, increases in emission taxes, and the rise in the cost of fossil and nuclear fuels.

Solar installations are often supported by federal and state subsidies, and many power companies allow these systems to be connected to their electricity grid so that the excess power generated can be sold to the power company for a credit (the electric meter is running backward), which "electricity credit" can be used later when solar energy is not available. In Europe, the highest amount of support paid is in Germany. In the United States, California supports each installation with about $25,000. In New York, the state pays 49–70% of the cost; in New Jersey, support is $3,800/kWp; and in Connecticut, $25,000 per installation.

1.5 Hydrogen Technology

The memory of the Hindenburg accident (1937) is fading, and as the safety record of hydrogen—based on its safe use in space exploration and in industry—becomes more widely known, it is also becoming accepted as a safe means of storing chemical energy. This trend has been further encouraged by the lessons learned from accidents, such as the one that occurred in 2008 on Interstate 84 in Connecticut, where a trailer truck carrying hydrogen plunged down the embankment. If the truck had carried gasoline, we know what would have happened—a huge fireball. However, because hydrogen does not form pools on the ground, but rather escapes into the atmosphere, there was no fire and no injuries were caused by the hydrogen.

When released, hydrogen quickly diffuses (3.8 times faster than natural gas) into a non-flammable concentration. It also rises 6 times faster than natural gas at a speed of almost 45 mph (20m/s). When it burns, due to the absence of carbon and the presence of heat absorbing water vapor, the fire produces much less radiant heat than a hydrocarbon fire. This reduces the risk of secondary fires. If only hydrogen is present, an explosion cannot occur. An oxidizer, such as oxygen, must be present in a concentration of at least 10% pure oxygen or 41% air. Hydrogen can be explosive at concentrations of 18.3% to 59% while gasoline can present a more dangerous potential, because it can explode at much lower concentrations, 1.1% to 3.3%. For more information on hydrogen safety, codes and standards, visit the following Web sites:

http://www1.eere.energy.gov/hydrogenandfuelcells/pdfs/doe_h2_safety.pdf

http://www.eere.energy.gov/hydrogenandfuelcells/codes.

TABLE 1.43

Properties of Hydrogen

Boiling point at atmospheric pressure	20.3°K (−253°C, −423.2°F, 36.5°R)
Storage pressure of compressed hydrogen gas	350 to 800 bar (5,000 to 12,000 psig)
Density (specific weight) of liquid hydrogen (LH_2)	71 kg/m³ (0.59 lb/gal, 0.265 kg/gal, 4.4 lb/ft³, 156 lb/m³, 2.0 kg/ft³)
Latent heat of vaporization of liquid hydrogen (LH_2)	107.6 kcal/kg (194 Btu/lb, 427 Btu/kg, 30.2 Btu/l, 114 Btu/gal, or 857 Btu/ft³)
Higher heating value (heat of combustion)	33,768 kcal/kg (39.3 kwh/kg, 61,000 Btu/lb, 134,000 Btu/kg, 141,500 kJ/kg, 36,000 Btu/gal, 12,000 kJ/m³)
Autoignition temperature of stoichiometric mixture	570°C (1,065°F)
LHV of oxyhydrogen (HHO)	241.8 kJ for every mol of H_2 burned

H_2 can be made by running solar-generated electric current through water (electrolysis). The solar energy that is stored in H_2 can be recovered by oxidizing it in a fuel cell, which can be viewed as an electrolyzer running backward. Here, the H_2 recombines into pure water, while generating electricity and heat. Freely available and inexhaustible solar energy is the fuel of this process, and H_2 is the means of storing it. This process is carbon emission free, because its only combustion product is distilled water. I for one believe that hydrogen will become the global energy carrier by the end of the 21st century and solar–hydrogen technology will be the basis of the coming third Industrial Revolution.

1.5.1 Properties of Hydrogen

H_2 was first recognized as a substance in 1766 by Henry Cavendish. It is the least dense gas, colorless, odorless, and tasteless. It is flammable and is slightly soluble in water. H_2 is a mixture of the para and ortho forms, which differ in their electronic and nuclear spins. At room temperature, atmospheric hydrogen gas (GH_2) is a mixture of 75% para-hydrogen and 25% ortho-hydrogen. Some of the properties of H_2 are given in Table 1.43.

As listed in Table 1.43, the heating value of 1 kg of liquid hydrogen (LH_2) is 134,000 Btu, approximately the same as the heating value of a gallon of gasoline (127,654 Btu). On a weight basis, the energy content of H_2 is 3.4 times that of gasoline. On a volume basis, LH_2 requires three times the volume to store the same amount of energy as does gasoline.

GH_2 is usually stored and transported at very high pressures, whereas LH_2 is stored at very low temperatures. The temperature of cryogenic LH_2 under atmospheric pressure is under −253°C (−423.2°F). The pressure of compressed GH_2 ranges from 350 to 800 bar (5,000 to 12,000 psig or 34.5 to 82.7 mPa).

Because of the lack of an international standardization of units, Table 1.44 provides some conversion factors among the various units used in defining the process properties of H_2.

TABLE 1.44

Conversion among Units of Pressure, Temperature, and Energy

Pressure	1 atm = 14.7 psi = 1.03 bar = 0.1013 mPa 1,000 psi = 68.05 atm = 68.95 bar = 6.89 mPa
Temperature	$°C = (°F − 32)/1.8 = °K − 273.16$ $°F = 1.8°C + 32 = °R − 459.67$
Energy	kWh = 3,412 Btu = 1.34 HP-hr = 0.36 mJ 1,000 Btu = 0.2931 kwh = 0.393 HP-hr = 0.36 mJ
Heat of combustion (Higher Heating Value—HHV)	Hydrogen: 61,000 Btu/lb = 134,000 Btu/kg = 39.3 kWh/kg Gasoline: 20,750 Btu/lb = 127,654 Btu/gal = 45,650 Btu/kg = 13.4 kwh/kg

In addition to being stored as a high-pressure gas or as a cryogenic liquid, H_2 can also be absorbed in solids such as metal hydrides (sodium borohydride) and in metallic "sponges" (zirconium, platinum, lanthanum). As of today, only small quantities of H_2 can be stored in solids by one of the "reversible solid" storage processes. Although it is the safest method of storage, it is still in the developmental stage. One such developmental process is planned to be used by Toyota in its Prius model, where it is planned to have the H_2 stored in metal powder using the technology of ECD Ovonics.

1.5.2 Hydrogen Generation

In nature, H_2 is produced by plants, including algae. Vegetation uses sunshine as the energy source to break down water into H_2 and O_2. The catalyst used by nature to decompose water and generate H_2 is chlorophyll. Once decomposed, the H_2 reacts with CO_2 to produce glucose. During this process, plants emit O_2 while taking CO_2 from the atmosphere. Animals do the opposite; they inhale O_2 and exhale CO_2. When the ratio of plant and animal life on the planet is balanced, the atmospheric concentration of CO_2 is constant.

Hydrogen can be made from water by electrolysis or by the photoelectrochemical processes. The second is just being developed and is in its early research state of development. Therefore, these photoelectrochemical processes are very inefficient at this time. If the reader is interested in the state of the art of this technology, the work of T. E. Mallouk at Penn State University and the reports of government scientists John Turner or Mark Spitzer can be consulted.

Today, the United States is using some 9 billion cubic feet of H_2 a day in the petrochemical, food, and rocket propulsion industries. Around 98% of the bulk H_2 is produced by steam reformation of natural gas (e.g., methane). Methane is reacted with water vapor over a catalyst to form carbon monoxide (CO) and H_2. H_2 can also be made from ethanol (alcohol), biomass, fossil fuels, or organic waste by the process of "reforming." Most of the currently operating H_2 production plants depend on reforming natural gas. This is a process that emits CO_2 while consuming a nonrenewable fossil

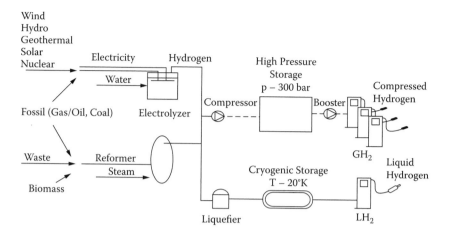

FIGURE 1.45

Hydrogen can be generated from hydrocarbons in steam reformers or by the electrolysis of water. Hydrogen can be stored either as a cryogenic liquid or as a high-pressure gas.

fuel. Therefore, when H_2 is made from fossil fuels, this process contributes to global warming.

As illustrated in Figure 1.45, a variety of energy sources and raw materials can be used to generate H_2. Of these sources, the only ones that are inexhaustible and pollution free are solar, geothermal, and wind energy.

1.5.3 Solar–Hydrogen Production Efficiency

When determining how much electricity is needed to produce H_2 by solar energy, the energy requirements of generation (electrolysis), compression, liquefaction, storage, and transportation all have to be considered and added up. The energy content of 1 kg of H_2 is 39.3 kWh. In order to generate 1 kg of H_2 by the electrolysis of water, about 50 kWh of electric energy is required. Therefore, the efficiency of H_2 generation is about 66%.

Once a kilogram of H_2 is produced, it is either compressed or liquefied before storage or distribution. If handled in the high-pressure gas form, about 3 kWh of energy is required for its compression and 2.5 kWh is required for its transportation over each 100 km distance. Therefore, a total of about 6 kWh is required to compress and transport the gas over a distance of 100 km. This energy corresponds to about 15% of the higher heating value (HHV) of the gas. As the transportation distance increases, this percentage also rises. Therefore, when transportation over long distances is required, it is more economical to transport the H_2 in liquid form by trucks, rails, or ships.

If handled as a cryogenic liquid, about 12 kWh is required to liquefy each kilogram of H_2 and about 1 kWh is needed to store and transport it, for a total of about 13 kWh, which is about 33% of the HHV of the liquid. In Table 1.46, some approximate data are provided on the costs and efficiencies of H_2 processing steps.

TABLE 1.46

Equipment Efficiencies and Energy Costs of Hydrogen Production

Process Step	Energy Cost in (kWh/kg)	Energy Used as a % of the HHV of H_2 (39.3 kWh/kg)	Equipment Efficiencies	Partial List of Suppliers
Electrolysis*				Proton Energy, Hgenerators™, Hydrogenics Corp.,
Theoretical	33	83	100%	H2Gen Innov., Teledyne
Small unit (actual)	75–100	191–254	Under 50%	
Large plant (actual)	50	127	About 66%	
Compression	4	7–16	84%–1-stage	Ariel, CompAir, Dresser-Rand, Fluitron, Greenfield,
			90%–multi-stage	Hydro-Pac, Neuman & Esser, PDC Machines, Inc., RIX
Transport kg/100 km			Best means:	Linde, Air Products, Praxair
Gas	2.5	6.0	pipeline, truck,	
Liquid	1.0	1.0	ship or rail	
Liquefaction + storage				Linde, Air Products, Praxair, BOC, Air Liquide
Range of sizes	10–20	13–60	J-T: 35% to 60%	
Avg. large plant	15	38	Brayton: 80%	
Totals for avg. large plants:**				
Liquid (LH₂)	~ 66	168	44% to 57%	Linde, Air Products, Praxair, BOC, Air Liquide
Compressed Gas	~ 56	142		

* The energy cost of electrolysis in a large (1.0 gW) plant, is about 50 kWh/kgH₂. Therefore, if 1 gWh of electric energy per hour is introduced into the electrolyzers, they will produce about 20 tons of low pressure H₂ gas per hour.

** The total energy cost of producing a kilogram of H₂ liquid in a large (1.0 gW) plant is about 66 kWh/kgH₂. Therefore, if 1 gWh of electric energy is provided hourly to the plant, it will produce about 15 tons of LH₂ every hour.

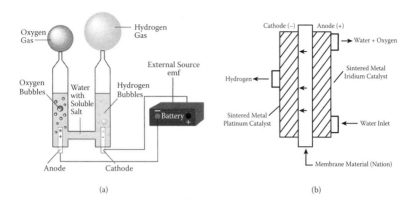

FIGURE 1.47
(See color insert following page 140.) The process of electrolysis can use liquid or solid electrolyte. (a) Liquid electrolyte. (b) Solid polymer membrane electrolyte (SPE).

1.5.4 Electrolysis of Water to Produce Hydrogen

In 1820, Faraday discovered electrolysis by passing electricity through water, thereby generating H_2 at the negative electrode (cathode) and O_2 at the positive electrode (anode). Thus, he found that H_2 can be produced from water by electrolysis (Figure 1.47). Electrolysis is an endothermic reaction that requires energy for its occurrence (up arrow in the upper left of Figure 1.48a), whereas the combustion of H_2 into water is exothermic (down arrow in Figure 1.48a).

The charge required to deposit 1 mol of any material at the cathode of an electrolytic cell is 96,484 coulombs (Faraday's constant). Therefore, it takes 237.13 kJ (224.79 Btu = 0.066 kWh) of electricity to synthesize 1 mol of H_2. When H_2 is generated at ambient temperatures, the environment contributes 48.7 kJ of thermal energy to that process for a total of 286 kJ.

The weight of 1 mol of H_2 is 2 g, and the energy content of 1 gal of gasoline approximately equals the energy content of about 500 mol of H_2. Therefore, the electricity needed to obtain 1 kg of H_2 (the amount of H_2 that has the same energy content as 1 gal of gasoline) is $[(237.13 \times 500 \times 0.948)/3,413] = 32.9$ kWh. Because electrolyzers use and PV solar collectors generate DC electricity, hydrogen can be generated from water without DC to AC conversion.

It takes the same amount of energy to split water into H_2 and O_2 as the energy obtained by oxidizing H_2 into water. The only difference between the two operations (between the process that takes place in fuel cells and electrolyzers) is that electrolysis increases the entropy and, therefore, not all the energy needs to be supplied in the form of electricity, as the environment contributes 48.7 kJ of thermal energy. Therefore, only 237.1 kJ of electric energy is needed to make 1 mol of H_2 from water (Figure 1.48b). On the other hand, in the case of fuel cell operation, the oxidation of 1 mol of H_2 will generate a total of 286 kJ of energy. Of that total 237.1 kJ is in the form of electric energy plus 48.7 kJ of thermal energy (heat).

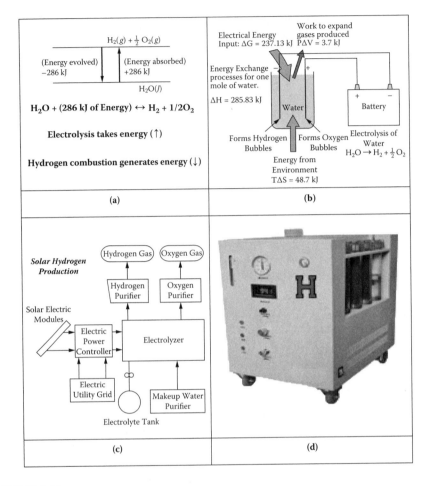

FIGURE 1.48
(See color insert following page 140.) The thermodynamics of converting water into hydrogen and the electrolysis equipment used in that process. (a) Thermodynamics of electrolysis. (b) Electrolysis equipment schematics. (c) Components of a solar–hydrogen generator. (d) Hydrogen gas generator. (Courtesy of HGenerators™.)

One mol of water produces one mol of GH_2 plus a half-mol of O_2 gas, both in their normal diatomic forms. As illustrated in Figure 1.48c, the electricity needed to convert water into H_2 and O_2 can be obtained from any electricity source, including solar converters or wind turbines. In other applications, such as in nuclear submarines, electrolysis can also be used to generate O_2 from water.

One commercially available hydrogen generator is shown in Figure 1.48d. It uses a prefluorinated solid polymer membrane electrolyte and can generate 0.65 kg/d of H_2 from 2.9 l/d of distilled water. In this design, after the H_2 and O_2 are separated, the O_2 is vented, and the H_2 is dried in a desiccant cartridge. H_2 pressure and flow are both regulated, and alarms are provided to signal such unsafe conditions as high pressure.

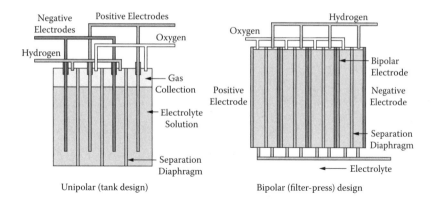

FIGURE 1.49
(See color insert following page 140.) Unipolar (left) and bipolar (right) alkaline electrolyzers.

HGenerators™' generator produces GH_2 at pressures adjustable from 0 to 4 bar (0–60 psig) while consuming 0.12 l of distilled water and 2 kW of electricity per hour. The cost of this generator is $20,000. If a distilled water supply is not available, the additional cost of a 10 L/h distiller is about $5,000.

There are two types of water electrolyzers: alkaline and polymer (e.g., PEM). The alkaline designs can be unipolar or bipolar (Figure 1.49). In the unipolar design, the anode and cathode are separated by a semipermeable diaphragm (membrane) and are immersed in an alkaline solution. The diaphragm allows only salts to pass through and prevents the gases from mixing. The H_2 and O_2 that are produced are dried downstream of the electrolyzer to remove water vapor. These units operate at lower current densities and temperatures.

In bipolar electrolyzer stacks, the face of the electrode to the left can be negative, whereas its other side, facing the next cell to the right can be positive. Naturally, these electrodes are separated and their electrical connections are provided by a metal separator plate (separation diaphragm). These units require less floor space and can operate at higher temperatures and pressures.

In proton exchange membrane (PEM) or solid polymer electrolyte (SPE) electrolysis, the electrolyte is replaced by an ion-exchange resin. These units are compact, provide high current densities, but are more expensive, and because of the corrosive nature of the electrolyte, require special construction materials.

Larger hydrogen-generating plants are marketed by Hydrogenics Corporation. These units have generating capacities from 1 to 120 Nm^3/h (2–240 kg/d) of H_2 at pressures from 10 to 25 bar (145–362 psi). The power consumption of these units is around 4.9 kWh/Nm^3 (54.8 kWh/kg), and the transportable generator is suitable for industrial outdoor installations.

As was summarized in Table 1.46, today electrolyzers operate at around 66% efficiency, whereas the ultimate target set by the DOE is 75%. Efficiency is defined as the ratio of the energy content of the generated H_2 to the energy required to produce that H_2. In the laboratory, the DOE efficiency target has already been reached using new, porous electrode materials. Another goal of the research and development effort in progress is to replace the expensive platinum electrodes with less expensive ones.

Today, the cost of producing H_2 from water by electrolysis is more expensive than making it by reforming natural gas, but the cost of natural gas is rising, and the cost of electrolyzers (and fuel cells) will drop when mass production starts. The new nano catalysts will also contribute to reducing electrolyzer costs, because they more than double the H_2 output by increasing the electrode surfaces (QuantumSphere Inc.). This new technology is available for only alkaline and not for PEM electrolyzers.

See Figure 4.2 for a description of the reversible fuel cell (RFC) and Section 2.12.5 for the controls of this device, which in one direction operate as an electrolyzer and in the opposite direction as a fuel cell.

1.5.5 Oxyhydrogen (HHO) Process

Once hydrogen is generated by the electrolyzer, its chemical energy content can be recovered in the form of heat or electricity. If that energy is needed in the form of electricity, a fuel cell can do the conversion (see Section 1.3.5). If the energy is needed in the form of heat, the hydrogen can be ignited and oxidized in air or heated to about 570°C (1,065°F) where, if a stoichiometric mixture at normal atmospheric pressure is present, autoignition occurs.

When its autoignition temperature is reached, oxyhydrogen will burn if the mixture is between about 4% and 94% hydrogen by volume. After ignition, this mixture converts to water vapor and releases 241.8 kJ of energy (LHV) for every mol of H_2 burned. The temperature of the flame is about 2,800°C (5,072°F) if the mixture is pure and stoichiometric. This temperature is about 700°C higher than when the hydrogen is burned in air, but drops when hydrogen and oxygen are mixed in excess of this ratio, or if mixed with nitrogen.

In some of the nontechnical media, the oxyhydrogen process has been described as an original method of making energy from water with the emission being also water. This is not true. This process is well known and the energy obtained in the resulting flame is less than the energy introduced at the electrolyzer (see Section 1.5.4), and if that pay energy is made by burning fossil fuel, nothing is gained. Therefore, if electrical energy is to be obtained from the chemical energy of hydrogen, fuel cells are preferred.

The one area where oxyhydrogen combustion is desirable is where high flame temperatures are required, such as in welding. In all other applications the most efficient system is to use fuel cells and the least expensive configuration is to use my new reversible fuel cell design (Section 1.3.5.4), which can operate both in the electrolyzer and the fuel cell modes and uses free solar energy to drive the electrolyzer.

1.5.6 Hydrogen Compression

Hydrogen can be stored as a gas compressed to 350–1,000 bar (about 5,000–15,000 psig) or as a cryogenic liquid. The energy cost of compression varies with the discharge pressure, compressor size, and the number of compression stages (Table 1.46). If a single-stage compressor is used, the energy required to compress H_2 is about 16% of the HHV of the compressed H_2. Multistage units with intercoolers are more efficient; they consume only 8–12% of HHV. To compress the gas, both large* and small[†] capacity compressors are available.

For example, the energy required to compress the H_2 from 1 to 200 bar at a rate of 1,000 kg/h in a 5-stage compressor is 7.2% of the HHV of the gas. If H_2 is compressed from 1 to 800 bar, the energy cost is about 10%. In terms of electricity consumption, compressing the gas from 7 to 470 bar (from 100 to 7,000 psig) requires about 3 kWh/kg of H_2. For these high-pressure applications, diaphragm compressors are most often used (see Figure 1.50).

These compressors are free from leakage to the same extent as are static seal systems. The leak-free integrity of the system is maintained even in the event of a diaphragm or diaphragm seal failure.

Polymer-coated metal diaphragm compressors are used to compress high-purity gases to eliminate leakage. In these designs, the metal diaphragm isolates the process gas from the hydraulic fluid to ensure purity. A self-energizing static o-ring seals the head assembly and provides both positive seal and low bolt torque. Any leakage that occurs owing to seal or diaphragm failure is retained in the head assembly and can be automatically detected to initiate maintenance.

These metal diaphragm compressors are available with capacities up to 680 m^3/h (400 cfm) and discharge pressures up to 2,000 bar (30,000 psi). Their drive motor sizes range from 1 hp to 200 hp (0.75–150 kW). A 1,000 bar (15,000 psi), 400 rpm 2-stage compressor with a 5 hp motor drive will compress 14 m^3/h (8 cfm) of H_2, whereas 10 hp unit will compress 85 m^3/h (50 cfm) and a 200 hp–680 m^3/h (400 cfm).

In high-pressure applications (70–8,250 bar = 1,000–120,000 psi) and in the capacity range of 3–1,200 kg/d (1–350 scfm), intensifiers are often added to the compressor. In these oil-free, nonlubricated gas pistons,[‡] the pressure of a hydraulic fluid moves the piston as it compresses the GH_2 (Figure 1.51). Both the flow and the discharge pressure of the H_2 are controlled by the hydraulic drive. This way, the rate at which the electrolyzer generates the H_2 is matched to the H_2 flow in the compressor.

Figure 1.51 illustrates the double-ended piston design of the intensifier, which can boost the pressure of the H_2 from 5 bar (70 psi) to 1,000 bar (15,000 psi). This pressure increase is standard for filling GH_2 storage tanks or transport vehicles. In stage 1 of the intensifier operation, the high-pressure

---

* Suppliers: Ariel, Dresser-Rand, Neuman & Esser, and Burckhardt Compresion.
† Suppliers of smaller (fuel station size) units: CompAir, Fluitron, GreenField Compression Inc., Hydro-Pac, Neuman & Esser, PDC Machines, Inc., and RIX Industries.
‡ Hydro-Pac Inc. (HP) is the manufacturer.

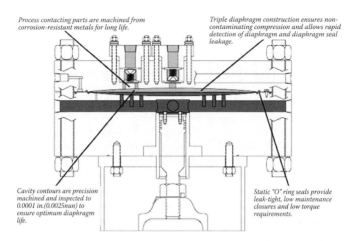

Process contacting parts are machined from corrosion-resistant metals for long life.

Triple diaphragm construction ensures non-contaminating compression and allows rapid detection of diaphragm and diaphragm seal leakage.

Cavity contours are precision machined and inspected to 0.0001 in.(0.0025mun) to ensure optimum diaphragm life.

Static "O" ring seals provide leak-tight, low maintenance closures and low torque requirements.

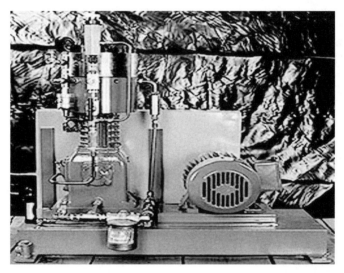

FIGURE 1.50
(See color insert following page 140.) Leakage-free diaphragm compressors. (Top) Pressure Products Industries (PPI) has furnished a number of petrochemical plants and hydrogen fueling stations with compressors (Hyundai in South Korea, Powertech in Canada, Senju in Japan, GM in Michigan, Hydrogenics in Canada). (Bottom) In the Fluitron diaphragm compressor there are no dynamic seals or packing glands. All of the seals to the atmosphere are static gasket or o-ring seals. (Photo courtesy of Fluitron Inc., Ivyland, PA.)

hydraulic fluid moves the motive cylinder to the left, thereby compressing the low-pressure H_2 in the first-stage cylinder and transferring this medium-pressure H_2 to the second-stage cylinder.

During stage 2 of the compression cycle, the high-pressure hydraulic fluid moves the motive cylinder to the right, thereby pulling fresh low-pressure H_2 into the first-stage cylinder while compressing the medium-pressure H_2 in the second-stage cylinder and discharging it as high-pressure H_2.

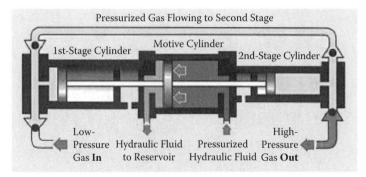

Stage 1 of the Intensifier Compression Cycle

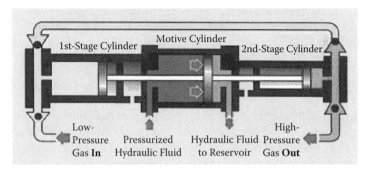

Stage 2 of the Intensifier Compression Cycle

FIGURE 1.51
(See color insert following page 140.) Intensifier with double-ended piston can boost the hydrogen pressure from 5 bar (70 psig) to 400 bar (5,800 psig) or from 70 bar (1,000 psig) to 1,000 bar (15,000 psig). (Courtesy of Hydro-Pac, Inc.)

The controls on the hydrogen compressor systems include both inlet and outlet pressure controls, motor starters, and temperature and flow controls for the cooling water. In addition, sequencing controls are provided. As illustrated in Chapter 4, Figure 4.1 (gatefold), for the optimized operation of the solar–hydrogen plant, it is necessary to integrate these controls with those of the solar collectors, electrolyzers, liquefiers, and fuel cells into a single system.

1.5.7 Liquefaction of Hydrogen

At atmospheric pressure, H_2 condenses at −423°F (−253°C). As the pressure rises, the condensation temperature also increases. The liquefaction process utilizes a number of heat pumps in series to reduce the H_2 temperature from ambient to the liquefaction temperature. The efficiency of this compression–condensation process ranges from 35 to 60% as a function of size and the refrigerant used in the heat pumps. One of the most efficient refrigerants is a mixture of helium and neon (Quack cycle). The energy consumption of liquefiers theoretically ranges from 5 to 15 kWh/kg

FIGURE 1.52
Hydrogen liquefier. (Courtesy of Linde Co.)

of liquefied hydrogen (LH₂). A small liquefier produced by Linde is shown in Figure 1.52.

Storage and transportation of H₂ are more practical in the liquid than in the gaseous form. This is because a gallon of LH₂ weighs 0.59 lb, whereas a gallon of atmospheric GH₂ weighs only 0.0007 lb. Consequently, an LH₂ trailer can carry about 4,000 kg LH₂, whereas a high-pressure gas trailer can transport only 300 to 400 kg of GH₂. When storing LH₂, some of the liquid will vaporize owing to heat infiltration (H₂ boil-off), which will reduce the volume available for storing the liquid.

The latent heat of vaporization of LH₂ is 450 kJ/kg (0.125 kWh/kg or 194 Btu/lb). The HHV (maximum heat of combustion) of H₂ is 141,500 kJ/kg (39.3 kWh/kg or 61,000 Btu/lb). Because the HHV of gasoline is 127,654 Btu/gal, the HHV of a kilogram of LH₂ (134,200 Btu) is approximately the same as that of a gallon of gasoline. This means that, on a volume basis, the energy content of gasoline is 3.5 times higher than that of LH₂, but on a mass basis, the energy content of LH₂ is 3 times that of gasoline.

Liquefaction plants are both energy and capital intensive. LH₂ suppliers in the United States include Linde, Air Products, Praxair, BOC, and Air Liquide. The daily capacities of liquefiers range from 6 to 35 tons (5,400–32,000 kg/d). The total cost of H₂ liquefaction is the sum of capital cost (65%), energy cost of production (30% = 12 kWh/kg), and operation and maintenance (5%). The proportion of capital cost drops as plant size increases. Energy efficiency can be improved by the use of expanders, high-speed centrifugal compressors, and by newer technologies such as magnetic and acoustic refrigeration.

1.5.8 Liquefaction Plants

A hydrogen liquefier consists of piston compressors that usually can operate at three pressure levels and a liquefier cold box. The cold box includes a precooler which uses liquid nitrogen as the coolant. The next portion of the system cools the hydrogen further as it passes through a number of heat pumps in which the working fluids are also hydrogen. There can be eight or more heat exchangers (with catalysts to convert H_2 from the para to the ortho form) and associated compressors and expansion turbines.

The liquefaction process uses a number of heat pumps in series to reduce the H_2 temperature from ambient to the condensation temperature of H_2. A heat pump is a device (such as a household refrigerator) that moves heat from a lower to a higher temperature by the introduction of work. This work is applied by a compressor, which increases the pressure of the refrigerant vapor. This working fluid can be the H_2 itself or can be other fluids such as a mixture of helium and neon. In this circulation, the heat of evaporation is taken from the H_2 stream being cooled by the working fluid (also H_2), which is at atmospheric pressure. The heat removed in the evaporator is taken by the working fluid to the condenser, where it is released as the fluid is condensed at the higher compressor discharge pressure. Thus, the heat pump removes the heat from the lower temperature (H_2 in the evaporator) and transfers it into the higher-temperature condenser.

If, in the heat pumps, the energy of compression is not recovered but is wasted in letdown valves (as the pressure of the working fluid is reduced to the low pressure of the evaporator (Joule–Thomson cycle), the liquefaction efficiency will be low (35–60%). This range of efficiencies is a function of the liquefier size and refrigerant used. If the letdown valves are replaced by turbo expanders (Brayton cycle), which recover some of the compression energy during pressure letdown, and if helium or neon refrigerants are used, the efficiency can theoretically reach 80–90%.

For a plant capacity of 100 kg/h of LH_2 production, the electric energy requirement is about 60 mJ/kg (17 kWh/kg). The specific energy input drops as plant size increases to a minimum of about 40 mJ/kg (11 kWh/kg). As the HHV of H_2 is 141.5 mJ/kg (39.3 kWh/kg), the energy cost of liquefaction at a 100 kg/h capacity is about 42%. In small units, the energy cost of production can exceed the HHV of the H_2, and even in large plants (1,000 kg/h capacity), it can be around 35%.

In the process of liquefaction, one must also consider the inversion temperature (−361°F or −183°C or 90°K) of H_2, because the behavior of this gas changes (inverses) at that temperature. Below the inversion temperature, when the pressure is reduced, the H_2 temperature will drop. Above that temperature the opposite occurs: a drop in pressure causes a rise in temperature. Therefore, in the process of liquefaction, H_2 first has to be cooled below its inversion temperature—by such means as cooling with LN_2—before the Joule–Thomson effect can be utilized.

Refrigeration down to 80°K is therefore provided by LN_2. The next step of refrigeration from 80 to 30°K is carried out using the Brayton cycle, in which high-pressure H_2 is expanded in a number of turbo expanders in series. From 30°K to liquefaction, the Joule–Thomson cycle is used, where high-pressure gas is throttled to low pressure to provide further cooling.

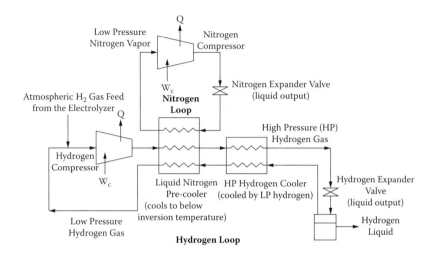

FIGURE 1.53
Simplified flow sheet of the hydrogen liquefaction process using liquid nitrogen for precooling and including an expander valve.

A simplified schematic of this "precooled liquefaction process" is shown in Figure 1.53. Before H_2 can be liquefied, it must be purified and converted from the normal to the para form. Normal H_2 is a mixture of 75% ortho and 25% para molecules. In an ortho molecule, the electrons rotate in the *clockwise* direction around both atoms, whereas in the para molecule the electrons *in one atom rotate clockwise and in the other, counterclockwise.* LH_2 is 99.8% para and only 0.2% ortho. Catalysts (which can be placed in the heat exchangers) are often used to speed up this conversion. The thermodynamic (theoretical) heat of conversion from the normal to the para form of H_2 is 0.146 kWh/kg.

Figure 1.54 illustrates a hydrogen liquefaction process developed by Praxair. This process uses both external refrigeration and LN_2 precooling. It has three sets of heat exchangers, two compressors, two expander turbines (for energy recovery), two accumulators, and a number of control and letdown valves. Other liquefaction process configurations have been developed under the names of Hydrogen Claude, Helium Brayton, combined reverse-Brayton Joule–Thomson (CRBJT), Neon Brayton, Neon with cold pump, etc. They can use more exchangers in series, more expanders and compressors, and naturally have different performance characteristics. As a consequence of all the R&D work, further improvements in the hydrogen liquefaction process are expected.

The energy consumption of liquefier plants ranges from 10 to 15 kWh/kg of LH_2 produced. Some estimates suggest that with optimization of the process, this energy consumption could be reduced to 5 kWh/kg, but these claims have not been proved. As was shown in Table 1.46, about 38% of the HHV energy content of the produced LH_2 is used up in the liquefaction and in the short-distance transportation to the users. This compares to about 15%

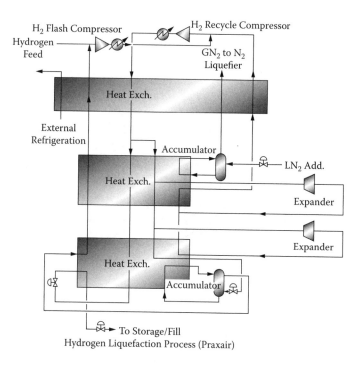

Hydrogen Liquefaction Process (Praxair)

FIGURE 1.54

(See color insert following page 140.) A hydrogen liquefaction process developed by Praxair using external refrigeration, a liquid nitrogen precooler, three sets of heat exchangers, two compressors, and two expander turbines for energy recovery. (Courtesy of Praxair Inc.)

if the H_2 is not liquefied but only compressed to 500–1,000 bar and transported over a distance of less than 100 km.

1.5.9 Transportation and Storage

The generation of high-pressure GH_2 requires less energy (10–15% of the hydrogen's energy content) than does the production of LH_2 (38%). Long-term storage of LH_2 is also a source of further losses. In conventional automotive applications without special seals, the loss due to leakage amounts to about 30% per month. These losses are disadvantages, but on the other hand, it is less expensive to store and transport H_2 in the liquid form.

Because of its lower density, transportation of GH_2 through pipelines requires more energy than does transportation of natural gas. Transportation of compressed GH_2 by trucks is inefficient, because the trucks can hold only about 400 kg of H_2. Therefore, a busy gas station could require 10–20 deliveries of GH_2 each day, whereas if LH_2 is used, a single delivery would suffice.

Safety is another important consideration. The experience gained by the daily production, transportation, and use of 9 billion cubic feet of H_2 (in the United States) in space exploration and other industries made the storage, transportation, and handling of H_2 reasonably safe. The design of hydrogen

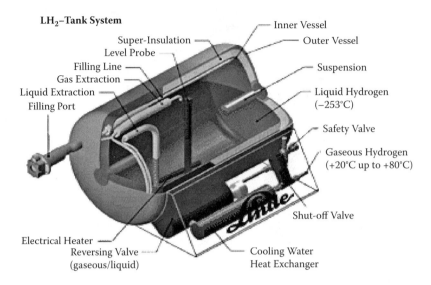

FIGURE 1.55
(See color insert following page 140.) Liquid hydrogen storage tank. (Courtesy of Linde Co.)

tankers and trucks is similar to those of liquefied natural gas (LNG) storage vessels and tankers. H_2 is usually stored in double-walled vessels (Figure 1.55); the space between the walls is filled with superinsulation or is evacuated.

The efficiency of fuel cells is much higher than that of internal combustion engines and therefore LH_2 fuel tanks on an average car need not hold more than about 10 kg of hydrogen. For truck and bus applications, several such tanks, containing about 60 kg of LH_2 can be used to provide a range of 500 km (about 300 miles). DyneCell tanks of Dynetek Industries are designed for 5,000 psig (350 bar). Toyota's FCHV vehicles use four 10,000 psig (700 bar) cylindrical tanks, which suffice for a driving range of over 400 miles.

One of the important aspects of liquid hydrogen storage is "thermal endurance" during "dormancy," which refers to the heat transferred from the environment into the liquid hydrogen fuel tank of a parked car. Tests at the Lawrence Livermore National Laboratory determined that if LH_2 is stored in cryo-compressed fuel tanks, the heat leak rate into a 151 liter tank (rated at 5,000 psig – 34.5 mPa) is about 10 Watts at 20–30°K, if the car is left parked for 2 weeks. A full tank provides a driving distance of over 650 miles. The liquid hydrogen fuel tank they designed for BMW held the cryogenic liquid for 6 days without venting. Their goal is extending "no-vent storage" to 15 days.

The operation of hydrogen tank farms and hydrogen transportation systems (Figure 1.56) is similar to that of LNG tank farms. Because H_2 will react violently with O_2, it is absolutely essential to prevent O_2 infiltration into the hydrogen storage tanks. The advantage of H_2 over LNG is that it is lighter than air. Therefore, if a leakage occurs, H_2 will quickly rise into

FIGURE 1.56
Hydrogen generation, storage, and bulk distribution systems. (Courtesy of Praxair, Inc.)

the atmosphere while LNG, propane, or gasoline will accumulate on the ground, or pool or soak into clothing or into the ground, where it can be ignited. In addition, when gasoline burns, glowing hot soot particles carry the heat and can become secondary ignition sources. In contrast, H_2 burns cleanly without a residue.

BMW tested its hydrogen tanks in a series of collisions that caused both fire and tank ruptures. In most of these tests, H_2 was found to be safer than gasoline or diesel fuels. The simulated 55-mph crash tests left the cars totaled, but the newly designed hydrogen tanks remained intact. Although pressurized hydrogen tanks could explode, enormous impact forces would be required and, therefore, they are reasonably safe.

In tests conducted at the College of Engineering at Miami University, 3,000 ft³/min of H_2 was leaked from the hydrogen storage tank of a vehicle and was set alight. Over the course of the burn, the temperature inside or outside the vehicle did not reach intolerable levels; it stayed at levels that are reached when a vehicle is parked in the sun.

1.5.10 Hydrogen-Assisted Fossil Power Plants

The transition from the present fossil fuel–based economy to the coming solar–hydrogen–based one will be gradual and slow. During this period, fossil power plants will continue to operate, but hydrogen will gradually start contributing to their efficiency improvement.* If solar–hydrogen sources and fossil power plants are physically close to one another, it is possible to send the O_2 (generated by the hydrogen electrolyzers) to enrich the combustion air to the fossil fuel burners and reduce the loss of efficiency that results from heating the 79% nitrogen in the air "just for the ride" from the boiler through the stack to the atmosphere.

---

* The one described here is the concept of William R. Lowery, PE of El Dorado, Arkansas.

TABLE 1.57

Efficiencies of Fossil and Hydrogen
Fueled Transportation and Power Plant
Processes

Industry	Efficiency
Transportation	
Gasoline—IC	25%
Diesel—IC	35%
Hydrogen—IC	38%
Hydrogen—Fuel Cell	45% to 60%
Electric Power Plants	
Sub-Critical Fossil	36%–38%
Super and Ultra-Critical Fossil	42%–48%
Combined Cycle Fossil	55%–60%
Standard Fuel Cell	45%–60%
Fuel Cell with Heat Recovery	80%

H_2 can also be used to capture carbon monoxide and CO_2 in the combustion products. CO_2 can be converted into methanol. Each mol of CO_2 will require 3 mol of H_2 to generate 1 mol of methanol (CH_3OH). After recovering this methanol, it can be used as a fuel or fuel supplement, thereby reducing the need for burning coal, natural gas, or other fossil fuels, whereas the emission of CO_2 will also be reduced.

1.5.11 Efficiencies of Fossil and Hydrogen Fuels

Theoretically, electrolysis of 1 kg of H_2 requires 32.9 kWh/kg, which at 66% efficiency corresponds to about 50 kWh/kg. Assuming that the liquefaction and the transportation of H_2 takes another 15 kWh/kg, the total energy cost of producing a kilogram of LH_2 is about 65 kWh/kg. The cost of a kilogram of LH_2 in the United States is about $5 and in Europe about $8. A kilogram of LH_2 has the same energy content as a gallon of gasoline, which in the United States costs about $4 and in Europe about $6–$8, and both are rising.

As can be seen from Table 1.57, the efficiency of fuel cells is about double that of gas-burning internal combustion (IC) engines. Therefore, if fuel cells can be made at the same cost as IC engines, hydrogen fuel cell–based transportation is already cost competitive with gasoline-based transportation. Unfortunately, at this point in time, fuel cells are much more expensive, but that is likely to change as soon as mass production starts.

Subcritical fossil power plants operate at 36%–38% efficiencies. The efficiency of supercritical designs is in the low- to mid-40% range, and the efficiency of the new "ultracritical" designs, operating at pressures of 30 mPa (4,350 psig) and using dual-stage reheat, can reach about 48%. Fossil power plants can also use gas turbines in conjunction with a steam boiler "bottoming" cycle. The efficiency of combined-cycle plants can approach 60% in large (over 500 mW) plant sizes.

As shown in Table 1.57, if the heat generated by the fuel cell is also recovered, the FC-based power plant efficiency can be nearly twice that of subcritical fossil power plants. In addition to the improvement in efficiency, FC-based power generation is also superior to fossil fuel–based power generation, because there are no expenses associated with carbon recapture; the only emission is distilled water.

1.5.12 Hydrogen Generation and Handling Costs

In a solar–hydrogen power plant, the first block in the chain of equipment consists of solar collectors. The electricity generated by these collectors is sent to the electrolyzers, which convert this electric energy into the chemical energy of H_2. As was shown in Table 1.46, the theoretical electrical energy required to produce 1 kg of hydrogen from water is about 33 kWh/kg. This means that theoretically it is enough to invest 83% of the HHV of a kg of hydrogen (39.3 kWh/kg) to make it by electrolysis. Today the most efficient electrolyzers have an efficiency of around 66% of the theoretical and the target set by the DOE is to increase that to 75%. The actual energy cost is naturally more than the theoretical and therefore the efficiency of actual units is less. As a function of size, the energy cost of electrolyzing a kilogram of hydrogen in actual units ranges from 50 to 100 kWh/kg. The installed capital cost of smaller electrolysis equipment is around $3,000/kW. This price drops as the size increases.

The DOE Hydrogen Program of the United States has made significant progress since its initiation in 2003. By 2007 it had reduced the cost of producing a "gallon gasoline equivalent (gge)" quantity of hydrogen—which is about 1 kilogram from natural gas—from $5.00/gge in 2003 to $3.00/gge and has a goal of reducing it to $2.00/gge by 2015. Considering that this was achieved using "pay fuel" (natural gas), it is safe to assume that the cost will be less when "free" solar energy is used.

After the electrolysis step, the H_2 is compressed. The energy cost of compressing the GH_2 to 500 bar (7,250 psig) is about 8% of the H_2's HHV and to compress it to 1,000 bar (14,500 psig) is about 16%, if a single-stage compressor is used. If multistage compression with intercoolers is utilized, the energy cost drops to 12% of the HHV of H_2. If H_2 is to be used as a cryogenic liquid, the energy consumption of the liquefier plants ranges from 10 to 15 kWh/kg of LH_2 produced.

Once the H_2 is available either as a pressurized gas or as a cryogenic liquid, it still has to be stored and transported to the users. In addition to transportation and storage, there are also infrastructure expenses. A National Renewable Energy Laboratory (NREL) reference provides dollar cost figures per kilogram of H_2.* These total costs include energy, freight, labor, and capital costs and have been developed for truck, rail, ship, and pipeline transportation of both gas and LH_2 at various flow rates and over a variety of

---

* Wade A. Amos at the National Renewable Energy Laboratory (NREL) prepared extensive data on this; see "Cost of Storing and Transporting Hydrogen" NREL/TP-570-25106.

distances. The truck or rail transportation of LH_2 is the least expensive, about $1.50/kg, if the amounts to be transported are small (under 45 kg/h) and the distance exceeds 800 km (500 mi). For large rates (4,500 kg/h) over the same 800 km distance, the cost is about $1/kg if LH_2 is transported by truck or rail or if gas is transported by pipeline.

If the transportation distance exceeds 100 km (62 mi) and the transportation is by truck or rail, the energy cost of compressed gas (at 200 bar—about 3,000 psig) is about 6% of the HHV. If, on the other hand, the H_2 is transported as a liquid, the transportation cost is about 1% of the HHV. It should be noted that although the transportation cost for LH_2 is lower, the storage cost is higher (about 5% HHV), because of the need for cooling.

At filling stations, the GH_2 pressure is usually increased from 200 to 400 bar (3,000–4,500 psig). This compression increases the energy cost of the total operation by about 2% HHV. Naturally, in addition to the operating energy costs, there will also be infrastructure, depreciation, operating, and maintenance costs.

Hydrogen storage costs with today's technology are much higher than those of gasoline storage, because of embrittlement of the metallic components when contacting H_2 and the daily boil-off (up to 15% in small tanks and a fraction of that in large ones). Because of the high costs of metal hybride or carbon-fiber tanks ($1,000 to $10,000/kg LH_2) and safety considerations. The gas station of the future might be all-electric, with large fuel cells used only to generate electricity to recharge the batteries on an all-electric fleet of cars. Only the future will tell. The choice of fuel cell cars or all-electric cars will depend on the prices of fuel cells and batteries of the future when both are mass produced. According to some estimates, transportation by fuel cell–operated electric cars will cost 10¢/mi to operate. This compares with today's gasoline cost in the United States of about 15¢/mi, if we do not consider the costs associated with their emission of CO_2.

The purchase price of LH_2 (made from natural gas) is about $5–$8/kg, whereas that of high-pressure GH_2 is about $25/kg. Naturally, when H_2 is made using "free" solar energy from water, instead of from costly hydrocarbons while using fossil fuel-generated electricity instead of solar energy, these costs should drop. As 1 kg of H_2 has the same energy content as a gallon of gasoline, which in Europe costs around $6 to $8 and in the United States from $3 to $4, the $5 to $8/kg price for H_2 in the United States is not yet competitive. On the other hand, if one considers that fuel cell efficiency is about twice that of gasoline engines and hydrogen does not generate carbon emissions, the price is more competitive.

1.6 Renewable Energy Economy

The "solar–hydrogen economy" of the future is expected to consist of both highly distributed and highly centralized components. The distributed

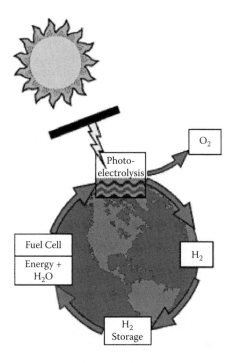

FIGURE 1.58

In the solar–hydrogen–based energy cycle, the fuel is inexhaustible solar energy which is first converted into electricity used to split water by electrolysis into oxygen and hydrogen. The oxygen is released while the hydrogen is distributed and used as fuel. When hydrogen is burned the combustion product is distilled water. (Source: DOE H2 Network.)

component will consist of the grid-connected "energy-free" residences and industries of the future. In addition to being energy efficient, these homes and office buildings will probably have solar roofs and use electric cars for transportation. If the solar collectors are mass produced, energy-free residences will probably be cost effective in all grid-supported parts of the world where the yearly insolation exceeds 1,500 kWh/m².

The other components of the renewable energy economy will probably be large solar–hydrogen power plants built near the equator in arid high-insolation regions. If there is no grid in the area, these plants will produce LH_2 that will be distributed around the world in a similar manner as is LNG today. Figure 1.58 schematically illustrates the functioning of this renewable energy economy.

1.6.1 "Zero-Energy" Homes

In addition to savings resulting from the use of more efficient appliances, low-energy light bulbs, more efficient heating and cooling equipment, double windows, and better insulation, most homes can also be free of outside energy sources. The technology already exists to make our homes energy free in all areas of the planet where the solar insolation exceeds 1,500 kwh/m². When nanosolar roof shingle collectors and reversible fuel cells (RFCs) are mass produced, "zero energy" homes can extend to most of Europe and the United States. Nanosolar roof shingles will double the efficiency of today's solar collectors and cut their costs to less than half, while my RFC design will make solar energy continuously available by storing the excess during the day in the form of hydrogen and then using it at night. In addition, if the residents have plug-in electric cars, the excess energy can also be used to recharge their batteries. If they don't yet have electric cars, they can sell the excess energy back to the electric utility over the grid, while their electric meter is running backward. If this is done, the United States will not only regain its global leadership and stop wasting its economic resources on energy wars, but can create the economic boom of the century, by producing and installing millions of solar roofs and RFCs around the country and the world.

The yearly energy needed for heat and hot water of a well-insulated home in the northeast United States is 150 million Btus = 44,000 kWh, which equals the energy content of about 1,000 gallons of oil. According to a 2007 study of the McKinsey Global Institute, the total residential energy consumption in the United States is 21.3 Q, and this total could be halved by installing double windows, improving sealing and thermal insulation, and using lighting and appliances with better efficiency. Of the total residential energy consumption, 8.5 Q is used to operate appliances, 7.5 Q for heating/cooling, 3.2 Q for hot water generation, and 2.1 Q for lighting. The study concludes that by 2020, residential efficiency improvements could reduce the energy consumption of American homes from 21.3 Q to 13.5 Q.

In June 2007, for the 1,289 kWh of electricity used in our home in Connecticut, we paid $245.50, which corresponds to a rate of 18.9¢/kWh. The electricity in Connecticut comes from a mix of nuclear and fossil-fueled power plants. This 18.9¢/kWh rate covered only the bill from the power company. It did not include the cost of protecting the energy supplies of the United States (including military expenses) and did not cover the price we all pay for global warming.

In comparison to the cost of fossil–nuclear energy, solar power plants generate electricity at 12–20¢/kWh. When using solar energy, the fuel is free and inexhaustible. Another potential of solar energy is economical: if it was decided to cover 10 million American homes with solar roofs, this decision alone would trigger the biggest economic expansion of the decade (if not the century?).

Energy consumption of private homes in the Southwest and many other suitable areas of the country (see Figures 1.29 and 1.30) could become "energy free," if solar roofs were installed. If these homes were also connected to the electric grid, when the generated solar electricity was in excess of what was needed, it could be stored on the grid and used when solar energy is not available. In these areas, the "zero-energy" homes could also provide the energy needed for plug-in hybrids or electric cars. Actually, if solar roofs were installed on 10 million homes, the resulting energy savings would more than exceed the present imports from the Persian Gulf, not to mention the jobs that this construction boom would create.

Larger installations could also use solar laminates and solar shingles. Such projects already exist. For example, GM has a 1 mW solar installation in Rancho Cucamonga, California. Many other such systems have been installed, and new designs, such as the ultrathin (Nanosolar) shingles, are under development, and promise drastic reductions in costs.

In the northern regions of the United States, the renewable energy supplies are insufficient to operate energy-free households and solar energy has had to be complemented by conventional energy supplies. To supplement solar energy, geothermal or wind energy can also be used. Naturally, today we are a long way from having 10 million zero-energy homes, but the process of conversion has already started. Many electric power companies already accept the excess electricity generated by renewable energy sources into their grids so that a home's electric meter is running backward when the excess energy is being sent to the grid. Figure 2.128 describes the components of such an installation, including the inverter needed to convert the direct current (DC) solar power to the alternating current (AC) grid. One of the many inverters on the market is SMA Technology's "sunny boy," which has been used in nearly 1 million installations and is available in sizes from 400 Wp up to mWp capacities.

One major problem with the global electricity system is the lack of continentwide electric grids that could serve to both distribute and store electricity. This "net metering" capability (Section 1.4.3.1) is essential to make the energy of intermittent energy sources (solar, wind, tide, etc.) continuously available.

This is beneficial to both parties, because most of the time excess solar generation coincides with peak demand periods when air-conditioning loads are high and, therefore, this solar excess reduces the peak demand of the utility. This is profitable for the utility because meeting that peak by fossil energy is very expensive. Thus, when more solar energy is being generated than is needed by the home, the peak power requirement of the system is reduced. The zero-energy household can later use this accumulated electricity credit to obtain free electricity, when (at night or during the winter) solar or wind energy is not available.

Other opportunities to economize also exist. For example, if a wind turbine is used in an energy-free household and the wind is blowing mostly at night when electricity is cheap, one can wait until it becomes more valuable by storing the generated electricity in batteries and sending it to the grid during peak demand periods.

When specifying the size of the solar collectors for a zero-energy home, it is advisable to double the maximum expected requirement so that sufficient excess electricity will be generated in the summer to "zero out" the electricity bill for the entire year.

1.6.2 Zero-Energy Home Costs

Today the average American generates 21 tons (46,000 lb) of carbon dioxide a year (some of the "green" cities in California and elsewhere have reduced this to 9–13 tons/capita), while the global per capita average is 4 tons (8,800 lb) a year. Yet, most of the private homes in Europe or in the United States could take advantage of inexhaustible, clean, and free forms of energy (solar, wind, and geothermal). Their use could completely eliminate the emission of greenhouse gases.

Naturally, although solar and other renewable sources of energy are freely available, the initial investment needed to convert them into electricity is high. Many governments subsidize the use of renewable energy systems so that the full cost of their installation need not be paid by the owner of the home. For individual home installations in the United States, contractors can be found on: http://findsolar.com/index.php?page=findacontractor. To estimate the cost of a specific solar installation, including state and federal support in the United States, go to: http://findsolar.com/index.php?page=rightforme.

The payback periods of alternative energy installations range from 5 to 20 years, with solar hot water systems being the least expensive and solar–hydrogen systems the most. The payback period for installing a photovoltaic (PV) electricity-generating system in California is about 15 years. This number is based on a home with a monthly electricity bill of $100, an installed system cost of $50,000, and a rebate plus tax credit of $20,000. If the monthly electricity bill is $250, the payback period drops to about 8 years, and if one also considers the increase in the value of the home, the payback period can drop to about 4 years. These payback periods were calculated on the assumption that the electricity cost in the area is 12¢/kWh, and it will not rise. As was mentioned earlier, this cost in my household in Connecticut is 18.9¢/kWh and rising.

In 2006, the unit cost of solar electricity from a typical 5 kilowatt (kWp) rooftop system was 30¢/kWh in Germany, 19¢/kWh in Spain and 22¢/kWh in California. By 2010, Photon Consulting estimates that solar electricity will be produced for 18¢/kWh in southern Germany, 12¢/kWh in Spain, and 13¢/kWh in California. In the United States the typical capital costs including system installation were approximately $3,600/kWp in 2007, with some installations costing only $3,000/kWp. These prices are expected to drop further as government subsidies rise and mass production starts.

The generating capacity of solar collector systems used for private homes ranges from 5 to 10 kWp.* The yearly energy bill of an average household in the north-central United States is about $4,500 ($2,500 for heat and hot water, $2,000 for electricity). (See http://www.findsolar.com/index.php?page=rightforme to obtain more data on the energy use of American households.)

In Europe, the highest amount of government support for household renewable energy installations is paid in Germany, whereas in the United States, state support of solar installations is the highest in California: about

---

* This is the maximum amount of electric power that a PV module is able to generate when it receives 1,000 W/m² of solar irradiation at 25°C. Recently, people in the industry started to refer to 1 kWp as "1 sun."

Installation of photovoltaic collectors in Colorado

Installing roof shingles that are coated with PV cells made of amorphous silicon

Solar home in New York

Solar home in Montréal

FIGURE 1.59
Grid connected "zero out" solar homes can be energy free even in regions of lower insolation if sufficient collector areas are provided. They can generate electricity only (shown) or a combination of electricity and hot water.

$25,000/household installation. A specific example of a 2.4 kW system that costs $81/month for 15 years with a zero down payment with installation and maintenance included is described on www.solarcity.com. In New York, the state pays 49–70% of the total cost; in New Jersey, the support is $3,800/kWp; and in Connecticut, $25,000 per installation. For a comprehensive summary of state, local, utility, and federal incentives that promote renewable energy and energy efficiency, refer to http://www.dsireusa.org/index.cfm?=altersystems.com.

In the United States, around 90% of the existing solar collector installations generate only hot water, whereas 10% also generate electricity. Today, there are only 50,000 solar homes (Figure 1.59) in the United States, although their numbers increased over tenfold during the last decade. As experience is gained and production volume rises (mass production is just beginning), the cost of collectors is dropping by about 5% per year.

Renewable energy homes can also use household wind turbines. These units are usually small, having a generating capacity of 2–10 kWh/d at a wind speed of 10 mph. When used only to charge batteries, a 10 kWh/d unit costs about $8,000; if an inverter is also included so that the system can be tied to the grid, the cost rises to about $12,000. Suppliers include Aircon (Germany), Energie PGE (Canada), Iskra Wind Turbine (U.K.), and Wind Energy Solutions (Holland).

There are already a few solar and geothermal combination homes in operation. One example is Michael Strizki's home in New Jersey, which cost $500,000. One of the first solar–hydrogen–generating installations can be found at Kings Point, New York, at the U.S. Merchant Marine Academy. Others include one designed by Advanced Solar Products on the Cayman Islands and the rest of them are on Stewart Island, Washington, in Indonesia, and in Friburg, Germany.

Another likely feature of the private homes of the future could be the capability for their continuous optimization and wireless remote access. Energy optimization can be provided by computer software packages that continuously monitor the cost effectiveness of operating the household. This might include maximizing the use of the least expensive form of energy when multiple energy sources exist. In this case, the system could continuously calculate the cost of alternative modes of operation, and for example, could make the decision on the cost effectiveness of selling or buying energy from the grid, adjusting thermostat set points, or cycling electric appliances, swimming pool pumps, etc., on and off.

This optimizing home-supervisory system could also be used for home monitoring if integrated with miniature cameras, smoke, and security alarms, which will automatically contact the owner in case of an emergency. The supervisory system could also allow the owner to view various parts of the home or change the conditions such as reprogramming the settings of thermostats or the on–off cycling periods of equipment, all by wireless mobile phone.

1.6.3 The Solar–Hydrogen Demonstration Plant

The costs and competitiveness of producing, storing, and distributing solar–hydrogen have been debated and estimated, but have not been determined. I believe that the time for studies and debates is over, and it is time to build the world's first solar–hydrogen demonstration plant to obtain hard, irrefutable data on costs and performance. Experience with the demonstration plants will help identify and eliminate technical bottlenecks, optimize processes, and generate standardized specifications that could later be used for the mass production of the equipment. Chapter 4 of this book describes such a demonstration plant. The four main equipment blocks of this solar–hydrogen power plant are shown in Figure 1.60.

After the solar–hydrogen demonstration plants have operated for a few years, methods will be found to increase their efficiencies, reduce their operating costs, and take advantage of mass production and free market competition to optimize their first costs. If the best scientific talent is mobilized, it

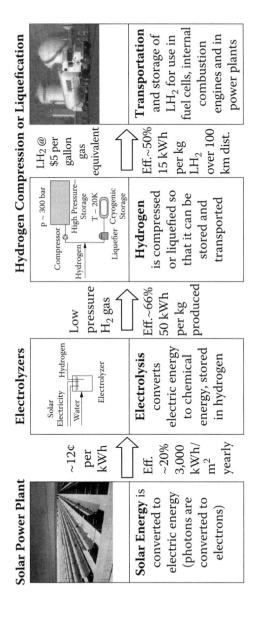

FIGURE 1.60

The four main equipment blocks of the solar–hydrogen power plant. (Four percent of the Sahara can supply the total energy needs of the planet or 15% of the Mojave Desert can supply the total electricity needs of the United States.)

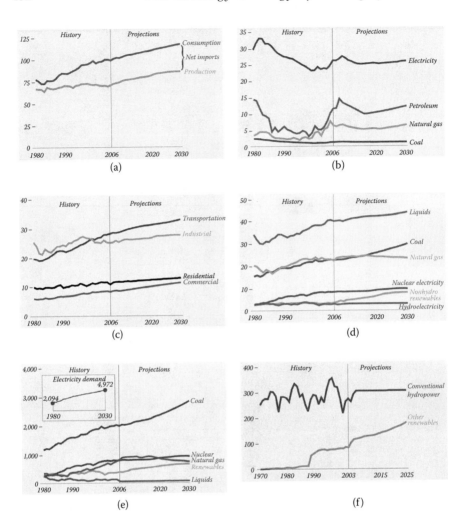

FIGURE 1.61
Data from the early release of the Energy Information Administration (EIA) of the United States for the year 2008. For additional details see: http://www.eia.doe.gov/oiaf/aeo/electricity.html. (a) Total energy production and consumption, 1980-2030 (quadrillion Btu). (b) Energy prices, 1980–2030 (2006 dollars per million Btu). (c) Delivered energy consumption by sector, 1980–2030 (quadrillion Btu). (d) Energy consumption by fuel, 1980–2030 (quadrillion Btu). (e) Electricity generation by fuel, 1980–2030 (billion kilowatt-hours). (f) Grid-connected electricity generation from renewable energy sources, 1970–2025 (billion kilowatt-hours).

is estimated that by 2020 the cost of solar electricity can be reduced below 10¢/kWh and the cost of 1 kg of H_2 (equivalent of a gallon of gasoline) can be lowered to about $3.

If one considers the present projections of most American government sources, the above expectations will seem overly optimistic. For example, the early release of the Energy Information Administration (EIA) of the United States energy outlook shown in Figure 1.61 provides estimates that suggest a much slower transition to the coming renewable energy economy and much higher renewable energy costs than I anticipate. Yet, it does not matter what I or they anticipate, what matters is that we should finally start dealing with facts instead of anticipations and that will not happen until the world's first solar–hydrogen power plant is built and its data collected.

1.7 Costs and Efficiencies

Today, renewable energy costs are often given inaccurately. Some are based on facts, others on wishful thinking or advocacy, but a few are based on the proven operation of full-sized power plants and actual experience. The very reason why a full-sized (1 gW) solar–hydrogen demonstration plant should be built is because its total costs of construction and operation should be established accurately. We do not have such data, because no such plant has ever been built; nobody has ever obtained competitive bids on such quantities of equipment.

We know that production volume and the associated mass production do affect costs. For example, during the last decade, the cost of solar collector cells (PVs) halved, and the production cost of solar electricity dropped to about 12¢/kWh. If the claims of Nanosolar (an ultrathin-film solar collector manufacturer) turn out to be correct, collector costs might drop drastically. With a reduction in solar collector costs, the cost of solar–hydrogen has also been dropping. Today, the cost projections for large-capacity LH_2 plants are around $3 to $5 per kg.

Today's fossil–nuclear energy economy is both exhaustible (Figure 1.1) and inefficient. As listed in Table 1.57, when fossil fuels are burned in internal combustion engines, almost three quarters of their energy content is wasted (gasoline efficiency: 25%, diesel: 35%), whereas the efficiency of hydrogen fuel cells is 50–60% and eventually might approach 80%. Similarly, fossil-based power plant efficiencies are also low. In subcritical fossil fuel power plants, more than half of the fuel's energy content is wasted, as the plants are only 36–38% efficient.

It is generally assumed that electricity from renewable sources is more expensive than that from fossil or nuclear plants (Table 1.62). This is not necessarily an accurate assessment. If we compare the first costs of new fossil power plants, which include carbon-recapturing or carbon-sequestering features or we consider the total cost of new nuclear plants (including waste disposal, decommissioning, insurance, and safety-related expenses), they both cost $5–$6 billion for a full-sized, 1.2 gW plant. Similarly, the cost of a coal-fired power plant rises drastically when good pollution protection and carbon dioxide recovery processes are added. Therefore, as shown in Table 1.63 the cost of a 1.2 gW fossil power plant (consisting of two 0.6 gW standard plants) is about $5 billion. Until the first full-sized renewable energy demonstration

TABLE 1.62

The Wholesale Costs of Electricity from Various Power Plants

Power Plant Fuel	*Wholesale** Cost of One kWh of Electricity Produced in 2007	*Wholesale** Cost of One kWh of Electricity if $50/ton Carbon Emission Charge Is Mandated
Pulverized coal	6.15¢	10.4¢
Nuclear	7.1¢	7.1¢
Coal gas	6.5¢	10.4¢
Natural gas	6.6¢	8.4¢
Wind	9.6¢	9.6¢
Biomass	10.7¢	11.2¢
Solar thermal	12.0¢	12.0¢

Note: Data according to the *New York Times*, 11/7/07 and Laura Martin of DOE/IEA.
* The estimates in the 2nd column are for plants coming on line in 2014. The costs in this table do not include the transportation and distribution costs which on a national average amount to 4–5¢/kWh. In addition, the cost of the summer peak electricity can be three times as much as the yearly average.

power plant is built, we will not know for sure what it will cost, but I would not be surprised if it costs less than its fossil or nuclear counterparts.

The *wholesale* cost of electricity from conventional (fossil or nuclear) power plants varies from 5¢ to 10¢ for a kilowatt-hour, but this wholesale price rises up to 18¢/kWh by the time it is delivered to users. In addition to the wholesale cost of generation, there are transportation and distribution costs as well. The total for us in Connecticut in June, 2007 came to 18.9¢/kWh. In 2007 the national average "retail" prices by end-user sector were as follows:

Residential	10.7¢/kWh
Commercial	9.9¢/kWh
Industrial	6.4¢/kWh
Transportation	10.4¢/kWh
All Sectors	9.2¢/kWh

The cost of electricity is also a function of the time of its use, because the summer "peak price" can be two or even three times the "night time" electricity cost. Actually the transportation cost of electricity is higher than that of hydrogen.*

Some of the alternative energy technologies (wind, geothermal) can already compete with these prices, whereas others come close to being competitive (solar: 10¢–18¢).

---

* "The Hydrogen Defense Against Climate Catastrophe" by D. S. Scott, 2007.

TABLE 1.63

The First Costs of Various Types of Power Plants and the Wholesale Costs of Electricity Produced

Type	Typical Plant Sizes	First Costs per Noted Plant Size	First Costs in $/W	Production Cost (¢/kWh)
Fossil*				
Coal outdoor	0.6 (1.2) gW	$1 (2) billion	$1.7	Coal—5.0
Coal indoor	0.6 (1.2) gW	$1.5 (3) billion	$2.5	Gas—8.0
Coal w/recapture	0.6 (1.2) gW	$3+ (6+)* billion	$5	Oil—10.0
Nuclear	1.2 gW	$5–$6 billion	$5–$6	6.0
Solar				
SEGS & DSG	30 mW	$75–$105 million	$2.50–$3.50	10–17
	150 mW	$ 300 million	$2	12
	700 mW	$1.3 billon	$1.8	10
ICCS	200 mW	$0.8 billion	$4–$5	18
PV	40 mW	$250 million	$7	14
Stirling	25 kW	$150,000	$6	12–14
	500 mW	$2.5 billion	$5	12
Solar tower	200 mW	$0.8 billion	$4	10
Solar satellite	1 gW	$17 billion	$17	?
Wind				
Residential	3 kW	$7,000 & up	$4.50	5–9
Power plant	1 gW	$2.5 billion	$2.50	5–9
Ocean	100 mW	$333 million	$3.3	?
Geothermal	100 kW	$60,000–$150,000 (highly variable)	$0.6–$1.50	5
Fuel Cells	100 kW	$12,000	$8	15
	2 mW	$6	$3**	10

* A 0.6 gW outdoor coal burning power plant with only the minimum required pollution controls costs $1 billion ($1,500/kW). If it is enclosed and provided with carbon sequestering or carbon recapturing, the cost rises to over $2 billion ($3,000/kW) and if a coal-to-diesel fuel plant is built (which does not exist today), the cost can be $3 to $5 billion ($4,500/kW to $7,500/kW).

** DOE estimates that if production volume of fuel cells reaches 500,000, their first costs could drop by an order of magnitude.

As of today, renewable electricity prices are coming down and the cost of fossil fuel–based electricity is rising. As shown in Table 1.42, solar electricity is already competitive with the price of natural gas–generated peak power, and as shown in Table 1.61, carbon emission charges will further improve their competitiveness. By the time the first full-scale demonstration plant is built (using solar–thermal, concentrating PV, or Stirling-based collector technology), solar electricity is likely to be cost effective.

Table 1.57 provided some data on the efficiency of various power plants. Table 1.63 gives the approximate sizes of typical electric power plants in the United States and their approximate first and production costs.

1.7.1 Overall Costs of Energy

Sir Nicholas Stern, former chief economist at the World Bank, estimated that by 2020 the cost of continued reliance on fossil fuels will reach 20% of the gross world product (GWP), which today is approaching $50 trillion. Naturally, in addition to the economic cost of inaction ($10 trillion/yr by 2020), there are social, environmental, and "energy war" costs.

This is not to say that action is cost-free. In June 2008 the International Energy Agency (IEA) estimated that a total of $45 trillion (or 1.1% of the GWP yearly) should be invested by mid-century to prevent energy shortages and greenhouse gas emissions from slowing the global growth of the economy. If the goal of halving emissions is to be achieved by 2050, while using exhaustible fossil and nuclear energy, the IEA estimates that 50 fossil power plants need to be converted to carbon capturing and 32 new nuclear plants need to be built annually. Therefore, the cost of inaction (just in economic terms) is 20% GWP and the cost of just maintaining the status quo while using up our exhaustible energy reserves is 1.1% of the GWP.

The cost of a **permanent** solution, the gradual, but total conversion to a solar–hydrogen economy should be compared to these GWP numbers. We know that we receive from the Sun as much energy in less than an hour as we use all year, and we also receive a similar amount of energy from the moon in the form of tides. We also know that the first cost of solar–hydrogen power plants is less than their nuclear or fossil counterparts (Table 1.63 and Table 5.1) and once built, their fuel is free. To achieve the complete conversion of the planet to renewable energy use (650 Q by 2100) would require a rate of conversion of about 7 Q per year (230 solar–hydrogen plants per year at a cost of about $4 billion each) or a total cost of less than $1 trillion/year, which is about 2% of today's GWP. The actual total cost would also be affected by two other factors. Costs would be increased by the installation of the hydrogen distribution infrastructure and would be reduced by the mass production of the needed equipment. In addition, the passage of time would improve on today's efficiencies and the jobs created by the conversion would create an unprecedented economic expansion, easily doubling the GWP.

The GDP of the United States is nearly $15 trillion ($13.2 trillion in 2006). The yearly budget of the federal government is about $3 trillion ($2.7 trillion in 2006). This budget includes a military budget* which is over $0.6 trillion.

If we estimate the yearly American electricity consumption by 2030 as 5 trillion kWh (Table 1.4 gives historic data up to 2006) and plan to make all

---

* For 2007, the military budget rose to U.S.$439.3 billion. This does not include many military-related items that are outside of the Defense Department budget, such as nuclear weapons research, maintenance and production (~$9.3 billion, which is in the Department of Energy budget, Veterans Affairs (~$33.2 billion), or the wars in Iraq and Afghanistan (which are largely funded through extra-budgetary supplements). Conversely, the military budget does allocate money for dual-use items, such as the development of infrastructure surrounding U.S. military bases. Altogether, military-related expenses totaled approximately $626.1 billion in 2007 and are higher in 2008. In addition to this sum, the energy-related military expenses also increase the budget deficit and reduce the purchasing power of the dollar.

electricity from renewable energy by 2030, we would need to build 570 solar–hydrogen power plants over 22 years, or 26 plants a year. If each plant costs $4 billion (Table 1.63 and Table 5.1), that would require a yearly investment of about $0.1 trillion or less than 0.7% of today's American GDP. I just mention, but do not elaborate on such estimates, because the whole premise of this book is to convince the reader that as of now we do not have hard numbers, and a demonstration plant needs to be built to obtain them. I am also convinced that with the passage of time and with mass production, competition and human inventiveness, prices will go down just as these factors lowered the cost of computers over the last decades.

It is also likely that as the cost of solar–hydrogen drops, the costs of fossil and nuclear energy will rise. Robert Stavins of Harvard University estimates that if a $100/ton charge is placed on CO_2 emissions, that would increase the cost of coal-based electricity by 400% and natural gas–based electricity by 100%, making the cost of solar-, wind-, and geothermal-based electricity more than competitive. Table 1.62 lists the effect of a $50/ton carbon emission charge on the unit costs of electricity from a variety of power plants.

Past experience shows that as the equipment needed for new technologies starts to be mass produced, its prices drop. The cost of wind power–generated electricity has already been reduced to one quarter of that of the first installations. An ultrathin-film solar collector manufacturer (Nanosolar) claims that it will soon market collectors at costs that are severalfold less expensive than today's PV prices. We do not know if that particular claim is correct or not, but we know that time is on our side. Therefore, it is realistic to expect that as markets expand, the costs of mass-produced renewable energy devices will also drop and the cost of transition to a solar–hydrogen economy over several decades will become not only affordable, but will also create jobs and an economic boom. We should remember that drastic changes can occur rapidly; after all, a century ago electricity was a luxury that only 3% of the households had. The same will occur with renewable energy over the 21st century.

If solar energy was to be collected to meet the total energy consumption of humankind, about 4% of the area of the Sahara would need to be covered with collectors (the global electricity demand could be met by covering about 1% of the area). The cost of these collectors, hydrogen generators, and worldwide infrastructure required for a solar–hydrogen-based economy is not known, but we do know that it is extremely high. On the other hand, if this cost is spread out over 50 years or a century and one considers both the jobs created by this effort and the savings that result from it, the cost is manageable and the investment makes sense.

One can safely assume that the equipment efficiencies would improve and the unit costs would be cut, possibly by a factor of two in each decade. One can also assume that the cost of inaction will be much higher. For example, updating the old refineries and modernizing the global oil infrastructure would cost $3 trillion today.* This does not even include the costs of new oil

---

* Estimated by Prof. Wolfgang Reitzle, president and CEO of Linde.

and natural gas pipelines (already being built), the costs associated with the facilities required to exploit the oil shale deposits, or the conversion of old power plants to less polluting ones and the building of new ones.

The implications of a shift to the solar–hydrogen economy are profound. For the first time since the beginning of industrialization, humankind would be freed from the dependence on fossil fuels, thereby ending not only global warming but also the geopolitical nightmare that has preoccupied national security planners for many decades. This transition can begin slowly. We can first concentrate on conservation and on improving the efficiency of existing energy users, but later the conversion from the fossil-based to the solar–hydrogen-based economy must start.

The pace and direction of energy use are determined not just by technological developments, but also by social, economic, and environmental forces and by the response of industries, governments, and society as a whole. We know what needs to be done, we have the technological tools to do it, and we still have the time to complete this transformation.

1.8 Conclusion

All fossil and nuclear fuels are exhaustible. There is no safe nuclear or clean fossil energy. Even biofuels are undesirable when they interfere with the food supply. The only free, inexhaustible, and nonpolluting energy source is the Sun, and the best way to store its energy is in hydrogen. The continued reliance on fossil fuels will not only accelerate global warming, but is also likely to result in "energy wars" and could cause the collapse of the global economy. Yet, none of these need to occur. There is still plenty of time to gradually convert to an inexhaustible and clean solar–hydrogen-based economy. Therefore, the 21st century has to be the century of transition from a fossil–nuclear to an inexhaustible and clean solar–hydrogen–based energy economy.

This conversion will be just as important as was the introduction of electricity 100 years ago. This transition will require the mobilization of scientific talent on the scale of the Manhattan Project and will require the same level of leadership that was displayed in the rebuilding of Europe or the landing on the moon. Once this effort is begun, humankind's confidence in the future will return.

The first steps on that road of transition should include a substantial carbon emission tax, mandatory emission limits, increasing the minimum mileage requirements on cars, providing a market for alternative energy products (for example, by installing solar collectors on government and military buildings), and shifting government subsidies from the fossil and nuclear industries to support solar, geothermal, and wind installations. Just in the United States, the installation of solar roofs on 10 million homes could save more energy than the total oil import from the Persian Gulf, not to mention the jobs that effort would create.

It is unacceptable to wait for the market forces (cap-and-trade, increased fuel costs) to start reducing emissions, because by doing just that and without introducing new technologies, global economic growth will be stifled. Instead of the market forces, we must apply large-scale public funding to develop and demonstrate the feasibility and cost effectiveness of renewable energy technologies.

Another ingredient of this transformation is the conversion of the electric power grid into an integrated and nationwide one, so that all green energy generators can use it for storage and for connecting millions of solar roofs, wind turbines, and electric cars into the grid. There is also time to build complete solar–hydrogen demonstration plants in the parts of the world where there is no grid and to build the global infrastructure to transport the H_2 from the arid regions near the equator to the users.

The fuel used to make solar–hydrogen is free (sunshine) and unlimited, the raw material for H_2 is water, and the emission when burning the H_2 in fuel cells, internal combustion engines, or in power plants is distilled water. The cost of building the solar–hydrogen plants will be known once the demonstration power plant described in this book is built. It might turn out that this cost is already competitive; but whatever it is, we know that it will drop by an order of magnitude when the mass production of ultrathin-film solar collectors and reversible fuel cells is started.

We do not need more conferences, debates, or articles. What we need is to prove that the technology for moving into the age of renewable energy does exist, and it is safe, and economical. To prove that, we not only need solar roofs and solar–hydrogen demonstration plants but also statesmen. We need leaders who are not worried about lobbies and reelection, but care about the planet, the future of our grandchildren, and are dedicated to lead the third Industrial Revolution. The United States has been the leader of the world before, so she can and should lead in implementing this new Marshall Plan to transform the planet.

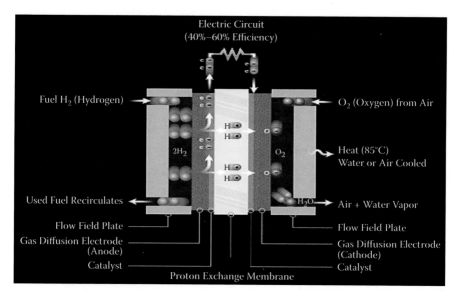

FIGURE 1.27
PEM fuel cell. (Courtesy of Ballard Power Systems Inc.)

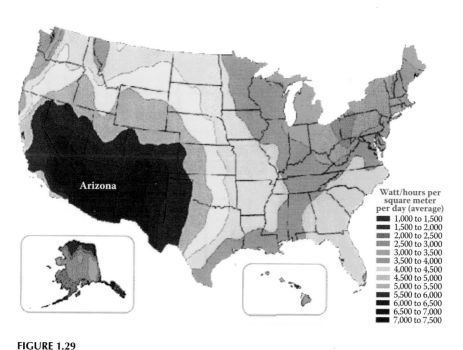

FIGURE 1.29
Annular average of daily solar energy received (insolation) in different locations in the United States.
(Courtesy of the National Renewable Energy Laboratory (NREL)/U.S. Department of Energy.)

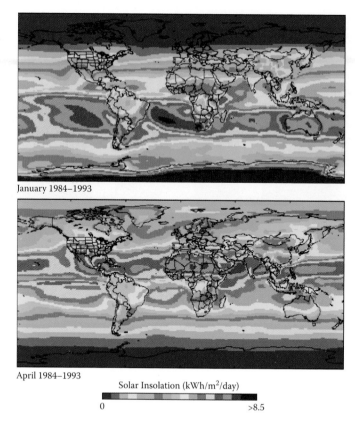

January 1984–1993

April 1984–1993

Solar Insolation (kWh/m²/day)

0   >8.5

FIGURE 1.30

Global map of insolation in units of kWh/m²/d. (Source: Earth Observatory, NASA.)

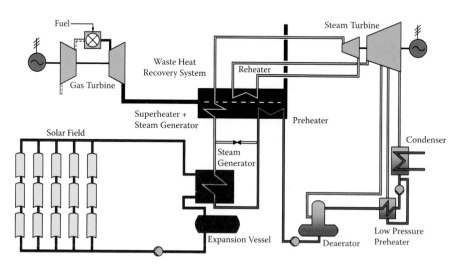

FIGURE 1.33

The integrated solar/combined cycle system (ISCCS) design, a thermal solar collector field is integrated with a steam- and gas turbine-based electric power generation. (Courtesy of FLAGSOL, Cologne.)

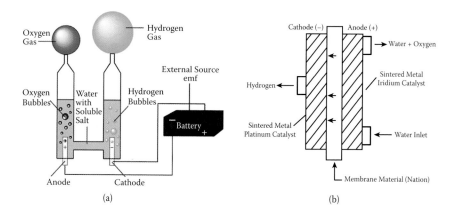

FIGURE 1.47
The process of electrolysis can use liquid or solid electrolyte. (a) Liquid electrolyte. (b) Solid polymer membrane electrolyte (SPE).

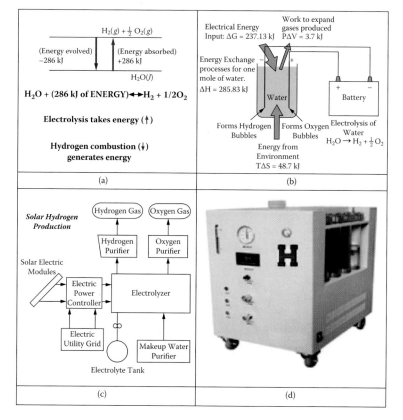

FIGURE 1.48
The thermodynamics of converting water into hydrogen and the electrolysis equipment used in that process. (a) Thermodynamics of electrolysis. (b) Electrolysis equipment schematics. (c) Components of a solar–hydrogen generator. (d) Hydrogen gas generator. (Courtesy of HGenerators™.)

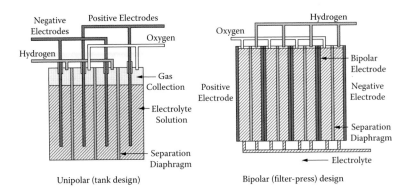

FIGURE 1.49
Unipolar (left) and bipolar (right) alkaline electrolyzers.

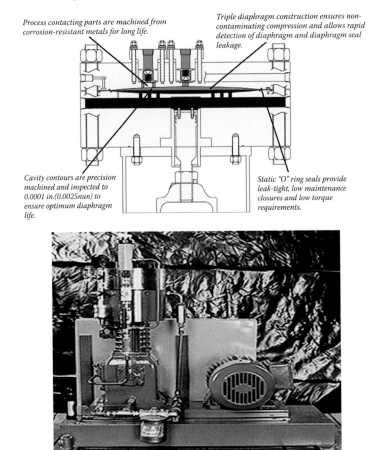

Process contacting parts are machined from corrosion-resistant metals for long life.

Triple diaphragm construction ensures non-contaminating compression and allows rapid detection of diaphragm and diaphragm seal leakage.

Cavity contours are precision machined and inspected to 0.0001 in.(0.0025nun) to ensure optimum diaphragm life.

Static "O" ring seals provide leak-tight, low maintenance closures and low torque requirements.

FIGURE 1.50
Leakage-free diaphragm compressors. (Top) Pressure Products Industries (PPI) has furnished a number of petrochemical plants and hydrogen fueling stations with compressors (Hyundai in South Korea, Powertech in Canada, Senju in Japan, GM in Michigan, Hydrogenics in Canada). (Bottom) In the Fluitron diaphragm compressor there are no dynamic seals or packing glands. All of the seals to the atmosphere are static gasket or o-ring seals. (Photo courtesy of Fluitron Inc., Ivyland, PA.)

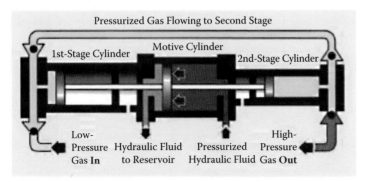

Pressurized Gas Flowing to Second Stage

1st-Stage Cylinder

Motive Cylinder

2nd-Stage Cylinder

Low-Pressure Gas **In**

Hydraulic Fluid to Reservoir

Pressurized Hydraulic Fluid

High-Pressure Gas **Out**

Stage 1 of the Intensifier Compression Cycle

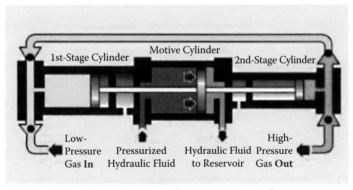

1st-Stage Cylinder

Motive Cylinder

2nd-Stage Cylinder

Low-Pressure Gas **In**

Pressurized Hydraulic Fluid

Hydraulic Fluid to Reservoir

High-Pressure Gas **Out**

Stage 2 of the Intensifier Compression Cycle

FIGURE 1.51
Intensifier with double-ended piston can boost the hydrogen pressure from 5 bar (70 psig) to 400 bar (5,800 psig) or from 70 bar (1,000 psig) to 1,000 bar (15,000 psig). (Courtesy of Hydro-Pac, Inc.)

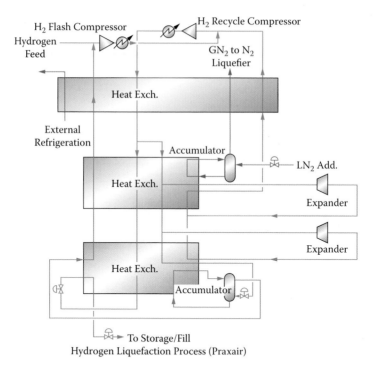

Hydrogen Liquefaction Process (Praxair)

FIGURE 1.54

A hydrogen liquefaction process developed by Praxair using external refrigeration, a liquid nitrogen precooler, three sets of heat exchangers, two compressors, and two expander turbines for energy recovery. (Courtesy of Praxair Inc.)

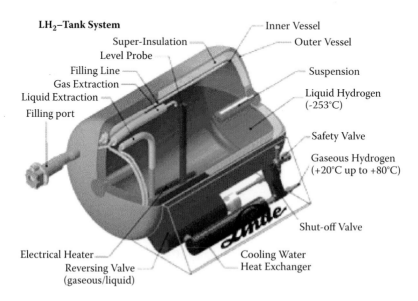

FIGURE 1.55

Liquid hydrogen storage tank. (Courtesy of Linde Co.)

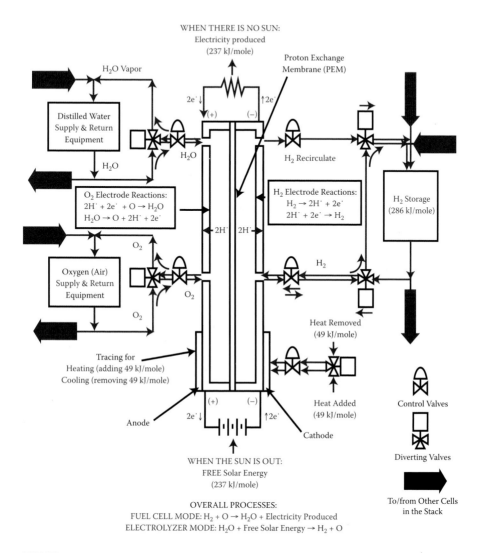

FIGURE 4.2

My reversible fuel cell (RFC) design will make the intermittently available solar energy continuously available by storing the excess solar energy when the sun is out and using this stored energy to make electricity when it is not. Compared to the cost of today's fuel cells, mass production and optimizing control algorithms (not shown) will probably lower the cost of RFCs by an order of magnitude. Small RFCs are most likely to be used in "zero energy" homes, if electric storage is not available and the electric cars are to be charged at night. Once large RFCs are also available, they can be used in full size power plants.

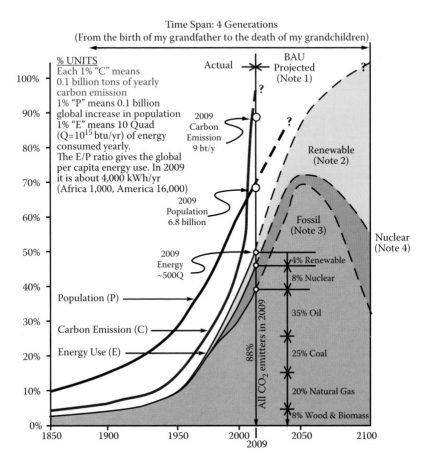

FIGURE 5.3

Global energy trends showing actual data up to 2009 and projections after that. The projection represents a "business as usual" case which assumes that the transition to renewable energy will be very slow and during that period no energy wars, climate or nuclear disasters will occur. (Note 1): If this BAU scenario is followed, it is likely that 20% of the land area of the planet will be flooded and civilization as we know it will end. (Note 2): If the equivalent of 1–2% of the GWP was invested annually in the conversion to using renewable energy and if no new fossil or nuclear plants were built, by 2100 the transition to a free, clean, safe and inexhaustible energy economy could be completed. (Note 3): All fossil fuels are exhaustible and carbon emitting. There is no such thing as "clean coal." Continued investments in these technologies take funds away from converting to renewable energy. (Note 4): No new nuclear power plants should be built because none are safe, their waste disposal is unresolved, uranium is exhaustible and breeder reactors use weapons-grade plutonium fuel which is readily usable for building "dirty" nuclear bombs.

2

Control and Optimization of Energy Conservation and Renewable Energy Processes

2.1 Introduction

The first chapter of this book described the technology needed to convert to the post-oil renewable energy economy. This chapter will show that the required know-how to integrate, operate, and optimize the needed processes and to build full-size renewable energy demonstration plants, which are needed to prove their feasibility, already exists. My goal here is to convince the engineering and scientific communities that it is time to stop writing articles and holding debates. The time has come to stop talking and to start building, so that we can establish the facts concerning the costs and feasibility of renewable energy demonstration plants.

The first step on the road to energy independence is not a complete conversion from fossil to renewable energy; the first step is energy conservation. This means an increase in the energy efficiency of the presently operating conventional processes. This chapter describes the state of the art of energy conservation technology for optimizing the traditional processes. The implementation of these strategies will immediately and drastically reduce the global consumption of energy.

These reductions can amount to about 20% in the industrialized world and to even more in the third world. For example, in China, the energy used to air-condition the buildings is twice what it needs to be. In general, Chinese industry is 30 to 50% more energy intensive and 100% more polluting than it is in Western industry. This chapter describes the devices and strategies that can improve the efficiency of energy conservation in buildings, boilers, chillers, compressors, pumping, distillation systems, etc.

The control systems serving the automation and optimization of solar collectors, wind turbines, fuel cells, geothermal, hydrogen, and other renewable energy processes are also described. These strategies are relatively new; some of them are original—I have invented them—and are described for the first time. The renewable energy processes differ from the traditional industrial ones, because some of them operate at very high temperatures (such as the concentrating and thermal solar collectors), some fuel cells require very small flows and pressure differentials, and hydrogen processes operate at very low temperatures (liquid hydrogen [LH_2]) or at very high pressures (hydrogen gas [GH_2]). This chapter will describe reliable, compact, and accurate means of precisely regulating these and many other processes.

For example, the thermal solar collector power plants are integrated with conventional boilers, turbines, and generators, and the geothermal systems depend on the optimization of conventional water-pumping and heat pump stations. In addition to control, this chapter will also describe optimization techniques, utilizing modern modeling techniques and multivariable controls. These controls are different from the traditional ones, because they treat flows, temperatures, levels, and pressures only as constraints of a control envelope; whereas the purpose of the envelope is to maximize efficiency and profitability, and minimize energy consumption (or maximize energy production).

2.2 Boiler Control and Optimization

Boilers are available in two basic designs: a fire tube, in which water circulates in tubes heated externally by fire and a water tube, in which hot gases from fire pass through the tubes in the boiler. Fire-tube boilers are generally limited in size to approximately 12,000 kg/h (25,000 lb/h) and to about 20 bar (250 psig). Their size and pressure limitations preclude their use in large industrial facilities or in power plants.

Steam boilers are used by the electric power plants to produce steam for power generation and by manufacturing and industrial processes to produce steam for both power generation and process heating and energy conversion. A typical coal-fired utility boiler might produce 1.5 million kg/h (3 million lbs/h) of superheated steam at 150 to 200 bar (2000 psig to 3000 psig). Industrial boilers are commonly referred to as cogeneration or combined heat and power (CHP) applications. Here the emphasis will be on the description of boiler optimization strategies, whereas for a detailed description of boiler controls (Figure 2.1) the reader is referred to Chapter 8.6 of the second volume of the *Instrument Engineers' Handbook*.

A fully loaded large boiler that is clean and properly tuned (with blow-down losses and pump and fan operating costs disregarded) is expected to have the following efficiencies:

On coal: 88%, with 4% excess oxygen (O_2); 89%, with 3% excess O_2

On oil: 87%, with 3% excess O_2; 87.5%, with 2% excess O_2

On gas: 82%, with 1.5% excess O_2; 82.5%, with 1% excess O_2

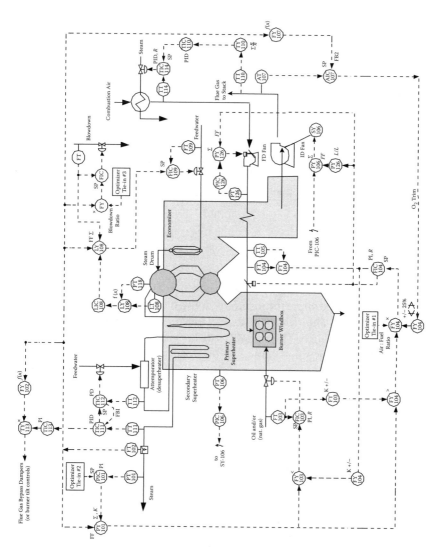

FIGURE 2.1
This diagram shows good boiler controls without optimization.

A 1% loss in efficiency on a 45,360 kg/h (100,000 lb/h) boiler will increase its yearly operating cost by about $25,000. A 1% efficiency loss can result from a 2% increase in excess O_2, or from about a 28°C (50°F) increase in flue gas temperature. The purpose of optimization is to continuously maximize the boiler efficiency, as variations occur in the load, fuel, ambient, and boiler conditions. The various goals of optimization include the following:

1. Minimize excess air and flue gas temperature
2. Minimize steam pressure
   a. Turbines thereby open up turbine governors
   b. Reduce feed pump discharge pressures
   c. Reduce heat loss through pipe walls
3. Minimize blowdown
4. Measure efficiency
   a. Use the most efficient boilers
   b. Know when to perform maintenance
5. Provide accountability
   a. Monitor losses
   b. Recover condensate heat

The first three methods of optimization are achieved by closed-loop process control and can be superimposed upon the overall boiler control system shown in Figure 2.1. The tie-in points for these optimization strategies are also shown in that figure. The benefits of the last two methods (efficiency and accountability) are not obtained in the form of closed-loop control signals, but they do contribute to better maintenance and better understanding of heat losses and equipment potentials.

2.2.1 Excess Air Optimization

Figure 2.2 plots the various boiler losses as a function of air excess or efficiency. The sum total of all the losses is a curve with a minimum point. Any process that has an operating curve of this type is an ideal candidate for optimization, because once the minimum loss point is determined, the optimizer can shift the operating conditions until that point is reached.

As shown in Figure 2.2, the radiation and wall losses of a boiler are relatively constant, and most of the heat losses occur through the stack. Under air-deficient operations, unburned fuel leaves, and under air-excess conditions, heat is lost as the unused O_2 and its accompanying nitrogen are heated up and then discharged into the atmosphere. The goal of optimization is to keep the total losses to a minimum. This is accomplished by minimizing both excess air and the stack temperatures.

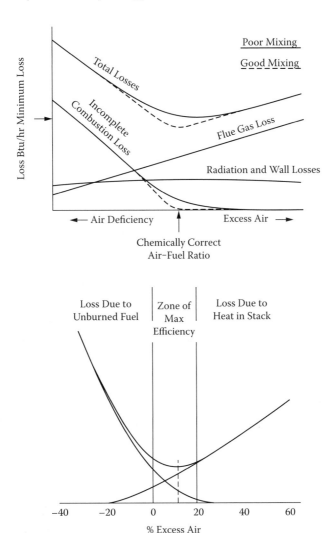

FIGURE 2.2
Boiler losses can be plotted as a function of excess air (top). The minimum of the total loss curve of a boiler is the point where optimized operation is maintained (bottom). Most efficient operation of a boiler occurs when the amount of excess air in the stack balances the losses of unburned fuel.

The minimum loss point in Figure 2.2 is not where excess O_2 is zero, but a minimum positive value. This is because no burner is capable of providing perfect mixing. This is why the theoretical minimum loss point is to the left of the actual one in the bottom part of Figure 2.2. This actual minimum loss or maximum efficiency point is found by lowering the excess O_2 as far as possible, until opacity or carbon monoxide (CO) readings indicate that the minimum has been reached. At this minimum loss point, the flue gas losses (heat lost through the stack) and the unburned fuel losses are the same.

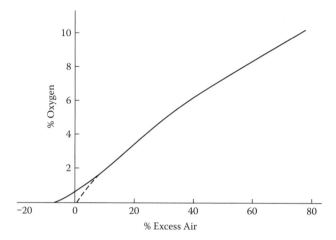

FIGURE 2.3

The amounts of oxygen in the flue gas and excess air are related and, therefore, the rate of the combustion air supplied (Figure 2.1) to control them.

Figure 2.3 shows the composition of the flue gas as a function of the amount of air present. The combustion process is usually operated so that enough air is provided to convert all the fuel into carbon dioxide (CO_2), but not much more. This percentage of excess O_2 is *not* a constant. It varies with boiler design, burner characteristics, fuel type, air infiltration rates, ambient conditions, and load.

The percentage of excess air must be increased as the load drops off. This is because at low loads turbulence is reduced and lowers the efficiency of mixing between the fuel and the air. This loss of mixing efficiency is compensated for by increasing the percentage of excess O_2 admitted at low loads. Because each boiler has its own unique personality, the relationship between load and excess O_2 must be experimentally determined. Because of the difference in mixing efficiency, gases require the lowest percentage and solid fuels the highest percentage of excess O_2 for complete combustion. The optimum excess O_2 percentages for gas, oil, and coal are around 1, 2, and 3%, respectively.

The analyzers available for the detection of CO_2, CO, and excess O_2 are discussed in Chapter 3. Excess air can be correlated to O_2, CO, CO_2, or combustibles present in the flue gas. The most sensitive indicator of flue gas composition is CO. As shown in Figure 2.2, optimum boiler efficiency can be obtained when the losses due to incomplete combustion *equal* the effects of heat loss through the stack. These conditions prevail at the "knee" of each curve. Whereas the excess O_2 corresponding to these knee points varies with the fuel, the corresponding CO concentration is relatively constant.

In Figure 2.1, the set point for the air–fuel ratio is designated as the #1 input of the optimizer controls. Similar optimizer inputs are provided to minimize (for heating) or maximize (for power generation) the steam pressure and minimize the blowdown flow.

2.2.2 Multivariable Envelope Control

Multiple measurements of flue gas composition can further improve boiler efficiency.

With microprocessor-based systems, it is possible to configure a control envelope, such as that shown in Figure 2.4, where several control variables are simultaneously monitored, and control is switched from one to the other, depending on which limit of the envelope is reached or violated.

For example, assuming that the boiler is on CO control, the microprocessor will drive the CO set point toward maximum efficiency, but if in so doing the opacity limit is reached, that will override the CO controller and will prevent the opacity limit from being violated. Similarly, if the microprocessor-based envelope is configured for excess O_2 control, it will keep increasing boiler efficiency by lowering excess O_2 until one of the envelope limits is reached. When that happens, control is transferred to the violated constraint parameter (CO, HC [hydrocarbon], opacity, etc.), and all variables are "herded" to stay within the envelope defined by these constraints. These limits are usually set to keep CO under 400 ppm and opacity below #2 Ringelmann, and below the local regulations for HC and NO_x.

Model-based boiler optimization schemes have proved successful in many power plant and industrial boiler applications. Successful NO_x reduction through this kind of optimization can avoid or postpone large capital expenditures for low NO_x burners, over-fire air modifications, and selective catalytic reduction/selective noncatalytic reduction (SCR/SNCR).

Depending on the type and size, a fossil-fired generating unit may contain hundreds of highly correlated, nonlinear, and time-varying variables. Shown in Figure 2.3 is the optimal region for a traditional excess-air-based efficiency optimization system. Through model-based optimization, the maximum efficiency zone shown in Figure 2.2 can be shifted to the right and the CO curve can be shifted to the left. This would open up more room for improved optimization compared to strategies that rely only on excess air control.

The models are a set of equations that relate key performance measurements to the major control variables influencing combustion efficiency. Owing to the complexity and uncertainty of the analytical models that are derived from physical principles, empirical models, based entirely upon the plant data, are typically used for practical boiler optimization control.

Modeled relationships can take the form of a step response, impulse response, state-space representation, or a neural network (see Section 2.6.17). If a linear form is desired, the model is usually linearized around some operating point. Another option is to produce a series of linear models, each representing a specific operating condition (usually load level). The obtained model can be used for solving a static optimization problem to find out the optimal operating point. The "optimal" criterion can be user selectable.

Fuel quality, boiler loading, heat exchanger surface fouling, ambient condition, and aging of equipment will all cause the process to drift and affect the model accuracy. Adaptive tuning computations can be built in to take care of

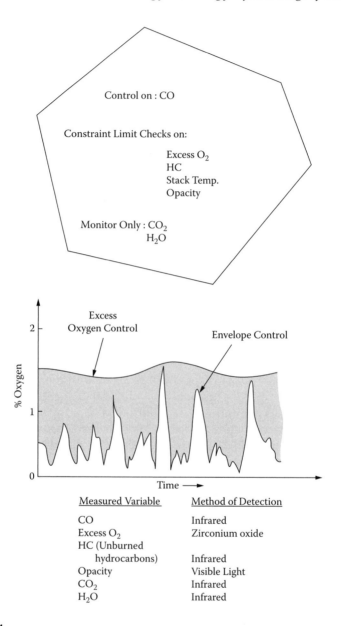

FIGURE 2.4
Multivariate envelope-based constraint control can lower overall excess oxygen. This can be achieved by monitoring both carbon dioxide and water and by performing constraint limit checks on excess oxygen, hydrocarbons, stack temperature, and opacity.

known quantifiable relations (e.g., variation of dead time with load). Online training, incorporating an adaptive learning algorithm, can be used to automatically train models in real time, combining recent results with the initial and historical training data.

Closed-loop multivariable boiler control has to be planned and performed carefully because plant operators are not traditionally willing to reduce air–fuel ratios due to concerns about CO and other symptoms associated with O_2-deficient combustion. Model predictive control (MPC) is by far the most widely used technique for conducting multivariable boiler optimization and control. Forms of MPC that are inherently multivariable and that include real-time constrained optimization in the design are best suited for boiler application.

2.3 Building Conditioning Optimization

Building optimization in the United States can save 25 to 50% (100% in China) of the energy used to air-condition residential and office buildings. This is on top of the savings resulting from the use of more efficient appliances, low-energy light bulbs, more efficient heating and cooling equipment, double windows, better insulation, and the installation of solar roofs and other renewable energy sources. Here, the focus will be on the optimization of the total space conditioning system and its automatic reconfiguration as the seasons change. The emphasis will be placed on systems in which air is the carrier of heat or cooling into the conditioned spaces.

The airhandler is the basic unit operation of space conditioning. It is used to keep the occupied spaces at the desired temperature and humidity. Figure 2.5 illustrates the U.S. and American Society of Heating, Refrigerating and Air-Conditioning Engineers (ASHRAE) "comfort zones." In addition to supplying or removing heat or humidity from the conditioned space, the airhandler also provides fresh air makeup. Depending on the climate, type of space, and the energy sources, the yearly cost of space conditioning can be up to $100 per square meter of floor space ($10 per square foot).

Airhandlers today are frequently controlled the same way as they were 20 or 30 years ago. For this reason, optimization can often cut their energy consumption in half—a savings that can seldom be achieved in any other unit operation. The optimization goals include the following:

Let the building heat itself

Use free cooling or free dying

Benefit from gap control or zero energy band (ZEB)

Eliminate chimney effect

Optimize start-up timing

Optimize air makeup (CO_2)

Optimize supply air temperature

Minimize fan energy use

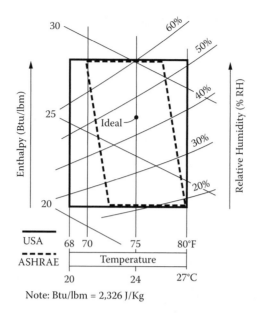

FIGURE 2.5
"Comfort zone" is defined in terms of temperature and humidity.

Automate the selection of operating modes

Minimize reheat

Automate balancing of air distribution

Figure 2.6 illustrates the main components of an airhandler system, including fans, heat exchanger coils, dampers, ducts, and instruments. The system operates as follows: Outside air is admitted by the outside air damper (OAD-05) and is then mixed with the return air from the return air damper (RAD-04). The resulting mixed air is filtered, heated, or cooled, and humidified or dehumidified as required. The resulting supply air is then transported to the conditioned zones (groups of offices) by the variable-volume supply fan station.

Any combination of temperature and humidity conditions within the envelopes on Figure 2.5 is considered to be comfortable. Therefore, as long as the space conditions fall within this envelope—referred to as ZEB—there is no need to spend money or energy to change those conditions. This concept is very cost-effective.

Airhandlers can be operated in a variety of "modes" including start-up, occupied, night, purge, emergency, and switching from summer to winter mode and vice versa. Conventional systems are switched according to the calendar, whereas optimized ones recognize that there are summerlike days in the winter and winterlike hours during summer days. Seasonal mode switching is therefore totally inadequate.

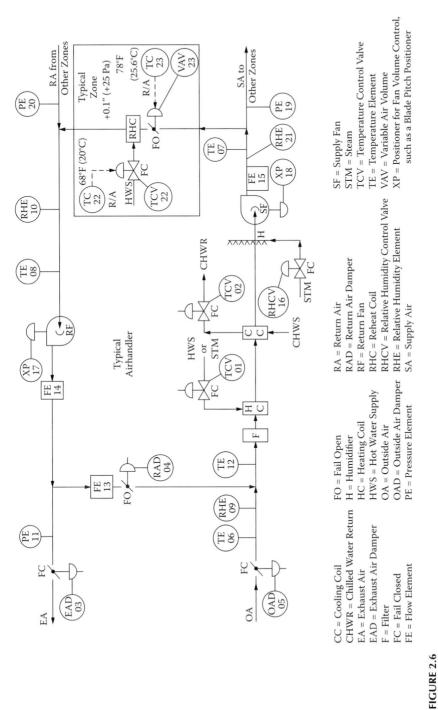

CC = Cooling Coil
CHWR = Chilled Water Return
EA = Exhaust Air
EAD = Exhaust Air Damper
F = Filter
FC = Fail Closed
FE = Flow Element

FO = Fail Open
H = Humidifier
HC = Heating Coil
HWS = Hot Water Supply
OA = Outside Air
OAD = Outside Air Damper
PE = Pressure Element

RA = Return Air
RAD = Return Air Damper
RF = Return Fan
RHC = Reheat Coil
RHCV = Relative Humidity Control Valve
RHE = Relative Humidity Element
SA = Supply Air

SF = Supply Fan
STM = Steam
TCV = Temperature Control Valve
TE = Temperature Element
VAV = Variable Air Volume
XP = Positioner for Fan Volume Control,
such as a Blade Pitch Positioner

FIGURE 2.6
A typical major airhandler has these components and controls.

2.3.1 Self-Heating Buildings

There are a variety of thermostats including night–day, set-back, heating–cooling, summer–winter, limited control range, slave or submaster, multi-stage, dew-point, and smart thermostats. A recent addition to the available thermostat choices is the ZEB design, which conserves energy by not using any when the room is within comfortable limits (Figure 2.5). This zero energy band and the system operation are illustrated in Figure 2.7.

When in a zone (a section of a building served by an airhandler), the comfort level is within acceptable limits, and the use of "pay energy" is no longer justified. Allowing the zones to float between such limits can substantially reduce the operating cost of the building. The added benefit of this approach is that during the winter—when heating is required in the rooms that have windows (perimeter spaces), while the interior spaces are still generating heat and need to be cooled—the building can become self-heating by moving the heat from the interior to the perimeter (Figure 2.8). This can result in periods during which buildings operate without using any pay energy.

In regions in which winter temperature does not drop below –12°C (10°F), ZEB control can eliminate the need to pay for heat altogether. In regions farther north, ZEB control can lower the yearly cost of heating by 30 to 50%.

When a zone is within the comfort gap, its reheat coil is turned off and its VAV box is closed to the minimum flow required for air refreshment (Figure 2.6). When all the zones are inside the ZEB, HW (hot water), CHW (chilled water), and STM supplies to the airhandler are all closed and the fan is operated at minimum flow. When all other airhandlers are also within the ZEB, the pumping stations, chillers, cooling towers, and HW generators are also turned off.

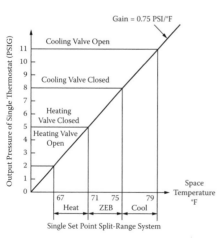
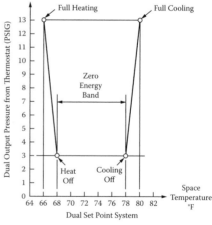

FIGURE 2.7

ZEB control schemes include the single set point split-range approach (shown on left) and the dual set point approach (shown on right).

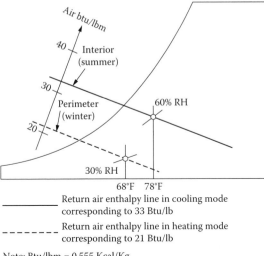

Return air enthalpy line in cooling mode
corresponding to 33 Btu/lb

‒ ‒ ‒ ‒ ‒ ‒ ‒ ‒ Return air enthalpy line in heating mode
corresponding to 21 Btu/lb

Note: Btu/lbm = 0.555 Kcal/Kg

FIGURE 2.8
The building can be made self-heating, because the return air can transport about 10 Btu/lbm
to the perimeter spaces, which in the winter, because of the windows, do require heat.

2.3.2 Elimination of Chimney Effects

In high-rise buildings in the winter when the cold air on the outside is heavier
than the conditioned air on the inside, the chimney effect tends to pull in
ambient air at ground elevation, and this cold air adds an additional load to
the building's heating system. Eliminating the chimney effect can lower the
operating cost by approximately 10%.

Figure 2.9 shows the controls required to eliminate the chimney effect. The
key element of this control system is the reference riser, which allows all
pressure controllers in the building to be referenced to the barometric pressure of the outside atmosphere at a selected elevation. Using this pressure
reference allows all zones to be operated at the same pressure of about 25 Pa
(0.1 in. H_2O) pressure and permits this constant pressure to be maintained at
both ends of all elevator shafts.

If the pressure is the same on all floors of a high-rise building, there will
be no pressure gradient to motivate the vertical movement of the air, and
the chimney effect will be eliminated. A side benefit of this control strategy is the elimination of all drafts or air movements between zones, which
also minimizes the dust content of the air. Another benefit is the capability
of adjusting the "pressurization loss" (out-leakage of air) of the building by
varying the setting of PC-7, 8, and 9.

If the building is maintained at a constant reference pressure, this will
result in higher pressure differentials on the windows on the higher floors
as the barometric pressure on the outside drops. Therefore, the pressure differentials on the windows will rise, requiring stronger windows.

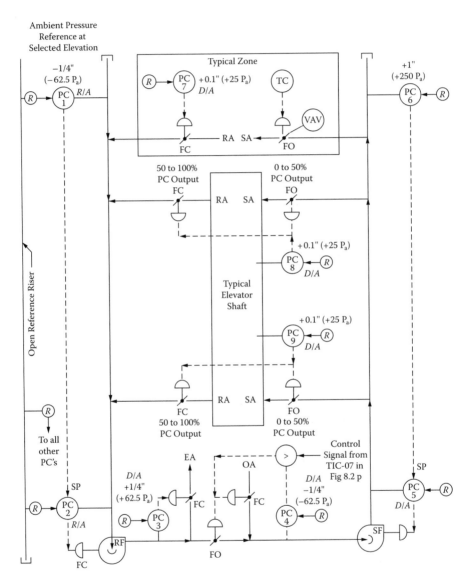

FIGURE 2.9
Chimney effects in high-rise buildings can be eliminated by using the proper pressure controls.

2.4 Chiller and Heat Pump Optimization

Heat pumps play a critical role in both renewable energy processes and in the general field of energy conservation. In the renewable energy processes, heat pumps are used to liquefy hydrogen and to increase the temperature of geothermal water. In conventional industrial and building conditioning processes, heat pumps (chillers) are used to generate chilled water and to air condition buildings. The optimization of chillers can reduce their energy consumption by about 33%.

The unit of refrigeration is the ton, and the units used include the following:

Standard ton	200 Btu/min (3520 W)
British ton	237.6 Btu/min (4182 W)
European ton (frigorie)	50 Btu/min (880 W)

Heat pumps serve to transport heat from a lower to a higher temperature level. They do not make heat; they just transport it. This is similar to a pump transporting water from a lower to a higher elevation. The required energy input of a heat pump is a function not only of the amount of heat to be transported but also of the temperature difference across which it is transported. A 1°C (1.8°F) reduction of this temperature difference will lower the yearly operating cost by 2.7% (1.5%).

Figure 2.10 illustrates how heat pumps can transport heat from a lower to a higher elevation and thereby can cool an already cold temperature substance, such as LH_2. The heat pump removes Q_l amount of heat from the cold process at the cost of investing W amount of work and delivers Q_h quantity of heat to the warm reservoir. In the lower part of Figure 2.10, the idealized temperature entropy cycle is shown for the chiller. The cycle consists of two isothermal and two isentropic (adiabatic) processes:

1 → 2: Adiabatic process that occurs in the expansion valve

2 → 3: Isothermal process taking place in the evaporator

3 → 4: Adiabatic process that occurs in the compressor

4 → 1: Isothermal process taking place in the condenser

The isothermal processes in this cycle are also isobaric (constant pressure). The efficiency of a heat pump is defined as the ratio between the heat removed from the process (Q_l) and the work (W) required to accomplish this heat removal.

$$\beta = \frac{Q_1}{W} = \frac{T_L}{T_H - T_L}$$

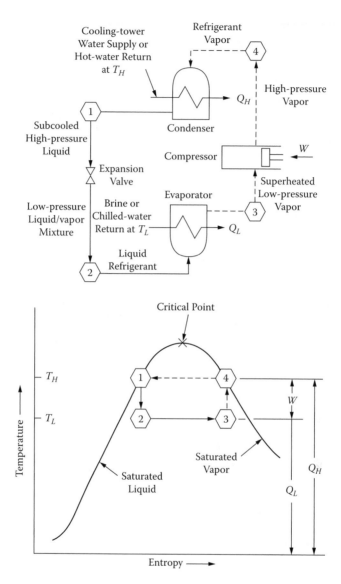

FIGURE 2.10
The top portion of this figure shows the main components of a refrigeration system. The bottom portion describes the refrigeration cycle, which consists of two isothermal and two isentropic processes.

Because the efficiency of a chiller can be much more than 100%, it is usually called the *coefficient of performance* (COP). If a chiller requires 1.0 kWh (3,412 Btu/h) to provide a ton of refrigeration 211.2 kW (12,000 Btu/h), its coefficient of performance is said to be 3.5. This means that each unit of energy introduced at the compressor will pump 3.5 units of heat energy into the cooling tower water.

2.4.1 Minimum Cost of Operation

Optimization can double the coefficient of performance; one unit of energy can remove 5, 6, or even 7 units of heat energy from the process if the chiller controls are optimized. The main goals of optimization are (1) minimizing the temperature difference, (2) minimizing pumping energy input, and (3) recovering let-down energy. In order to fully optimize a cooling system, in addition to the chiller, it is necessary to also optimize cooling tower controls, water distribution, and the pumping systems. Figure 2.11 describes the basic controls of an industrial chiller system, showing the optimum values of the set points of the four temperature control loops.

The yearly cost of operating the total cooling system can typically be broken down as follows:

*M*1 (fans)	10%
*M*2 (cooling tower [CT] pumps)	15%
*M*3 (compressors)	60%
*M*4 (chilled [CH] pumps)	15%

The proportion of *M*1 increases in warm weather regions. *M*2 and *M*4 increase when water transport lines are long and the proportion of *M*3 is lowered as the maximum allowable chilled water temperature rises. However, regardless of these proportions in a particular installation, the goal of optimization is to find the minimum chilled water and cooling tower water temperatures that will minimize the total cost of operation.

The optimum set point for both TIC-3 and TIC-4 are the allowable maximums. As *M*3 is about 60% and *M*4 about 15% of the total operating cost, a 1°C (1.8°F) decrease in the ΔT on the chilled water side (evaporator ΔT) will lower the yearly cost of operation by about $(1.8 \times 60)/100 + (11 \times 15)/100 = 2.7\%$.

Finding the optimum cooling tower water temperatures is more complicated, because changing the cooling tower water supply and return temperatures will increase some costs while it will lower some others, as shown in Figure 2.12. For example, an increase in the cooling tower water supply temperature (T_{ctws}) will increase the tower's approach and the total cost of chiller operation but will lower the operating cost of the cooling tower fans. Therefore, the optimum cooling tower water temperature is the temperature that can meet the particular load at the minimum total cost of all operating equipment (*M*1 + *M*2 + *M*3).

Figure 2.12 illustrates that this is a function of the load and of the weather. When the conditions are such that a lowering of the approach (or T_{ctws}) would require a large increase in fan operating costs (*M*1″), the minimum point on the total cost curve (A_0'') .will be above the minimum attainable approach (A_0'). Conversely, if the approach can be reduced with a small increase in fan operating costs (*M*1′) the optimum operating point for the total system will correspond to the minimum T_{ctws} temperature and therefore would be at the minimum attainable approach (A_0'). Whichever total

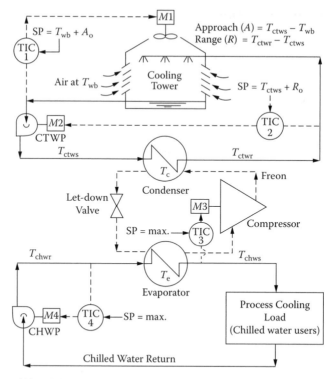

Abbreviations:

A_o : Optimum Approach
R_o : Optimum Range
SP   : Set Point
T_{wb} : Wet Bulb Temp.
T_{ctws} : Cooling Tower Water Supply Temp.
T_{ctwr} : Cooling Tower Water Return Temp.
T_{chws} : Chilled Water Supply Temp.
T_{chwr} : Chilled Water Return Temp.
CTWP: Cooling Tower Water Pump
CHWP: Chilled Water Pump
T_c : Freon Temperature in Condenser
T_e : Freon Temperature in Evaporator

FIGURE 2.11
In order to optimize a cooling system, the cooling towers, pumping stations, chillers, and process equipment should be treated as an integrated single system.

cost curve ($\$'$ or $\$''$) is applicable, its minimum point becomes the set point for TIC-1 in Figure 2.11.

A number of other optimization strategies will be discussed later in this chapter. The savings resulting from replacing the letdown valves with expander turbine generators will be discussed in Section 2.13. The optimization of multistage chillers will be covered in connection with hydrogen liquefaction in Section 2.15.3, and coolant distribution controls will be covered under pump optimization in Section 2.17.2.

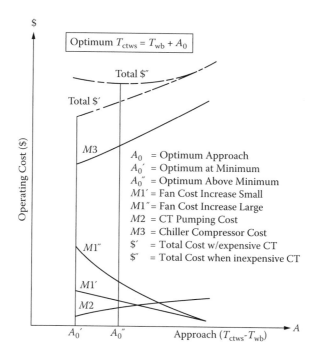

Figure 2.12
The empirically determined operating costs are shown as a function of the approach at two cooling loads. The optimum approach at each load is the one that corresponds to the minimum total cost of operation.

Additional details on dozens of chiller optimization configurations can be found in Chapter 8.13 of Volume 2 of the *Instrument Engineers' Handbook*.

2.4.2 Reconfiguration and Storage

The selection of the most efficient configuration of a cooling system depends on the cooling load, ambient conditions, and utility costs. When the load is high, *pay cooling* is unavoidable, but as the demand for cooling drops, it is possible to consider *free cooling* equipment configurations. Figure 2.13 illustrates the equipment and valving required to implement four modes of chiller operation.

As defined in Table 2.14, Mode #1 is the state of normal mechanical refrigeration operation, also referred to as pay cooling, because in this mode the chiller compressor is on. The other three modes are referred to as free cooling modes because the compressor is off. In these modes, the cooling tower water (CTW) can either directly cool the process (the chilled water pump [CHWP] is off) or indirectly cool it. In Table 2.14 direct free cooling is Mode #4. During indirect free cooling, the CTW does not reach the process, but only cools the chilled water (ChW), and therefore the CHWP is on. In these indirect free cooling modes, the CTW will cool the ChW either in a

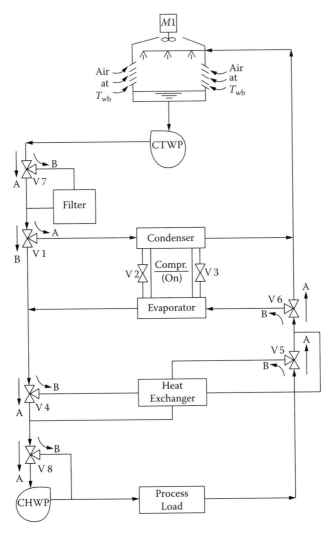

FIGURE 2.13
This mechanical refrigeration system can be automatically reconfigured to operate in any of the four modes including the three 'free cooling' configurations, which are listed in Table 2.14.

separate heat exchanger (Mode #3) or through the idle chiller (Mode #2). If the chiller is used, this Mode #2 operation is called the *thermosiphon mode*. In this mode the cooling is obtained through refrigerant vapor migration in the evaporator.

Heat pump operating costs can also be minimized by operating the chillers when the cost is least expensive, namely, at night when both the ambient temperature and the cost of electricity are low. Optimization can also be served by meeting partial loads by operating the chillers part of the time at maximum efficiency. Both of these strategies require storage, so that when the chillers are off, the chilled water or brine will continue to be available. As illustrated

TABLE 2.14

Equipment Status in Four Operating Modes of a Refrigeration System

Equipment	Mode 1 (Mechanical refrigeration)	Mode 2 (Vapor migration based indirect free cooling)	Mode 3 (Indirect free cooling by the use of heat exchanger)	Mode 4 (Direct free cooling with full filtering)
Compressor	On	Off	Off	Off
Cooling-tower pumps	On	On	On	On
Chilled-water pumps	On	On	On	Off
Valve V1	A	A	B	B
Valve V2 and 3	Closed	Open	Closed	Closed
Valve V4	A	A	B	A
Valve V5	A	A	B	A
Valve V6	A	B	A	A
Valve V7	A	A	A	B
Valve V8	A	A	A	B

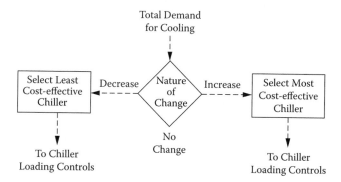

FIGURE 2.15
Optimized, computer-based load allocation will direct load increases to the most cost-effective chiller, while sending the load decreases to the least cost-effective chiller.

in Figure 2.15, the logic of starting and stopping individual chillers in a multi-chiller installation should always consider efficiency, so that the most efficient idle unit is started and the least efficient active unit is stopped.

By applying all the optimization strategies available, the energy cost of operation of many heat pump systems can be cut in half.

2.4.3 Cooling Towers

The cooling water used in most chillers is received from cooling towers (CTs) and as was shown in Figure 2.11, chillers and CTs are optimized as a single unit. In CT optimization it is important that both the water pump and the airflow fan be variable speed. Naturally, the goal of cooling tower optimization is to maximize efficiency (the amount of cooling per unit of operating cost invested).

The cost of fan operation can be reduced by allowing the cooling tower water temperature (T_{ctws} in Figures 2.11 and 2.16) to rise, thereby increasing the approach ($T_{ctws}-T_{wb}$) at which the tower operates. As shown in Figure 2.16, as the approach, and therefore T_{ctws}, rises, the temperature difference ($T_{ctwr} - T_{ctws}$) across the chiller condenser is reduced. This will cause an increase in water flow, and consequently the pumping costs ($M2$) will rise.

If the actual total operating cost of the CT is plotted against the approach (as in the right side of Figure 2.16), the fan cost ($M1$) tends to drop and the compressor and pumping costs ($M2$ and $M3$) tend to rise with an increase in approach. Once a total operating cost curve is obtained, the optimum approach can be found as the minimum point on that curve and that becomes the set point of TIC-1 on Figure 2.11.

As there is an empirical relationship between the values of approach and range, once the optimum approach (A_0) is determined, the corresponding optimum range (R_0) can be obtained from the empirically derived approach versus the range curve (on the right of Figure 2.16), which is used as the optimum set point for TDIC-1 in Figure 2.16.

2.5 Compressor Optimization

Compressor controls play an important role in the optimization of renewable energy processes. Their applications are discussed in detail in connection with hydrogen processing (see Chapter 1, Section 1.5.5), geothermal power plants (see Section 1.3.4), and heat pumps (see Section 2.4).

The three main compressor types are rotary, reciprocating, and centrifugal. The compressor drives can be constant or variable speed, and can be driven by electric motors, steam turbines, gas turbines, gasoline, or diesel engine drives. For a particular application, the type of compressor is selected by considering the required capacity and discharge pressure.

As was shown in Figures 1.50 and 1.51, the designs most often used on high-pressure hydrogen services are the hydraulic-fluid-operated diaphragm compressors and the reciprocating intensifiers. When the hydrogen is to be compressed to 1,000 bar (15,000 psi) or higher, only reciprocating compressors can be considered. The discharge pressure and flow rate delivered by the reciprocating compressors and intensifiers are both controlled by modulating the hydraulic fluid flow.

The flow and discharge pressure ranges of the various compressor designs are shown in Figure 2.17.

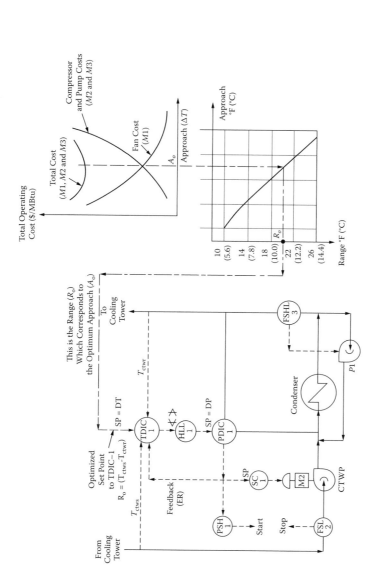

FIGURE 2.16

The cooling tower flow rate is modulated to keep the ΔT of the chiller's condenser (R_o, which is the range of the cooling tower) at the value that corresponds to the optimum value approach.

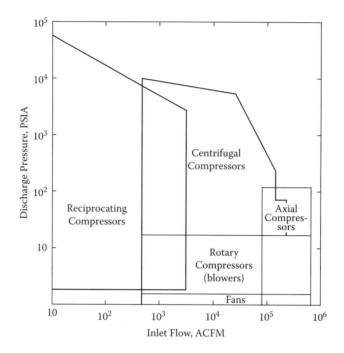

FIGURE 2.17
The range of flow capacities and discharge pressures that the different compressor designs can handle.

Positive-displacement compressors pressurize gases through confine-ment. Dynamic compressors pressurize them by acceleration. The axial compressor moves the gas parallel to the shaft; in the centrifugal compres-sor, the gas receives a radial thrust toward the wall of the casing where it is discharged. The axial compressor is better suited for constant flow applications, whereas the centrifugal design is more applicable for constant pressure applications. This is because the characteristic curve of the axial design is steep, and that of the centrifugal design is relatively flat. Axial compressors are more efficient; centrifugal ones are better suited for dirty or corrosive services.

2.5.1 Centrifugal Compressor Control

Centrifugal compressors are capable of generating only a few hundred bars of discharge pressure, but can handle much higher volumes than can the reciprocating compressors.

The centrifugal compressor is a machine that converts the momentum of gas into a pressure head. The compressor pressure ratio (P_D/P_I) var-ies inversely with mass flow (W). For a compressor running at constant speed (ω), constant inlet temperature (T_1), and constant molecular weight, the discharge pressure may be plotted against mass flow (Curve I in

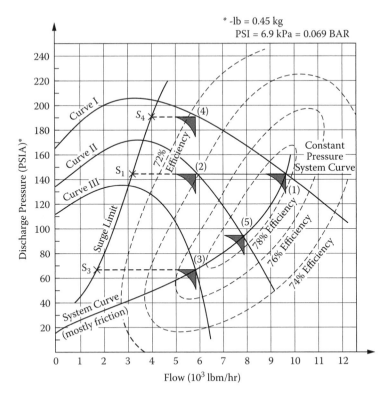

FIGURE 2.18
The operating point is located where the centrifugal compressor curves cross the system curve of the process. The system curve can be a constant pressure one (horizontal line), a mostly friction one, or any other.

Figure 2.18). The compressor operates at the point where the compressor curve crosses the system curve of the process. In the illustrated case the design point (1) is located in the maximum efficiency region (78%) at design flow and pressure.

Compressor loading can be reduced by throttling a discharge or a suction valve, by modulating a prerotation vane, or by reducing the speed. Of these options, it is only the variable-speed operation that does not waste energy. Curve II in Figure 2.18 could correspond to a compressor speed of 80% and Curve III to 60% speed of operation. Discharge throttling is the least and speed modulation is the most energy-efficient method of throttling. Suction throttling is in between, and it gives a little better turndown than discharge throttling; but it still results in the wasting of energy. When speed is modulated this energy is not introduced in the first place.

If the discharge pressure is constant, flow tends to vary linearly with speed. If the discharge head is allowed to vary, it will change with the square of flow and, therefore, with the square of speed as well. This square relationship between speed and pressure tends to limit the speed range of compressors to the upper 30% of their range. Electric governors give better turndowns and are quicker and simpler to interface with surge or computer controls. It is desirable to make the speed control loop response as fast as possible. On turbines, this goal is served by the use of hydraulic actuators, and in case of electric drives, the motor response is usually increased by the use of tachometer feedback.

If the compressor serves a constant pressure process (point [1] on Figure 2.18) by speed variation and the process demand is reduced to point (2), the compressor speed has to be reduced to 80% (Curve II) where the operating efficiency is only 74%. In order to accomplish the same capacity reduction in a "mostly friction" system, it is necessary to reduce the compressor speed to 60% (Curve III), so that the compressor will operate at the new operating point (3) at 77% efficiency.

If the same compressor is modulated by discharge throttling and the flow is reduced in a mostly friction process, the compressor must follow Curve I and therefore operate at point (4) at 72% efficiency. However, the mostly friction system curve at this capacity requires only the discharge pressure corresponding to point (3) and therefore the pressure difference between points (2) and (3) will be wasted in the form of a pressure drop across the discharge valve.

2.5.2 Surge Control

The phenomenon of momentary flow reversal is called *surge*. During surging, the compressor discharge pressure drops off and then is reestablished on a fast cycle. Intense surges are capable of causing complete destruction of compressor parts, such as blades and seals. Figure 2.18 illustrates the surge line. At the beginning of surge, the total flow drops off within 0.05 seconds, and then it starts cycling rapidly at a period of less than 2 seconds. If the flow cycles occur faster than the control loop can respond to them, this cycling will pass through undetected as uncontrollable noise. Therefore, fast sensors and instruments are essential for the surge control loop, which serves to temporarily increase the compressor loading by venting or recycling some of the vapors.

Figure 2.19 illustrates the basic controls required to keep the compressor out of surge by quickly introducing an artificial load as the surge line is approached within some limit. In order to do this, the response speed of both the flow sensor and the valve must be very fast.

The surge controller (FIC) in Figure 2.19 is a direct-acting controller with proportional and integral action and with anti-reset windup (ARW) features. It is a flow controller with both the proportional band and the integral settings minimized (PB ~50%, 1 to 3 s/repeat) and with a set point which is about 10% higher than the flow at which surge starts. The aim of an ARW feature is to hold the FIC output under normal conditions at around 105%, so that the valve is closed but the signal is just lingering above the 100% mark without saturation.

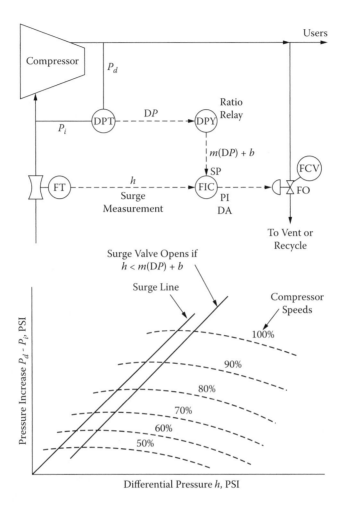

FIGURE 2.19
On a plot of P vs. h, the surge curve becomes a straight line.

In digital controllers, it is desirable to set the sample time at about one quarter of the period of surge oscillation, or at 0.3 seconds, whichever is shorter. If a flow-derivative backup control is used, the sample time must be 0.05 seconds or less, because if it is slower, the backup system would miss the precipitous drop in flow as surge is beginning.

2.5.3 Optimized Load Following

When a compressor is supplying gas to several parallel users, the goal of optimization is to satisfy all users with the minimum investment of energy. The minimum required header pressure is found by the valve position controller (VPC) in Figure 2.20. It compares the highest user valve opening (the

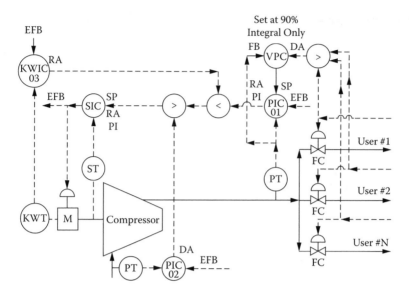

FIGURE 2.20
Protective overrides can be added to optimized load-following controls, so that the system is protected against excessively low pressures on the suction side of the compressor or from overloading the compressor's motor drive.

needs of the most demanding user) with a set point of 90%, and if even the most open valve is not yet 90% open, it lowers the header pressure controller (PIC-01) set point.

This supply–demand matching strategy cannot only minimize the use of compressor power but also protects the users from being undersupplied, because it protects all supply valves from being fully open. The VPC is an integral-only controller, with its integral time set for about 10 times that of the PIC. This guarantees that the VPC will act more slowly than all the user controllers, thus giving stable control even if the user valves are unstable. The external feedback (FB) protects the VPC from reset windup when its output is limited or when the PIC has been switched to manual.

In addition to following the load, protective overrides prevent the development of excessively low suction pressures (PIC-02), which could result in drawing oil into the compressor and from overloading the drive motor and thereby tripping the circuit breaker (KWIC). In order to prevent reset windup when the controller output is blocked from affecting the SIC set point, external feedback (EFB) is provided for all three controllers.

2.5.4 Multiple Compressor Optimization

Compressors can be connected in series to increase their discharge pressure (compression ratio) or they can be connected in parallel to increase their flow capacity. When driven by different shafts, they require separate antisurge

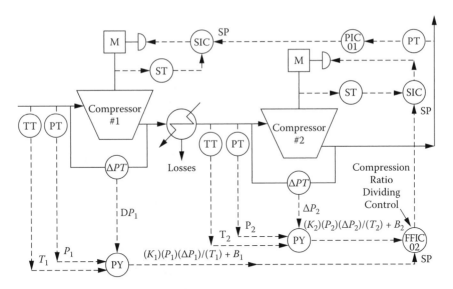

FIGURE 2.21
When two compressors operate in series, one can be dedicated to maintain the total discharge pressure (PIC-01) from the pair, while the speed of the other can be manipulated to keep them at equal distance from their surge curves. This is achieved by FFIC-02 controlling the distribution of the total compression ratio between them.

systems, although an overall surge bypass valve can be made available to all surge controllers through the use of a low signal selector. This eliminates the interaction that otherwise would occur between surge valves, as the opening of a bypass around one stage not only would increase the flow through the higher stages but would also decrease it through the lower ones.

If streams are extracted or injected between the compressors, which are in series and therefore have unequal flows, a control loop needs to be added to keep both compressors automatically away from their surge lines by distributing among them the total compression ratio that must be provided (Figure 2.21). In hydrogen-processing plants, it is common to have two or more compressors operate in series. In this system, PIC-01 controls the total discharge pressure by adjusting the speed of compressor #1, whereas the speed of compressor #2 is set by FFIC-02. Both compressors are thus kept at equal distances from their respective surge lines.

Controlling two or more compressors operating in parallel requires load distribution controls. Figure 2.22 illustrates how two compressors can be proportionally loaded and unloaded, while keeping their operating points at equal distance from the surge curve. Improper load distribution is prevented by measuring the total load (summer #9) and assigning an adjustable percentage of it to each compressor by adjusting the set points of FFIC-01 and FFIC-02. The load distribution can be computer-optimized by calculating compressor efficiencies (in units of flow per unit power) and loading the units in the order of their efficiencies.

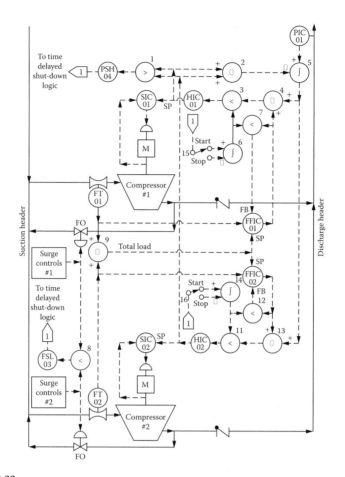

FIGURE 2.22

A flow-balancing bias can be superimposed on direct pressure control. This control system can distribute the load between the two machines in the order of their efficiencies.

The pressure controller (PIC-01) directly sets the set points of SIC-01 and 02, whereas the balancing controllers (FIC-01 and 02) slowly bias those settings. This is a more stable and responsive configuration than a pressure-flow cascade, because the time constants of the two loops are similar. A high-speed integrator (item #5) is used on the summed speed signals to assure a correspondence between the PIC output signal and the number of compressors (and their loading). The integrator responds in a fraction of a second and therefore does not degrade the speed of response of the PIC loop.

The compressors are automatically started and stopped as the load varies. When the total flow can be handled by a single compressor or when any of the surge valves open, FSL-03 triggers the shutdown logic interlock circuit #1 after a time delay. Automatic starting of an additional compressor is also initiated by interlock #1 when PSH-04 signals that one of the compressors has reached full speed. The ratio flow controllers (FFIC-01 and

02) are protected from reset windup by receiving an external feedback signal through the low selector #7 (or #12), which selects the lower of the FFIC output and the ramp signal.

Interlock #1 is also provided with *rotating sequencer logic,* which serves to equalize run times between machines and protects the same machine from being started and stopped too frequently. A simple approximation of these goals is achieved if the machine that operated the longest is stopped and the one that was idle the longest is started.

2.6 Control and Optimization Theory

Here, a summary is provided of the general aspects of process control. These basic concepts are applicable to all control systems, regardless of the variable or the process being controlled. Naturally, this has to be an abbreviated treatment of the subject and therefore, if the reader is interested in an in-depth treatment, turning to the second volume of the *Instrument Engineers' Handbook* is recommended.

2.6.1 Basics

The fly-ball governor was the first known automatic control system. It was installed on Watts' steam engine over 200 years ago in 1775. It detected the speed of the rotation of the shaft and automatically opened up the steam supply when a drop was registered in that speed.

Today, we understand that in order to control and optimize a process, we must understand its "personality," because the controller must be correctly matched to the process it controls. The key personality traits of controlled processes are their resistance, capacitance, and dead time.

Pipelines and equipment are mostly resistance processes. When the flow is turbulent, the resistance is a function of the square root of pressure drop. A change in flow results in an immediate and proportional change in pressure drop. The amount of change is a function of the process gain, also called *process sensitivity.*

Most processes include some form of capacitance or storage capability. These capacitance elements can provide storage for materials (gas, liquid, or solids) or storage for energy (thermal, chemical, etc.). Thermal capacitance is directly analogous to electric capacitance and can be calculated by multiplying the mass of the object (W) with the specific heat of the material it is made of (Cp). The gas capacitance of a tank is constant and is analogous to electric capacitance. The liquid capacitance equals the cross-sectional area of the tank at the liquid surface, and if the cross-sectional area is constant, the capacitance of the process is also constant at any head.

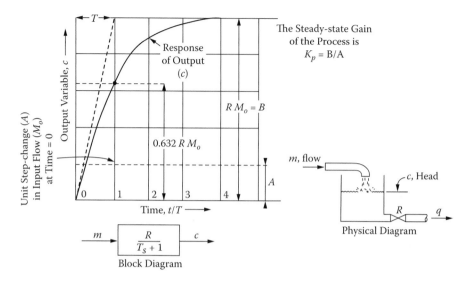

FIGURE 2.23

A single time-constraint process consists of a capacitance and a resistance element. The time it takes for the controlled variable (c) to reach 63.2% of its new steady-state value is the value of the time constant.

Resistance and capacitance in combination are the most common in industrial processes involving heat transfer, mass transfer, and fluid flow operations. The combined effect of supplying a capacity through a resistance is time retardation (time constant). Combining a capacitance-type process element (tank) with a resistance-type process component (valve) results in a single-time-constant (τ) process. If the tank was initially empty and then an inflow was started at a constant rate of m, the level in the tank would rise as shown in Figure 2.23 and would eventually rise to the steady-state height of $c = Rm$ in the tank.

If two tanks are connected in series, the system has two time constants operating in series. The response curve of such a process is slower than that of the single-time-constant process, because the initial response is retarded by the second time constant. Figure 2.24 illustrates the responses of processes having up to six time constants in series. As the number of time constants increases, the response curves get slower (the process gain is reduced) and the overall response gradually changes into an S-shaped reaction curve.

A first-order system coupled with dead time (transportation lag) is a good model for many process systems. The dead time (L or t_d) is the time that has to elapse before the output first starts to respond to a change in the input. The effect of a change in steam rate on the water temperature at the end of the pipe will depend not only on the resistance and capacitance effects in the tank but will also be influenced by the length of time necessary for the water to be transported through the pipe. The effect of dead

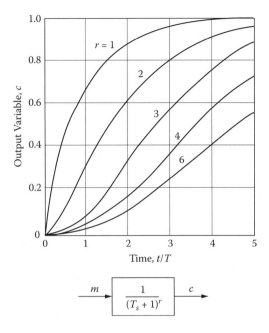

FIGURE 2.24
The responses of processes slow as the number of their (equal size) time constants in series increases.

time can be compared to driving a car (the process) with closed eyes. Dead time is the worst enemy of good control, and the process control engineer should minimize it.

The various means of reducing dead time are usually related to reducing transportation lags. This can be achieved by increasing the rates of pumping or agitation, reducing the distance between the measuring instrument and the process, eliminating sampling systems, etc. When the nature of the process is such that the dead time exceeds the time constant,

$$\left(\frac{t_d}{\tau} \right)$$

ratio exceeds unity, the traditional PID control must be replaced by periodic adjustments, called *sample and hold controls* or *Smith predictors.*

2.6.2 Degrees of Freedom and Loop Gain

Typically, the variable chosen to describe the state of the system is termed the *controlled variable* and the variable chosen to control the system's state is termed the *manipulated variable.* To fully understand the personality of a process, one must also know the number of variables that can be independently controlled in that process. The maximum number of independently acting automatic controllers that can be placed on a process is defined as the degrees of freedom (df) of that process. Mathematically, the number of degrees of freedom is defined as:

$$df = v - e$$

where:

> df = number of degrees of freedom of a system
> v = number of variables that describe the system
> e = number of independent relationships that exist among the various variables

For example, a train has only one degree of freedom because only its speed can be varied; a boat has two, an airplane three. When looking at industrial processes, the determination of their degrees of freedom becomes more complex and cannot always be determined intuitively. When the process is more complex, as in the case of binary distillation, the calculation of the degrees of freedom also becomes more involved. Figure 2.25 lists 14 variables of this process, but they are not all independent.

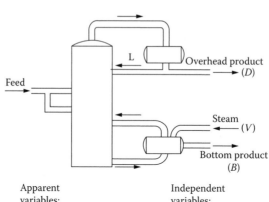

Apparent variables:	Independent variables:
C_1 = Overhead temperature	
C_2 = Overhead pressure	2
C_3 = Overhead composition	
C_4 = Overhead flow rate	1
U_1 = Bottom temperature	
U_2 = Bottom pressure	2
U_3 = Bottom composition	
U_4 = Bottom flow rate	1
U_5 = Feed temperature	
U_6 = Feed pressure	2
U_7 = Feed composition	
U_8 = Feed percent vapor	1
U_9 = Feed flow rate	1
m = Steam flow rate (heat input)	1
	11

FIGURE 2.25
In a binary distillation process the number of independent variables is 11 and the number of degrees of freedom is 9.

There being two components and two phases in the following streams at the bottom, feed and overhead, Gibbs's law states that only two of the three variables (pressure, temperature, and composition) are independent. Therefore, the number of independent variables is only 11. The number of defining equations is two (the conservation of mass and energy), and, therefore, the number of degrees of freedom for this process is $11 - 2 = 9$. Consequently, not more than nine automatic controllers can be placed on this process.

The process gain indicates how much a process property (output) changes per unit of input change. The input can be a flow through a control valve and the output can be a process property, such as temperature. The process gain frequently depends on the load or operating point. For example, the gain changes when the cross section of a vessel varies (this is the case with horizontal cylindrical tanks), or if the load is modified (for example, the gain of heat transfer processes will drop with rising load). These variations introduce nonlinearities, making the process gain highly dependent on the operating point.

Dynamic gain of a sinusoidal input is the ratio of its amplitude to that of the output. Sustained oscillation results if the loop gain is 1.0, and quarter-amplitude damping (the amplitude of a sinusoidal error drops to one quarter in each cycle) results if the loop gain is 0.5. The goal when tuning most process control loops is to obtain quarter-amplitude damping. This will result if the product of all the gains in the loop comes to 0.5 (Figure 2.26). This end result is achieved through tuning, which is the process of finding the controller gain, which will make the overall gain product 0.5.

The loop gain is the product of all the gains in the loop, including sensor, controller, control valve, and process. In a properly tuned loop, the product of all these gains is 0.5. What makes tuning difficult is that the process gain often varies with process load. For example, in heat transfer processes, as the load rises, the process gain drops. One way to compensate for this effect is to install an equal percentage (=%) control valve, which increases its gain as the load rises, so the total loop gain remains relatively unaffected.

2.6.3 Feedforward Control

Most automatic control strategies are a combination of feedback (closed-loop) and feedforward (open-loop) control components. Feedback control maintains a desired process condition by measuring and comparing it to a set point and initiating corrective action based on the difference between the desired and the actual conditions. Feedforward control provides fast compensation to disturbances if its "model" is accurate. The main limitation of feedforward is our inability to prepare perfect process models or to make perfectly accurate measurements.

Because of these limitations, pure feedforward would accumulate the errors in its model and would eventually "self-destruct." The main limita-

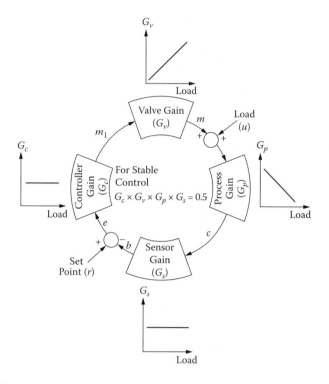

FIGURE 2.26
If the process gain varies with load, such as the case of a heat transfer process, the gain product of the loop can be held constant by using a valve whose gain variation with load compensates for the process gain variation.

tions of feedback control are that feedback cannot anticipate upsets and can only respond after they have occurred, and that it makes its correction in an oscillating, cycling manner.

It has been found that combining feedback and feedforward is desirable in such a way that the imperfect feedforward model corrects for about 90% of the upset as it occurs, whereas feedback corrects for the remaining 10%. With this approach, the feedforward component is not pushed beyond its abilities, whereas the load on the feedback loop is reduced by an order of magnitude, allowing for much tighter control.

Filters are needed to eliminate noise from such signals as the controlled variable. If the noise in the measurement signal is not reduced, it can pass through the controller and cause cycling and eventual damage to the control valve or other final control element. Noise tends to be of a higher frequency and is likely to change more rapidly than does the controlled process variable. The task of a filter is to block the rapidly changing component of the signal but pass the slowly changing component, so that the filter output will have less noise than the raw signal does.

Filters are likely to be more effective on fast processes and can complicate or limit the response of a PID controller. One way to compensate for this is to tune the controller with the filter inserted into the feedback loop.

2.6.4 Process Reaction Curves

The dynamic behavior of processes (pipe–vessel combinations, heat exchangers, transport pipelines, furnaces, boilers, pumps, compressors, turbines, and distillation columns) can be described using simplified models composed of process gains, dead times, and process dynamics.

Dead time can result from measurement lag, analysis, and computation time, communication lag or the transport time required for a fluid to flow through a pipe. Figure 2.27 illustrates the response of a control loop to a step change, showing that the response started after a dead time (t_d) has passed and reaches a new steady state as a function of its time constant (τ), defined in Figure 2.23. When material or energy is physically moved in a process plant, there is a dead time associated with that movement. This dead time equals the residence time of the fluid in the pipe. Note that the dead time is inversely proportional to the flow rate. For liquid flow in a pipe, the plug flow assumption is most accurate when the axial velocity profile is flat, a condition that occurs when Newtonian fluids are transported in turbulent flow.

The time constants of pressure and differential pressure measurements are on the order of 0.1 seconds. Temperature measurement time constants are usually between 1 and 10 seconds. Composition measurements (analyzers) are even slower, varying from 5 seconds to 10 minutes.

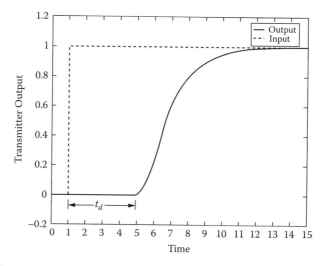

FIGURE 2.27
The response of a process having both dead time and a time constant.

Signal processing delays and periods in digital systems are around 0.1 seconds. Depending on the length and diameter of the tubing between the I/P converter and control valves, pneumatic control valves can introduce considerable delays. Electric actuators or valve positioners can considerably reduce these delays. The stroking times of conventional (pneumatic) valves vary from 2 seconds to 1 minute, whereas high-speed electric and hydraulic actuators can reduce them to the millisecond range.

A complete mathematical description of the process can be constructed using mass and energy balances, engineering relations, valve equations, etc. It is a difficult and time-consuming job to develop a dynamic process model. An approximation of the dynamic process model can be obtained by using the reaction curve method. This experimental technique is based on applying a change in the manipulated variable (input of the process) with the loop open (controller in manual) and recording the response of the controlled variable (output of the process). In the Laplace domain, most of the industrial processes can be represented by a first-order plus dead-time model

$$G(s) = \frac{K}{\tau_s + 1} e^{-t_d s}$$

where:

K is the steady-state gain
τ is the time constant
t_d is the dead time

The parameters of the process model, K, τ, t_d are obtained by using the following procedure:

1. The control loop is opened by switching it to manual.
2. This is done when the controlled variable (system output) is at a constant value and no disturbances or other upsets are allowed to occur while the reaction curve is developed.
3. A step change is applied to the manipulated variable (controller output, which is an input to the process). The step changes are usually 5 to 10%. The step time should be long enough for the manipulated variable (system input) to reach a new steady state.
4. The manipulated variable (system input) response is recorded to provide good visibility on both the amplitude and time scales.

The evaluation involves drawing a tangent to the initial process response curve at the point of maximum slope and noting the point where this tangent intercepts the final steady-state value. In this case, the process time constant is the time between the tangent intercepting the original and

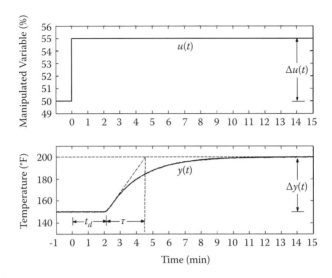

FIGURE 2.28
Reaction curve evaluation to determine dead time and time constant of a first-order plus dead-time process.

the new steady-state lines (Figure 2.28). The dead time is measured as the time it takes between applying the step change and the beginning of the time constant. Therefore, the process gain (K) is the ratio $\Delta y(\frac{1}{2})/\Delta u(t)$.

Sometimes industrial processes contain pure integration elements. This is the case with controlling level in a tank, when the manipulated variable is the feed flow rate into the tank. This is not a self-regulating process, because any step change in feed flow will cause the controlled variable (level) to increase or decrease linearly with time. The process model of an integrating process is obtained by applying a pulse test (not a step test), because the step test would produce an output curve that will change linearly, without reaching a steady-state value. As shown in Figure 2.29, the gain of the integration element can be calculated when a pulse step is applied, as

$$K = \frac{\Delta y}{\Delta ut_r}$$

2.6.5 Proportional Control

A controller compares its measurement (y) to its set point (r), and based on the difference (e = error) sends an output signal (m – manipulated variable) to the final control element (e.g., control valve) to eliminate the error. The control mode options include on–off, floating, proportional (P), integral (I), differential (D), and many others. The proportional mode considers the present state of the process error, the integral mode looks at the past history of the error, and the derivative mode anticipates the future values of the error

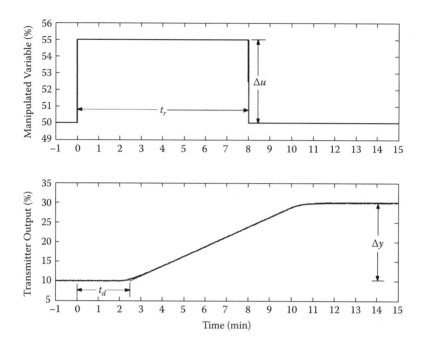

FIGURE 2.29

A pulse test is used to establish the dynamic characteristics of an integrating plus dead-time process.

and acts on that prediction. A controller is called *direct-acting* if its output increases when its measurement rises, and is called *reverse-acting* if its output decreases when its measurement rises.

On–off control is implemented by a switch that keeps the manipulated variable at its maximum (or minimum) value when the controlled variable is below (or above) its set point. In most practical applications, there is a narrow band that the error must exceed before a change will occur. This band is known as the *differential gap.*

For self-regulating processes with little or no capacitance, a single-speed floating controller can be used, such as a single-speed reversible motor. The controller is usually provided with a neutral zone (dead zone), and when the measurement is within that zone, the output of the controller is zero. This dead zone is desirable, because otherwise the manipulated variable would be changing continually in one direction or the other.

The response of the proportional (P) mode (as the name suggests) is proportional to the error. The proportional setting K_c, is called the proportional gain, frequently expressed in terms of percent proportional band, PB, which is inversely related to the proportional gain ($K_c = 100/PB$). "Wide bands" (high *PB* values) result in less sensitivity (lower gains), and "narrow bands" (low *PB* percentages) result in more "sensitive" controller response. The output of a proportional controller (m) is

$$m = e(100/PB) + b$$

where:

      e = the deviation from set point or error

      PB = the proportional band $(100/K_c)$

      b = the live zero or bias of the output, which in analog electronic loops is usually 4 mA.

The proportional controller cannot consider the past history or the possible future consequences of an error trend. The gain in DCS control packages is usually adjustable from 0 to 8, whereas in analog controllers it can usually be adjusted from 0.02 to about 25.

The main limitation of plain proportional control is that it cannot keep the controlled variable on set point, because it cannot bring the error to zero. This is because it can only change its output, when there is an error. The resulting deviation is called the offset (Figure 2.30).

2.6.6 Integral Control Mode

The integral (I) control mode (sometimes called *reset mode*, because after a load change it returns the controlled variable to set point and eliminates the offset) generates an output (m) according to the equation:

$$m = \frac{1}{T_i} \int edt + b$$

The term "T_i" is the integral or reset time setting of the controller. If the bias (b) is zero, this mode acts as a pure integrator, the output of which reaches the value of the step input during the integral time. The integral mode eliminates the offset of plain proportional control because it continuously looks at

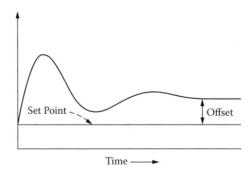

FIGURE 2.30
After a permanent load change, the proportional controller is incapable of returning the process back to the set point and an offset results. The smaller the controller's gain, the larger the offset will be.

the total past history of the error by continuously integrating the area under the error curve and eliminates the offset by forcing the addition (or removal) of mass or energy, which should have been added (or removed) in the past.

When the error is constant, and therefore the proportional correction is also constant, the integral correction is rising at a constant rate because the area under the error curve is still rising. When the error is dropping, the integral correction is still rising because the area under the error curve is still rising. When the error reaches zero, the integral correction is at its maximum. It is this new signal level going to the control valve that serves to eliminate the offset.

The unit used in setting the integral time is "repeats/minute" or "minutes/repeat." This means that so long as an error exists, the integral mode will keep repeating the proportional correction after the passage of each integral time. The shorter the integral time, the more often the proportional correction is repeated (the more repeats/minute), and thus the more effective the integral contribution is to the output. In DCS systems, the integral setting of control loops can usually be set from 0 to 300 repeats per minute, or from 0.2 seconds to about 60 minutes or more in units of minutes/repeat.

The fact that integral looks at the past history of the error and integrates it is useful when the loop is operational, but it can be a problem when the loop is idle. For example, when a process is shut down, the controller does not need to integrate the errors under the error curves, because if it does, it will eventually saturate, and its output will either drop to zero or rise to the maximum value of the air or power supply. Once saturated, the controller will not be ready to take control when called upon to do so. In all such installations the controller must be provided with either external reset, which protects it from ever becoming idle, or with an ARW feature.

The process responses to a load change are shown in Figure 2.31 for a variety of controller mode combinations.

In selective and cascade control loops, external feedback is the most-often-applied solution. Here, instead of looking at its own output, which can be blocked, the integral mode of the controller looks at an *external feedback* signal (such as the opening of the valve), which cannot be blocked. In surge control or reactor heat-up applications, the chosen solution usually is to use the slave measurement as the external reset signal to prevent saturation.

2.6.7 Derivative Control Mode

The derivative mode anticipates the future values of the error and acts on that prediction. This third control mode became necessary as the size of processing equipment increased and, correspondingly, the mass and the thermal inertia of such equipment also became greater. The purpose of the derivative mode is to predict the process errors before they have evolved and take corrective action in advance. The derivative action is not used by itself, but only in combination with the proportional mode. The output of a PD controller is as follows:

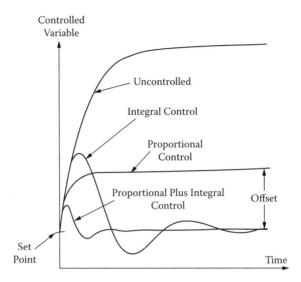

FIGURE 2.31
The different responses of a process variable to a load change in the process when it is uncontrolled and also when it is under P, I, and PI control.

$$m = K_c \left[e + T_d \frac{d}{dt} e \right] + b$$

When the error is rising, the derivative contribution is positive, and when the error is dropping, it is negative. The unit of the derivative setting is the derivative time (T_d). This is the length of time by which the D-mode "looks into the future." In other words, if the derivative mode is set for a time T_d, it will generate a corrective action that is the same size, which is the proportional mode would have generated one derivative time (T_d) later. The longer the T_d setting, the further into the future the D-mode predicts and the larger its corrective contribution is.

The derivative (or rate) settings are in units of time and can be adjusted from a few seconds to up to 10 h or more. Because the derivative mode acts on the rate at which the error signal changes, it can also cause unnecessary upsets because, for example, it will react to the sudden set point changes made by the operator. It will also amplify noise, and will cause upsets when the measurement signal changes occur in steps, as in case of periodic measurements. Therefore, in such situations it should either be avoided or the controller be reconfigured so that the D-mode acts only on the measurement and not the error.

Excessive noise and step changes in the measurement can also be corrected by filtering out changes that occur faster than the maximum speed of response of the process. DCS systems, as part of their software library, are provided with adjustable filters on each process variable.

A special-purpose control action used on extremely fast processes is the so-called *inverse derivative* mode. The output of the inverse derivative mode is inversely proportional to the error's rate of change. It is used to reduce the gain of a controller at high frequencies and is therefore useful in stabilizing a flow loop. Inverse derivative can also be added to a proportional-plus-integral controller to stabilize flow and other loops requiring very low proportional gain for stability. Because inverse derivative is available in a separate unit, it can be added to the loop when stability problems are encountered.

2.6.8 PID Control

Figure 2.32 describes the responses of the individual control modes, and it also shows the responses of various mode combinations to a number of disturbance inputs. Similarly, Figure 2.33 shows the controller responses to set point step changes.

Table 2.34 tabulates the working P, I, and D values in noninteracting, interacting, and parallel PID algorithms. Interacting behavior was unavoidable in the past when pneumatic controllers were used, because their three settings were physically interconnected. The interacting controller had the advantage that the working derivative time (T_d) could never exceed one quarter of the working integral time T_i. This contributes to safety, because if it were possible to make $T_d > T_i/4$, the controller action would reverse and that could cause accidents.

Nowadays, the different DCS and PLC suppliers use a variety of PID control algorithms, including the three most widely used ones listed in Table 2.34.

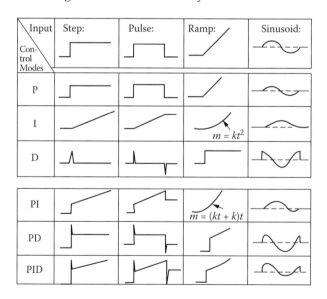

FIGURE 2.32

Responses of the P, I, D control modes and of the PI, PD, and PID controllers to a variety of input disturbances.

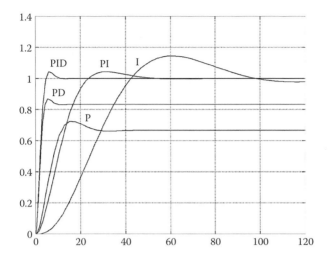

FIGURE 2.33
The unit-step responses of P, I, PD, and PID controllers.

It should be understood that, on the one hand, all of these algorithms are usable but, on the other hand, their settings will mean different things and that the noninteracting and parallel algorithms do need software protection against an operator accidentally setting T_d to exceed $T_i/4$.

2.6.9 Digital and Sample-and-Hold Algorithms

When PID algorithms are implemented digitally, what used to be integration becomes summation, and what used to be differentiation becomes difference. The scan period of DCS systems is fixed at around 0.5 seconds or is selectable for each loop from under 0.1 to over 30 seconds. As digital controllers do not continuously evaluate the measurements, but look at them intermittently, this increases the dead time of the loop by two "scan periods" which are needed to calculate the "present" and the "previous" error.

There are two basic types of digital algorithms in use. One is called positional, meaning that the full output signal to the valve is recalculated every scan period:

$$m = K_c \left(e + \frac{1}{T_i} \sum_o^n edt + \frac{T_d}{\Delta t} \Delta e \right) + b$$

where, Δt is the scan period, Δe is the change in the error since the previous scan, and

$$\sum_o^n e$$

is the sum of all previous errors between time zero and time n.

TABLE 2.34

Conversions and Relationships between the Different PID Algorithms

Types of Algorithms	Standard or Noninteracting	Interacting	Parallel
Equation for Output (m) =	$K\left(e + \dfrac{1}{T}\int edt + T_d\dfrac{de}{dt}\right) + b$	$K_c\left(e + \dfrac{1}{T_i}\int edt + T_d\dfrac{d}{dt}e\right) + b$	$K_c e + \dfrac{1}{T_i}\int edt + T_d\dfrac{de}{dt} + b$
Block Diagram Representation			
Working PB ($100/K_c$)	PB = PB$_0$	$PB = \dfrac{PB_0}{1 + T_{d0}/T_{00}}$	PB = PB$_0$
Working T_i (minutes/repeat)	$T_i = T_{I0}$	$T_i = T_{d0} + T_{I0}$	$T_i = (T_{I0})(K_c)$
Working T_d (minutes)	$T_d = T_{d0}$	$T_d = \dfrac{(T_{d0})(T_{I0})}{T_{I0} + T_{d0}}$	$T_d = T_{d0}/K_c$

Where: PB$_0$, T_{I0}, and P_{d0} are the proportional, integral, and derivative settings on the controller.

The other commonly used algorithm is the velocity algorithm. In that case, the value of the previous output (m) is held in memory, and only the required change in that output is calculated:

$$\Delta m = K_c \left(\Delta e + \frac{1}{T_i} \sum_{0}^{n} e\, \Delta t + \frac{T_d}{\Delta t} \Delta(\Delta e) \right)$$

where $\Delta(\Delta e)$ is the change in the change in the error between the previous and the present scan period.

The positional algorithm is preferred when the measurement is noisy, because it works with the error and not the rate of error change when calculating its proportional correction. Velocity algorithms have the advantages of providing bumpless transfer and less reset windup (0 or 100%), and are better suited for controlling servomotor-driven devices. Their main limitations include noise sensitivity, likelihood to oscillate, and lack of an internal reference.

When the dead time of a control loop exceeds its time constant, the conventional PID algorithms cannot provide acceptable control. This is because the controller cannot distinguish between a nonresponding manipulated variable and the transportation lag (dead time) in the system. For such applications the sample-and-hold type algorithms are used. These algorithms are identical to the previously discussed PID algorithms except that they activate the PID algorithm for only part of the time. After the output signal (m) is changed to a new quantity, it is sealed at its last value (by setting the measurement of the controller equal to its set point), and it is held at that constant value until the dead time of the loop is exhausted.

When the dead time has passed, the controller is switched back to automatic and its output (m) is adjusted based on the new measurement it "sees" at that time. A timer sets the period for which the controller is switched to manual, and that period is adjusted to exceed the dead time of the loop. This way, the sample-and-hold controller has less time to make the same amount of correction as the conventional PID would and therefore its integral setting must be increased in proportion to the ratio of cycle to automatic time periods (Figure 2.35).

Control strategies in digital systems are implemented by an organization of function blocks available from the manufacturer's function block library. Signals from one block to another are designated either graphically or in questionnaire format; other configuration choices are given in response to a questionnaire. A minimum set of configuration questions is given in Table 2.36. Most systems would have many more questions covering loop description, engineering ranges, alarm limits, etc.

2.6.10 Open-Loop PID Tuning

The ideal personality of an ocean liner captain is not the same as that of a supersonic jet pilot. Similarly, the nature of controllers must also match

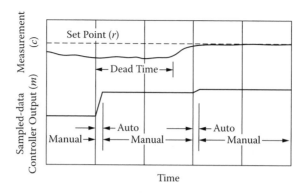

FIGURE 2.35

Sample and hold controllers are periodically switched from manual to automatic and back again. They are used mostly on dead-time processes.

TABLE 2.36

PID Configuration Items (Minimum Set)

Block Identification	Block Tag
Block type	PID
Cascade?	Yes or No
If yes, set point source	Primary block tag
Set point track PV in manual?	Yes or No
Process variable source	AI block tag
Direct or reverse acting	DA or RA
Derivative mode on measurement or error	M or E
Proportional mode on meassurement or error	M or E
Controller gain	K_c
Integral time, minutes/repeat	T_I
Derivative time, minutes	T_D

the dynamic characteristics of the processes they control. This matching is called "tuning," aimed at a control loop that can correct for all types of upsets (including set point changes and load disturbances) quickly and accurately. With the proliferation of tuning software packages, this subject cannot be fully covered here and the reader is referred to Sections 2.35 through 2.38 of the *Instrument Engineers' Handbook* for further details.

The purpose of a controller is to keep the controlled variable as close to its set point as possible. Meeting this objective is a function of the nature of the process, the nature of the disturbances, and of tuning. Disturbances arise from three different sources: load, set point changes, and noise. The controller has no impact on noise, other than possibly amplifying it and passing it onto the final actuator, which can cause excessive wear and ultimate failure.

Set point and load changes affect the behavior of the control loop quite differently, owing to the dynamics in their path. There are no dynamics involved with changing the set point, unless intentionally placed there for purposes of filtering the set point. However, there are always dynamics in the load path. A controller tuned to follow set point changes tends to respond sluggishly to load variations, and a controller tuned to correct for load disturbances tends to overshoot when its set point is changed.

Set point response is of no importance to the majority of control loops; actually, the only loops in a continuous plant which must follow set point changes are flow loops. Controllers that can be transferred "bumplessly" between manual and automatic modes (most do) allow simulating a load change by using that feature. This is done by waiting until the loop is at steady state and on set point (zero deviation). When steady state has been achieved, the next step is switching to the manual mode and stepping the output by the desired amount in the desired direction, and immediately (before a deviation develops) transferring back to the automatic mode. This procedure can be followed for all but the fastest loops, such as flow loops. For them, a step in set point is acceptable, both because flow loops must follow set point changes and because, for them, set point tuning gives acceptable load response.

When a process is at steady state and it is upset by a step change, it usually starts to react after the dead time (Figure 2.37). After the dead time, most processes will reach a maximum speed (reaction rate), then the speed will drop (self-regulating process) or the speed will remain constant (integrating process).

Knowing the change in controller output (ΔCO) and having determined the dead time ($t_d = L_r$) and the reaction rate (R_r), the controller settings are calculated by using the equations in Table 2.38, where Ziegler–Nichols recommends using the ratio of the controller output divided by the product of the slope, and the dead time to calculate the proportional gain. If the reaction rate (slope) is high, then the controller gain must be small, because the process is sensitive and it reacts quickly. If the dead time is long, the controller gain must be small because the process response is delayed and therefore the controller cannot be aggressive. If one can reduce the slope and the dead time of any process, it will be easier to control. The integral and derivative are calculated as a function of dead time.

One of the advantages of open-loop tuning over the closed-loop tuning technique is its speed, because one does not need to wait for several periods of oscillation during several trial-and-error attempts. The other advantage is that one does not introduce oscillations into the process with unpredictable amplitudes. The disadvantages are also multiple. The open-loop test is not as accurate as the closed-loop one, because it disregards the dynamics of the controller. Another disadvantage is that the S-shaped reaction curve and its inflection point are difficult to identify when the measurement is noisy or if a small step change was used. Therefore, it is advisable to refine

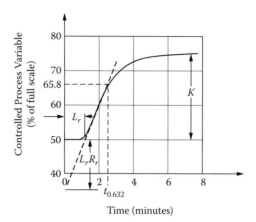

FIGURE 2.37

Reaction curve of a self-regulating process, caused by a step change of one unit in the controller output. $L_r = t_d$ is dead time, R_r is reaction rate, and K is process gain.

TABLE 2.38

Equations for Calculating the Ziegler–Nichols Tuning Parameters for an Interacting Controller

Type of Controller	P (gain)	I (minutes/repeat)	D (minutes)
P	$\dfrac{\Delta CO}{R_r{}^* t_d}$	—	—
PI	$0.9 * \dfrac{\Delta CO}{R_r{}^* t_d}$	$3.33\, t_d$	—
PID	$1.2 \dfrac{\Delta CO}{R_r{}^* t_d}$	$2 t_d$	$0.5\, t_d$

these settings once the system is operating by retuning the loop using the closed-loop method.

The tuning settings based on the process reaction curves obtained by the open-loop tuning method, in addition to the Ziegler–Nichols method (Table 2.38), can also be selected by other methods. Figure 2.39 compares the load responses and Figure 2.40 compares the set point responses of these methods.

To remove oscillations in a control loop, hence to increase the robustness, it is necessary to give up some performance. By reducing the proportional gain in a control loop, the robustness will be increased and the oscillations will be reduced.

Digital control loops differ from continuous control loops because the continuous controller is replaced by a sampler. In such cases, the open-loop tun-

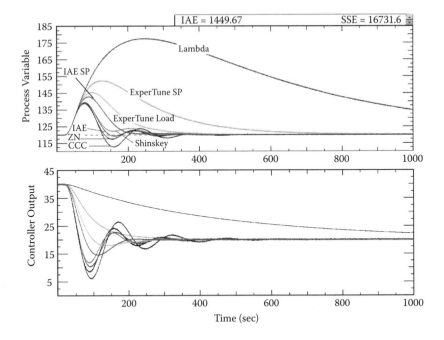

FIGURE 2.39
Load responses of the different tuning techniques (example).

ing methods presented previously may be used, considering that the dead time used is the sum of the true process dead time and one-half of the sampling time.

2.6.11 Closed-Loop PID Tuning

In 1942 Ziegler and Nichols first proposed "the ultimate method," because its use requires the determination of the ultimate gain (sensitivity) and the ultimate period of the loop, when it is in sustained, undamped oscillation. The settings determined by this method are obtained on the basis of load disturbance and are not suitable for set point changes The ultimate gain K_u is the maximum allowable gain of a proportional controller, which will result in a loop gain = 1.0, corresponding to the state of sustained oscillation. The ultimate period (P_u) is the time period of one sinusoidal in that oscillation.

A control loop can be tuned for any damping ratio (Figure 2.41). The most common goal is to obtain quarter-amplitude damping. To do that, the total loop gain must be at 0.5. This means that the product of the gains of all the components in the loop—the process gain, the sensor gain, the transmitter gain, the controller gain, and the control valve gain—must be at 0.5. When the loop is in sustained, undamped oscillation, the gain product of the loop is 1.0 and the amplitude of cycling is constant.

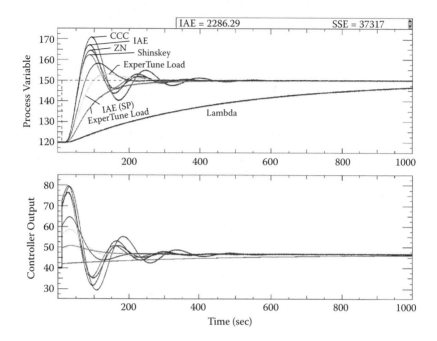

FIGURE 2.40
Set-point responses of the different turning techniques (example).

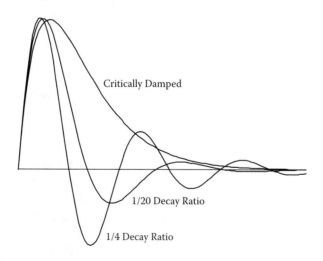

FIGURE 2.41
Responses to equal maximum deviation when the controllers are tuned for the noted three decay ratios.

The period of oscillation of a closed loop depends on the loop dead time. The period of oscillation in flow loops is 1 to 3 seconds; for level loops, it is 3 to 30 seconds (sometimes minutes); for pressure loops, 5 to 100 seconds; for temperature loops, 0.5 to 20 minutes; and for analytical loops, it ranges from 2 minutes to several hours. A proportional loop oscillates at periods ranging from two to five dead times. PI loops oscillate at periods of three to five dead times, and PID loops at around three dead time periods.

Table 2.42 gives the recommended settings for processes with little or no dead time. As the dead time to time constant ratio rises, the integral setting becomes a smaller percentage of the oscillation period. For noninteracting PID loops, as dead time rises to 20% of the time constant, (I) drops to 45% and (D) to 17% of the period. At 50% dead time, (I) = 40% and (D) = 16%. When the dead time equals the time constant, (I) = 33% and (D) = 13%, finally, if dead time is twice the time constant, (I) = 25% and (D) = 12%.

The steps in this tuning sequence are as follows: (1) set integral to maximum and derivative to zero (or minimum), (2) set the gain or proportional band at $K_c = 1$ ($PB = 100\%$) if no better information is available, and (3) once the process variable is stable, move the set point up or down 2% for half a minute and then return it to its original value. If the response is a dampened sinusoidal, increase the controller gain (narrow the PB) and if the response is undamped (run away), reduce the gain. The test is repeated until sustained oscillation is obtained, at which point K_u and P_u are read. Once the values of K_u and P_u are known, one might use the recommendations of Ziegler–Nichols (Table 2.42) or the recommendations that also consider the dead time to time constant ratio (Figures 2.39 and 2.40). No tuning method is perfect, and "fudge factors" based on experience can be very valuable.

This closed-loop tuning method considers the dynamics of all system components and therefore gives accurate results at the load where the test is performed. Another advantage is that the readings of K_u and P_u are easy to read, and the period of oscillation can be accurately read even if the measurement is noisy. On the other hand, when tuning unknown processes, the amplitudes of undamped oscillations can become unsafe or can take a long time to develop. For these reasons, other tuning techniques have also been developed and are described in the *Instrument Engineers' Handbook*.

TABLE 2.42

Tuning Parameters Based on the Measurement of
K_u and P_u Recommended by Ziegler–Nichols for a
Noninteracting Controller

Type of Controller	P (gain)	I (minutes/repeat)	D (minutes)
P	$0.5\,K_u$	—	—
PI	$0.45\,K_u$	$P_u/1.2$	—
PID	$0.6\,K_u$	$P_u/2$	$P_u/8$

The cause of control loop cycling can be that they interact with other loops. When loops interact, it is necessary to make sure that their response speeds are not even similar. To avoid oscillation, one should select response speeds that differ by a factor of 3 to 10, depending on the degree of interaction. In tuning interacting loops, one would place the downstream loops in manual while tuning the "upstream loop," and once the upstream loop's speed of response is determined, use a multiple of that to set the downstream controllers.

Once the "upstream loop" is tuned (for quick response), it stays in automatic and is treated as if it were part of the process, while the first downstream loop is tuned (for moderate response). If there is a second downstream loop it is tuned (for slow response), while the other two are in automatic and are considered to be part of the process.

2.6.12 Tuning by Computer

The tuning steps can all be performed automatically under computer control. The pattern recognition type self-tuning controllers do this, by applying the "open-loop" method of tuning during their "pretune phase" and by using the "closed-loop" method to evaluate the loop performance during upsets. Computers can also perform "model-based" tuning. Several good commercially available packages exist. These computer programs include nonlinear simulation tools, linearizing routines, predictor and estimation techniques, linear quadratic Gaussian methods, and time series and frequency response techniques.

The maintenance of control systems is equally important as is a good workable design. Many good control schemes fall into disuse because of a lack of proper fine-tuning, failure to adapt to changing process requirements, or poor maintenance. As time progresses, the operating processes change in the plant.

In addition to keeping the controllers tuned, other methods are available to improve the quality and reliability of process measurements. Overall process balance calculations and the use of predictor/estimator filters (e.g., Kálmán filters) can help to improve the quality of measurements. These better-quality measurements are contributing to better control of performance, which will be discussed in more detail in the following subsections.

Table 2.43 lists the equations to be used in determining the proportional, integral, and derivative settings for a PID controller tuned for load disturbance using the various criteria listed in Table 2.44. The selection of the performance criteria can be done manually or automatically (per programming).

2.6.13 Cascade Control

Cascade loops consist of two or more controllers in series and have only a single, independently adjustable set point, that of the primary (master) controller. The main value of having secondary (slave) controllers is that they act as the first line of defense against disturbances, preventing these upsets from entering and upsetting the primary process, because the cascade slave

TABLE 2.43

Controller Tuning Criteria Selection

$$ISE - 1 = \int_{0}^{\infty} [c(t) \times c(\infty)]^2 \, dt$$

$$ISE - 2 = \int_{\theta_0}^{\infty} [c(t) - c(\infty)]^2 \, dt$$

$$ISE - 3 = \int_{0}^{\infty} \left[\frac{c(t) - c(\infty)}{c(\infty)} \right]^2 dt$$

$$IAE - 1 = \int_{0}^{\infty} |c(t) - c(\infty)| \, dt$$

$$IAE - 2 = \int_{\theta_0}^{\infty} |c(t) - c(\infty)| \, dt$$

$$IAE - 3 = \int_{0}^{\infty} \left| \frac{c(t) - c(\infty)}{c(\infty)} \right| dt$$

$$ITAE - 1 = \int_{0}^{\infty} |c(t) - c(\infty)| t \, dt$$

$$ITAE - 2 = \int_{\theta_0}^{\infty} |c(t) - c(\infty)| t \, dt$$

$$ITAE - 3 = \int_{0}^{\infty} \left| \frac{c(t) - c(\infty)}{c(t)} \right| t \, dt$$

detects the upsets and immediately counteracts them, so that the primary loop is not even aware that an upset occurred.

In order for the cascade loop to be effective, the slave should be more responsive (faster) than the master. The slave's time constant should be one quarter to one tenth that of the master loop and the slave's period of oscillation should be one half to one third that of the master loop. The goal is to distribute the time constants between the inner (slave or secondary) and outer (master or primary) loops, while making sure that the largest time constant is not placed within the inner loop. When that occurs, such as in the case when the valve has a positioner (the slave) on a fast flow or liquid pressure controller (the master), stability will be sacrificed because the slave (valve)

TABLE 2.44

Tuning Equations for PID Controller to
Correct for a Load Disturbance

$$K_c = \frac{A}{K_u}\left(\frac{t_u}{\tau}\right)^u$$

$$\frac{I}{T_i} = \frac{A}{\tau}\left(\frac{t_0}{\tau}\right)^B$$

$$T_d = \tau A\left(\frac{t_0}{\tau}\right)^B$$

Constants			
Criterion	Controller Mode	A	B
ISE	Proportional	1.495	−0.945
	Reset	1.101	−0.771
	Rate	0.560	1.006
IAE	Proportional	1.435	−0.921
	Reset	0.878	−0.749
	Rate	0.482	1.137
ITAE	Proportional	1.357	−0.947
	Reset	0.842	−0.738
	Rate	0.381	0.995

has the largest time constant in the loop. Therefore, in such cases one would try to either speed up the valve or avoid the use of cascade loops.

Providing external reset for the cascade master from the slave measurement is always recommended. This guarantees bumpless transfer when the operator switches the loop from slave control to cascade control (Figure 2.45). The internal logic of the master controller algorithm is such that as long as its output signal (m) does not equal its external reset (ER), the value of m is set to be the sum of the ER and the proportional correction ($K_c(e)$) only.

An addition to the noted advantages is that the set point of the secondary controller can be limited. In addition, by speeding up the overall cascade loop response, the sensitivity of the primary process variable to process upsets is also reduced, whereas the secondary loop can reduce the effect of control valve sticking or actuator nonlinearity. The primary or outer control loop of a cascade system is usually a PI or PID controller. A properly selected secondary will reduce the proportional band of the primary controller.

Adding a cascade slave to a fast loop can destabilize the primary if most of the process dynamics (time lags) are within the secondary loop. The most common example of this is using a valve positioner in a flow-control loop. The

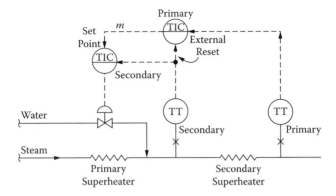

FIGURE 2.45
External reset is provided for the primary controller to prevent integral windup when the secondary controller is in manual.

valve positioner is a secondary controller—a proportional controller usually provided with a fixed band of about 5% (gain of 20)—serving to overcome the effects of changes in stem friction and line pressure. Friction produces hysteresis, which, without a positioner, would degrade performance, particularly where liquid level or gas pressure is being controlled with integral action in the controller. The combination of the natural integration of the process, reset integration in the controller, and hysteresis, can cause a "limit cycle" that is a constant-amplitude oscillation.

Flow as a secondary cannot only overcome the effects of valve hysteresis, but also insures that line pressure variations or badly selected valve characteristics will not affect the primary loop. For these reasons, in composition control systems, flow is usually set in cascade. Cascade flow loops are also useful in feedforward systems. Flow controllers invariably have both proportional and integral modes. If their proportional band exceeds 100%, they must have an integral mode.

In reactor temperature control applications, a slave controlling the jacket outlet temperature is recommended so that the dynamics of the jacket is transferred from the primary to the secondary loop. In temperature-on-temperature cascade systems, such as shown in Figure 2.45, the secondary controller should have little or no integral.

2.6.14 Digital Configuration of Cascade Control

Control strategies in digital systems are implemented by an organization of function blocks from the manufacturer's function block library. Signals from one block to another are designated either graphically (or in questionnaire format); the signals which must be passed are as follows:

- Primary process measurement from an analog input (AI) block to the primary PID

- Secondary process measurement from an AI block to the secondary PID
- Set point from the primary controller output to the secondary PID
- Controller output from the secondary PID to an analog output (AO) block

It is also desirable to pass status information between the blocks. For example, if the secondary block is manual or local automatic, the primary controller should be aware of this. Furthermore, the set point of the secondary should be passed back to the primary controller for initialization purposes; thus when the secondary is returned to cascade automatic, no bump will be created in the primary controller output.

The interblock communication can be implemented several ways, depending on the manufacturer. One way is to configure specific forward and backward connections as shown in Figure 2.46. The forward connections are as follows:

AI "OUT" to PID "IN"—floating point signal values

PID (primary) "OUT" to PID (secondary) "IN"—floating point signal value

PID (secondary) "OUT" to AO "IN"—floating point signal value

PID (secondary) "TCO" to PID (primary) "TCI"—Boolean signal

PID (secondary) "TRO" to PID (primary) "TRI"—Boolean signal

Whenever the secondary controller is in manual or on local set point, the status is communicated backward to the primary controller via the TCO–TCI communication link. At the same time, the set point of the secondary controller is communicated backward to the primary controller via the TRO–TRI communication link. The status notification puts the primary PID into an initialization mode. This forces the primary controller output to be the same as the secondary controller set point. (The secondary controller should be configured so that its set point tracks its measurement whenever it is in manual.)

A minimum set of configuration questions is given in Table 2.36. Most systems would have many more questions covering loop descriptions, engineering ranges, alarm limits, etc.

2.6.15 Ratio Control

Ratio controls are used primarily for blending ingredients into a product such as the addition of a gasoline additive to gasoline, in order to maintain the required octane number of the product, a number that may or may not be measured. The load, or wild flow, as it is called, may be uncontrolled, controlled independently, or manipulated by other controllers that respond to the variables, such as pressure, level, etc. Maintaining a ratio R between ingredients A and B can be done by manipulating the set point of stream B to keep it equal to RA (Figure 2.47).

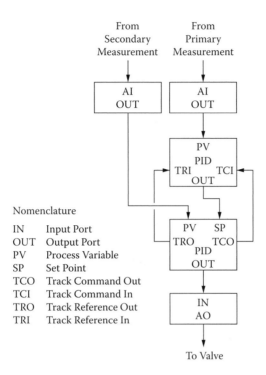

FIGURE 2.46
Function block structure for cascade control.

An alternate method is to calculate the ratio R from the individual measurements of flows A and B, and use this calculated ratio as the measurement input into a manually set ratio controller (RIC). In this case the loop gain varies with both the ratio R and the flow B. Because the loop gain varies inversely with flow B, this can cause instability at low rates. Therefore, the use of equal-percentage valve characteristics is essential to overcome this danger (Figure 2.48).

No matter what portion of the ratio control loop is implemented in hardware or in software, a computing element must be used whose scaling requires some consideration. The ratio station (FY in Figure 2.50) normally has a gain range of about 0.3 to 3.0. The primary flow signal F_p, in percent of scale, is multiplied by the gain setting to produce a set point for the secondary flow controller (F_s) in percent of scale. The true flow ratio must take into account the scales of the two flow meters. The setting of the ratio station is related to the true flow ratio by: $R = $ (true flow ratio)(scale of F_p)/(scale of F_s).

When head-type flow meter signals are used and the scale is linear, such scale should use the square root of R in order to be meaningful. As shown by Table 2.49, the available range of ratio settings is seriously limited when using squared flow signals.

Figure 2.50 shows a combination of cascade and ratio controls of blending fluids A and B and sending the mix into a blend tank. In this process, the

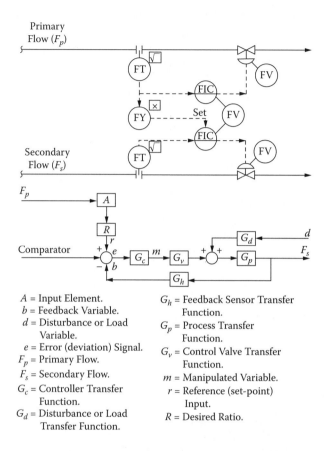

A = Input Element.
b = Feedback Variable.
d = Disturbance or Load
     Variable.
e = Error (deviation) Signal.
F_p = Primary Flow.
F_s = Secondary Flow.
G_c = Controller Transfer
     Function.
G_d = Disturbance or Load
     Transfer Function.

G_h = Feedback Sensor Transfer
     Function.
G_p = Process Transfer
     Function.
G_v = Control Valve Transfer
     Function.
m = Manipulated Variable.
r = Reference (set-point)
     Input.
R = Desired Ratio.

FIGURE 2.47
If the ratio calculation is made outside the secondary flow control loop, its setting does not change the dynamics or response of the loop.

liquid level in the blend tank is affected by total flow, hence the liquid level controller sets flow *A*, which in turn sets flow *B* proportionately. Whenever the slave of a cascade system is a head-type flow controller, the square root must be removed (linear flow signal used) in order to maintain stability.

Conversely, composition is not affected by the absolute value of either flow, only by their ratio. Therefore, to make a change in composition, the AIC controller must adjust the ratio set point of the multiplier (FY). To minimize the interaction of the composition controller with the blend tank level control (through its manipulation of flow *B*), the smaller of the two streams should be controlled by level.

2.6.16 Model-Based Controls

The basic concept of a model-based control (MBC) is illustrated in Figure 2.51 where the "controller," C_I, is the inverse of a model of the process

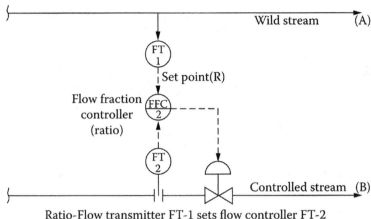

Ratio-Flow transmitter FT-1 sets flow controller FT-2

FIGURE 2.48
Controlled stream is manipulated to maintain a constant ratio (R) of controlled flow (B) to wild flow (A).

TABLE 2.49

Comparing the Actual Ratio Settings Required for Linear and Head-Type Flowmeters to Obtain the Same Flow Ratio

| | Actual Ratio (Gain) Setting Required to Achieve the Desired Ratio | |
Flow Ratio Desired	If Signals Are Linear	If Flow-Squared Signals Are Used and the Scale Is Linear
0.6	0.6	0.36
0.8	0.8	0.64
1.0	1.0	1.0
1.2	1.2	1.44
1.4	1.4	1.96
1.6	1.6	2.56

to calculate the manipulated variable (m) that should make the controlled variable (x) be on set point. The controller does not have PID components, and there is only one tuning parameter per controlled variable. MBC determines the process input that causes a desired process output response; this is the model inverse.

Few controller models are perfect; therefore, they require some form of feedback correction. In Figure 2.51 the difference (d) between the model (C_M) and the process output (x) is monitored and used to adjust a controller feedback (c), which usually is either a bias to the set point (x_{sp}) or a model coefficient. The controller consists of the three functions C_I, C_M, and C_A, enclosed by the dashed line. MBCs are generally classified by model type.

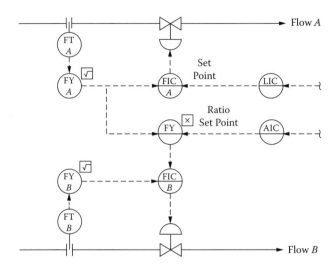

FIGURE 2.50
The level controller manipulates flow rate and the composition controller manipulates flow
ratio.

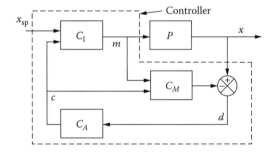

FIGURE 2.51
The three MBC functions are: inverse, model, and adjustment.

Model predictive control (MPC) was developed in the 1970s and 1980s
to meet control challenges of refineries. The advantages of MPC are most
evident when it is used as a multivariable controller integrated with an
optimizer. The greatest MPC benefits are realized in applications with
dead-time dominance, interactions, constraints, and the need for optimization. As opposed to a traditional control loop, where the controller
responds to a difference (error) between the set point and measurement,
the predictive controller uses a vector difference between the future trajectory of the set point and the predicted trajectory of the controlled variable
as its input (Figure 2.52).

MPC is used mainly for multivariable and highly interactive processes.
Processes are tested by applying a special pulse test sequence instead of a
single step. Then, a process model is built from process test data for every

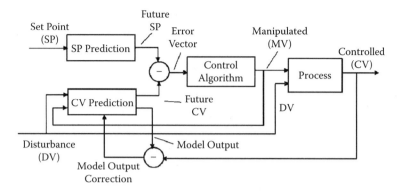

FIGURE 2.52
Model predictive control operation diagram. (From the Instrumentation, Systems, and Automation Society, 2002. All rights reserved. Used with permission of ISA.)

output that equals the number of collected samples. The number of collected samples in test data is normally significantly higher than the number of unknown model coefficients. Such equations are solved using the least-squares technique. The least-squares technique finds the coefficients' fit optimally for all equations in such a way that the total squared error for all equations is minimal.

Figure 2.53 illustrates a typical procedure of MPC application development consisting of the following basic steps:

1. Process analysis
2. MPC configuration development
3. Process testing
4. Process model development and controller generation
5. MPC simulation and tuning validation
6. MPC control evaluation and tuning adjustment

Model predictive control and optimization are the primary techniques for achieving high-performance unit operations. Modern MPC products, especially those integrated with DCS, are easy to apply and use. Good process understanding, however, is essential in setting the control objectives, designing, and commissioning an MPC application.

Prediction and adaptation are combined in model-based predictive control (MBPC). There are such applications as the better control of boiler chemicals as well as optimal dispatching of multiunit cogeneration facilities. More and more plants designed for base load service are being used to meet peaking power demand. Steam turbines designed to operate at steady loads are being cycled to meet peaking load. This can induce large thermal stresses in both the turbine and the boiler as a result of steep steam-to-metal temperature gradients that develop during rapid loading or unload-

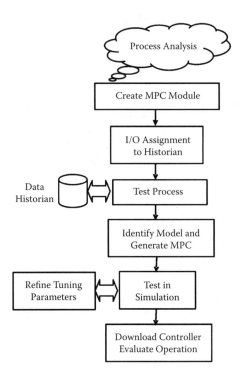

FIGURE 2.53
A typical procedure for MPC application. (From the Instrumentation, Systems, and Automation Society, 2002. All rights reserved. Used with permission of ISA.)

ing of the turbine. As shown in Figure 2.54, MBPC can be used to predict loads in power plant peak-shedding applications.

2.6.17 Model-Free Adaptive and Artificial Neural Network Control

Model-free adaptive (MFA) control does not require process models. It is most widely used on nonlinear applications because they are difficult to control, as there could be many variations in the nonlinear behavior of the process. Therefore, it is difficult to develop a single controller to deal with the various nonlinear processes. Traditionally, a nonlinear process has to be linearized first before an automatic controller can be effectively applied. This is typically achieved by adding a reverse nonlinear function to compensate for the nonlinear behavior so that the overall process input–output relationship becomes somewhat linear. It is usually a tedious job to match the nonlinear curve, and process uncertainties can easily ruin the effort.

The nonlinear MFA controller is a general-purpose controller that provides a more uniform solution to nonlinear control problems. Figure 2.55 illustrates how a multivariable MFA control system works with a two-input–two-output (2 × 2) system, which consists of two controllers (C_{11}, C_{22}, and

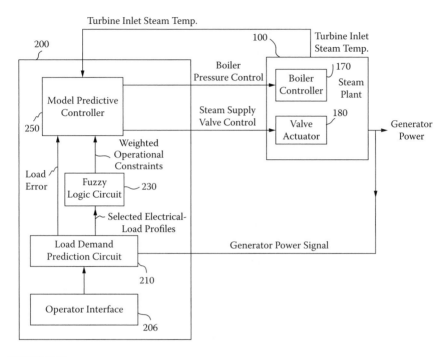

FIGURE 2.54
Patent for MBPC design used in peak-shedding applications to predict loads based on extended load profiles.

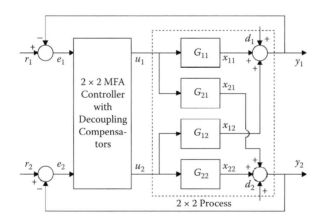

FIGURE 2.55
Two-input two-output MFA control system.

two compensators (C_{21}, C_{12}), and a process consisting of four subprocesses: G_{11}, G_{21}, G_{12}, and G_{22}.

In a simplistic sense, artificial neural networks (ANNs) may be thought of as a functional mapping of inputs connected to outputs through an interconnected network of nodes. The weights given to the node interconnections are varied to achieve the desired mapping. The advantages of neural networks lie in their ability to learn arbitrary function mapping with little or no prior knowledge about the function itself. Thus, they provide the capability to do "black-box modeling" of a process for which only input/output data is given. In addition, these networks have been shown to be robust, resilient, and capable of adaptive learning. The disadvantage of this model is that the knowledge is implicit in the network connection weights.

Many control systems are not amenable to conventional modeling approaches due to the lack of precise formal knowledge, strongly nonlinear behavior, a high degree of uncertainty, or time-varying characteristics. Computational intelligence, the technical umbrella of ANNs, has been recognized as a powerful tool which is tolerant of imprecision and uncertainty.

Among the techniques of computational intelligence, ANNs attempt to mimic the structures and processes of biological neural systems. According to the type of available information, four types of models are used:

1. *White-box or first-principle modeling.* A dynamic model for well-understood processes derived from mass, energy, and momentum balances.

2. *Fuzzy logic modeling.* A linguistically interpretable rule-based model, based on the available expert knowledge and measured data.

3. *Black-box or empirical modeling.* A model structure for processes when no physical insight is available and the parameters of the model are identified based on actual input–output data collected during operation.

4. *Gray-box modeling.* ANNs can learn complex functional relations by generalizing from a limited amount of training data. When part of the process is well understood, white- and black-box models can be combined and called *gray-box modeling.*

The common characteristic of fuzzy logic and neural networks is that one does not need to know anything about the mathematical model of the process in order to utilize them. In a way it is like the tennis player who can hit the ball without the in-depth knowledge of Newton's laws of motion and how these laws apply to the tennis process. A fuzzy logic controller just mimics the operator (the tennis player) in its responses.

Neural networks are similar to fuzzy logic insofar as the mathematical model relating the inputs to the outputs of the process need not be known. It is sufficient to know the process response (as in tennis). The major difference between fuzzy logic and neural networks is that neural networks can only be trained by data, but not with reasoning. This is different, because in fuzzy logic each controller parameter can be modified in terms of both its gain (its importance) and its functions.

ANNs can be looked upon as tools for examining data and building relationships. ANNs have found applications in the process industry for capturing relationships among measurements and modeling. Compared to the technology of expert systems, the technology of ANNs is immature and still evolving. Advantages of neural networks include the following:

- Their good fit for nonlinear models
- Ability to adapt, generalize, and extrapolate results
- Speed of execution in recall mode
- Ease of maintenance

However, neural networks have the following inherent disadvantages:

- Cannot handle constraints alone
- Cannot optimize
- Need lots of data
- Need lots of training (learning) sessions
- Are unpredictable for utilization in "untrained" areas
- Are not well understood and, therefore, are not yet widely accepted

A neural network processing element has many input paths. These input paths are individually multiplied by a weight and then summed. The output path is then connected to input paths of other processing elements through connection weights. These weights and connections form the "memory" or knowledge of the neural net. Because each connection has a corresponding weight, the signals on the input lines to a processing element are modified by these weights prior to being summed. Thus, the summation function is a weighted summation.

A neural network consists of many processing elements joined together. A typical network consists of a sequence of layers with full or random connections between successive layers. A minimum of two layers is required: the input buffer where data is presented and the output layer where the results are held. However, most networks also include intermediate layers called hidden layers. An example of such an ANN network is one used for the indirect determination of the Reid vapor pressure (RVP) and the distillate boiling point (BP) on the basis of 9 operating variables and the past history of their relationships to the variables of interest (Figure 2.56).

2.6.18 Adaptive and Optimizing Controls

Adaptive control can be thought of as a system that automatically designs a control system. The adaptation can result in automatically scheduling or

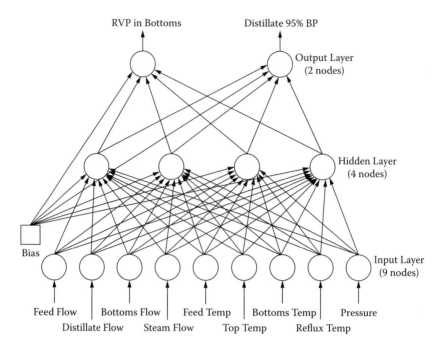

FIGURE 2.56
Example of a three-layer back-propagation network describing the application of physical property prediction.

changing controller gains in a feedback or feedforward manner, and can be used for self-tuning in pattern recognition or in model reference adaptive control (MRAC). Figure 2.57 illustrates an application for controller gain switching in a nonlinear process.

The process-gain characteristic of the process illustrated in Figure 2.57 is of a typical strong acid–strong base neutralization process. The control set point is at pH = 7. The process has extremely high gain at this point (i.e., a small amount of reagent change results in a dramatic change in pH). The required controller gain near neutrality must be small for stability. If the pH measurement were to move above or below this region, the low gain would result in a very sluggish response; therefore the gain is switched to a high value.

MRACs are composed of a reference model that specifies the desired performance; an adjustable controller whose performance should be as close as possible to that of the reference model; and an adaptation mechanism. This adaptation mechanism processes the error between the reference model and the real process to modify the parameter of the adjustable controller accordingly.

Pattern recognition self-adaptive controllers exist that *do not* explicitly require the modeling or estimation of discrete time models. These controllers adjust their tuning based on the evaluation of the system's closed-loop response characteristics (i.e., rise time, overshoot, settling time, loop damp-

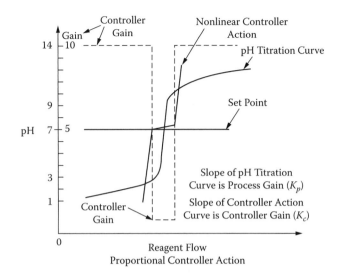

FIGURE 2.57
Variable breakpoint nonlinear control action is used for strong acid–strong base neutralization.

ing, etc.). They are microprocessor-based and usually heavily constrained with regard to the severity of allowable tuning parameter adjustments. They are gaining fair acceptance in operating plants.

The classical adaptive control scheme is shown in Figure 2.58. Its goal is to use online identification through artificial intelligence (AI), neural networks, and fuzzy logic to adapt the model to the actual process. AI and model predictive control (MPC) can tolerate inaccuracy and uncertainty in the model, and online training can continuously improve the model.

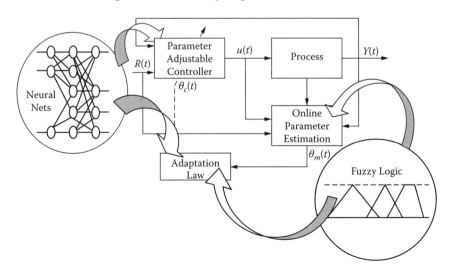

FIGURE 2.58
Intelligent identification and tuning in adaptive control.

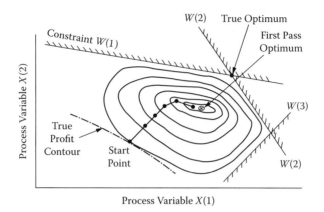

FIGURE 2.59
Gradient search using created response surface (first pass).

The goal of optimization is usually maximum profitability, which is most often obtained if one can minimize energy use and maximize production. Figures 2.2 and 2.12 already gave some examples of optimization strategies, which usually involve the finding of a minimum or maximum point on a curve or on a surface.

Any process in which the cost has two components with opposite slopes relative to the load will have a total operating cost curve with a minimum point. In other processes the optimum condition cannot be described by a curve (two dimensions) but requires the climbing and evaluation of multiple peaks. The methods of searching for minimum and maximum points include both continuous and sampling techniques.

Two types of nonlinear optimizers—the sectionalized linear program and the gradient search—have been successfully implemented in advanced computer control schemes.

The sectionalized linear program is especially useful for those processes that exhibit slight-to-moderate nonlinearities in their variable relationships.

An alternative to the sectionalized linear program is the use of a gradient or "hill-climbing" approach. In contrast to the linear program approach, which considers only one variable at a time in its serial and sequential search for the optimum, the gradient methods generally simultaneously perturb all the independent variables. Once the constraint has been recognized, the algorithm must attempt to move along this boundary.

A typical profit surface is shown in Figure 2.59. Note that a fictitious hill is created whose summit is within the feasible operating range. Standard gradient search methods are then used to find the peak. Obviously the peak is not at the true optimum as defined by the intersection of the constraints. If the true value of the objective function is greater than the value at the starting point, then a recentering is employed. The profit contours passing through the peak are assigned a value of zero. This, along with the zero-valued constraints, permits the creation of a second hill with a summit that is much closer to the true optimum. This recentering is repeated until the improvement in the true objective function is less than a desired value.

2.7 Control Valves

In thermal solar power plants, high-temperature oil flows need to be controlled; in hydrogen processes, very-low-temperature LH_2 flows have to be controlled. For these reasons, the emphasis in this treatment will be on control valve designs that are suited for these applications and on phenomena (such as noise and cavitation) that are common in these applications.

All control loops consist of three equally important components: the sensor, the controller, and the final control element, which usually is a throttling valve. Control valves provide a means of adjusting the flows of process or heat transfer fluids and thereby serve to stabilize the process's response to material and heat balance variations. They manipulate these flows by changing their openings and thereby modifying the amount of energy needed for the flow to pass through them. As a control valve closes, its pressure differential rises and therefore its flow is reduced. In order for a constant-speed pump to generate this increased pressure, it has to reduce its flow.

The pumping energy that is invested to overcome the valve differential is wasted energy. The amount of this waste is the difference between the pressure required to "push" (transport) the fluid into the process (see ΔP in Figure 2.60) and the pump curve of the constant-speed pump. Pumps are selected to be able to meet the maximum possible flow demand, and therefore, most of the time they operate at partial loads. Consequently, using control valves to manipulate the flow of constant-speed pumps wastes energy and thereby increases operating cost.

Therefore, in connection with flow control system design, the first decision is whether control valves or variable-speed pumps should be used. In case of the second, flow is reduced by reducing the speed, and therefore, instead of the valve burning up the unnecessarily introduced pump head, that energy is not introduced in the first place. This reduces the operating cost of the process, but increases the capital investment, because the cost difference between variable and constant-speed pumps is usually more than the cost of control valves.

2.7.1 Selection and Characteristics

If the cost–benefit analysis or user preference points toward the use of throttling valves, the next task is to select the right valve type for the application. As shown in Table 2.61, the various valve designs have different pressure

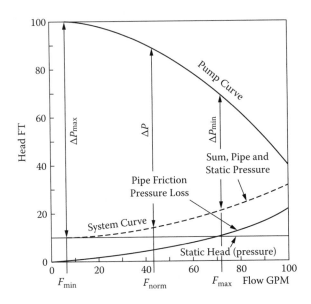

FIGURE 2.60
The difference between the pump discharge pressure curve and the system curve (which is the sum of the static head and the pipe friction loss) is the available valve differential.

and temperature ratings, costs, capacities ($C_d = C_v/d^2$, where C_d is the discharge coefficient, C_v the valve coefficient, and d the valve diameter), etc. One selects the appropriate valve type on the basis of these features, including the required valve characteristics, materials of construction, and compatibility with the process fluid.

When the type of valve has been selected, the next task is to determine the required size and characteristics. The valve characteristic determines the relationship between valve stroke (control signal received) and the flow through the valve. The gain of a control valve is the ratio between the percentage change in the control signal and the resulting percentage change in flow through the valve (G_v = [GPM]/[% stroke]). If the result of a 1% change in the stroke (control signal) results in the same amount of change (such as, say, [5 GPM]/[%]), no matter how open the valve is, the valve is a linear valve (straight line in Figure 2.62).

If a 1% change in the valve stroke results in the same percentage (percent of the flow at that opening) change, the valve characteristic is called *equal percentage* (=%). If the valve characteristic is equal percentage, the change in flow is a small quantity when the flow is small (valve is nearly closed), and it becomes a larger and larger quantity as the valve opens. In the case of quick opening (QO) valves, the opposite is the case; at the beginning of the stroke, the flow increases quickly, and toward full opening, very little. Therefore, the gain (G_v = GPM/%) of a linear valve is constant, the gain of an equal percentage valve increases at a constant slope, and the gain of a QO valve drops as the valve opens.

TABLE 2.61

Orientation Table for Selecting the Right Control Valves for Various Applications

Control Valve Types

Features & Applications	Ball: Conventional	Ball: Characterized	Butterfly: Conventional	Butterfly: High-Performance	Digital	Globe: Single-Ported	Globe: Double-Ported	Globe: Angle	Globe: Eccentric Disc	Pinch	Plug: Conventional	Plug: Characterized	Saunders	Sliding Gate: V-Insert	Sliding Gate: Positioned Disc	Special: Dynamically Balanced
Features:																
ANSI class pressure rating (max.)	2500	600	300	600	2500	2500	2500	2500	600	150	2500	300	150	150	2500	1500
Max. capacity (C_d)	45	25	40	25	14	12	15	12	13	60	35	25	20	30	10	30
Characteristics	F	G	P	F,G	E	E	E	E	G	P	P	F,G	P,F	F	F	F,G
Corrosive Service	E	E	G	G	E,G	G,E	G,E	G,E	F,G	G	G,E	G	G	F,G	G	G,E
Cost (relative to single-port globe)	0.7	0.9	0.6	0.9	3.0	1.0	1.2	1.1	1.0	0.5	0.7	0.9	0.6	1.0	2.0	1.5
Cryogenic service	A	S	A	A	A	A	A	A	A	NA	A	S	NA	A	NA	NA
High pressure drop (over 200 PSI)	A	A	NA	A	E	G	G	E	A	NA	A	A	NA	NA	E	E
High temperature (over 500°F)	Y	S	E	G	Y	Y	Y	Y	Y	NA	S	S	NA	NA	S	NA
Leakage (ANSI class)	V	IV	I	IV	V	IV	II	IV	IV	IV	IV	IV	V	I	IV	II
Liquids:																
Abrasive service	C	C	NA	NA	P	G	G	E	G	G,E	E,G	E,G	E,G	NA	E	G
Cavitation resistance	L	L	L	L	M	H	H	H	M	NA	L	L	NA	L	H	M
Dirty service	G	G	F	G	NA	E,G	F	G	E,G	E	G	G	G,E	G	F	F
Flashing applications	P	P	P	F	F	G	G	E	G	F	P	P	F	P	G	P
Slurry including fibrous service	G	G	F	F	NA	E,G	E,G	G,E	E,G	E	G	G	E	G	P	F
Viscous service	G	G	G	G	F	G	E,G	G,E	E,G	G,E	G	G	G,E	F	F	F
Gas/Vapor:																
Abrasive, erosive	C	C	F	F	P	G	G	E	E,G	G,E	E,G	E,G	G	NA	E	E
Dirty	G	G	G	G	NA	G	F,G	G	E,G	G	G	G	G	G	F	G

Abbreviations: A = Available, C = All-ceramic design available, E = Excellent, F = Fair, G = Good, H = High, L = Low, M = Medium, NA = Not available, P = Poor, S = Special designs only, Y = Yes.

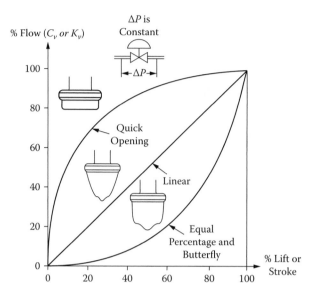

Valve Characteristics Selection Guide

Service	Valve ($\Delta P_{max}/\Delta P_{min}$) Under 2 : 1	Valve ($\Delta P_{max}/\Delta P_{min}$) Over 2 : 1 *but* Under 5 : 1
Orifice-type Flow	Quick-opening	Linear
Flow	Linear	Equal %
Level	Linear	Equal %
Gas Pressure	Linear	Equal %
Liquid Pressure	Equal %	Equal %
Temperature	Equal %	Equal %

FIGURE 2.62

The most common control valve characteristics (top) and the recommended valve characteristics selections (bottom) for a variety of applications.

Table 2.61 lists the inherent valve characteristics of the various types of control valves. The characteristics that are recommended are a function of the intended application, and are listed in the tabulation at the lower part of Figure 2.62. It should be noted that the listed valve characteristics assume that the valve pressure drop is constant. Unfortunately, in most applications (Figure 2.60), the available valve pressure differential is not constant but drops as the load (flow) increases. This is the reason why the recommended valve characteristics in Figure 2.62 are different if the ratio of the maximum to minimum pressure differential is above or below 2:1.

If the valve $\Delta p_{max}/\Delta p_{min}$ is under 2:1, the general recommendation is to use QO for orifice-type flow; linear for nonorifice flow, level, or gas pressure; and equal percentage for liquid pressure or temperature applications. If that ratio is above 2:1, use linear for orifice-type flow and equal percentage for all other applications.

If after start-up, the control loop tends to oscillate at low flows but is sluggish at high flows, one should switch the valve trim characteristics from linear to equal percentage. Conversely, if oscillation is encountered at high flows and sluggishness at low flows, the equal-percentage valve trim should be replaced with a linear one. Changing the valve characteristics can also be done (more easily) by characterizing the control signal to the actuator instead of replacing the valve trim.

One approach to characterizing an analog control signal is to insert a divider or a multiplier into the control signal line. By adjusting the zero and span, a complete family of curves can be obtained. A divider is used to convert an air-to-open, equal-percentage valve into a linear one, or an air-to-close linear valve into an equal-percentage one. Similarly, a multiplier can be used to convert an air-to-open linear valve into an equal-percentage valve or an air-to-close, equal-percentage valve into a linear one.

2.7.2 Distortion and Rangeability

Figure 2.63 illustrates the effect of the distortion coefficient (D_c) on the characteristics of a linear and an equal-percentage valve. As the ratio of the minimum to maximum pressure drop increases, the D_c drops and the equal-percentage characteristics of the valve shift toward linear and the linear characteristics shift toward QO. In addition, as the D_c drops, the controllable minimum flow increases, and therefore, the rangeability (the flow range within which the valve characteristic remains as specified) of the valve also drops.

The conventional definition of *rangeability* is the ratio between the maximum and minimum controllable flow through the valve. Minimum controllable flow (F_{min}) is *not* the leakage flow (which occurs when the valve is closed), but the minimum flow that is still controllable and can be changed up or down as the valve stroke is changed.

Using this definition, manufacturers usually claim a 50:1 rangeability for equal-percentage valves, 33:1 for linear valves, and about 20:1 for quick-opening valves. These claims suggest that the flow through these valves can be controlled down to 2, 3, and 5% of maximum. However, it can be seen in Figure 2.63 that the F_{min} rises as the distortion coefficient (D_c) drops and at a D_c of 0.1, for example, the 50:1 rangeability of an equal-percentage valve drops to about 10:1.

Therefore, the rangeability should be defined as the flow range over which the actual installed valve gain stays within ±25% of the theoretical (inherent) valve gain (GPM/% stroke). Figure 2.64 shows that the actual gain of an equal percentage valve, for example, starts to deviate more than 25% when the flow reaches about 65%. Therefore, in determining the rangeability of such a valve, the maximum allowable flow should be 65%. Actually, if one uses this definition, the rangeability of an equal-percentage valve is seldom more than 10:1 and the rangeability of a linear valve can be greater than that of an equal-percentage valve. Also, the rangeability of some rotary valves can be higher

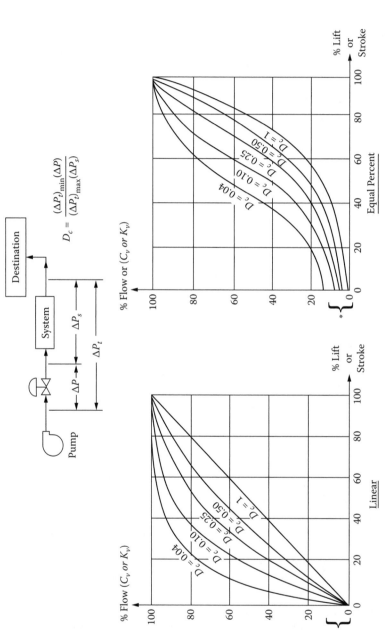

FIGURE 2.63

These figures illustrate the effects of the distortion coefficient (D_c) on inherently linear (left) and inherently equal-percentage valves (right), according to Boger.

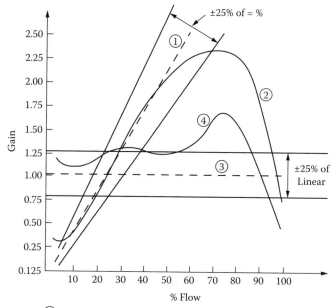

① Theoretical gain characteristics of equal % valve
② Actual, inherent gain characteristics of equal % valve
③ Theoretical gain characteristics of linear valve
④ Actual, inherent gain characteristics of linear valve

FIGURE 2.64
Theoretical vs. the actual characteristics of a 2 in. (50 mm) cage-guided globe valve, according to Driskell.

because their clearance flow tends to be lower, and their body losses near the wide-open position also tend to be lower than those of other valve designs.

In order to stay within ±25% of the theoretical valve gain, the maximum flow should not exceed 60% in a linear valve and 70% of full flow in an equal-percentage valve. In terms of valve *lift*, these flow limits correspond to 85% for an equal-percentage and 70% for linear valves.

2.7.3 Valve Gain and Stability

The gain of any device is its output divided by its input. The characteristics, rangeabilities, and gains of control valves are interrelated. For a linear (constant gain) valve, the gain (G_v) is the maximum flow divided by the valve stroke as a percentage ($F_{max}/100\%$). As was discussed in Sections 2.6.11 and 2.6.12, most control loops are tuned for quarter-amplitude damping. This amount of damping (reduction in the amplitude of succeeding cycles) is obtained by adjusting the controller gain ($G_c = 100/\%PB$) until the total loop gain (the product of the gains of all the loop components) reaches 0.5 (Figure 2.65).

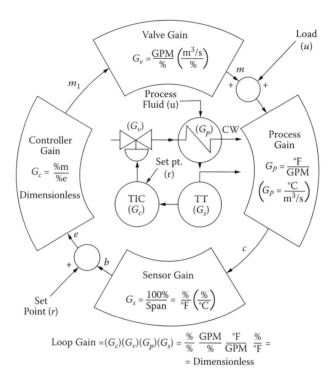

FIGURE 2.65
The loop gain is the product of the gains of the loop component. In a properly tuned loop (decay ratio of 1:4), this gain product should be constant at 0.5.

The gain components of a linear controller (plain proportional) and a linear transmitter (for temperature $G_s = 100\%/°F$) are both constant. Therefore, if the process gain ($G_p = °F/GPM$) is also constant, a linear valve is needed to maintain the total loop gain at 0.5 ($G_v = 0.5/G_pG_cG_s = $ constant).

If the transmitter is nonlinear (sensor gain increases with flow), as in the case of an orifice d/p cell, one can correct for that nonlinearity by using a nonlinear valve whose gain drops as the flow increases (quick opening). Conversely, in other cases the process gain (G_p) drops with load, as is the case with heat transfer processes, where, because the heat transfer area is fixed, the efficiency of heat transfer (process gain G_p) drops as the amount of heat to be transferred over the fixed area rises. To compensate for this nonlinearity (drop), the valve gain (G_v) must rise with load. Therefore, an equal-percentage valve should be selected for all temperature control applications (in lower part of Figure 2.62).

One effective method of keeping the valve gain (G_v) perfectly constant is to replace the valve with a linear flow control loop. The limitation of this cascade configuration (in addition to its higher cost) is that if the controlled process is faster than the speed of response of the flow loop, cycling will occur. This is because the slave—in this case, the flow control loop—in any cascade system must be faster than its master. The only way to overcome this cycling is to slow down (detune) the master by lowering its gain (increasing the proportional band), which in turn degrades control quality. Therefore, replacing the valve with a linear flow control loop should only be considered on such slow loops as temperature.

2.7.4 Positioners

The positioner is a high-gain (0.5 to 10% proportional band), proportional-only, valve stroke position controller. Its set point is the controller output signal. The main purpose of having a positioner is to guarantee that the valve does in fact move to the position that the controller requires.

The addition of a positioner can correct for changes in packing friction due to dirt buildup, corrosion, or lack of lubrication; variations in the dynamic forces of the process; or nonlinearities in the valve actuator. In addition, the positioner can allow for split-ranging the controller signal between several valves or can increase the actuator speed or thrust by increasing the pressure or volume of the actuator air signal. This is only the case when using pneumatic actuators. In case of constant-speed electric actuators, both the speed and the thrust are fixed. Positioners can also modify the valve characteristics by the use of cams and electronic function generators or by resident software.

A positioner will be helpful on most slow loops that are controlling analytical properties, temperature, liquid level, blending, slow flow, and large-volume gas flows. A controlled process can be considered "slow" if its period of oscillation is three times the period of the oscillation of the positioned valve.

Positioners are also useful to overcome the "dead band" that has been caused by valve stem friction. This occurs because whenever the direction of the control signal is reversed, the stem remains in its last position until the friction force is exceeded. Positioners eliminate this limit cycle by closing a loop around the valve actuator. Integrating processes such as liquid-level, volume (as in digital blending), weight (not weight-rate), and gas pressure are prone to limit cycling and can usually benefit from the use of positioners.

In the case of fast loops (fast flow, liquid pressure, small-volume gas pressure), positioners are likely to degrade loop response and cause limit cycling, because the positioner (a cascade slave) is not faster than the speed at which its set point (the control signal) can change. A controlled process is considered "fast" if its period of oscillation is less than three times that of the positioned valve.

Split-ranging control valves does not necessarily require the use of positioners, because one can also split-range the valves through the use of different spring ranges in the valve actuators. If the need is only to increase the speed or the thrust of a pneumatic actuator, it is sufficient to install an air volume booster or a pressure amplifier relay, instead of using a positioner. If the goal is to modify the valve characteristics on fast processes, this should not be done by the use of positioners, but by installing dividing or multiplying relays in the output of analog controllers or characterizing the output of digital ones.

2.7.5 Self-Diagnosing and Flow-Sensing Smart Valves

Improved performance can be provided by intelligent and self-diagnosing positioners and control valves. The detection and correction of the wearing of the trim, packing friction, air leakage in the actuator, and changes in valve characteristics can all be automated. If the proper intelligence is provided, the valve can compare its own behavior with its past performance, and when the same conditions result in different valve openings, it can conclude, for example, that its packing needs to be lubricated or the valve port is getting plugged. In such cases it can automatically request and schedule its own maintenance.

A traditional valve positioner serves only the purpose of keeping the valve at an opening that corresponds to the control signal. Digital positioners can also collect and analyze valve position data, valve operating characteristics, and performance trends, and enable diagnostics of the entire valve assembly. The control signals into smart positioners can be analog (4–20 mA) or digital (via bus systems). The advantages of digital positioners include better accuracy (0.1–1% versus 0.3–2% for analog), improved stability (about 0.1% compared to 0.175%), and wider rangeability (up to 50:1 compared to 10:1).

Smart valves should also be able to measure their own inlet, outlet, and vena contracta pressures; flowing temperature; valve opening (stem position); and actuator air pressure. Valve performance monitoring includes the detection of "zero" position and span of travel, actuator air pressure versus stem travel, and the ability to compare these against that of a new valve. Major deviations from the "desired" characteristic can be an indication of the valve stuffing box being too tight, the valve stem being corroded, or the actuator spring being damaged.

Additional features offered by smart valves might include the monitoring of packing box or bellows leakage by "sniffing," using miniaturized chemical detectors, checking seat leakage by measuring the generated sound frequency, or by comparing the controller output signal at "low flow" with the signal when the valve was new. Another important feature of digital positioners is their ability to alter the inherent characteristics of the valve by the use of software.

A control valve can also be viewed as variable-area flow meter. Therefore, smart valves can measure their own flow by solving the appropriate valve-sizing equation. For example, in the case of turbulent liquid flow applications, where the valve capacity coefficient

$$C_v = \frac{q}{F_p} \sqrt{\frac{G_f}{\Delta P}}$$

the flow (q) can be obtained by inserting the known values of C_v, G_f (specific gravity), ΔP, and the piping geometry coefficient (F_p).

Naturally, in order for the smart valves of the future to be able to accurately measure their own flow, they must be provided with sufficient intelligence to identify the applicable sizing equation for the particular process.

2.7.6 Sizing of Control Valves

For a complete treatment of the many aspects of control valve sizing, refer to the *Instrument Engineers' Handbook*. For some basic equations to calculate the required valve coefficients (C_v) for various liquid and gas flow applications using different valve types, refer to Tables 2.66 and 2.67.

2.7.7 Applications—High Pressure

The processing and transportation of hydrogen requires very-high pressure control valves, which must have high physical strength and special seals. For this reason, higher-strength materials are used with stems that have increased diameters, are short, and are well-guided (Figure 2.68). High-pressure services also require special attention to noise and vibration considerations.

High operating pressure frequently involves high pressure drops. With high flow rates and high pressure drops, a large amount of energy is dissipated in turbulence and radiated as noise. Liquid droplets may also develop and cause erosion and abrasion at the trim or the high-velocity jets leaving the valve can erode downstream piping. Very high forces are developed on the valve body and internal parts, and can cause valve instability. Materials resistant to erosion and abrasion include 440C stainless steel, flame-sprayed aluminum oxide coatings (Al_2O_3), and tungsten carbide.

On the stem, where the unit pressure between it and the packing is high, it is usually sufficient to chrome-plate the stem surface to prevent galling. Special "self-energizing" seals are used with higher-pressure valves above 700 bar (10,000 psig) service, so that the seal becomes tighter as pressure rises. Popular body seal designs for such services include the delta ring closure and the Bingham closure shown in Figure 2.68.

These designs depend on the elastic or plastic deformation of the seal ring at high pressures.

Special packing designs and materials are also required in high-pressure service because conventional packing would be extruded through the clearances. To prevent this, the clearance between stem and packing box bore is minimized, and extrusion-resistant material, such as glass-impregnated Teflon®, is used for packing. The stem thrust can also be excessive for globe valves. If globe valves are flow to close, high Δp can damage the seat or prevent the actuator from opening the valve. If they are flow to open, they might open against the actuator.

The thrust required to throttle large valves can exceed the force available from pneumatic actuators, which are often limited by an air supply pressure

TABLE 2.66

Orientation Table: Summary of Valve Sizing Equations (for U.S. Customary Units: gpm, SCFH, psi, °F, lbm/h, lb/ft³, etc.)

Selection Basis		Fluid State	
		Liquid	Gas or Vapor
Nonchoked, turbulent flow in liquids: $\Delta p < F_L^2(P_1 - F_F P_v)$	Volumetric flow in gpm or SCFH	$C_v = \dfrac{q}{F_P}\sqrt{\dfrac{G_l}{p}}$	$C_v = \dfrac{Q}{1360\,F_P P Y}\sqrt{\dfrac{G_g T Z}{x}}$ $Y = 1 - \dfrac{x}{3 F_k x_T};\ x = \dfrac{\Delta P}{P_1};\ F_k = \dfrac{k}{1.4}$
Nonchoked vapors and gases: $\Delta p/P_1 < x_T$	Mass flow in lbm/h	$C_v = \dfrac{w}{63.3\,F_P\sqrt{(\Delta p)\gamma_1}}$	$C_v = \dfrac{w}{63.3\,F_P Y\sqrt{x P_1 \gamma_1}}$
Choked flow due to cavitation or flashing in liquids, or choked flow in gases	Volumetric flow in gpm or SCFM	$C_v = \dfrac{q_{max}}{F_{LP}}\sqrt{\dfrac{G_f}{P_1 - F_F P_v}}$ $F_F = 0.96 - 0.28(P_v/P_c)^{1/2}$	$C_v = \dfrac{Q_{max}}{7320\,F_P P Y}\sqrt{\dfrac{M T_1 Z}{F_k x_T}}$
Choked flow due to sonic velocity in gases or vapors, or choked flow in liquids	Mass flow in lbm/h	$C_v = \dfrac{w_{max}}{63.3\,F_{LP}\sqrt{(P_1 - F_F P_v)\gamma_1}}$	$C_v = \dfrac{w_{max}}{19.3\,F_P P Y}\sqrt{\dfrac{T_1 Z}{F_K x_T M}}$

Piping effect for above equations	Not choked	$F_P = \left[1 + \dfrac{(\Sigma K)C_d^2}{890}\right]^{-1/2}$
	Choked	$F_{LP} = \left[\dfrac{1}{F_L^2} + \dfrac{K_i C_d^2}{890}\right]^{-1/2}$
Nonturbulent (viscous) flow	Volumetric flow in gpm	$C_v = \dfrac{q}{F_R}\sqrt{\dfrac{G_f}{\Delta p}}$	Laminar, i.e., nonturbulent, conditions generally do not occur in gases or vapors except for small flow valves.
	Mass flow in lbm/h	$C_v = \dfrac{w}{63.3\,F_R\sqrt{(\Delta p)\gamma_i}}$	

TABLE 2.67

Representative Values of Relative Valve Capacity Coefficients (C_d) and of Other Sizing Factors for a Variety of Valve Designs. The C_d Values Listed Are for Valves with Full Area Trims, When the Valve Is Fully Open

Valve Type	Trim Type	Flow Direction*	X_T	F_L	F_d	F_s	$C_d = C_v/d^{2**}$	K_c
GLOBE								
Single-port	Ported plug, 4 port	Either	0.70	0.90	0.48	1.0	9.5	0.65
	Contoured plug	Open	0.72	0.90	0.46	1.1	11	0.65
		Close	0.55	0.80	1.0	1.1	11	0.58
	Characterized cage, 4 port	Open	0.75	0.90	0.45	1.1	14	0.65
		Close	0.75	0.85	0.41	1.1	16	0.60
	Wing-guided, 3 wings	Either	0.75	0.90	0.58	1.1	11	0.60
Double-port	Ported plug	Either	0.75	0.90	0.28	0.84	12.5	0.80
	Contoured plug	Either	0.70	0.85	0.32	0.85	13	0.70
	Wing-guided	Either	0.75	0.90	0.41	0.84	14	0.80
Rotary	Eccentric spherical plug	Open	0.60	0.85	0.42	1.1	12	0.60
		Close	0.40	0.68	0.42	1.2	13.5	0.35
ANGLE	Contoured plug	Open	0.72	0.90	0.46	1.1	17	0.65
		Close	0.65	0.80	1.0	1.1	20	0.55
	Characterized cage, 4 port	Open	0.65	0.85	0.45	1.1	12	0.60
		Close	0.60	0.80	1.0	1.1	12	0.55
	Venturi	Close	0.20	0.50	1.0	1.3	22	0.21
BALL	Segmented (throttling)	Open	0.30	0.80	0.98	1.2	25	0.25
	Standard port (diameter ≈ 0.8 d)	Either	0.42	0.74	0.99	1.3	30	0.20
BUTTERFLY	60°, no offset seat	Either	0.42	0.70	0.5	0.95	17.5	0.39
	90°, offset seat	Either	0.35	0.60	0.45	0.98	29	0.32
	90°, no offset seat	Either	0.08	0.53	0.45	1.2	40	0.12

* Flow direction tends to open or close the valve, i.e., push the closure member away from or toward the seat.

** In this table, d may be taken as the nominal valve size, in inches.

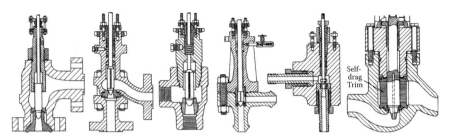

FIGURE 2.68
High-pressure valve designs.

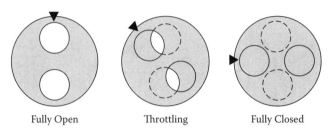

Fully Open   Throttling   Fully Closed

FIGURE 2.69
High-pressure process streams can be throttled by the positioned disc-type slide gate valve.

of 10 bar (150 psig). In such cases, high-pressure, high-performance, hydraulic actuators are used that function much like larger versions of the hydraulic power steering used on automobiles. They provide both very high speeds and power.

When smaller valves are needed (25–75 mm; 1–3″) variable chokes can be considered. They have a stationary and a movable disk, both with two holes. If rotated, the movable disk can progressively cover the two holes in the stationary one to throttle flow (Figure 2.69). The use of ceramic or tungsten carbide disks allows these choke valves to handle pressures up to 700 bar (10,000 psig). An angle version of this valve design is used for proportioning control, with an actuator capable of controlling the discharge flow at quarter-turn movement. Both linear and rotary type actuators can be used. The valve opening (the relationship between the discs) remains in the last position if power fails.

2.7.8 Applications—Noise

The noise sources in control valves include mechanical vibration (usually below 100 dBA); hydrodynamic noise caused by liquid turbulence, cavitation, or flashing (usually below 110 dBA); and aerodynamic noise (can reach 150 dBA). In control valve design, aerodynamic noise can be a major problem. Aerodynamic noise generation, in general, is a function of mass flow rate and the pressure ratio (p_1/p_2) across the valve. The point at which sonic speed is reached in the valve vena contracta is a function of the valve design.

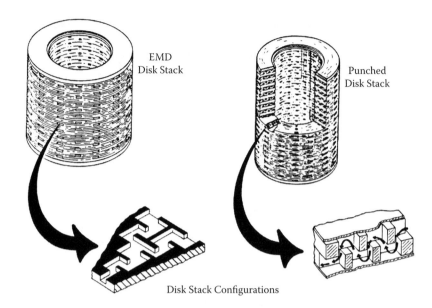

FIGURE 2.70
Special noise element design using labyrinth passages incorporated on plates. (Courtesy of Control Components, Inc.)

When sonic velocity is reached at the vena contracta (the minimum point on the pressure profile across the valve), the valves are said to be *choked*, because their capacity does not increase if the pressure ratio is increased while the upstream pressure is kept constant. Generally, choked valves are the sources of the highest noise levels, but subsonic flows can also generate high noise levels. Velocity of flow in the downstream pipe can also generate significant noise starting at pipe velocities of about Mach 0.4 to Mach 1.0. Noise-induced damage can drastically reduce valve service life, and in some cases it can cause valve or piping failures in a matter of minutes or hours.

Depending on the magnitude of the aerodynamic noise and assuming that massive valve damage is not a factor, valve noise treatment can be accomplished either by path treatment or source treatment. If the noise level exceeds 100 dBA, valve damage can only be reduced or eliminated by source treatment. Figure 2.70 shows the cross section of a multistage noise reduction trim that utilizes a combination of noise reduction strategies and reduces valve noise by up to 40 decibels. As will be discussed later, many of the features of low-noise control valves are similar to the features of the anticavitation valves.

The impingement of liquid droplets can be erosive if the velocity is great enough (>60 m/s, >200 ft/s) across the orifice, such as in applications involving high-pressure let-down of gas or vapor with suspended droplets. On high-pressure let-down applications the ideal valve to use is the dynamically balanced plug valve provided with a hard-faced plug.

Erosion caused by high exit velocities can also cause corrosion problems. Some metals do not corrode due to a self-regenerating protective surface film; however, if this film is removed by erosion faster than it is formed, the metal corrodes rapidly. For lining either noble metals or ceramics should be considered in such situations. The preferred arrangement for flashing service is to use a reduced port angle valve discharging directly into a vessel or flash tank.

2.7.9 Applications—High Temperature

Operating temperatures in excess of 250°C (450°F) are considered to be high temperatures. The maximum temperatures at which control valves have been successfully operated are up to 1400°C (2500°F). At elevated temperatures, the standard materials of control valve construction, such as plastics, elastomers, and standard gaskets must be replaced by more durable materials. Seating is always metal-to-metal, packing is semimetallic or laminated flexible graphite with spiral-wound stainless steel, and flexible graphite gaskets are necessary. Temperature cycling can cause thermal ratcheting and stress resulting in body or bolting rupture if the rate or frequency of temperature cycles is high.

It is important to select the trim materials and valve designs that will not experience sticking and gasket failure due to thermal expansion. If dissimilar trim materials are used, it is possible that one part will react to the high temperature more or faster than another, causing the components to gall. It is also important to allow the trim to grow axially in the valve body. Hanging the cage element from the top of the valve body does this. Some designs also incorporate internal springs or load rings to allow for the thermal expansion of the trim.

Similar to cryogenic applications, extension bonnets are used to protect the packing box parts from extremely high temperatures. Figure 2.71 gives an example of a jacketed valve design.

For the body, it is suggested that bronze and iron be limited to services under 200°C (400°F), steel to operation below 450°C (850°F), and the various grades of stainless steel, Monel®, nickel, or Hastelloy® alloys to temperatures up to 650°C (1200°F). For the valve trim, 316 stainless steel is used up to 400°C (750°F), 17-4 PH stainless steel up to 480°C (900°F), tungsten carbide up to 650°C (1200°F), and Stellite or aluminum oxide up to 980°C (1800°F). Up to 300°C (600°F) 316 stainless steel guide-posts in combination with 17-4 PH stainless steel guide bushings give acceptable performance. If the guide-posts are surfaced with Stellite, this combination can be extended up to 400°C (750°F) service. At higher temperatures, both the posts and the bushings must be Stellite.

The packing temperature limitation for most nonmetallic materials is in the range of 204 to 288°C (400 to 550°F), the maximum temperature for metallic packing is around 480°C (900°F), Teflon® should not be exposed to temperatures above 230°C (450°F), Graphoil (pure carbon) can be used from −240

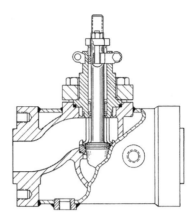

FIGURE 2.71
Steam-jacketed valve. (Courtesy of Flowserve Corp.)

to −399°C (−400° to −750°F) in oxidizing service and up to 650°C (1200°F) in nonoxidizing service, with an ultimate potential of 1650°C (3000°F). Solid rings or ribbons of Graphoil are also popular at higher temperatures, but they require more loading to energize the packing than does Teflon, and the resulting friction can cause stem lockup.

The bonnets are usually flanged and are extended on hot services, so as to bring the operating temperature of the packing closer to the ambient. The sliding stems can sometimes drag atmospheric contaminants or process materials into the packing, but this can be overcome by close-tolerance guide bushings or wiper rings. Packing contamination is less likely with rotary valves. For sliding stem valves, Teflon V-rings within extension bonnets are frequently selected and used up to 450°C (850°F). On high-temperature services, it can be effective to mount the bonnet below the valve. If the bonnet is mounted below the valve, no convection occurs and the heat from the process fluid is transferred by conduction in the bonnet wall only. If the bonnet is below the valve, the relationship between process and packing temperature is not affected by the phase of the process fluid. In case of ball or plug valves with double-sealing, it is important to vent the space between the seals to the line so that damage will not be caused by thermal expansion.

Heating jackets with steam or hot oil circulation are used to prevent the formation of cold spots in the more stagnant areas of the valve or where the process fluid otherwise would be exposed to relatively large masses of cold metal. Many standard globe pattern valves can be fitted with a jacket to allow heating or cooling as required (Figure 2.71). Often the manufacturer can provide these jackets, but where this is not available, there are firms that specialize in designing and installing such jackets on valves and other equipment. These special jackets can either be designed to weld to the valve as a permanent fixture (Figure 2.72) or as separate devices bolted or clamped to the valve body. In the latter case, it may be necessary to use a heat transfer

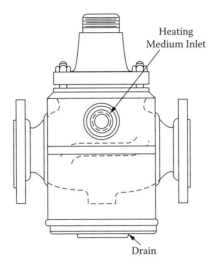

Heating
Medium Inlet

Drain

FIGURE 2.72
Special steam jacket for retrofit installation on valve.

paste between jacket and valve body to give efficient transfer by eliminating the air gap.

2.7.10 Applications—Cryogenic

Cryogenic service is usually defined as temperatures below −100°C (−150°F). Properties of some cryogenic fluids are listed in Table 2.73. Valve materials for operation at temperatures down to −268°C (−450°F) include copper, brass, bronze, aluminum, 300-series stainless steel alloys, nickel, Monel, Durimet, and Hastelloy. The limitation on the various steels falls between 0° and −150°F (−17 and −101°C), with cast carbon steel representing 0°F (−17°C) and 3.5% nickel steel being applicable to −150°F (−101°C). Iron should not be used below 0°F (−17°C).

Cryogenic valves should have small body mass and small heat capacity to provide a short cool-down period. The inner parts of the valve should be removable without removing the body from the pipeline, and if the valve is installed in a cold box, no leakage can occur inside this box. Conventional valve designs can be used for cryogenic service with the proper selection of construction materials and with an extension bonnet to protect the packing from becoming too cold. The extension bonnet is usually installed vertically so that the boiled-off vapors are trapped in the upper part of the extension, which provides additional heat insulation between the process and the packing. If the valve is installed in a horizontal plane, a seal must be provided to prevent the cryogenic liquid from entering the extension cavity.

When the valve and associated piping are installed in a large box filled with insulation ("cold box"), this requires an unusually long extension in order to keep the packing box in a warm area. The valve illustrated in Figure 2.74 can

TABLE 2.73

Properties of Cryogenic Fluids

	Methane	Oxygen	Fluorine	Nitrogen	Hydrogen	Helium
Boiling point (°K)	−259	−297	−307	−320	−423	−452
(°C)	(−162)	(−183)	(−188)	(−196)	(−253)	(−269)
Critical temperature (°F)	−117	−181	−200	−233	−400	−450
(°C)	(−83)	(−118)	(−129)	(−147)	(−240)	(−268)
Critical pressure (psia)	673	737	808	492	188	33
(bars absolute)	(46.1)	(50.5)	(55.3)	(33.7)	(12.9)	(2.26)
Heat of vaporization at boiling point (Btu/lbm)	219	92	74	85	193	9
(J/kg)	$(5.09 \cdot 10^5)$	$(2.14 \cdot 10^5)$	$(1.72 \cdot 10^5)$	$(1.98 \cdot 10^5)$	$(4.49 \cdot 10^5)$	$(2.09 \cdot 10^4)$
Density (lbm/ft³) gas at ambient conditions	0.042	0.083	0.098	0.072	0.005	0.010
(kg/m³)	(0.673)	(1.33)	(1.57)	(1.153)	(0.080)	(0.16)
Vapor density (lbm/ft³) at boiling point	0.111	0.296	—	0.288	0.084	1.06
(kg/m³)	(1.778)	(4.74)	—	(4.614)	(1.346)	(16.98)
Liquid density (lbm/ft³) at boiling point	26.5	71.3	94.2	50.4	4.4	7.8
(kg/m³)	(424.5)	(1142)	(1509)	(807.4)	(70.5)	(125)

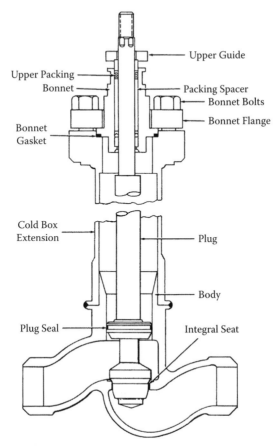

FIGURE 2.74
Cold-box valve with weld ends and welded bonnet. (Courtesy of Flowserve Corp.)

be used down to –270°C (–450°F). Where tight shut-off is required, Kel-F®
has been found satisfactory as a soft seat material, whereas other elastomer
materials will harden, set, or shrink and can cause leaking.

Because of the need for Charpy impact for the extremely cold service, the
materials are limited to bronze and austenitic stainless steels such as 304, 316,
and 316L. Body ratings through ANSI 600 and flange ends, either integral or
separable, are available depending upon manufacturer. Seat rings may be
integral hard-faced with Stellite.

The long, extended bonnet is provided with a plug stem seal to minimize
liquid "refluxing" into the bonnet and packing area, thereby minimizing the
heat loss due to conduction and convection. Cold boxes are commonly used
in hydrogen and in the air separation industry. Valves used in these applications feature bodies with welded extension necks and standard-length bonnets to allow in-place trim maintenance from outside the cold box. On LH₂
service these valves are often provided with vacuum jackets for additional

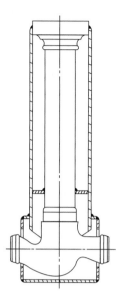

FIGURE 2.75
Vacuum jacketing of cryogenic valve.

insulation (Figure 2.75). The potential leakage problems are eliminated, because there are no gaskets inside the jacket.

For noncold-box applications, a cryogenic valve with an extension bonnet is used (Figure 2.76), allowing the packing box area of the control valve to be warmed by ambient temperatures, thus preventing frost from forming on the steam and packing box areas.

Actually, the small amount of liquefied gas passing into the bonnet vaporizes and provides a vapor barrier between the liquefied gas and the packing area. In addition, the pressure resulting from the vaporization of the liquid prevents additional liquid from passing into the bonnet area. Excess pressure vents back into the body.

When testing cryogenic valves for shutoff or in hydrostatic tests, the use of water-based tests should be avoided. If water tests were conducted, it is possible that moisture can be trapped inside the body or extension bonnet, which could ultimately form ice in the valve. For these types of applications, in order to prevent freezing, the proper test medium is usually helium.

2.7.11 Applications—Cavitation

Cavitation problems are common when handling cryogenic liquids. Cavitation can occur when the pressure at the vena contracta (P_{vc}) is less than both the vapor pressure (P_v) and the outlet pressure of the valve (P_2). Under such conditions, vapor bubbles can form at the vena contracta, and these bubbles can implode and release powerful microjets that will damage any metallic surface as the pressure rises downstream (Figure 2.77).

FIGURE 2.76
The construction of a cryogenic valve with extended bonnet. (Courtesy of Fisher Controls.)

Cavitation damage always occurs downstream of the vena contracta when pressure recovery in the valve causes the temporary voids to collapse. Destruction is due to the implosions that generate the extremely high-pressure shock waves in the substantially noncompressible stream. When these waves strike the solid metal surface of the valve or downstream piping, the damage gives a cinderlike appearance. Cavitation is usually coupled with vibration and a sound like rock fragments or gravel flowing through the valve. In case of flow-to-open valves, the destruction is almost always to the plug and seldom to the seat.

Table 2.67 listed, for the various valve designs, the liquid pressure recovery factors (F_L)—which are related to the ratio between the valve pressure drop and the difference between the inlet and the vena contracta pressure—and the cavitation coefficients (K_c)—which are the ratios between the valve pressure drop at which cavitation starts and the difference between the inlet and the vapor pressure—of the application. The allowable maximum Δp before cavitation begins is $\Delta p = K_c (p_1 - p_v)$. As the F_L and K_c values of the different valve designs drop, the probability of cavitation increases.

Because no known material can remain indefinitely undamaged by severe cavitation, the only sure solution is to eliminate it. The greatest damage is caused by a dense pure liquid with high surface tension (e.g., water or mercury). Methods to eliminate cavitation include the reduction

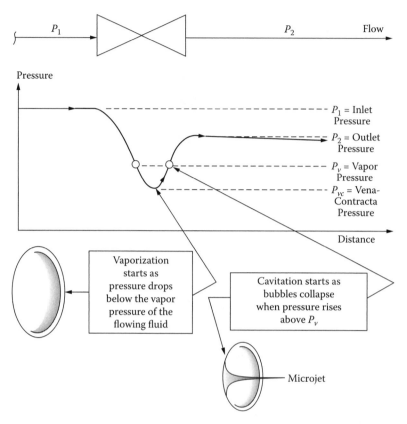

FIGURE 2.77
Cavitation occurs when the pressure rises downstream of the vena contracta. When it reaches the vapor pressure of the process fluid, the vapor bubbles implode and release powerful microjets that will damage any metallic surface in the area.

of operating temperature, which can sufficiently lower the vapor pressure. Similarly, increased upstream and downstream pressures, with Δp unaffected can relieve cavitation. Therefore, control valves that are likely to cavitate should be installed at the lowest possible elevation in the piping system and be operated at minimum Δp. Moving the valve closer to the pump will also serve to elevate both the up- and downstream pressures.

If cavitating conditions are unavoidable, then it is preferable to not only have cavitation but also some permanent vaporization (flashing) through the valve. This can usually be accomplished by a slight increase in operating temperature or by decreasing the outlet pressure. Flashing eliminates cavitation by converting the incompressible liquid into a compressible mixture.

When operating conditions cannot be changed, one can select a valve with a treacherous flow path, either of the multiple-port or multiple-flow-path variety. Valves that are most likely to cavitate are the high-recovery valves (ball, butterfly, gate) having low F_L and F_c coefficients (Table 2.67). Therefore, if cavitation is anticipated, low recovery valves with high F_c and F_L coefficients should be selected. Different valve designs react differently to the effects of cavitation, depending on where the bubbles collapse. If the focus is in midstream, materials may be unaffected.

Labyrinth-type valves avoid cavitation by a series of right-angle turns with negligible pressure recovery at each turn. The multistep valves at the bottom of Figure 2.78 can avoid cavitation by replacing a single and deep vena contracta, as would occur in a single-port valve, with several small vena contracta points as the pressure drop is distributed between several ports working in series. On the other hand, if P_v is greater than P_2 (described in Figure 2.78 as Condition B), this valve is likely to cavitate at an intermediate port. Therefore, these multistep valves are not recommended for flashing applications.

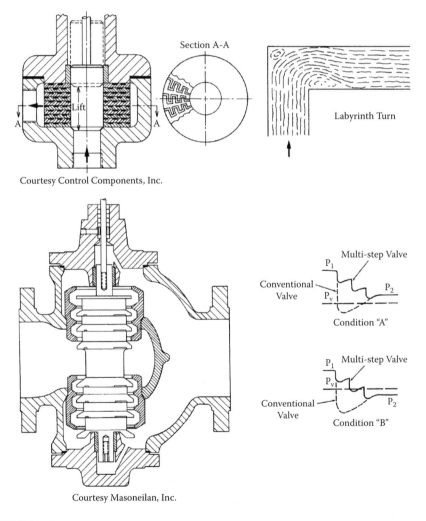

FIGURE 2.78
The labyrinth (top) and multistep (bottom) valve designs help to reduce the probability of cavitation.

Although no known material will stand up to cavitation, some will last longer than others. The best overall selection for cavitation resistance is Stellite 6B (28% chromium, 4% tungsten, 1% carbon, 67% cobalt). This is a wrought material and can be welded to form valve trims in sizes up to 3 inches (75 mm). Stellite 6 is used for hard-facing of trims and has the same chemical composition but less impact resistance. Correspondingly, the cost of Stellite 6 is lower than the cost of Stellite 6B.

In summary, the applications engineer should first review the potential methods of eliminating cavitation. These would include adjustment of process conditions, revision of valve type, or change of installation layout. If none of these techniques can guarantee the complete elimination of cavitating conditions, chokes or special anticavitation valves should be installed that can last for some reasonable period, even if some cavitation is occurring.

2.8 Cooling Tower Optimization

Cooling towers (CTs) are water-to-air heat exchangers that are widely used in renewable energy processes. They are simple devices, and possibly for that reason, their optimization is often neglected. Yet, the yearly operation of the cooling tower fans and pumps adds up to several hundred thousand dollars and can be cut in half by optimization. Optimization is achieved by meeting the variable cooling load of the plant by the minimum water and airflows, and by the cost-effective balancing of the distribution of the returning water among the tower cells.

The plant's demand for cooling water (CW) is variable, and meeting that load with constant-speed pumps is wasteful. When the demand for CW drops, the airflow can also be reduced. In optimized CTs both the water and the airflows are controlled by variable-speed devices. The goal of CT optimization is to maximize the amount of heat discharged into the atmosphere per unit of operating cost invested.

The cost of fan operation can be reduced by allowing the CT water temperature (T_{ctws} in Figure 2.79) to rise, thereby increasing the approach to the atmosphere's wet-bulb temperature ($T_{ctws}-T_{wb}$). The cooling tower cannot generate a water temperature that is as low as the ambient wet-bulb temperature, but it can approach it (hence the term "approach"). As the cost of operating the CT fans drops, the cost of pumping increases. The optimum approach is the one that will result in a minimum-cost operation. This ΔT is the set point of TDIC-1. As shown in the figure, the approach can be increased to a point where the tower operates by natural convection, the fans are off, and their operating cost is zero. Under most load conditions, however, this would not produce a low enough CW temperature for the process.

As the CT approach increases, and therefore T_{ctws} rises, the temperature difference across the process cooler (T_p-T_{ctws}) is reduced. This will cause the process temperature (T_p) to rise and its controller (TIC-4) to further open

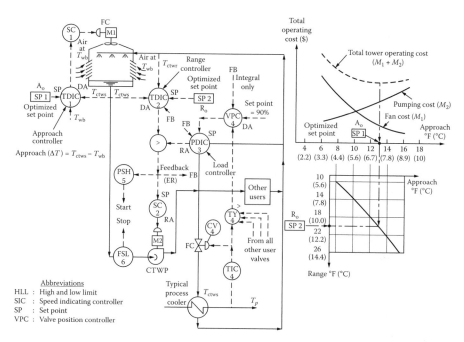

FIGURE 2.79
Optimization minimizes the unit cost of cooling by minimizing the operating speeds of both the cooling tower fans and pumps.

the coolant valve, CV-4. In order to reduce the process temperature, more and more water must be pumped. Therefore, an increase in approach will result in an increase in water transportation, and consequently, the pumping costs will rise. Therefore, the actual total operating cost is the sum of these expenses. Once a total operating cost curve is obtained, the optimum approach can be found at the minimum point on the total cost curve.

The data for the cost curves is empirically collected and is continually updated through the actual measurement of fan and pump operating costs. Consequently, for any combination of load and ambient conditions, there is a reliable prediction of optimum approach setting. Once the initial prediction is set, that setting can be refined by adjusting it in 0.5°F increments, if the total operating cost is lowered by such adjustments. As shown in Figure 2.79, if the optimum value of approach (SP-1) has been determined, the corresponding optimum range ($T_{cwtr} - T_{cwts}$ = SP-2) under the prevailing load conditions can also be obtained.

TDIC-2 is the range controller, which is set by the optimized set point of SP-2. This is the range value corresponding to the optimum approach. TDIC-2 throttles the water circulation rate and sets the speed of the CT water circulating pumps in such a way that none of the process user valves (CV-4) will be fully open. This is guaranteed by the cascade loop of PDIC-3 and valve position controller (VPC-4).

In larger power plants, a number of CT cells are used to cool the returning cooling water, and part of the optimization strategy is to fully utilize their

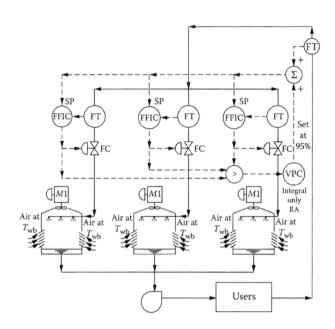

FIGURE 2.80

The distribution of the return water among several cooling tower cells can satisfy the dual goals of sending a preset percentage to each and to do that at a minimum pumping cost. The second goal is served by the valve position controller (VPC), which increases the measurement of all flow ratio controllers (FFICs) until the most open valve opens to 90%.

cooling capacity by automatically distributing the water flows among them. For optimum operation, the water flows to those cells that have fans at high speed should be equal and high, and those with fans operating at low speeds should receive water at equal and low flow rates; cells with their fans off should receive equal minimum flow rates.

Figure 2.80 shows how automatic water distribution balancing is achieved. Here, the total flow is used as the set point of the ratio flow controllers (FFICs), and the ratio settings reflect changes in fan speeds. Naturally, the total of the ratio settings must always be 1. Therefore, if one ratio setting is changed, all others should also be modified. The cost of operation is the energy cost of pumping, which will be minimum when the pressure drop through the distribution control valves is minimum. The valve position controller in Figure 2.80 keeps increasing the FFIC set points to keep this pressure drop to a minimum by opening all distribution valves until the most open valve reaches the desired 90 or 95% opening.

2.9 Distillation Optimization

During the global transition from the present oil-based economy to a clean and inexhaustible energy economy, the importance of distillation will increase because biofuel production also involves distillation and because

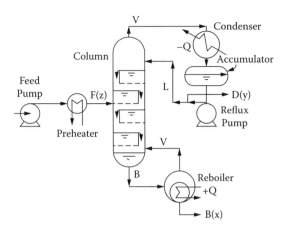

FIGURE 2.81
Illustration of a tray type distillation tower, where (without accumulation), the material balance is: $F = D + B$ and $D = V - L$. The mol fractions of the light key component in the bottoms, distillate and feed are identified as x, y, and z. For binary separation, $S = (y[1 - x])/(x[1 - y])$.

energy conservation requires the optimization of the existing processes. Globally, more than 80 million barrels of crude oil are refined daily. In the United States 146 refineries operate, employing over 65,000 people and producing a total value that exceeds $200 billion. Refineries are old (the last one built in the United States was 30 years ago), and spend 50 to 60% of their operating costs (i.e., excluding capital costs and depreciation) on energy, whereas the chemical industry spends only 30 to 40%. This difference shows the saving potential of implementing better control and optimization.

2.9.1 The Process

Distillation separates the components of a mixture on the basis of their boiling points and on the difference in the compositions of the liquids and their vapors. The product purity of a distillation process is maintained by the manipulation of the material and energy balances. Difficulties in maintaining that purity arise because of dead times, nonlinearities, and variable interactions.

A basic difference between distillation operations is in the handling of the heat removed by the condenser at the top of the column. The more common approach is to waste that heat by rejecting it into the cooling water. Figure 2.81 illustrates this configuration and identifies its main components. An alternate configuration (not shown) is to recycle the heat instead of wasting it. This is done by a heat pump (compressor), which moves the heat from the condenser into the reboiler.

The main distillation equipment is the *column* (also called *tower* or *fractionator*). It has two purposes: First, it separates a feed into a vapor portion that ascends the column and a liquid portion that descends. Second, it achieves intimate mixing between the two phases. The purpose of the mixing is to get an effective transfer of the more volatile components into the ascending

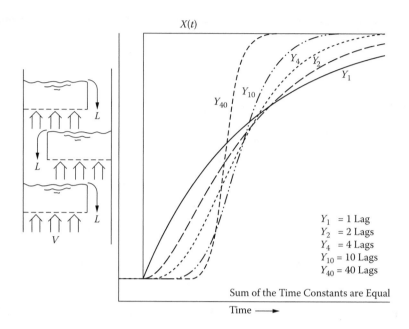

FIGURE 2.82

The contact between liquid and vapor is made intimate as the vapors ascend through the liquids which are held on each tray, as the liquid descends (left). The dynamics of a multiple tray column can be approximated as a second-order lag plus dead time.

vapor and a corresponding transfer of the less volatile components into the descending liquid.

The intimate mixing is obtained by either filling the column with lumps of an inert material (*packing*) or by the use of a number of horizontal plates, or *trays*, that cause the ascending vapor to be bubbled through the descending liquid (left side of Figure 2.82). The portion of the column above the feed is called the *rectifying* section and below the feed is called the *stripping* section.

Generally, trays work better in applications requiring high flows, because plate efficiencies increase with increased vapor velocities, and therefore increase the influence of the reflux to feed ratio on overhead composition. Column dynamics is a function of the number of trays, because the liquid on each tray must overflow its weir and work its way down the column. Therefore, a change in composition will not be seen at the bottoms of the tower until some time has passed.

These lags are cumulative as the liquid passes each tray on its way down the column. Thus, a 30-tray column could be approximated by 30 first-order exponential lags in series having approximately the same time constant. The effect of increasing the number of lags in series is to increase the apparent dead time and increase the response curve slope. Thus, the liquid traffic within the distillation process is often approximated by a second-order lag plus dead time (right side of Figure 2.82).

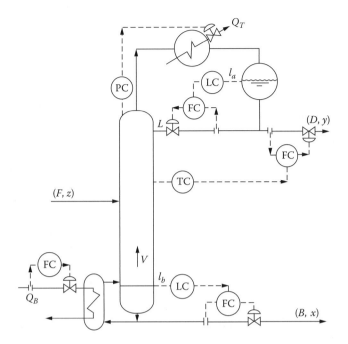

FIGURE 2.83
In this example, the five manipulated variables are so assigned to the five controlled variables that the heat input at the reboiler and the distillate composition are fixed and therefore the bottoms flow and composition are allowed to change with the variations in feed flow or composition.

2.9.2 Column Variables and Their Pairing

Controlled variables include product compositions (x,y), column temperatures, column pressure, and the levels in the tower and accumulator. Manipulated variables include reflux flow (L), coolant flow (Q_T), heating medium flow $(Q_B$ or V), and product flows (D,B) and the ratios L/D or V/B. Load and disturbance variables include feed flow rate (F), feed composition (z), steam header pressure, feed enthalpy, environmental conditions (e.g., rain, barometric pressure, and ambient temperature), and coolant temperature. These five single loops can theoretically be configured in 120 different combinations, and selecting the right one is a prerequisite to stability and efficiency.

Column pressure almost always is controlled by heat removal (Q_T). This loop closes the heat balance around the column, while the levels are controlled to close its material balance. Therefore, the key task is the assignment of the manipulated variables to the composition controllers. No matter how we make that selection, these two loops will "interact" (a change in one will upset the other). This is because whenever they change the openings of their control valves, the material and heat balance of the column will also change.

Figure 2.83 illustrates a possible end result of calculating the relative gain (RG) values. Here, it was concluded that fixing the production rate (heat input to the column $[Q_B]$) and controlling only the distillate composition, while allowing the bottoms composition to float, will give the most stable and efficient operation.

2.9.3 Composition Control

Conceptually, product quality is determined by the heat balance of the column. The heat removal determines the internal reflux flow rate, whereas the heat addition determines the internal vapor rate. These internal vapor and liquid flow rates determine the degree of separation between two key components.

If the feed composition and the column pressure are constant, temperature can be used as an indirect measure of composition. When the bottom product composition is being controlled, the temperature sensor is located in the lower half of the column and when overhead composition is controlled, in the upper half of the column. The temperature sensor should be located on a tray that strongly reflects changes in composition (Figure 2.84). When two compounds of relatively close vapor pressures are to be separated, two temperatures or a temperature difference can be used instead of a single sensor. This configuration can also be used to eliminate the effects of column pressure variations.

Analyzer controllers in a feedback configuration can only be considered when the dead time caused by analysis update is less than the response time of the process. The composition controller provides a feedback correction in response to feed composition changes, pressure variations, or changes in tower efficiencies.

In part (a) of Figure 2.85, the analyzer controller (ARC) uses the chromatographic measurement to manipulate the reflux flow by adjusting the set point to the reflux flow controller (FRC). Controllability of the process is degraded by the dead time between measurement updates.

As shown in part (b) of Figure 2.85, the *Smith-predictor* compensator provides a process model in terms of its time constant and dead time and thereby predicts what the analyzer measurement should be between analysis updates. When an actual analysis is completed, the model's prediction is compared

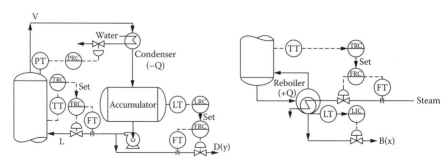

FIGURE 2.84

Distillate composition can be controlled by a cascade temperature master on the upper part of the column, which manipulates the reflux flow L (left). Similarly, the bottoms composition can be controlled by a cascade temperature master located on the lower half of the column, throttling the reboiler heat input (right).

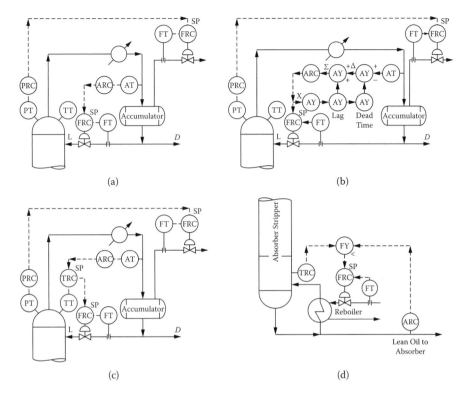

FIGURE 2.85
(a): Overhead composition controls by cascading reflux flow as the slave controller (FRC). (b): Shows the Smith predictor, which is the same as (a), but with dead time compensation added. (c): Overhead composition control by triple cascade of ARC to TRC to FRC. (d): Absorber bottoms composition control (ARC) cascaded to reboiler heat input (FRC) with temperature override (TRC).

to the actual measurement and the input to the controller is biased by the difference.

Part (c) in Figure 2.85 illustrates a triple cascade loop, where a temperature controller is the slave of an analyzer controller while the reflux flow is cascaded to temperature. Because temperature is an indicator of composition at constant pressure, the analyzer controller serves only to correct for variations in feed composition. Cascade loops will work only if the slave is faster than the master, which adjusts its set point. Another important consideration in all cascade systems (not shown in Figure 2.85) is that an external reset is needed to prevent the integral mode in the master from saturating, when that output is blocked from reaching and modulating the set point of the slave (when the slave is switched to local set point).

Part (d) of Figure 2.85 illustrates a limit control configuration where the analyzer controller is overruled by temperature when it reaches its high limit. The reason for this limit is energy conservation, because no additional stripping of the light component can be accomplished, once the boiling point of the impurity is exceeded. Therefore, even though an analyzer controller may call for more heat, this heat would only increase the bottoms' temperature without removing the impurity, thereby wasting heat.

Selective control configurations also require external feedback to protect them from reset windup. Part (d) has a combination of selective and cascade systems and in such configuration, the external reset (ER) signal (not shown) is taken from the measurement of the slave controller (FRC).

2.9.4 Pressure Control

In controlling the pressure of a column, the key pieces of equipment are the condenser and accumulator. First, the overhead vapors enter the condenser (partial or total), and next, the liquid condensate is collected in an accumulator vessel. Some of the accumulated condensate is returned to the column as reflux, while the remainder is withdrawn as overhead product (distillate).

Most distillation columns are operated under constant pressure, because at constant pressure temperature measurement is an indirect indication of composition. When the column pressure is allowed to float, the composition must be measured by analyzers or by pressure-compensated thermometers. The primary advantage of floating pressure control is that one can operate at minimum pressure, and this reduces the required heat input needed at the reboiler. Other advantages of operating at lower temperatures include increased reboiler capacity and reduced reboiler fouling.

A number of column pressure control options are shown in Figure 2.86. These options include the choice of controlling the column pressure by throttling the condenser water flow (a), throttling the venting of inerts (b), throttling flow of vapor distillate flow, if the condenser is "dry" (c,d), throttling hot gas bypass when the distillate product is a vapor (e), and throttling hot gas bypass when the distillate product is a liquid (f). Throttling hot gas bypass for fast response with and without inerts is described in parts (g) and (h).

2.9.5 Minimizing Pressure and Vapor Recompression

Operating the column at the minimum pressure minimizes the energy cost of separation. Lowering this pressure increases the relative volatility of distillation components and thereby increases the capacity of the reboiler by reducing operating temperature, which also results in reduced fouling. Reducing pressure also affects other parameters, such as tray efficiencies and latent heats of vaporization.

Minimum pressure operation can be achieved by adjusting the set point of the pressure controller so as to keep the condenser fully loaded at all times. This can be done by using valve position control (VPC), which keeps the condenser control valve in a nearly fully open position ([a] in Figure 2.87) or by refrigerant level control ([b] in Figure 2.87). The column pressure can also be minimized on the basis of condensate temperature ([c] in Figure 2.87).

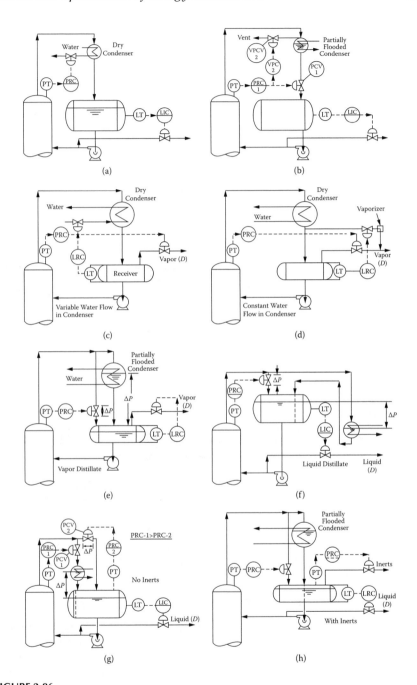

FIGURE 2.86
Pressure control system configurations for no inerts: (a) describes a process controlled by CW throttling, with inert purging; (b) by adjusting level of condenser flooding; (c) by CW throttling; (d) by vaporizer throttling; with vapor bypass throttling; (e) and vapor distillate; (f) and liquid distillate; (g) without inerts; (h) with inerts.

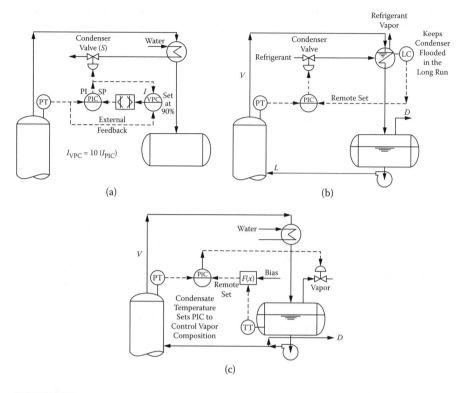

FIGURE 2.87

Optimization and vacuum control strategies: (a) minimizing (floating) pressure by maximizing coolant valve opening, (b) floating pressure control of partial condenser with vapor distillate, (c) floating pressure control when the distillate is both vapor and liquid.

Vapor recompression is another means of improving energy efficiency. As shown on the left in Figure 2.88, the overhead vapor from the distillation column is compressed to a pressure at which the condensation temperature is greater than the boiling point of the process liquid at the tower bottoms. This way, the heat of condensation of the column overhead is reused as heat for reboiling the bottoms. This scheme is known as vapor recompression.

2.9.6 Maximizing Feed Flow and Preheat

One of the best means of stabilizing the operation of a distillation column is to hold both the feed flow and feed temperature constant. If feed flow variations are unavoidable, the impact of these disturbances can be reduced by feedforward correction of the reflux flow (material balance). If the reflux flow is modified in the right proportion (*m*) and at the right time, the consequences of feed flow variations can be minimized.

In cases where both the demand for the product and the availability of feedstock are unlimited, increasing the throughput of the column maximizes profitability. In such installations a valve position controller can be

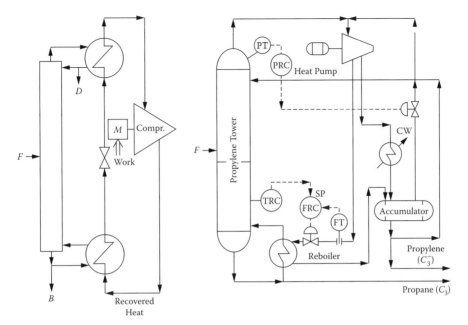

FIGURE 2.88
The vapor recompression system uses recovered heat (left). The pressure of such a distillation process can be controlled by modulating the speed of the compressor or the bypass around it (right).

cascaded to the feed flow controller in order to increase the feed rate until an equipment constraint is reached. If the constraint on maximum production is the cooling capacity of the condenser, the VPC will manipulate the feed rate to keep the back-pressure control valve always nearly open (Figure 2.89a). If the constraint is the heating capacity of the reboiler, a similar cascade VPC-based control configuration of the boiler heat input can be used to keep the reboiler fully loaded (Figure 2.89b).

Heat energy can be saved by using the hot bottom product to preheat the feed in an economizer (Figure 2.89c). In order to maximize the amount of heat recovered, a VPC is used as the cascade master of the feed temperature controller. The goal of optimization is to keep the bypassed flow at a minimum. Therefore, the VPC is usually set at about 10% of the valve opening.

2.9.7 Reflux Controls

Stable column operation is guaranteed by keeping the internal reflux of the distillation tower constant. Consequently, internal reflux controls are designed to compensate for changes in the temperature of the external reflux caused by ambient conditions. Figure 2.90a is controlled by a typical internal reflux control system (top) and the equations that need to be solved in calculating the required external reflux rate are shown at the bottom. This control system corrects for either an increase in overhead vapor temperature or a decrease in external reflux liquid temperature.

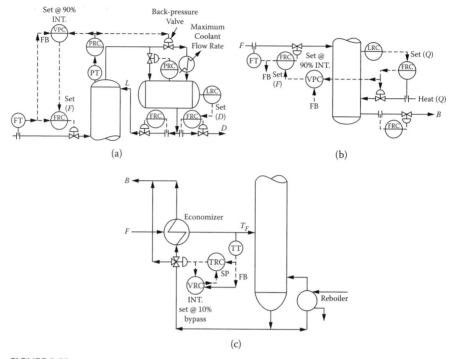

FIGURE 2.89
Feed flow optimization: (a) maximizing throughput by fully loading the condenser; (b) maximizing throughput against a reboiler constraint; (c) maximum recovery of the heat content of bottom product by economizer.

External instead of internal reflux control can be used in some cases when the external reflux flow (L) is controlled under the cascade control of accumulator level. To overcome the accumulator lag, the reflux rate, L, is manipulated in direct proportion to the distillate rate (D), rather than by waiting for the response of a level controller (Figure 2.90b).

2.9.8 Maximized Recovery or Constant Separation

In unit operations' control, the individual column variables are treated only as constraints and so long as the values of these constraints are within acceptable limits, the column is controlled (optimized) to maximize production rate, profitability, etc. Economics of individual fractionators may continually change throughout the life of the plant, because energy savings can be important at one particular time, whereas product recovery can be more important at other times.

A distillation column operating under constant separation conditions has one fewer degree of freedom, because its energy-to-feed ratio is constant. This means that, for each concentration of the key component in the distillate, a corresponding concentration exists in the bottoms. Therefore, if the concentration of a component in one product stream is held constant, that

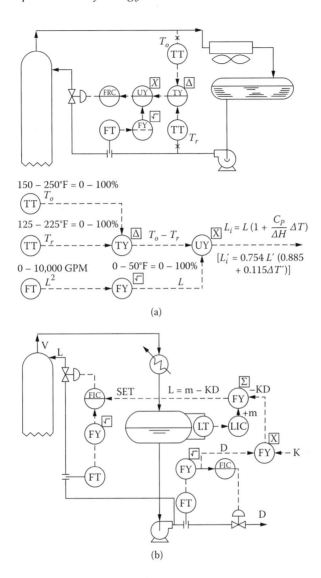

FIGURE 2.90
The controls required to keep the internal reflux flow of the column constant (a) and the controls needed to eliminate the effect of accumulator lag in controlling the external reflux to a column (b).

fixes it in the other. Figure 2.91 shows a feedforward control system for constant separation in which distillate is the manipulated variable. The block labeled "dynamics" serves to correct the transient response. Because the "dynamic personality" of the various distillation processes are different, a variety of dynamic compensators are shown in the figure.

If one product, such as the distillate (D), is worth much more than the other, the control system can be designed to maximize the production of the more

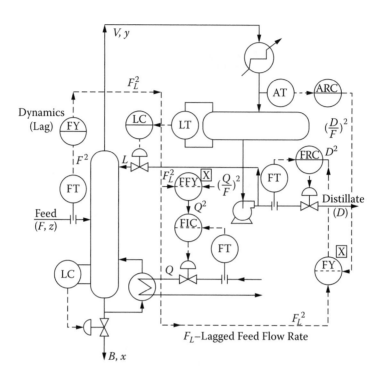

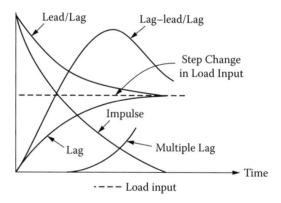

FIGURE 2.91

Feedforward control system that provides constant separation by manipulating the distillate flow (top). At the bottom, a variety of dynamic compensators are shown, which can be used to match the "dynamic personality" of the process.

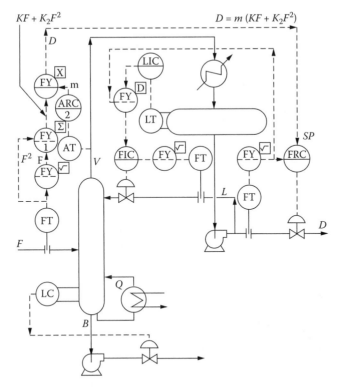

FIGURE 2.92
Control system that guarantees the maximum recovery of the more valuable product.

valuable stream. Figure 2.92 shows the controls of a maximum recovery system. In this configuration the boil-up (heat input rate) is constant, and consequently the distillate product flow is not linear with feed rate. In order to increase the response of the system (minimize accumulator lag), the reflux flow set point is adjusted by the distillate flow measurement.

2.9.9 Controlling Two Products

Whenever the feed composition is unpredictable, one must directly control the compositions of both products. The main benefit of dual composition control is minimized energy consumption. The main limitation is caused by the interactions between the two composition loops. On the left of Figure 2.93 an example of a feedforward dual composition control system is shown. In this configuration, the distillate flow is manipulated to control the distillate composition by maintaining the relationship:

$$D = F \left(\frac{z - x}{y - x} \right)$$

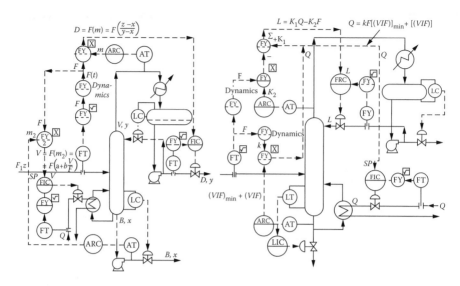

FIGURE 2.93
Configuration for controlling the composition of both products of a distillation column without much interaction (left) and with interaction (right).

Consequently, the bottom composition (x) has to be controlled by manipulating the energy balance of the column. The control system computes V based on the equation, $V = F(a + b[V/F])$, where $[V/F]$ equals the desired ratio of boil-up to feed.

Interaction is unavoidable between the material and energy balances in a distillation column. The severity of this interaction is a function of feed composition, product specification, and the pairing of the selected manipulated and controlled variables. It has been found that the composition controller for the component with the shorter residence time should adjust vapor flow, and the composition controller for the component with the longer residence time should adjust the liquid-to-vapor ratio, because severe interaction is likely to occur when the composition controllers of both products are configured to manipulate the energy balance of the column and thereby "fight" each other.

The decoupling scheme shown on the right side of Figure 2.93 serves to minimize the tendency of a change at one end of the column from upsetting the controller at the other end. The limitations of decoupling include that overrides can drive the loops to saturation when constraints are encountered.

2.9.10 Multiple Product and Multivariable Controls

Most biofuel separations involve multiple components and produce two or more liquid or vapor products. When there is a sidestream product in addition to the overhead and bottom products, an additional degree of freedom is available for the control system, because the overall material balance becomes $F = D + C + B$, where C is the sidestream flow rate. Therefore, two product streams can be manipulated to control, while the material balance can still be closed by the third. If the feed rate and column pressure are constant, five degrees of freedom exist: three composition specifications and two levels. These five controlled variables can manipulate the material and heat balance by throttling three product flows, and the loading of two heat exchangers (V and L), as is shown on the left side of Figure 2.94. Heat balance is often controlled by throttling the pump around reflux flows, as shown on the right side of Figure 2.94. The goal is to maximize the amount of heat that is transferred to the feed.

Multivariable models can be of the "white-box" or "black-box" type. White-box modeling is used for well-understood processes, such as distillation, where the knowledge of mass, energy, and momentum balances allow the development of accurate dynamic models. These internal model control (IMC) systems are useful in optimizing the process and in anticipating future events. "Black-box" or model-free controls (MFCs) include the artificial neural networks (ANNs), fuzzy logic, and statistical process control strategies. These algorithms are trained on the data obtained from the past operation of the controlled process. Their limitations include the relatively long learning period that they require and knowledge based on the past. Therefore, they are not well suited to anticipating future events, and if conditions change, they require retraining.

2.9.11 Model-Based and ANN Control

Once a process model has been established, it is possible to build the inverse of that model, which can be used as a controller. A simple internal model-based controller (IMC) is the Smith-predictor (Figure 2.85b), which is a first-order system with dead time combined with a PI controller.

Multivariable controls (MVCs) are particularly well suited for controlling highly interactive fractionators where several control loops need to be simultaneously decoupled. MVCs can simultaneously consider all the process lags, and apply safety constraints and economic optimization factors in determining the required manipulations to the process. The technique of multivariable control requires the development of dynamic models based on fractionator testing and data collection. Multivariable control applies the dynamic models and historical information to predict future fractionator characteristics. For towers that are subject to many constraints, towers that have severe interactions, and towers with complex configurations, multivariable control can be a valuable tool.

Dynamic matrix control (DMC) is also an MVC technique, but it uses a set of linear differential equations to describe the process. The DMC method obtains its data from process step responses and calculates the required manipulations utilizing an inverse model. Coefficients for the process dynamics are determined by process testing. During these tests, manipulated and load variables are perturbed, and the dynamic responses of all

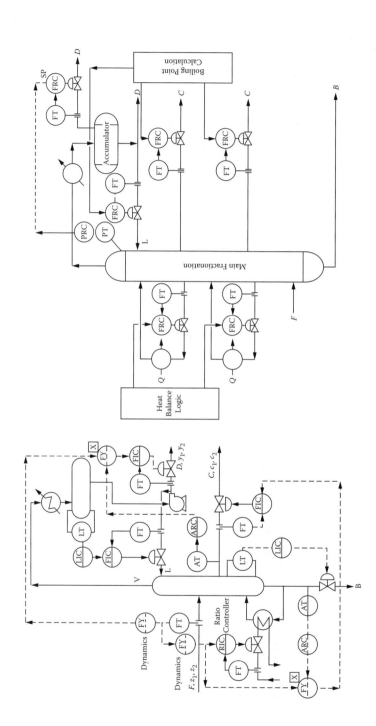

FIGURE 2.94

Multiproduct fractionator controls, where, after dynamic correction, the boil-up, side-draw and distillate flows are ratioed to the feed flow (left). On the right, the true boiling points are controlled by throttling the product flows, while heat balance is controlled by manipulating the reflux flows.

controlled variables are observed. This identification procedure is time consuming and requires substantial local expertise.

Figure 2.95 shows a three-layer, back-propagation ANN configuration, which predicts the manipulated steam and reflux flows of a column. The process model is stored in the ANN according its processing elements (nodes), which are connected, and by the importance that is assigned to each node (weight). The ANN is "trained" by example, and therefore, it contains the adaptive mechanism for learning from examples. During the "training" of these networks, the weights are adjusted until the output of the ANN matches that of the real process. Naturally, when process conditions change, the network requires retraining. The hidden layers help the network to generalize and even to memorize.

Because the neural network paradigm can accommodate multiple inputs and multiple outputs, an entire fractionator model can be built into a single controller. The neural controller can be thought of as a model-based control algorithm, whereby the neural network is used to obtain the inverse of the process model. As shown on the right of Figure 2.95, the back-propagation network can be trained to behave as an inverse model of the process, with load and controlled variables being input vectors and manipulated variables being output vectors.

Because the neural controller is an empirical model as opposed to a theoretical model, it is susceptible to errors if operated outside the conditions of the training set. Data for the training set need to be continually gathered, and the network retrained whenever novel conditions occur in order to increase the robustness of the neural controller throughout its life of operation.

2.9.12 Profitability-Based Optimization

The goal of optimization is safety at maximum profit, but this can only be done if the market value of each product is known. This is not the case when the products of a column are not final products but feed flows to other unit processes. When the product prices are unknown, it is still possible to perform optimization, but the optimization goal changes. The criterion in that case becomes the generation of the required products at minimum operating costs. This can be called an optimum with respect to the column involved, but only a "suboptimum" with respect to the plant of which the column is a part.

When the market values of the products are known, the column can be fully optimized, but additional variables must still be considered. These include the type of the market that exists for the products. If the market is limited, the goal is to generate the products at optimum separation and minimum operating cost. This cost varies as the feed flows and their compositions vary. When the market is unlimited and sufficient feedstock is available, the optimization task is more difficult, because one must determine both the optimum separation and the value of the feed streams. In this case the goal of optimization is either maximum loading or maximum energy efficiency.

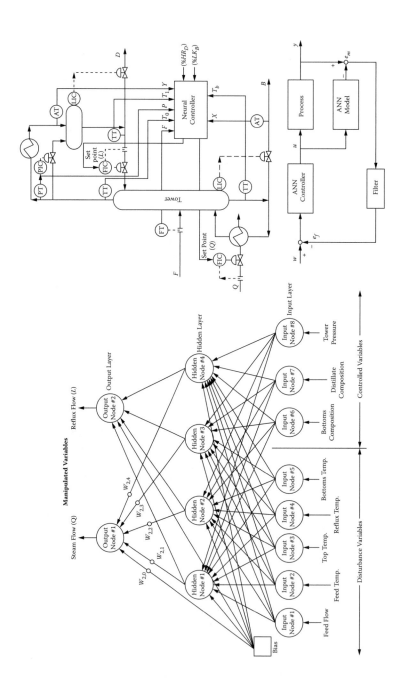

FIGURE 2.95

The configuration of a back-propagation neural network and its use as an internal model controller (IMC).

Optimization implies maximum profit rate. An objective function is selected, and manipulated variables are chosen that will maximize or minimize that function. Unit optimization addresses several columns in series or parallel. It is concerned with the effective allocation of feedstocks and energy among the members of that system. Plantwide optimization involves coordinating the control of distillation units, furnaces, compressors, etc., to maximize profit from the entire operation. All lower-level control functions respond to set points received from higher-level optimizers.

The advanced process control strategies that are most applicable to the optimization of the distillation process are usually based on white-box modeling, where the theoretical dynamic models are derived on the basis of the mass, energy, and momentum balances of this well-understood process. Although the optimization techniques described here can improve productivity and profitability by 25%, this goal will only be achieved if the distillation process is treated as a single and integrated unit operation and the variables, such as flows, levels, pressures, etc., become only constraints, and the controlled and optimized variables are productivity and profitability.

2.10 Distribution of Liquids or Gases

In renewable energy processes it is a common task to distribute heat transfer fluids among rows of solar collectors, gases among parallel compressors, or return water among cooling towers, and transport them with the minimum investment of pumping energy. The key to distribution optimization is the VPC.

The basic motivation for controlling on the basis of selecting the most open valve (MOV) is that by opening the user valves, the pressure drops, and therefore, the transportation energy requirement is reduced. The MOV control systems operate by opening all user valves until one of them approaches full opening. This way, all loads are satisfied, yet a minimum of transport energy is invested.

The advantages of MOV control are not limited to energy savings, they also include the reduction of valve maintenance due to high-pressure-drop-based abrasion and cavitation. Another advantage of MOV control has to do with the positioning accuracy of control valves, which is about ±1% of the valve stroke. This positioning error is a fixed quantity, which becomes a larger percentage of the operating stroke (and therefore of the flow through the valve) as the valve is throttled down. When the valve is 90% open, the error is $1/90 = 1.1\%$, whereas with a 10% open valve it is $1/10 = 10\%$.

Figure 2.96 describes a control configuration in which the VPC measures the opening of the most open user valve, and if it is under 90%, opens it up further by lowering the set point of the discharge pressure controller (PIC).

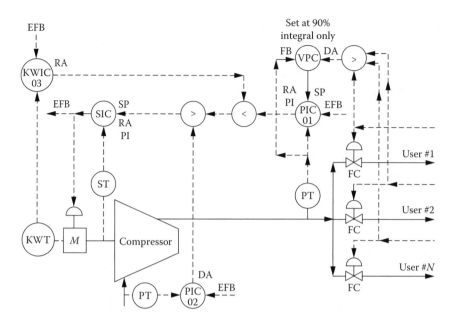

FIGURE 2.96

In this configuration, the flows to the individual users are controlled, while the compressor discharge pressure and therefore its power consumption is minimized.

A liquid flow distribution control system is illustrated in Figure 2.97. The extent to which the most open valve should be allowed to open depends on the nature of the process, because the higher that percentage, the less safety margin is available to overcome an upset. In well-maintained systems, which are operating with only a few valves, it is good practice to select the most open valve as the measurement for the VPC. In processes that are not well maintained (in heating, ventilation, and air conditioning [HVAC] or water treatment, where valves can be stuck in full open positions) or in processes involving the selection from among several hundred valves (fuel cells, aeration), selecting a single valve might not be the best choice. In these processes one might select five or even ten of the most open valves and average their openings to arrive at the VPC measurement.

In most fluid distribution applications, the use of MOV controllers to throttle variable-speed fans, blowers, or pumps can be expected to bring an energy saving of about 25%. Improvements, such as power factor correction or the use of high efficiency motors, can also contribute to energy reduction. Adjustable frequency drives (ADFs) are most often used, although direct current (DC) and pneumatic actuators are also still in use. DC units usually require high maintenance, whereas pneumatic units have high hysteresis. The actuators for the capacity controls of blowers should be fast and have positioners. Electrohydraulic actuators can meet this requirement.

When suction dampers or guide vane controls are used instead of variable-speed controls, reciprocal gain compensation is recommended due to the

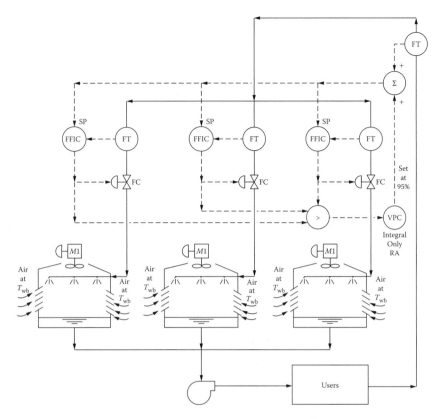

FIGURE 2.97
Water distribution can be controlled automatically while minimizing pumping costs.

variable gain of the final control element. If the distribution system consists of many valves, and the restart sequence after a power failure is initiated manually, it is desirable to provide master controls for the restart of all valves and other instruments by a single switch or other control device.

Distribution control system design should consider the response to power failure, sensor or actuator failure, valve plugging, and operator errors. The control system should also include the capability of both predicting and signaling the need for maintenance. Such self-diagnostics can be based on the comparison of valve opening and valve flow, and also on the comparison between control signal and valve opening.

2.11 Fan and Blower Optimization

The optimization of renewable energy processes, when applied to building conditioning, dryer, boiler, and cooling tower processes, necessitates the

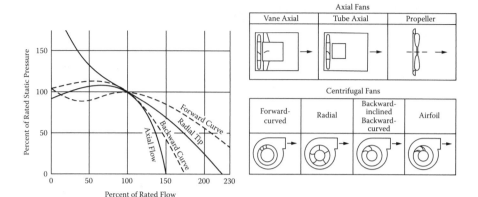

FIGURE 2.98
Different fan types exhibit different pressure-flow characteristics.

reduction of fan and blower energy consumption. Fans transport large volumes—up to 100,000 m³/m (1 million CFM)—of gases at discharge pressures of a few centimeters of water to 0.1 barg (a couple of psig). PD and centrifugal blowers generate higher pressures but smaller flows.

Fan designs are classified into axial and radial (also called centrifugal) flow types because of the difference between the nature of the flow through the blade passages. Figure 2.98 describes their pressure-flow characteristics. If the operating point is to the left of the maximum pressure point on the fan curve, it is likely that pulsation and unstable operation will occur. This maximum point on the fan curve is referred to as the *surge point* or *pumping limit.* All fans must always operate to the right of the surge point.

Under low-load conditions, the fan can be kept out of surge either by artificially increasing the load or by venting the gas that is not required by the process. The other, more energy-efficient way of eliminating surging is to replace suction or discharge dampers with blade pitch, speed, or vane control. The fan discharge pressure, in mostly friction-type transportation systems, is a parabolic function of flow. The operating point for the process is where the fan curve crosses the system curve.

When the flow and pressure requirements of the process are below the fan curve of the constant-speed fan, one way to bring them together is to introduce an artificial source pressure of drop, such as a suction or discharge damper, which, as it is throttled, modifies the fan curve. Although the introduction of this artificial pressure drop does result in the wasting of transportation energy, its advantage is that this throttling shifts the surge point to the left and thereby, at lower flows, allows for stable operation.

Tube-axial fans are provided with adjustable pitch blades that permit the balancing of the fan to match the varying process load either automatically or by infrequent manual adjustments. Vane-axial fans are also available with controllable pitch blades (that is, pitch that can be varied while the fan is in operation) for use when frequent or continuous flow adjustment is needed.

Throttling by varying the pitch angle retains high efficiencies over a wide range of conditions.

From the standpoint of power consumption, the most desirable method of control is to vary the fan speed to match reduced process loads. If the load does not change too frequently, belt-driven fan drives can be considered. The speed in such designs is adjusted by changing the pulley on the drive motor of the fan. When the process load varies often or when continuous fan flow modulation is desired, electrical or hydraulic variable-speed motors are required. From the standpoint of noise, variable speed is preferred to the variable-blade pitch design. On the other hand, both the variable-speed and the variable-pitch throttling designs are much quieter and more efficient than the discharge damper or suction vane-throttling-type systems.

When two fans are operated in parallel serving the same load and one of them is idle, its damper should be closed in order to prevent recirculation from the operating fan. When one of the two fans are shut down or started up, it is important to simultaneously correct the damper position on the operating damper so that the supply pressure to the process will not be upset.

The minimum cost of operation is achieved by first minimizing the number of fans that remain in operation and then by reducing the power consumption of the operating fans to the minimum. Figure 2.99 illustrates a control system that fulfills both of these goals. The lower part of the figure describes

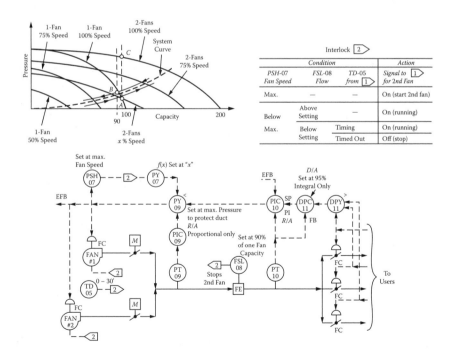

FIGURE 2.99
Optimized fan station controls include automatic fan cycling and a damper position to discharge pressure cascade loop to keep the most open user supply damper at 90% opening.

the control loops, whereas the operating fan curves are shown in the upper left and the interlock table is shown in the upper right.

In this system the second fan is started and stopped in response to load variations according to the logic in the interlock table in the upper right of the figure. The optimum discharge pressure is the minimum pressure that is still sufficient to satisfy all the users. This minimum pressure is found by observing the opening of the most open damper. If even the most open damper is not fully open, the supply pressure can be safely lowered, whereas if the most open damper is fully open, the supply pressure has to be raised. This supply–demand matching strategy not only minimizes the use of fan power but also protects the users from being undersupplied.

PIC-10 is the load-following controller. It compares the optimized set point with the actual header pressure and adjusts the fan speeds or the blade pitch angles of the fans. When its output signal reaches the maximum set on PSH-07, it starts another fan.

The role of PIC-09 is to provide overpressure protection at the fan discharge. When the pressure limit set on PIC-09 is reached, it takes over control from PIC-10 and protects the ductwork from being damaged.

2.12 Fuel Cell Controls

Fuel cell control is relatively simple for one cell; the difficulties come when multiple cells need to be controlled in parallel. The control challenge is similar to controlling a single cylinder in a conventional engine, to controlling 24 cylinders in parallel and, of course, no one has yet built a functioning 400-cylinder engine. Added to this challenge is the fact that, because all cells are connected in series in a stack, the performance of the worst cell also determines the performance of the best cell.

The operation of fuel cells has already been described in Section 1.3.5 (Chapter 1). Here, the emphasis will be on the control of these devices. Further research is required to reduce the cost of instruments for all fuel cell systems. For example, a complex fuel cell system can require upward of 100 flow control valves. Even if the cost is only $200 for a typical low-cost commercial valve, this cost can exceed the total cost of alternative electricity generation components by a sizable margin. Transition to high-temperature fuel cells pushes the valve price up as special materials are required, yet low cost is critical for commercial viability and salability of fuel cells if they are ever to move out of the laboratory and into general use.

Another source of problems are the thermal losses induced by instrumenting a typical high-temperature system. Thermocouple wiring and metallic capillary tubing to pressure and flow transducers can easily double the thermal losses of the system. Therefore, nonmetallic and thermally insulating sensor leads are required at reasonable costs for reliable operation at 1200°C.

Today's costs of tens of thousands of dollars per point for using such commercial instruments as optical strain gauges and thermometry systems are totally unacceptable.

It is essential that the cost of the control instrumentation required to operate fuel cells be reduced to the level that car manufacturers are paying for their sensors and controllers. This is essential to the commercialization of the current generation of fuel cell technology. The reduction in price must also be matched with several orders of magnitude improvement in reliability.

2.12.1 Fuel Cell Characteristics

Fuel cells are a means of converting chemical energy into electrical energy, and hence, to work. In theory, almost 100% conversion efficiency is possible as the energy conversion is via electrochemical reactions bypassing the losses inherent in thermal conversion. In practice, however, other limitations due to electrical losses and chemical concentration gradients reduce the achievable performance to between 30 and 85% efficiency. (See Section 1.3.5.)

In case of hydrogen-powered fuel cells, the previously mentioned advantages must be balanced with the consequences of the hydrogen production method used. If the hydrogen is made by reforming of natural gas, that is a rather wasteful process if the CO_2 is vented and therefore the energy content of the carbon is not utilized. Electrolytic systems that generate hydrogen from water are cleaner but their power consumption and power generation are essentially the same, so they are efficient only if the energy source of the production of electricity is renewable.

Because the basic fuel cell needs no mechanical drive, its operation is quiet and involves no frictional losses (Figure 2.100). These characteristics should make it possible to locate them near the final user, producing a more even distribution of the generation capacity. Auxiliaries, particularly fans and blowers, must be quiet; therefore, they should be well supported to prevent their motion and be provided with variable-speed drives. In addition, the feathering of the blade edges and the use of noise-reducing enclosures are recommended.

Fuel cells share a number of characteristics with heat exchangers. Similar forces limit the transfer of heat as those limiting the transfer of ions. Also, the exchange of ions is dependent upon the available surface area in contact with the process fluids and by the resistance to transfer of energy through the separator material. The transfer of energy also depends upon the movement and mixing of the molecular components of the fluids on either side of the exchange surfaces.

Reagent mixing can be enhanced by narrowing the passages between the electrodes or by placing obstructions into these streams to induce turbulence in the fluid flows. The downside of these solutions is that they increase pressure drop. In addition, if the dimensional tolerances cannot be accurately maintained or if the system model is not accurate, such obstructions can potentially interfere with the gas distribution between cells.

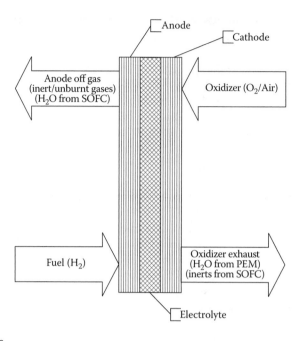

FIGURE 2.100
Illustration of one of the fuel cells in a stack of solid oxide fuel cells (SOFCs) where hydrogen fuel contacts an anode and the oxidizer acts on the cathode's surface.

Fuel cell control is conceptually simple. The fuel and oxidizer flows are regulated to ensure that the reagent flows match the required current generation, taking into account the cell efficiency factor. At low levels of current generation and low volumetric efficiencies, simple diffusion is adequate to ensure that some current output is generated. When higher power outputs are required, a control system is required to match the power generation with the power requirement of the process.

2.12.2 Oxidant and Fuel Flow Metering

In air-breathing cells, the airflow can be maximized to ensure the availability of O_2 by maintaining its partial pressure. The partial pressure achieved is subject to maximum capacity limitations and to limitations imposed by losses due to heating in high-temperature cells and due to pumping losses in other cell types. In the proton-exchange-type cells, this maximized airflow also sweeps the waste gases, principally water, away from the cathodes. In fuel cells running at high volumetric power levels, cold air bypass streams are provided to cool the cells if needed and to balance the temperatures between the cell stacks that are connected to a common manifold.

As illustrated in Figure 2.101 for a single cell in a stack, the fuel flows are generally metered more precisely because of their generally higher costs and are throttled to meet the current demand of the process being served.

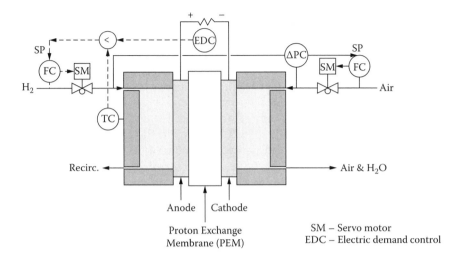

FIGURE 2.101
The basic controls of one cell in a fuel cell stack.

Fuel flow control in low-temperature cells is relatively simple. The control valves used are servomotor-driven quarter-turn valves, and the flow sensors can be the types used on gas furnaces, such as thermal dispersion type mass flow meters.

The previously used flow sensor option in research fuel cells were the integrated circuit (IC)-type differential sensors, which were limited by their low tolerance for water contamination. This was a serious limitation in reforming-type fuel cell applications, because the presence of water is essential for the operation of the system, and no protection against water contamination was provided in the smaller and cheaper IC sensors.

In spacecraft and similar closed-circuit fuel cell applications, partial pressure reductions are not a problem, other than for the production of wastewater. Therefore, constant supply pressure can be used, which greatly simplifies the control requirements, as the system becomes self-regulating on the demand side. In such applications, the waste can simply be blown down to a reservoir, based on time and current draw.

In high-temperature cells, special valve designs are required to handle the fuel gas, which is up to 1000°C. These valves are used to balance the gas flows between stacks and for stack isolation to permit continued operation when some of the stacks fail. High-temperature servomotor drives (Mitsubishi and Hitachi) can be used on these valves.

In most fuel cells a measurement of the cell voltages (based on the stack voltages) will indicate if the fuel or airflow is inadequate, as the cell voltage will drop substantially. In large stacks (20 or more cells), this stack voltage-based measurement can be too insensitive to be of much use. On the other hand, the measurement of the voltages generated by the individual cells can be too complex to be economically viable. In case of low-voltage systems, cost is not a serious problem, and individual monitoring is possible for up to 40 cells using commercially available products.

In high-temperature solid oxide fuel cells (SOFCs), fuel utilization can also be measured by the O_2 content in the exhaust fuel gas, using zirconium oxygen probes. Both cell voltage and fuel O_2 content measurement respond to the worst cell in the stack. Erroneous readings can also occur if a cell is cracked, as this will result in the same reading as an overloaded cell. In the case of overloading, the fuel needs to be increased, whereas in the case of cracking, the cell has to be isolated; this reading is insufficient for accurate diagnosis.

Both the cell voltage and the O_2 monitoring are negative correlations, because these measurements give an indication only when the utilization of O_2 drops. Although this is useful information when the load is rising, the detection of falling load conditions require another mechanism to ensure that fuel is not being wasted. Therefore, as a means of checking, it is also desirable to measure the cell performance on the basis of a fuel cell model.

Temperature monitoring of the stack exhaust is needed to determine whether the temperature rise is a consequence of cell overloading or of stack fire.

In a load-following control system, as the demand for current falls, the fuel feed flow is reduced until a predetermined minimum value is reached or until a fuel deficit is detected. If the fuel flow cannot be raised when a fuel deficit is detected, the inverter drawing power from the cell stack will reduce the current draw. If the fuel cell system is connected to a grid, which is also connected to a booster, this is not a problem, because more power can be drawn from the grid to make up the shortfall. If the grid is isolated, it is necessary to shed some of the load on the cells or by making up the shortfall from battery or other storage systems such as capacitor banks in the short term.

In multiple-stack installations, it is important to control the performance of each stack separately to ensure that one stack cannot discharge into another. This is necessary, because the manufacturing of identical stacks is just about impossible with the current means of manufacturing in the industry. This is particularly a problem for active anode SOFCs and molten carbonate cell designs, because the O_2 drawn through the cell electrolyte can oxidize and destroy the catalytic ability of the cell.

To ensure that backfeeds cannot occur, it is usual to design the stack electronics to act as a variable step-up DC/DC inverter with diode protection. In fuel cells current flow always equals gas flow. In other words, whenever current is drawn, gas must also move from one place to another. With PEM cells, the current drawn results in the movement of hydrogen to the oxidizer side, whereas in high-temperature cells, it results in moving O_2 into the fuel side. If the reagents are not available to remove these flows, reagents can appear on the wrong side of the cell, which can cause adverse reactions or, in extreme cases, can cause cell fires and explosions.

Further concerns are the balancing of the inlet pressures on both sides of the cells to ensure that the maximum differential pressures are not violated.

This is particularly a problem when thin-film electrolytes are used in PEM cells or with some SOFC systems. The absolute pressures can, however, be raised quite high as the strength of the outer casing is not limited by the requirement to enhance diffusion across a membrane, thereby potentially improving the power density of some fuel cell systems. This improvement has to be balanced against the energy cost of pressurizing the cell system. If fed from a high-pressure source like a gas bottle, this cost is negligible at the consumer's end but can be significant at the source.

2.12.3 Temperature and Auxiliary Controls

Temperature controls are used to override the power generation controls in fuel cells. The operating temperature of low-temperature cells can rise owing to a combination of increased current draw, cell impedance, and elevated gas inlet temperatures from the external reformers to raise the temperature of the cell system. This can degrade electrolyte and destroy the cell. In high-temperature cells, this heating effect can be useful, because at low temperatures it will increase the load on a cold cell to raise its temperature, reducing the load on hot stacks. Temperature controls are usually successful, when a combination of heat exchanger bypasses and auxiliary heater controls are used to maintain the temperature of loaded cells.

A further issue for high-temperature fuel cells is the melting temperature and related vapor pressure of the current-collecting cabling. Whereas most commercial high-temperature cable is nickel, its high resistance makes its use in fuel cell systems unwise and inefficient. Instead, solid silver cabling was used; however, the silver's vapor pressure limits its service to below 900°C with 850°C limit being more common. Recent developments in inexpensive cable insulation including ready availability of commercially produced silica and Nextel®-insulated thermocouple wiring, which solve this problem.

Figure 2.102 describes a high-temperature fuel cell system, using a gas fuel and a reformer to produce hydrogen. The fuel treatment controls described in the following subsections include the scrubbers, humidifiers, and reformers. In addition, the controls of the exhaust systems, the inverters, and the required shutdown controls will also be covered.

For all fuel cells, except those running on high-purity hydrogen, some form of fuel treatment is required. The main problem with fuel supplies intended for conventional combustion systems is the presence of minor contaminants containing ash-making chemicals and sulfur compounds. In fuel cell applications, the sulfur compounds form corrosive substances that poison the catalysts in the reformer stages and the fuel cell itself.

Therefore, scrubbing is often required. The scrubbing process is not tolerant of water contamination because it can cause the bed to collapse if exposed for prolonged periods to high water vapor pressures or to condensation. Alkaline fuel cells also require the removal of carbon oxides including CO_2.

After the scrubbing, water vapor must be added to the scrubbed mix, in order for the shift reaction to properly function in the reformer. The steam

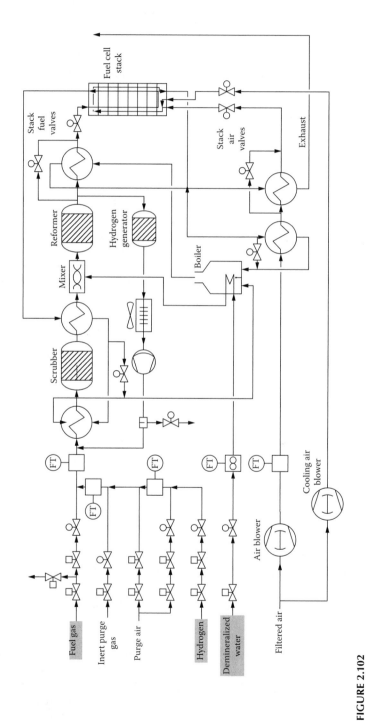

FIGURE 2.102
A simplified flow diagram of a high-temperature fuel cell system.

flow to the mixer is controlled on a mass flow–ratio basis to the gas flow, considering the gas composition and the degree of reformation required. The steam produced by the boiler is mixed with the fuel stream by in-stream mixers ahead of the reformer. Mixing the gas with the highly superheated steam also assists with the heating of the fuel gas before it enters the reformer.

Controlling the water feed to these boilers is a challenge in itself as the flow rates are low. For a 40 kW fuel cell system, the required water flow range is 0.25 to 10 g/s. Thermal dispersion flow meters can be considered. If conventional control valves do not prove adequate, pulsed solenoids can be used, with the flow being averaged to match the required ratio to the fuel gas.

All fuel sources except hydrogen require some reformation to provide a feed rich in hydrogen to the fuel cells and the scrubber. The reformer reaction requires temperatures that are higher than that of the scrubber stage. They are around 300 to 400°C for common iron/chrome catalysts, but temperatures of over 500°C have also been found to be effective. Low-temperature fuel cells require a further scrubber stage after the reformer to remove any CO using platinum group metals, in combination with cobalt, silver, iron, copper, and molybdenum. As most fuel cells do not produce complete conversion (other than hydrogen types), and most do vent unburned gas, some means of handling the waste fuel is required. Reformer and exhaust controls are not required for fuel cells operating on hydrogen.

2.12.4 Safety and Shutdown Controls

The fuel cell's control system should also provide the controls for a safe shutdown. The emergency shutdown sequence starts with the isolation of the fuel gas supply and is continued with "inerting" of the fuel cell system to snuff out any fires that could be caused by leaks, and to prevent the possibility of the fuel cell becoming an ignition source. Conventionally, this is done by applying a nitrogen blanket, which is supplied from a high-capacity liquid nitrogen tank. High-temperature cells that are located in or near hazardous areas may require more drastic design features, such as using liquid CO_2 to cool the cell to below the auto-ignition temperature of 400°C (destroying the fuel cells in the process). This occurs in under 2 minutes, and as such, it complies with requirements for a fire suppression system and removes the need for the nitrogen supply.

A more normal shutdown sequence would first flush the fuel cell, reformer, and scrubber with nitrogen or CO_2 (if it is safe for the cell design). This step is followed by a slow bleed of air and nitrogen to repassivate the fuel cell and reformer under temperature control. If this is not done gradually, the reformer can reach temperatures high enough to violate its containment (meltdown) and become unrecoverable. The reformer performance does decrease after this treatment, but over 90% of its capacity can be retained. Similar problems are present when reforming the fuel cells themselves.

High-temperature fuel cells are governed by the same gas safety regulations as are fired heaters and boilers, because they, too, operate at above the autoignition temperature of the fuel gas. Some gas reformers are also in the same category. Therefore, double block and bleed arrangements are required for commercial systems in all but the smallest installations, and rapid removal of internal ignition sources are required after a forced shutdown, in addition to meeting the system purge requirements that are set in the relevant standards.

2.12.5 Reversible Fuel Cell (RFC) Controls

In the solar–hydrogen demonstration power plant I have designed, the functions of the electrolyzers and of the fuel cells are combined into reversible units (RFCs). These dual-state fuel cells are expected to be much lighter than the combined weights of separate electrolyzers and fuel cells. The RFC during the day would operate in the electrolyzer mode converting solar energy into the chemical form (hydrogen), while at night it will switch into its fuel cell mode and will convert the chemical energy stored in hydrogen back into electricity (see Figure 4.2 in Chapter 4, which is reproduced below).

It takes the same amount of energy to split water into hydrogen and oxygen as the energy obtained when oxidizing hydrogen into water. The only difference between the two operations is that electrolysis increases the entropy and, therefore, not all the energy needs to be supplied in the form of solar electricity as the environment contributes an additional 48.7 kJ/mol of thermal energy. Inversely, when the RFC is operated in the fuel cell mode, part of the energy in the fuel, which is released as the hydrogen is oxidized, will be released as heat. Therefore, the electrolysis mode of operation can require heating and fuel cell mode cooling (or heat recovery).

In a solar–hydrogen power plant, when excess solar energy is available, the RFC is switched into the electrolyzer mode to split water into hydrogen and oxygen. The hydrogen is collected at about 3 bar (45 psig) of pressure and is either liquefied or is compressed to some high pressure (about 1,000 bar = 15,000 psig) and sent to storage.

On the other hand, when solar electricity is insufficient and needs to be supplemented, the RFC is switched into the fuel cell mode of operation where the oxidation of one mol of hydrogen will generate 237.1 kJ/mol of electrical energy plus 48.7 kJ/mol of thermal energy. This waste heat can also be used for heating buildings or for preheating boiler feedwater.

The role of process control is critical in operating RFCs. The complexity of the control challenge can be appreciated if we view a 400-cell RFC stack as 400 pumps operating in parallel and we realize that switching the RFC from one mode of operation to the other is like shooting down one process in a chemical reactor while starting up another. Fortunately, the switchover does not need to occur quickly, but once the processes are running, they are fast. For example, it takes only a couple of seconds between starting the charging of hydrogen to the fuel cell and beginning to generate electricity or vice versa in the electrolyzer mode (Figure 2.102).

In addition to the electric controls that control the grid connection and the conversion between direct and alternating current circuits, a massive

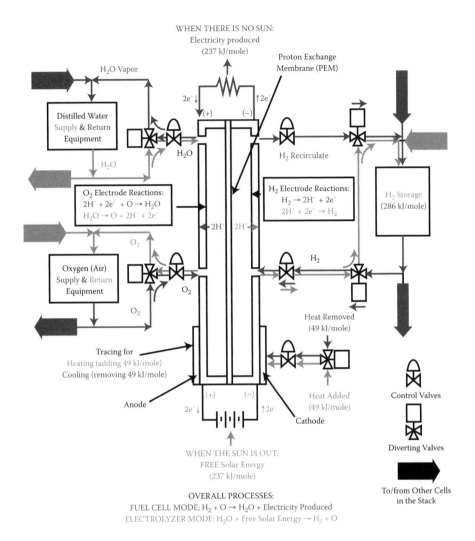

WHEN THERE IS NO SUN:
Electricity produced
(237 kJ/mole)

Proton Exchange
Membrane (PEM)

H₂O Vapor

Distilled Water
Supply & Return
Equipment

H₂O

H₂O

H₂ Recirculate

O₂ Electrode Reactions:
$2H^+ + 2e^- + O \rightarrow H_2O$
$H_2O \rightarrow O + 2H^+ + 2e^-$

H₂ Electrode Reactions:
$H_2 \rightarrow 2H^+ + 2e^-$
$2H^+ + 2e^- \rightarrow H_2$

H₂ Storage
(286 kJ/mole)

O₂

Oxygen (Air)
Supply & Return
Equipment

O₂

H₂

O₂

Heat Removed
(49 kJ/mole)

Tracing for
Heating (adding 49 kJ/mole)
Cooling (removing 49 kJ/mole)

Anode

Control Valves

Heat Added
(49 kJ/mole)

Cathode

Diverting Valves

WHEN THE SUN IS OUT:
FREE Solar Energy
(237 kJ/mole)

To/from Other Cells
in the Stack

OVERALL PROCESSES:
FUEL CELL MODE: $H_2 + O \rightarrow H_2O$ + Electricity Produced
ELECTROLYZER MODE: H_2O + Free Solar Energy $\rightarrow H_2 + O$

FIGURE 4.2
(See color insert following p. 140.) One cell of my reversible fuel cell (RFC) design, using the basic proton electrolyte membrane (PEM) fuel cell.

quantity of measurements and controls is required. These involve switching between the heating and cooling mode controls as the FRC operation is reversed. Both of these modes naturally require high rangeability and fast and accurate temperature controls. It is also necessary to carefully control the pressures of the oxygen and hydrogen streams that are entering (in the FC mode) or leaving (in the electrolyzer mode). The oxygen and hydrogen pressures have to be identical so that the PEM diaphragms do not experience excessive pressure differences.

In addition to electric, pressure, and temperature controls, the loads (the rates of hydrogen or electricity generation) also need to be controlled. These requirements are based either on the excess solar electricity availability (electrolyzer mode) or on the electricity requirement (FC mode). The load controls in both modes of operation require fast and accurate flow controls. In the fuel cell mode, the hydrogen fuel flow has to be controlled, while in the electrolyzer mode the water supply flow needs to be controlled. Controls are also needed to direct the generated distilled water to its destination (FC mode) and the generated oxygen to its destination (electrolyzer mode). This destination can be the air supply to a fired heater or boiler (to increase its oxygen concentration), if such unit exists on the site.

All the above described instruments will have to be mass produced when mass production of the RFCs starts. These instruments will have to be miniaturized, accurate and inexpensive, just like the units used in the automobile industry (a single modern car has 500 sensors). Therefore, in the transition period, when we change from the fossil/nuclear to the inexhaustible and free solar energy economy, process control will play a leading role. Good sensors and fast and stable controls are needed to make the RFCs small, light, and inexpensive. I am sure that the process control profession will meet these challenges and thereby will play not only a key role in this, the third industrial revolution, but will also gain the respect it deserves as the most important field of engineering technology.

2.13 Geothermal Controls

As has already been discussed in Section 1.3.4, geothermal energy is renewable and is available continuously. The temperature of thermal springs can reach 350°C (662°F). If the groundwater temperature exceeds 150°C (302°F), "flash steam" power plants can be built; if the water temperature is between 100 and 150°C (212 and 302°F), "binary cycle" power plants can be operated. When the groundwater temperature is less than 100°C (212°F), "geothermal heat pumps" (GHPs) can be used to heat buildings in the winter and cool them in the summer.

Geothermal energy systems are often integrated with solar power plants, so that energy production can continue during the night hours and other periods when the sun is not shining. One such combined control system—including energy storage and total plant integration—is described in detail in Chapter 4 of this book. This last part describes in detail the design of controlling the solar–hydrogen demonstration plant.

2.13.1 Direct and Indirect Geothermal Pumping

The temperature difference between the desired temperature of the conditioned space and the actual underground soil or groundwater temperature (Figure 1.24) can be used for direct cooling in the summer and direct heating in the winter. This temperature difference can be utilized in "open" or "closed" systems. In the case of an open system, the groundwater (or pond water) is pumped directly into the heat transfer equipment of the conditioned building (top left of Figure 1.24) and after that it is returned underground.

In a closed system, city water is continuously circulated in a loop of tubing that has been laid in a ditch (top right of Figure 1.24), underground, or in a pond. Closed systems are often preferred, because the quality of city water is usually better than the quality of natural water bodies and therefore systems using city water require less maintenance.

The controls of both the "open" and the "closed" direct systems are very simple, because they consist of only a differential thermostat and a pump. The differential thermostat starts and stops a constant-speed pump whenever the detected temperature difference is favorable. In more sophisticated systems, it adjusts the speed of variable-speed pumps (or selects a small or a larger constant-speed pump for operation). What ΔT is "favorable" is a function of not only the "availability" of free cooling or free heating, but also of the cost of pump operation, which increases the ΔT thermostat set point.

When the measured ΔT is favorable, but is insufficient to condition the building on its own, the geothermal energy resource needs to be supplemented to meet the total energy need. This can be done by using conventional means, such as supplementing it by starting a furnace in the winter (or air conditioner in the summer).

2.13.2 "Binary Cycle" Geothermal Heat Pumps (GHPs)

When the previously discussed temperature difference is small or negative (say, the groundwater is cooler in the winter than the thermostat setting on the conditioned space), "direct" operation is no longer possible. In such cases, the heat content of the groundwater cannot just be pumped into the building, because first its temperature has to be elevated. This requires the use of an intermediate heat pump loop, hence the name GHP (geothermal heat pump) for this configuration. As illustrated on the upper left of Figure 1.24, a "direct" geothermal system can be converted into a GHP by the insertion of a heat pump into the circulating loop.

In Chapter 4 of this book, where I provide the detailed design of the world's first 1,000 mW solar–hydrogen demonstration power plant, I included and described the controls of a GHP system, which provide (in addition to the solar energy storage) an alternate source of electricity, during the night or at other times when solar energy is not available. For the details of this optimized control system, please refer to Figure 4.1 (gatefold).

Solar and geothermal systems can also be combined when the size of the system is small, such as in cases of "energy-free" private homes, where the combined goal of the solar and geothermal systems is to continuously meet the electricity needs of the home. Figure 2.103 illustrates such a system. In that system the main load variable is the insolation (the solar heat input to the hot water collectors). The changes in insolation are reflected by the

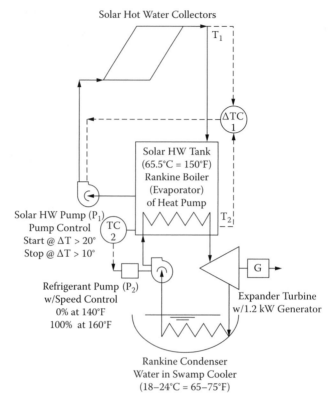

FIGURE 2.103
Geothermal heat pump (GHP) and solar hot water system combined to continuously meet the
electricity needs of a home.

temperature of the solar hot water. Therefore, a simple control scheme is to
look at both the temperature difference between the temperature at the solar
collector (T_1) and the solar water tank (T_2), and start the water pump (P_1)
when $T_1 > T_2$ by over, say, 20°C and stop it when this difference drops to 10°C
using a differential thermostat ΔTC-1.

The same holds true for the Rankine cycle refrigerant pump (P_2), operating between the solar hot water tank and the pond (swamp). Because the
condenser temperature (swamp temperature) is more or less constant, one
need not detect the ΔT and can just look at the boiler (solar water tank temperature—T_2) and keep P_2 running whenever the temperature T_2 is above
60°C (140°F). The temperature controller TC-2, brings P_2 to full speed as the
T2 temperature reaches, say, 70°C (160°F).

If one wants to maximize the efficiency of the system, one can use temperature difference control, but it costs more and is usually not justified for
a private home. If such controls are used, the measured variable would be
the ΔT between the boiler and condenser waters, and one would prepare
a three-dimensional plot, where the coordinates are ΔT, refrigerant flow,
and net electricity production. The *net* electricity produced is the difference

between the power that is generated minus pumping power invested. The resulting three-dimensional surface will have a maximum point for each ΔT. Therefore, one would read the flow corresponding to that maximum and set the pump speed to match it.

For a power plant or other large system (a geothermal power plant using groundwater or a lake as its heat source), this optimization strategy might be justified, because one might obtain another 10% of power, but in small household systems, the effort required does not "pay."

2.13.3 "Flash Steam" Systems

In the case of "flash steam" power plants, the steam is either generated directly by the production wells or the wells produce hot water from which steam can be separated to drive conventional steam turbine generators. The size of these plants ranges from 100 kW to 150 mW. Figure 2.104 illustrates the optimizing control system for such a geothermal facility.

Here, a variable-speed pump transfers the hot water from the production well into the steam separator. The speed of the pump is set by the tank level. (Variable-speed pump station controls are discussed in Section 2.17). The level control signal is corrected for steam pressure variations by multiplying the two. This is called a two-element feedwater system.

The high-pressure steam from the separator is sent to the steam turbine under speed control, and the generated electricity is sent to the grid or to other users. The low-pressure steam is condensed by cooling it with cooling water. The flow of the cooling water is modulated by the turbine exhaust pressure controller.

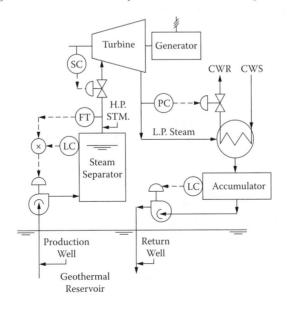

FIGURE 2.104
Geothermal flash steam power plant controls.

The lower this pressure, the more energy the turbine will generate, so if there are no other considerations (such as the limited availability of cooling water), the optimum operation occurs when this valve is fully open, if the incremental energy gain exceeds the increased pumping cost of cooling water. The condensate is collected in an accumulator tank and is returned under level control (by gravity or pumping, depending on elevations) into the return well.

For additional details on turbine controls and optimization, refer to Section 2.19.

2.14 Heat Exchanger Optimization

The transfer of heat is one of the most basic and best-understood unit operations. Heat can be transferred between the same phases (liquid–liquid, gas–gas) or phase change can occur on either the process side (in the case of condensers, evaporators, and reboilers) or the utility side (in the case of steam heater) of the heat exchanger.

Heat transfer, being one of the best-understood processes, has a good potential for modeling and optimization. In addition to energy savings, through heat exchanger control and optimization, monitoring is also needed to obtain plantwide energy audits and to provide historical trend and efficiency information. Trend data can be useful in deciding when to clean heat transfer surfaces, to select optimum cooling water temperatures, or steam-supply pressure and superheat values.

2.14.1 Degrees of Freedom, Gain, Time Constants

The degrees of freedom of a process define the maximum number of independently acting automatic controllers that can be placed on it. The available degrees of freedom are determined by subtracting the number of system-defining equations from the number of process variables. The temperatures and flows are considered to be variables; specific and latent heats are parameters. A steam condenser for example has four variables, which are the two flows (cooling water and steam) and the two temperatures, the inlet and outlet temperatures of cooling water. So there are four process variables and one defining equation, the conservation of energy. Therefore, this system has three degrees of freedom; thus, a maximum of three automatic controllers can be placed on it.

In a liquid–liquid heat exchanger, there are six variables (four temperature and two flow variables) and one (conservation of energy) defining equation, resulting in five degrees of freedom. In a steam-heated reboiler or in a condenser cooled by a vaporizing refrigerant (assuming no superheating or supercooling), there are only two flow variables and one defining equation. Consequently, they have only a single degree of freedom, and therefore, only one automatic controller can be used. In the majority of installations, fewer controllers are used than the available degrees of freedom, but every once in a while mistakes are made by using too many controllers and thereby over-defining the process.

Because the heat transfer surface of a heat exchanger is fixed, the more heat it needs to transfer, the less efficient it becomes, and therefore, the lower the process gain (G_p) of the heat transfer process will be. Therefore, as the load rises and the G_p drops, the valve gain (G_v) should be rising to keep the loop gain constant. A valve having this characteristic is called *equal percentage* (see Section 2.7.1). The loop gain is the product of the process, sensor, control-ler, and valve gains: $(G_p)(G_s)(G_c)(G_v)$. The gain product of a well-tuned control loop should be 0.5, if a decay ratio of 4:1 is desired.

The time constant of any process is the result of its capacitance and resis-tance. Usually, the heat exchanger outlet temperature is the controlled vari-able, and the flow rate of the heat transfer fluid is the manipulated variable. The time constant of an exchanger is a function of the mass and the specific heat of the tube material, the mass flow, and the specific heat of the process and utility streams and their heat transfer coefficients.

Dead time is also called transportation lag, because it is the time required for fresh heat transfer fluid to displace the contents of the exchanger and its associated piping. The dead time is the worst enemy of control, because until it has expired, a change in the heat transfer fluid flow (or temperature) will not even begin to have an observable effect. For a heat exchanger, the dead time is usually between 1 and 30 seconds. When the equipment is correctly designed, the dead time is much less than the time constant.

2.14.2 Tuning the Control Loop

The tuning of the controller is the process in which the dynamics of the controller is matched to the dynamics of the process. During tuning, the controller gain for example is adjusted until the loop gain (the gain product of the loop components) is about 0.5. Control loops can be tuned by consider-ing the dynamics of only the process (open-loop method) or by evaluating the response of the complete loop (closed-loop method). Because the period of oscillation of all control loops (P) is 3.5 to 4.0 times their dead time, the integral and derivative settings of PID controllers are generally adjusted as a function of the dead time of the process.

For noninteracting control loops with zero dead time, the integral setting (minutes per repeat) is about 50% and the derivative, about 18% of the period of oscillation (P). As dead time rises, these percentages drop. If the dead time reaches 50% of the time constant, I = 40%, D = 16%, and if dead time equals the time constant, I = 33% and D = 13%. When tuning the feedforward control loops, one has to separately consider the steady-state portion of the heat transfer process (flow times temperature difference) and its dynamic compensation. The dynamic compensation of the steady-state model by a lead/lag element is necessary, because the response is not instantaneous but affected by both the dead time and the time constant of the process.

2.14.3 Liquid–Liquid Heat Exchangers

In a liquid–liquid exchanger, the total heat transferred (Q) from the hot process fluid to the cooling water is dependent on the overall heat transfer coefficient (U), the heat transfer area (A), and the log mean temperature difference (ΔTm). Therefore, any of these can be manipulated to control Q.

The dead time of a heat exchanger equals its volume divided by the flow rate through it. As process flow increases, the process dead time is reduced and the loop gain is also decreased. If the controlled variable (Th_2 in Figure 2.105) is differentiated with respect to the coolant flow (manipulated variable F_c), the steady-state gain of the process is given by:

$$\frac{dTh_2}{dFc} = \frac{C_{pc}}{F_h C_{ph}}$$

where C_{pc} is the specific heat of the cold fluid and C_{ph} is the specific heat of the hot fluid. If $C_{pc} = C_{ph}$, the steady-state gain equals $1/F_h$. Therefore, at lower flows, a unit adjustment of a valve opening will have more effect, because the process fluid will spend more time in the exchanger and as the process gain increases, the loop becomes more sensitive and more prone to cycling. Conversely, as the load rises, the loop becomes more sluggish, as the residence and dead times are reduced and the process gain is lowered.

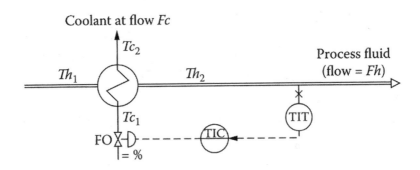

FIGURE 2.105
The feedback-type temperature control of a process cooler involves the control of a nonlinear and variable gain process.

As the process gain of all heat exchangers varies with load, it is not possible to tune the TIC if all other loop gains remained constant. Therefore, using an equal-percentage valve is a good solution if the temperature difference through the exchanger ($Th_1 - Th_2$) is constant. If it is not constant, feedforward compensation of the gain is required, or if sluggish behavior at high loads is acceptable, the TIC can be tuned for the minimum load (maximum process gain).

It is generally recommended that positioners be provided for the control valves to minimize the effects of valve friction (hysteresis and dead band) effects. In the majority of installations, a three-mode controller would be used for heat exchanger service. The use of the derivative or rate action is essential in slower (long time lag) systems or when sudden changes in heat exchanger throughput are expected. Because of the relatively slow nature of these control loops, the proportional band setting usually approaches 100%. This means that the valve will be fully stroked only as a result of a substantial deviation from the set point. To correct this offset, the use of the integral control mode is also required.

In order to obtain good control response (a maximum of 1 second distance velocity lag), the temperature sensor should not be more than 1 m (3 ft) from the exchanger outlet. Also, in order to detect a temperature change, first heat must be transferred into the thermal bulb through its fixed area. This time lag varies from a few seconds to minutes, depending on the area and mass of the bulb and on the process fluid being detected. Measurement of gas temperatures at low velocity involves the longest time lags, and measuring water (or dilute solutions) at high velocity results in the shortest time lags.

The addition of a thermowell will further increase the total sensor lag. When thermowells are used, it is important to eliminate any air gaps between the bulb and the socket. One method of reducing time lag is by miniaturizing the sensing element. For example, the accurate detection of high-temperature gases (500 to 2000°C) at low velocities of 1 to 2 m/s (3 to 6 ft/s) is a problem, even for thermocouples. This is because conductance through the lead wires and radiation both tend to alter the sensor temperature faster than the low velocity gas flow can resupply the lost heat. In such applications (fuel cell controls are prime examples), optical fiber thermometry is a good choice.

Thermocouples are usually not accurate enough for the precise measurement of temperature differences and are not fast enough to detect high-speed variations in temperature. Resistance temperature detectors (RTDs) can detect temperature differences of 5°C (10°F) at a measurement error of ± 0.02°C (±0.04°F). If higher-speed response is desired, thermistors or infrared detectors should be considered.

In the conventional control loop, the measurement lag is only part of the total time lag of the control loop. For example, an air heater might have a total lag of 15 minutes. Of this lag, 14 minutes is contributed by the *process lag*, 50 seconds by the bulb lag, and 10 seconds by the control valve lag. Bypass control is often applied to circumvent the dynamic characteristics of heat exchangers, thus improving their controllability. Bypass control can be achieved by the use of either one three-way valve or two two-way valves.

2.14.4 Heating with Steam

The steam-heated exchanger is also nonlinear (Figure 2.106). The steady-state gain of such an exchanger is the derivative of its outlet temperature with respect to steam flow, having the dimension of °F(lb/h):

$$K_p = \frac{dT_2}{dF_s} = \frac{\Delta H_s}{FC_p}.$$

Therefore, its process gain varies inversely with flow. Its time constant, dead time, residence time, and period of oscillation also vary with flow.

Therefore, the TIC loops have a tendency to become unstable at low and sluggish at high loads. In order to eliminate cycling, these loops are often tuned at minimum load, resulting in sluggish response at higher loads. One way to compensate for the drop in process gain as the flow increases is to use an equal-percentage valve whose gain increases with load. This is sufficient if the temperature rise $(T_2 - T_1)$ is constant. Otherwise, feedforward compensation is needed.

$$Q = F_s \,\Delta H_s = F \, C_p \, (T_2 - T_1)$$

Where Q = Heat-Transfer Rate
F_s = Steam Mass Flow
ΔH_s = Latent Heat of Vaporization
F = Feed Rate
Cp = Heat Capacity of Feed
T_0 = Steam Supply Temperature
P_1 = Steam Supply Pressure
P_2 = Steam Valve Outlet Pressure
P_s = Condensing Pressure
T_1 = Inlet Temperature
T_2 = Outlet Temperature
ΔTm = Log Mean Temperature Difference
T_s = Condensing Steam Temperature

F_s = 500 to 2500 lb/hr (227 kg/hr to 1134 kg/hr)
P_1 = 200 PSIA (1.38 MPa)

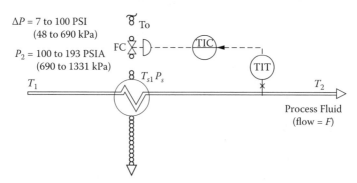

FIGURE 2.106
The feedback control of a steam-heated exchanger and its characteristic equation.

The condensing pressure is a function of the load when the temperature is controlled by throttling the steam inlet. Once this pressure drops below the sum of the trap back pressure plus trap differential, it is no longer sufficient to discharge the condensate; it will start accumulating in the exchanger, covering up more and more of the heat transfer area, which results in a corresponding increase in condensing pressure. When this pressure reaches the pressure needed to discharge the trap, the condensate is suddenly blown out and the effective heat transfer surface of the exchanger increases instantaneously. This can result in noise, hammering, and cycling as the exchanger surface is covered and uncovered.

This cycling can be eliminated by mounting the control valve in the condensate pipe, but this creates new problems, because when the load decreases, the process is slow; steam has to condense before the condensate level is affected, and when the load increases, the process is fast, because blowing out liquid condensate is fast. With such "nonsymmetrical" process dynamics, control is bound to be poor. A better option is to use lifting traps to prevent condensate accumulation. These pumping traps will make temperature control possible even when the heater is under vacuum, but will not improve the problem of low rangeability, and the possible use of two control valves in parallel can still be necessary.

Improved control can be obtained by using condensate level controllers or by bypassing some of the heated process stream. As the required level is a function of the load, it can be adjusted automatically. Bypass control using one 3-way or two 2-way valves also circumvents the transient characteristics of coolers; it also creates an additional degree of freedom. Therefore, the steam feed, for example, can be throttled to maintain a constant condensing pressure, which eliminates problems associated with condensate removal. A limitation of this approach is that when all the process flow is sent through the bypass, boiling may result in the stagnant exchanger.

2.14.5 Condenser and Vaporizer Controls

Condenser controls have already been discussed in connection with optimizing the pressure controls of distillation columns. Condensers can also be operated by keeping the condensate temperature constant by throttling the cooling water flow through the condenser. This can cause a high temperature rise on the water side, which will cause fouling if the water is not chemically treated. As was shown in Figure 2.86 and discussed in Section 2.9.4, this problem can be overcome by controlling the condensate level.

Vaporization controls depend on the operating pressure. If vacuum is required to vaporize the process fluid and if steam jets are used to create that vacuum, it is recommended that a pressure controller be installed on the steam inlet to maintain the optimum pressure required by the ejector. For processes in which load variations are expected, the operating costs can be lowered by installing a larger and a smaller ejector and automatically switching to the small unit when the load drops off, thereby reducing steam demand.

When steam-heated reboilers are used, only one degree of freedom is available; therefore, only one controller can be installed without overdefining

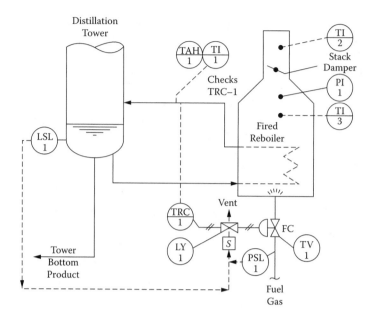

FIGURE 2.107
The feed heater of a refinery.

the system. This one controller usually throttles the rate of steam addition. Minimum condensing pressure considerations are the same as have been discussed earlier in connection with liquid heaters. When several vaporizers operate in parallel, it is important to balance the load distribution between them, as was described in Section 2.10.

In fired reboilers with natural circulation, one must guard against overheating, because most tower bottoms will coke or polymerize if subjected to excessive temperatures for some length of time. As shown in Figure 2.107, the flow of tower bottoms is usually not measured, because the liquid is near the flash point, and because it is usually of a fouling nature, tending to plug most flow elements. Proper circulation is guaranteed by hydraulic design.

Return temperature is controlled by throttling the fuel gas flow. The furnace draft can be manually set by the operator adjusting the stack damper. The major dangers in this type of furnace operation include the interruption of process fluid flow or the stoppage of fuel flow. If tower bottom flow is lost, overheating of the tubes may result. Momentary loss of fuel can also be dangerous because flames can be extinguished, and on the resumption of fuel flow, a dangerous air–fuel mixture can develop in the firebox. Under either of these conditions, the fuel flow must be stopped automatically.

2.14.6 Advanced and Optimizing Controls

Override controls are used to protect against fouling of the heat transfer surfaces, when the water outlet temperature exceeds 50°C (122°F) or to prevent steam or process pressures from reaching undesirable limits. It is also possible to apply override controls in an envelope configuration, so that a number of limits (temperature, level, valve opening, etc.) can all override the normal operation when their limits are reached.

In heat exchanger applications, cascade loops are configured so that the master detects the process temperature and the slave detects a variable, such as steam pressure, that may upset the process temperature. The cascade loop, responds immediately and corrects for the effect of the upset before it can influence the process temperature. The cascade master adjusts the set point of the slave controller to assist in achieving this. Therefore, the slave must be much faster than the master. A rule of thumb is that the time constant of the primary controller should be ten times that of the secondary, or the period of oscillation of the primary should be three times that of the secondary. One of the quickest (and therefore best) cascade slaves is the simple and inexpensive pressure regulator.

In a feedback configuration the controlled variable (temperature) has to be upset before correction can take place. *Feedforward* is a mode of control that corrects for a disturbance before it can cause an upset. Figure 2.108 illustrates feedforward control of a steam heater. The feedforward portion of the loop detects the major load variables (the flow and temperature of the entering process fluid) and calculates the required steam flow (W_s) as a function of these variables. When the process flow increases, it is matched with an equal increase in the steam flow controller set point. Because instantaneous response is not possible, dynamic correction by a lead-lag element is provided.

The feedback portion of the loop has to do much less work in this configuration, as it only has to correct for minor load variables, such as heat losses to the atmosphere, steam enthalpy variations, and sensor errors. The feedback and feedforward portions of the loop complement each other. The feedforward portion is responsive, fast, and sophisticated, but inaccurate. The feedback portion is slower, but is capable of correcting the upsets caused by unknown or poorly understood load variations, and it is accurate.

The process gain of heat transfer varies inversely with the process flow (W). If the temperature rise ($T_2 - T_1$) across the exchanger is also variable, even the use of equal-percentage valves cannot correct for this nonlinearity in the process. In that case, the only way to keep the process gain constant is to use the feedforward system shown in Figure 2.108. Here, a reduction in process flow causes a reduction in the gain of the multiplier, which cancels the increase in process gain. Thus, the feedforward loop provides gain adaptation as a side benefit. The result is constant total loop gain, and therefore, stable loop behavior.

Similarly to cascade control, load upsets or supply disturbances are corrected by feedforward control before they can upset the main variable, and therefore contribute to stable operation and to fast recovery from upsets. In addition to feedforward, more sophisticated model-based strategies are also available, as discussed in Sections 2.6.16 through 2.6.18.

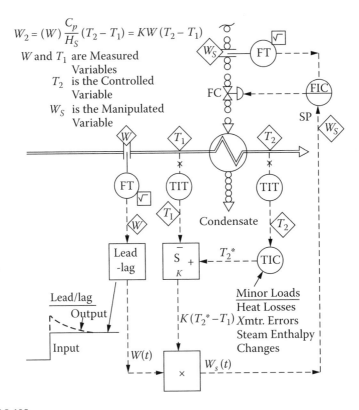

FIGURE 2.108
In feedforward optimization of steam heaters, major load variations (T_1 and W) are corrected by the feedforward portion of the loop, leaving only the minor load variables for feedback correction.

When the process might require either cooling or heating, the control loops operate on a *split-range signal*. When the process temperature is above the desired set point, first heating is reduced (between 50 and 100%), and if that is not sufficient to bring the process temperature down to set point, the signal will further decrease, thereby fully closing the heat supply and causing the coolant valve to start opening. At a 0% signal, the total cooling capacity of the system is applied to the recirculating oil stream, which at that time flows through the cooler without bypass.

The limitations of such a split-range operation include that near 50%, the system can be unstable and cycling, and when the signal is between 50% and 100%, the shell side of the exchanger can become a reservoir of cold heat transfer fluid. This upsets the control system twice: once, when the system just begins to heat, and once when the cold fluid has been completely displaced and the outlet temperature suddenly rises. Finally, most of these systems are nonsymmetrical in that the process dynamics (lags and responses) are different for the cooling and heating portions. These limitations can be overcome by the use of cascade loops and by overlapping of the two valve positioners. The resulting sacrifice of heat energy can be justified by the improved control obtained.

2.15 Hydrogen Process Optimization

The control and optimization of high-pressure and low-temperature processes require the same control strategies as any other industrial process. The difference is only in the selection of the construction materials, which are used in the sensors and control valves. In fact, process control tends to be less sophisticated in hydrogen applications.

A good example is the control of the Space Shuttle engines, where each of the three engines generates 200,000 kgf (400,000 lbf) lifting force and each weighs 3,200 kg (7,000 lb). These engines can operate at extreme temperatures as the LH_2 fuel is at $-253°C$ ($-423°F$), and when burned, its combustion temperature reaches $3,300°C$ ($6,000°F$). Yet, the controls used are not that sophisticated at all.

On–off control is used to keep the vapor pressure within limits in both the fuel (hydrogen) and oxidant (O_2) tanks. The LH_2 storage tank vapor space, for example, is held between 32 and 34 psig. Below 32 psig a GH_2 supply valve is opened to the tank; at 33 psig this valve closes, and at 35 psig a relief valve is opened. The same type of on–off control is applied to keep the O_2 tank pressure between 20 and 22 psig. Similarly, to control the propellant flows, the 17 in. control valves are pneumatically operated.

As will be seen here, the control and optimization of solar-generated hydrogen processes are also similar to the strategies that have been used for decades. The main steps in the production of hydrogen from solar energy are as follows: (1) the generation of hydrogen from water, by electrolysis, (2) compression, (3) liquefaction, (4) storage and transportation, and (5) utilization of this fuel in power plants, IC engines, and fuel cells. The control and optimization of fuel cell and combustion processes have already been covered (Sections 2.2 and 2.12). The others will be discussed here.

2.15.1 Electrolyzer Controls

As was discussed in detail in Chapter 1, Section 1.5.4, passing electricity through an aqueous electrolyte solution generates hydrogen at the negative electrode (cathode) and O_2 at the positive electrode (anode). There are a number of electrolyzer designs on the market, and they all are made up of parallel cells. Each cell is split in two by a diaphragm, and in each half there is an electrode; an anode in one half and a cathode in the other. One popular

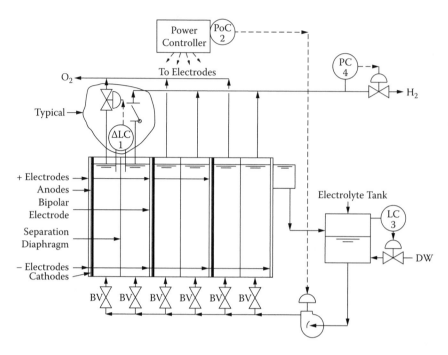

FIGURE 2.109
Hydrogen from water and solar energy are generated by optimized electrolyzer.

design uses "bipolar" electrodes (Figure 2.109), which serve not only as the separation walls between the cells, but also provide a negative electrode surface to one cell and a positive electrode surface for the next.

The optimization goals for the electrolyzer control system are to maximize hydrogen production at minimum cost of operation while protecting the separation diaphragms from damage. Diaphragm damage can occur if pressures of the generated O_2 and hydrogen (on the two sides of the diaphragm) are not identical, and therefore, the pressure difference between them exceeds the design limit of the diaphragm. On the other hand, if that difference is reduced and the diaphragms can be made thinner, both the cost and the weight of the electrolyzer are reduced. Therefore, one of the optimization goals is to minimize the ΔP across the diaphragm. (This is also the case with fuel cells.) When several hundred cells are operated in parallel, this design differential across the diaphragms has a substantial influence on the system costs.

In today's electrolyzers and fuel cells, ΔP controllers are used for this purpose, and therefore, the diaphragms are thicker, and the units are bulkier and heavier than they need to be. Figure 2.109 shows a better, more sensitive control system, which looks at the difference in electrolyte levels. ΔLC-1 can detect differences of a couple of millimeters between the half-cell levels and can modulate the O_2 outflow to match the pressure on the cathode side. The cathode side pressure is a function of the set point of the common hydrogen pressure controller (PC-4) and the drops across the check valves, connecting each half-cell to the hydrogen header.

Another goal of electrolyzer optimization is to make sure that both the available electric power (PoC-2) and the electrolyte flow are evenly distributed among the cells, because good electrolyte mixing and high velocity are both required to maintain cell efficiency. Even distribution of the electrolyte can be achieved by manually or automatically adjusting the openings of balancing valves (BVs) and making sure that the valve in the path of the most resistance (usually, but not always the cell, which is furthest from the electrolyte pump) is fully open, so pumping costs are minimized.

In order to maximize electrolyzer efficiency, the available solar energy has to be equally distributed by the power controller (PoC-2) among the cell electrodes and the rate of electrolyte circulation has to be matched to the electrolyzer loading. The other contribution to efficiency is minimizing pumping costs, which is achieved by the use of variable-speed pumps and by circulating only as much electrolyte as the power distribution controller (PoC-2) requires to maximize efficiency.

2.15.2 Hydrogen Compression

The hydrogen generated by the electrolyzers is usually at a pressure of 3 to 25 bar (40 to 360 psig). In order to fill GH_2 storage tanks or transport vehicles, the gas from the electrolyzer has to be compressed to about 1,000 bar (15,000 psig). As was discussed in detail in Section 1.5.5, diaphragm-type compressors and multiple-stage hydraulic intensifiers are used to generate these higher pressures. Figure 2.110 illustrates one of these compressor packages.

The control and optimization of compressors have already been described in Section 2.5. That description is also applicable to hydrogen compressors, particularly the controls and optimization of multiple compressors. Because of the high pressures required, compressors are often operated in series. Multiple compressors are also used in parallel, because when solar energy is the source of electricity for operating the electrolyzers, the volume of hydrogen generated will vary greatly as insolation varies. Therefore, in order to maximize the efficiency of compressing the highly variable flow of hydrogen, several compressors need to be operated in parallel.

Controlling two or more compressors operating in parallel and having identical characteristics would be relatively simple. It is very difficult, if not impossible, to find two compressors having identical performance characteristics. Slight variations in flow can fully load one compressor and cause the other to operate with wasteful recycling. The control scheme shown in Figure 2.111 alleviates that problem.

Figure 2.112 illustrates how two compressors can be proportionally loaded and unloaded. Because of age, wear, or design differences, no two compressors are identical. A change in load will not affect them equally, and each should therefore be provided with its own antisurge system. Another reason for individual surge protection is that check valves are used to prevent

FIGURE 2.110
Volumetric flow is regulated by throttling a diaphragm-type control valve in this nonlubri-
cated, vertical, three-stage, two-cylinder hydrogen compressor. (Courtesy of Koehler and
Hoerter GmbH.)

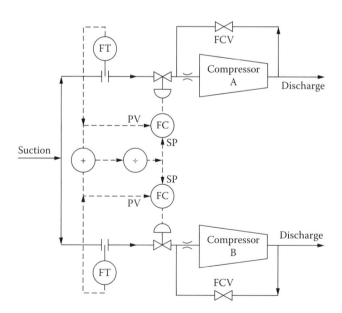

FIGURE 2.111
When operating multiple compressors in parallel, the total flow (load) can be distributed
among the machines that one is fully loaded and the other handles the variations in demand.

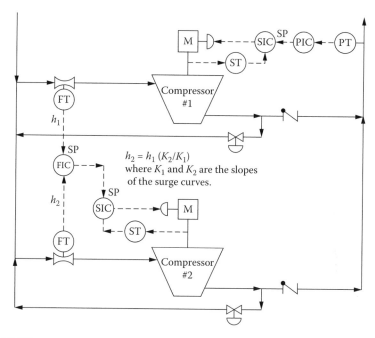

FIGURE 2.112
Two parallel compressors can be loaded to keep both of them at equal distance from their surge curves.

backflow into idle compressors. Therefore, the only way to start up an idle unit is to let it build up its discharge head while its surge valve is partially open. If this is not done and the unit is started against the head of the operating compressors, it will surge immediately. The reason why the surge valve is usually not opened fully during start-up is to protect the motor from overloading.

Improper distribution of the load is prevented by measuring the total load and assigning an adjustable percentage of it to each compressor. The load distribution can be computer-optimized by calculating compressor efficiencies (in units of flow per unit power), and loading the units in the order of their efficiencies. The same goal can be achieved if the operator manually adjusts the ratio settings. This is a more stable and responsive configuration than a pressure-flow cascade, because the time constants of the two loops are similar.

2.15.3 Hydrogen Liquefaction Optimization

In the process of hydrogen liquefaction, one must consider the inversion temperature (−361°F or −183°C or 90°K) of hydrogen, because the behavior of this gas changes (inverses) at that temperature. Below the inversion temperature, when the pressure is reduced, the hydrogen temperature will drop (above that temperature, the opposite occurs; a drop in pressure causes a rise in tem-

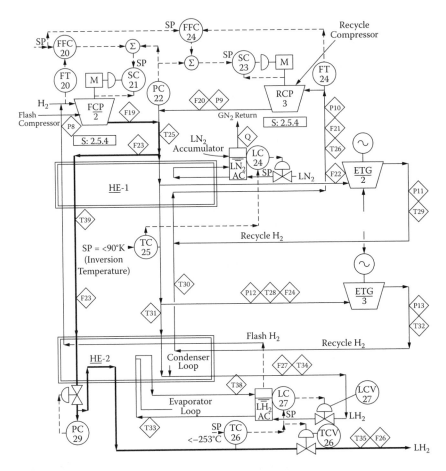

FIGURE 2.113

The control and optimization of a nitrogen precooled hydrogen liquefaction process.

perature). Therefore, in the process of liquefaction, hydrogen first has to be cooled below its inversion temperature—by such means as cooling with liquid nitrogen—before the Joule–Thomson effect can be utilized. Figure 2.113 describes the control and optimization of a precooled liquefaction process.

In Figure 2.113, only the main control loops are shown; they will be explained in the following text. Not shown are the basic compressor controls for surge protection, dynamic compensation of the control loops, and auto–manual switching. These can be found in other figures in this book. The corresponding section numbers are given in rectangles below the particular piece of equipment shown in Figure 2.113. It should also be noted that the control loops of the liquefaction process are highly interacting and their decoupling systems are not shown in Figure 2.113, but the general strategies for control loop decoupling are described in Section 2.6.

Before hydrogen can be liquefied, it must be purified and converted from the normal to the "para" form. LH$_2$ is 99.8% para and only 0.2% ortho. Catalysts, which can be placed in the heat exchangers (HEs), are often used to speed this conversion. The "thermodynamic" (theoretical) heat of conversion from the normal to the para form of hydrogen is 0.146 kWh/kg. The higher heating value of hydrogen is 39.3 kWh/kg.

Figure 2.113 describes the simplified controls of a hydrogen liquefaction process. This particular process uses both nitrogen precooling and hydrogen heat-pump cooling. Other liquefaction process configurations have been developed under the names of Hydrogen Claude, Helium Brayton, combined reverse-Brayton Joule–Thomson (CRBJT), Neon Brayton, Neon with cold pump, etc. Each is being evaluated. As a consequence of all the R&D work, further improvements are expected in the hydrogen liquefaction process.

The GH_2 from the electrolyzers (usually at about 5 bar = 72.5 psig) is distributed among the power plant's liquefiers and high pressure compressors as a function of the demand for high pressure gas ($HPGH_2$) and LH_2 by flow ratio controllers (FFC-20 and FFC-24). These ratio controllers measure the total hydrogen flows and receive their set points from the plantwide production optimizer controller that decides what part of the generated solar electricity is to be used to make hydrogen, what portion of that should be gas, and what portion should be liquid or sent to the methanol plant. This liquid portion is then distributed among the liquefiers so that the most efficient units are always in operation.

The two key temperatures of the liquefaction process (inversion and liquid product temperatures) are controlled by TC-25 and TC-26. TC-25 modulates the level set point of LC-24 of the LN2 accumulator, and TC-26 adjusts the level set point of LC-27 of the LH_2 accumulator. The LH_2 (or nitrogen) supplies to the evaporators are controlled by cascade loops that adjust the levels in the accumulators, which in turn vary the heat transfer area, and therefore, the rate of evaporation.

The flash compressor (FCP-2) speed (SC-21) is set by a loop cascade in which the sum of the output signal of the ratio controller (FFC-20) and the discharge pressure controller (PC-22) is the set point of SC-2 in accordance with the control details shown in Figure 2.22. The speed controller (SC-23) of the recycle compressor (RCP) is adjusted by the discharge pressure controller PC-22 and FFC-24 as described in detail in Figure 2.22.

The heat pumps operate by moving the heat from the low-temperature (low-pressure) hydrogen product liquid flowing through the evaporator into the higher-temperature (higher-pressure) condenser. This is done because the evaporation of the coolant (in this case it is also hydrogen) can cool the hydrogen product stream, and because it is evaporated at that low pressure. Then, after being compressed, it gives up that heat as it is condensed at that high pressure. In Figure 2.113 the condenser and evaporator loops are noted in HE-2. Some of the energy of compression is regained in the expander turbine generators (ETGs). They recover the energy corresponding to the pressure difference between the high pressure hydrogen (or nitrogen) vapors from the condenser and the low pressure evaporator.

2.16 Power Plant Optimization

During the transition from the fossil-based to the renewable-based energy economy, it is important to reduce carbon emission and to increase the efficiency of conventional power plants. One way to do that is to recapture the CO_2 and convert it to methanol, by the addition of hydrogen as shown in the gatefold.

Subcritical fossil power plants operate at 36 to 38% efficiencies. Supercritical designs have efficiencies in the low- to mid-40% range, and the new ultracritical designs operate at pressures of 30 mpa (megapascal) and use dual-stage reheat at efficiencies up to 48%. Fossil power plants can also use gas turbines in conjunction with a steam boiler "bottoming" cycle. The efficiency of a combined cycle plant can approach 60% in large-size (500 + mWe) power plants.

Cogeneration is a mode of operation in which the plant produces both heat (steam or hot water) and electricity. In a combined cycle power plant, electricity is produced by two types of turbines, gas and steam. The gas turbine is operated by the combustion products of the fuel (Brayton cycle), whereas the steam turbine (Rankine cycle) is operated by the steam generated by the heat content of the exhaust gases leaving the gas turbine. In a traditional power station, consisting of a fired boiler, full condensing steam turbine and electric generator major losses occur in the steam condenser, which wastes more than 45% of the total thermal energy. Therefore, when the total electric power generated is constant, the overall power plant efficiency can be improved by lowering the inlet steam quantity to the condenser.

2.16.1 Cogeneration and Combined Cycle

Cogeneration means that there are two marketable products: electricity and heat. The heat is usually in the form of steam and is sold to process industries or for civil applications (district heating); thus, the latent heat of the steam is not wasted, but utilized. The steam user should be near the power plant in order to reduce steam transportation losses and the demand for steam should be continuous. The steam can be obtained from back-pressure steam turbines or from extraction in the intermediate- and low-pressure (IP or LP) stage of the turbine.

Smaller combined-cycle power plants (up to 60–100 mW) are particularly suitable for cogeneration, because they can be located very close to the industrial steam user plant.

In many cases the steam is generated at two pressure levels, and the heat recovery steam generator (HRSG) is designed to serve the steam users. In some installations, if the high- and low-pressure (HP and LP) steam characteristics are suitable for the steam turbine, flexible operation is obtained by sending the excess steam to the turbine or by drawing steam from the turbine supply. Figure 2.114 shows a steam distribution control system where

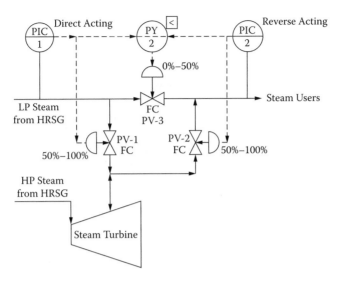

FIGURE 2.114
Control system used to distribute the steam generated by an HRSG in such a way that the variable demand for steam can be met by either sending the excess to the steam turbine or by supplementing it from the steam turbine.

the steam generated by the HRSG is sent to the steam users while sending the remaining low-pressure (LP) steam to the steam turbine. In case no steam is required by the steam users, PV-1 is fully open, and PV-2 and PV-3 are closed.

In a combined-cycle power plant, electricity is produced by two turbines, a gas and a steam turbine. The term *combined cycle* comes from the fact that the combustion gas turbine operates according to the Brayton cycle and the steam system operates according to the Rankine cycle. As shown in Figure 2.115, the dual-shaft combined-cycle plant consists of a gas turbine (GT)

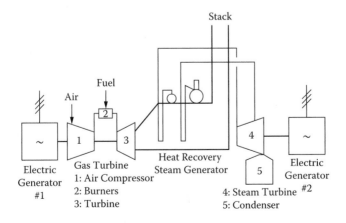

FIGURE 2.115
The main components of a dual-shaft combined-cycle power plant.

with its associated electric generator, an HRSG, a steam turbine (ST) with its associated condenser and electric generator plus auxiliaries such as a demineralization/polishing process, a fuel gas or fuel oil system, and a closed-circuit cooling water system. The GT exhausts into the HRSG, where the heat content of the flue gas produces steam, which is fed to the ST.

The combined-cycle configuration is preferred for base load applications, i.e., to operate continuously at full power or very close to it, even though sometimes the combined-cycle configurations can also be used to meet peak loads. During transients (either an increase or decrease of the total power), the ratio of the generated power between the GT and the ST can be different, because the GT has a much quicker dynamic response (seconds) than the HRSG and ST assembly (minutes). This fact has normally little impact on the operation of the power station, because a combined cycle is always operating in nearly steady-state conditions.

If the load increase is large, it is possible that after a first step of 5%, the rate of change in the GT needs to be limited to 2–5% per minute. This is because the ST is unable to accept a sudden change in steam characteristics and quantity. The net heat rate for combined-cycle units >150 mW is in the range of 5,700 to 6,800 Btu/kWh (6,015 to 7,175 kJ/kWh), whereas for 60 mW it is in the range of 6,500 to 7,000 Btu/kWh (6,860 to 7,385 kJ/kWh).

2.16.2 Level and Steam Temperature Controls

Boiler controls have already been described in Section 2.2, so their discussion here will be limited and oriented toward power generation. The level control of the steam drum of an HRSG is very similar to that of fired boilers, except that up to 30% of nominal steam flow, a single-element controller is usually used. Above 30%, the loop is bumplessly transferred to a three-element control (Figure 2.116).

The feedwater control valve can be located upstream or downstream of the economizer. Several HRSG manufacturers do not allow the economizer to steam and therefore the control valve is installed downstream of the economizer. The valve in these installations is quite critical because it can be subject to flashing and cavitation. Some units are designed to maintain a minimum flow through the economizers to prevent this, and the excess water is dumped from the steam drums to a condensate recovery system. Other HRSG suppliers accept a little steaming in the economizer and install the control valve upstream of the economizer with an accurate heat transfer design to reduce steaming during some transient conditions. The valve, in this case, is much less critical as it handles "cold" water without flashing or cavitation problems.

The valve body size is frequently determined not only on the basis of the *required valve capacity* (C_v), but also by considering the maximum acceptable flow velocity (approximately 30 ft/s [9 m/s]).

Steam temperature is usually controlled with desuperheating valves installed between the first and second superheaters, similarly to the arrange-

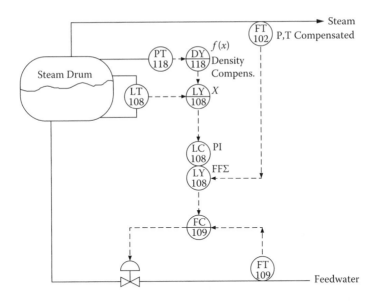

FIGURE 2.116
Three-element feedwater system, plus density compensation with drum pressure.

ment in regular boilers (Figure 2.117). If the superheaters are split into more banks in parallel, good practice requires a desuperheating valve per each bank. The control function is normally implemented by a cascade loop in which the master controller (TIC-111) senses the final steam temperature, and the slave senses the temperature after the desuperheater before the steam enters the second superheater. The temperature used as the measurement of the slave controller (TT-112) should be located about 60 ft (20 m) downstream of the desuperheater to allow the water droplets to vaporize, which is necessary to obtain a correct measurement and to prevent mechanical damage to the protecting thermowell. Sometimes a feedforward action by steam flow is added (FY-102) to the control system.

Condensate preheater (economizer) temperature control is unique to HRSGs and is not used on conventional fired boilers. Its purpose is to (1) prevent external condensation on the tubes of the preheater, which could cause external corrosion and (2) to obtain the correct inlet temperature to the deaerator.

The correct inlet temperature for the deaerator is obtained by mixing the hot condensate downstream of the preheater with cold condensate. This inlet temperature is maintained by a temperature controller around the preheater, which detects the mixed condensate's inlet temperature to the deaerator and is set at a little lower (15°F [8°C]) than the boiling temperature.

Burner management controls, complete with flame scanners, purging, and cooling should be provided to ensure boiler safety. Supplementary firing should be limited to a fraction of the gas turbine load. The applicable safety codes are a function of the presence of fuel gas in the proximity. Precautions include the use of double stuffing glands with bleeds between the glands on

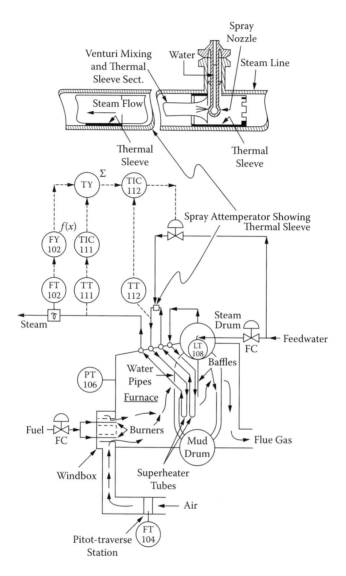

FIGURE 2.117
Cascade configuration controlling the desuperheater control valve.

the fuel gas valves. The intermediate bleeds from such valves are collected and vented to the atmosphere at safe location.

2.16.3 Gas Turbine and Electric Generator Controls

The control of steam turbines will be discussed separately in Section 2.19. Therefore, here the focus will be on the optimization of gas turbine (GT) operation. The GT system consists of three main parts: the air compressor (axial type), the burners, and the turbine itself. The mechanical power generated by the turbine is partly (about 50%) used to drive the air compressor, so that the net generated power is the difference between the full-generated power and the power required by the air compressor.

The characteristics of gas turbines are given at ISO (International Standards Organization) conditions, which assumes no losses (i.e., no pressure drop in the inlet and outlet ducts), and operation at 59°F (15°C), 14.696 psia (101.325 kPa), and 60% relative humidity. As the air temperature rises, the efficiency drops, whereas with a drop in temperature it rises. The current generation of GTs has an efficiency of 33 to 38%, referred to the low heating value of the fuel, when running in open cycle (i.e., exhausting to the atmosphere). This efficiency is close to the overall efficiency of a traditional power station. The heat rate for heavy-duty turbines in sizes greater than 100 mW is in the range of 9,000 to 10,000 Btu/kWh (9,500 to 10,550 kJ/kWh), whereas for sizes of about 25 mW, the heat rate is about 10,000 to 12,000 Btu/kWh (10,550 to 12,650 kJ/kWh). For aeroderivative turbines in sizes of about 40 mW, it is in the range of 8,200 Btu/kWh (8,650 kJ/kWh).

In the single-shaft arrangement, there is only one electric generator, driven by both the gas turbine and the steam turbine. This means that the electric power plant is simpler, because there is only one step-up transformer and one bay to connect to the grid.

There are two possible single-shaft configurations, shown in the top and bottom of Figure 2.118. One is to locate the gas turbine and the electric generator at the two ends of the shaft (top), and the other is to locate the steam and gas turbines at the two ends of the shaft (bottom).

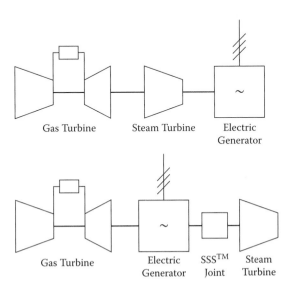

FIGURE 2.118
Single-shaft power plant configurations can locate the gas turbine and the electric generator at the two ends of the shaft (top) or can locate the steam and gas turbines at the two ends (bottom).

The configuration at the bottom requires a clutch or a special joint (e.g., SSS™) located between the generator and the steam turbine (ST). This is needed for the generator to be able to rotate during start, driven by the GT, with the ST stopped until the steam conditions are suitable for starting it with its own start-up procedure. Sometimes, when an even number of GTs are used, it is possible to have two GTs feeding only one ST. This configuration is known as 2+1. This simplification reduces the investment costs and reduces the electrical installation, with the additional benefit of having similar electrical components.

The GT is supplied with its own governor that takes care of all safety and control functions, including antisurge control of the compressor, IGVs (inlet guide vanes) control, burner control, start-up and shut-down sequences, excessive vibrations, etc. All safety functions of the GT are performed by the turbine governor that normally has a "2oo3" (two out of three) or a "1oo2D" (one out of two diagnostics) configuration, which is usually independent of the control functions. The control functions are performed in simple, redundant, or 2oo3 configurations that ensure the degree of safety and meet the availability requirement, which should be consistent with the size and criticality of the plant.

The governors are provided with their own operator interface and with processors that store information on the behavior of the turbine, including the sequence of events leading to a shutdown. The governor includes comprehensive self-diagnostics that allow easy maintenance while the GT is running, and still keeping all of the protections active. The GT runs in temperature control mode when it is at base load. When it is ramping up, it is in speed control, with the temperature of the first row of blades limiting the generated power.

Steam turbine controls are discussed in Section 2.19, so they are only briefly mentioned here. The largest steam turbines today are in the range of 150 mW. Only when one ST is fed by two HRSGs (in a 2+1 configuration) can the steam turbine's rated power reach 250 to 300 mW. The steam turbine is fed with HP superheated steam, is designed to accept IP and LP steam from the HRSG, and also can extract full steam flow at IP level in case of reheating. The vacuum condenser can be water- or air-cooled. The present trend is in favor of air-cooled condensers, particularly if the water availability is limited. The use of air-cooled vacuum condensers reduces the size of the cooling water towers required for the plant (still needed for cooling the rotating machinery) if closed-circuit cooling is selected. This approach also eliminates the need for large flows of cooling water and the thermal pollution of rivers, lakes, or the sea, if open-circuit cooling is used. The air-cooled condensers slightly decrease the efficiency of the steam turbine, mainly in hot climates.

When the plant is located in built-up areas, there might also be an atmospheric condenser that receives the steam, which is vented during the start-up to decrease the noise level at the plant. The auxiliary circuits of the steam turbine are the lube oil, control fluid, and vacuum system for the condenser.

The vacuum system can be equipped with steam ejectors or with liquid ring pumps (vacuum pumps), or a combination.

The electric generator is a three-phase unit and can be air- or hydrogen-cooled. The excitation system is static and the voltage/cosϕ control is obtained via the AVR (automatic voltage regulator) that is hardwired to the DCS or can be serially linked to the DCS, which is a less frequently used arrangement. In large groups, at start-up the generator is fed with variable frequency and acts as the launching motor. The lubricating system is normally common with the driver's system.

2.17 Pumping Station Optimization

The lifetime operating cost of a pumping station is about 100 times its first cost, because pumping is a very energy-intensive process. For this reason, the returns on pump optimization can be substantial. Pump optimization includes the goal of introducing only the energy that is needed to transport the fluid, but no more. The elimination of energy waste and the providing of good supply–demand matching, will not only lower operating costs, but will also reduce maintenance. The full optimization of pumping stations—including automatic start-up and shutdown—will not only reduce operating costs but will also eliminate human errors and increase operating safety.

2.17.1 Pump Curves, NPSH, and Cavitation

A pump is a liquid transportation device, which must develop enough pressure to overcome the hydrostatic and frictional resistance of the process as it delivers the required fluid. These resistance components are unique characteristics of the process served and can be described by system curves. The system curve of a process relates the pressure (head) required and the amount of fluid flow that is being delivered.

The characteristics of the system, which is served by a pump or pumping station, can be represented by a head-capacity system curve (Figure 2.119). The head at any one-flow capacity is the sum of the static and the friction heads. The static head does not vary with flow rate, as it is only a function of the elevation or back pressure against which the pump is operating. The friction losses are related to the square of flow and represent the resistance to the flow caused by pipe and equipment friction.

A generalized equation, describing the system curve of a process is $P = H + F_f(Q^x)$, where P is the required to pump discharge pressure, H is the static or elevation head, F_f is the friction factor, Q is the flow rate, and x is an exponent that varies between 1.7 and 2.0; usually, 2.0 is used.

Figure 2.119 illustrates the typical pump curve of a single impeller pump. It shows both the head-capacity curve of a centrifugal pump and the system

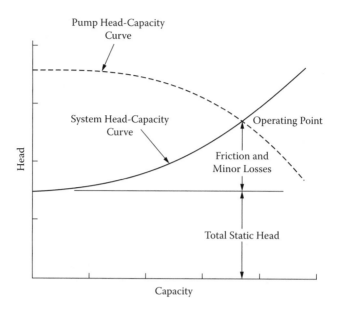

FIGURE 2.119
The system curve crosses the pump curve at the operating point.

curve of a process. The operating point is the point at which the pump and system curves cross each other. If the process flow is controlled by a control valve, a new system curve is generated by the added pressure drop through the valve. This added friction loss wastes energy that can be saved by not using valves, but changing the pump speed.

Figure 2.120 shows a three-dimensional plot—(a) pressure, (b) flow, and (c) speed—where the system curves form one surface (surface A) and the pump curves form another surface (surface B). The intersection of surfaces A and B is the operating line of the variable-speed pump.

Flow control via pump speed adjustment is less common than the use of throttling with valves, because most AC electric motors are constant-speed devices. If a turbine drive is used, speed control is even more convenient. However the advent of the pulse-width modulated (PWM) adjustable speed drive with sensorless flux-vector control has brought adjustable speed (AS) pumping into the mainstream of everyday applications.

The shape of the system curve determines the saving potentials of using variable-speed pumps. All system head curves are parabolas, but they differ in steepness and in the ratio of their static head to friction drop. The value of variable-speed pumping increases as the system head curve becomes steeper. Therefore, in mostly friction systems, the savings will be greater.

The cause of cavitation is that the pumped liquid flashes to vapor at one point inside the pump, where pressure is below the vapor pressure, and as the spinning impeller throws the liquid and vapor outward, the vapor bubbles collapse as the pressure rises above the vapor pressure. When the col-

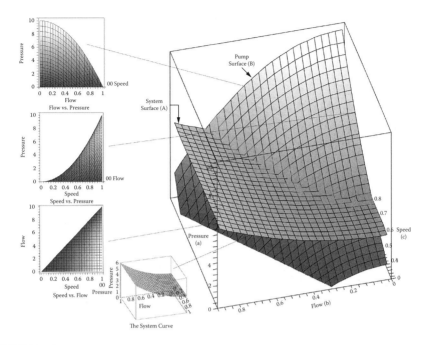

FIGURE 2.120
The variable speed pump operates on the line where the surface formed by the system curves intersects with the surface formed by the pump curves.

lapsing bubbles reach the pump wall, they collide with it, releasing extreme force. This gives rise to the characteristic sound of cavitation, and also to the consequent erosion, usually of the impeller.

If a pump cavitates, one might try to eliminate or minimize it by removing restrictions at the pump suction. If the pump cavitates at high flows, a second pump should be started at a lower flow. If cavitation occurs at low flows, one might turn off the pump. An extreme option is to inject a compressible gas into the impeller. This reduces pump efficiency and capacity, but it can eliminate cavitation.

When liquids are being pumped, it is important to keep the pressure in the suction line above the vapor pressure of the fluid. The available head measured at the pump suction is called the *net positive suction head available* (NPSHA). At sea level, pumping 15°C (60°F) water with the impeller about 1 m below the surface, the NPSHA is about 9.1 m (30 ft). It increases with barometric pressure or with static head, and decreases as vapor pressure, friction, or entrance losses rise. Available NPSHA is the characteristic of the process and represents the difference between the existing absolute suction head and the vapor pressure at the process temperature. The required net positive suction head required (NPSHR), on the other hand, is a function of the pump design (Figure 2.121). It represents the minimum margin between suction head and vapor pressure at a particular capacity that is required for pump operation. Cavitation can occur at suction pres-

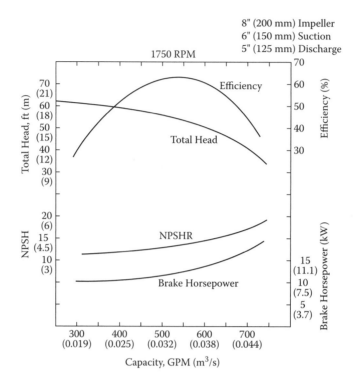

FIGURE 2.121
Typical characteristic curves of a single-impeller centrifugal pump.

sures exceeding the NPSHR of the pump, and therefore, if one wants to be positive, actual cavitation testing is recommended.

2.17.2 Pumping Station Optimization

Multiple pumps operate in parallel and are used if the process flow rangeability exceeds the throttling capability of a single pump. Booster pumps are installed in series and are used to increase the total discharge pressure of the station. Individual centrifugal pumps have a rangeability of about 4:1, which can be obtained by either speed control or by discharge throttling.

When two or more pumps operate in parallel, the combined head-capacity curve is obtained by adding up their individual capacities at each discharge head, as illustrated in Figure 2.122. The total capacity of the pump station is found at the intersection of the combined head-capacity curve with the system head curve. When constant-speed pumps are used in parallel, the added increments of pumping can be started and stopped automatically on the basis of flow.

When two or more pumps operate in series, the total head-capacity curve is obtained by summing up the pump heads at each capacity. Series pumping is most effective when the system head curve is steep. Multiple pumps in

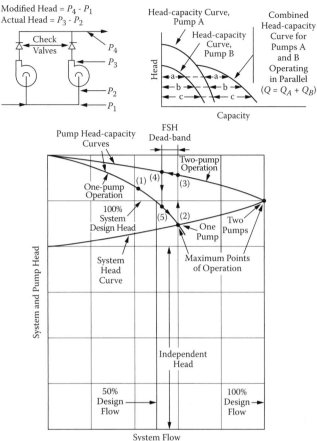

Modified Head = $P_4 - P_1$
Actual Head = $P_3 - P_2$

Check Valves

P_4
P_3
P_2
P_1

Head-capacity Curve, Pump A

Head-capacity Curve, Pump B

Head

←a→ ‑‑‑ ←a→
←b→ ←b→
←c→ ←c→

Capacity

Combined Head-capacity Curve for Pumps A and B Operating in Parallel $(Q = Q_A + Q_B)$

Pump Head-capacity Curves

FSH Dead-band

System and Pump Head

One-pump Operation
(1) (4)

Two-pump Operation
(3)

100% System Design Head
(5) (2)

One Pump

Two Pumps

System Head Curve

Maximum Points of Operation

Independent Head

50% Design Flow

100% Design Flow

System Flow

Two pumps, each with a capacity of 50% design flow at 100% design head

FIGURE 2.122
Pump turndown and rangeability can be increased by operating two or more pumps in parallel.

series are preferred from an operating cost point of view, but the capital cost investment of a single two-speed pump is lower.

When constant-speed pumps are used, the booster pump can be started and stopped automatically on the basis of pressure. In this case, an adjustable dead band is provided in the low pressure switch (PSL), which prevents the on–off cycling of the booster pump. The width of the dead band is a compromise; as the band is narrowed, the probability of cycling increases, whereas the widening of the band results in extending the periods during which the booster is operated unnecessarily.

A pumping system is optimized when it meets the process demand for liquid transportation at minimum pumping cost, and does that in a safe and stable manner.

In Figure 2.123, the pump station consists of a variable-speed and a constant-speed pump. PDIC-01 maintains a minimum of 10 psid (69 kPa) pres-

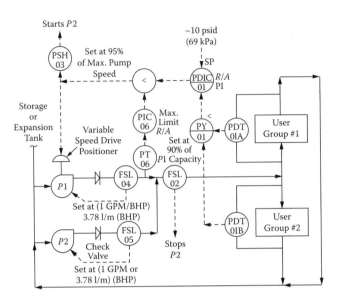

FIGURE 2.123
Optimization controls of a pump station consisting of a constant- and variable-speed pump.

sure difference between supply and return liquid pressures of each group of users. When the variable-speed pump approaches its maximum speed, PSH-03 will automatically start the constant-speed pump (P2). When the load drops down to the set point of FSL-02, this second pump is stopped.

One important recommendation to remember is that extra increments of pumps are started by pressure but stopped on flow. The pump speed is set to keep the pressure drop across the lowest user above some minimum limit. If the header supply pressure instead of the pressure drop across the load is to be controlled, the pressure transmitters should be located on main raisers or headers and should not be near major on–off loads.

One of the best methods of finding the optimum pump discharge pressure is illustrated in Figure 2.124. Here the optimum discharge pressure is selected to keep the most open user valve at a 90% opening. As the pressure rises, all user valves close; as the pressure drops, they will all open. Therefore, opening the most open valve to 90% causes all others to be opened also. This keeps the pump discharge pressure and the use of pumping energy at a minimum. As the valve position controller (VPC-02) lowers the set point of PC-01 and thereby opens the user valves, it not only minimizes the valve pressure drops but also reduces valve cycling and maintenance.

In order to make sure that the pressure controller (PC-01) set point is changed slowly and in a stable manner, the valve position controller (VPC-02) is provided with integral action only, and its integral time is set to be about 10 times that of PC-01. In order to prevent reset windup when PC-01 is switched to manual or local control from cascade, the valve position controller is also provided with an external feedback signal off the pump speed.

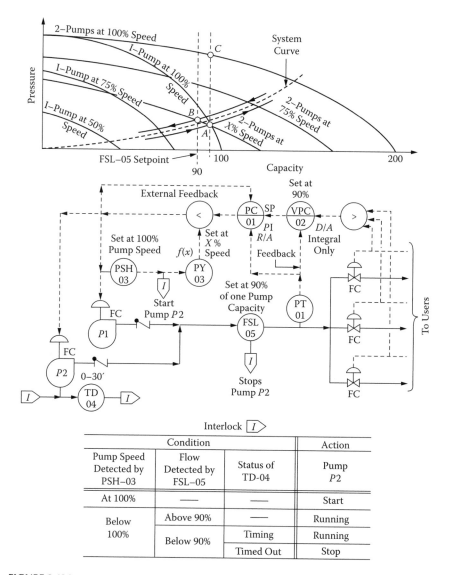

FIGURE 2.124

Illustration of a control system that optimizes the energy consumption of a pumping station, consisting of two variable-speed pumps, by keeping the most-open user valve at near 90% opening.

The pump station consists of two variable-speed pumps. When only one pump is in operation and the output of PC-01 approaches 100%, PSH-03 is actuated and the second pump is started. When both pumps are in operation and the flow drops to 90% of the capacity of a single pump, the second pump is stopped if this condition lasts longer than the setting of TD-04. The purpose of the time delay (TD-04) is to make sure that the pump is not started and stopped too often.

However, if the speed-control signal was unchanged when the second pump was started, an imbalance would occur, because the pressure and the corresponding pump speed of both pumps would instantaneously jump from point A to C. In order to eliminate this temporary surge in pressure, PY-03 is introduced. This is a signal generator, which, upon actuation by interlock 1, drops its output to x. This x corresponds to the required speed for the two-pump operation at point A. Once both pumps are operating smoothly and the load drops back to the point where it can be met by a single pump, the low-flow switch FSL-05 stops the second pump.

2.17.3 Calculating the Savings

Efficient pump starting and stopping requires finding the optimum pump selection by summing the required power consumption for any pump combination at any load, and implementing new combinations that could meet the total flow requirement at a lower power consumption. By the use of this model predictive approach, the pump combination that consumes the least amount of power can be identified and implemented.

To determine the savings resulting from variable-speed pumping, the relationship between the demand for flow and the power input required to meet that demand can be plotted. Figure 2.125 shows these curves for both constant-speed (control valve throttling) type and variable-speed type pumping systems. The difference between the two curves is the savings resulting from optimization. Once the savings curve has been established, the next step is to determine the operating cycle. The operating cycle identifies the percentages of time when the load is 10%, 20%, etc., up to 100%. Knowing the

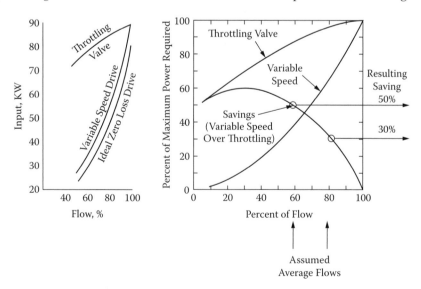

FIGURE 2.125
The savings generated by variable-speed pumping increase as the load drops off.

total horsepower of the pumps and the cost of electricity makes it possible to convert the resulting percentages into yearly savings.

The adjustable-speed centrifugal pump optimization techniques will reduce the energy consumption of pumping stations by 12% or more, depending on the nature of the load served. They will also reduce pump wear commensurately. Combined with surge-free pump starts and stops, and with their inherent predictive maintenance capability, these strategies can also improve the overall plant operation.

It is expected, that in the not too distant future, unit operations such as pumping will no longer be controlled by individual loops for flow, pressure, etc., but will be controlled by pump station controllers, which would be supplied with the software required so that the user only needs to plug in the system curve and the characteristics of each pump and to push the start button. Such unit operation software packages would include the logic, optimization, self-diagnostics, auto-start/stop, and other features to fulfill the needs of pumping stations and other unit processes. It is high time that we stopped controlling flows, pressures, and temperatures and started controlling and optimizing unit operations.

2.18 Solar Collector Farm Controls

For all collector designs, the control and optimization of solar farms include the positioning of the collectors to maximize collection efficiency. This is done by following the sun. For solar hot water generators, the controls include the water flow distribution controls (balancing) and the pumping rate optimization controls. For photovoltaic collectors, inverter and grid connection controls are needed, whereas for thermal systems the optimization involves both hot oil pumping/distribution and the boiler/steam turbine generator controls.

2.18.1 Solar Hot Water System Controls

Figure 2.126 illustrates a "passive" solar hot water collector. In this system there is no pump, and the water moves only when somebody in the house asks for hot water by opening a user valve. At that point the cold water inflow through the valve (V4) moves cold water into the collector, transfers the heated water from the solar collector into the solar hot water tank (V5), and sends the hot water from the tank to the user (V7). This system has no controls as such. If there is no usage and the sun is out, it is possible for the water to overheat in the collector or the tank. In that case, thermal expansion will increase the pressure, and when it reaches the settings of the pressure safety valves (PSVs), these safety valves will open, and both water and heat will be relieved and wasted.

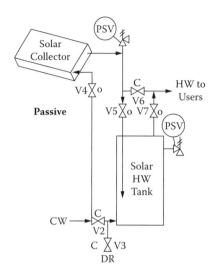

FIGURE 2.126
Passive solar hot water collector.

Figure 2.127 illustrates the "direct" (top) and "indirect" (bottom) hot water collector designs. Both are controlled by differential temperature switches (ΔTs), which start pump circulation only if the temperature difference between the hot water leaving the solar collector and the cold water entering the storage tank exceeds the setting of these switches. In the "direct" design (at the top of Figure 2.127) there is only one circulating pump, which is started and stopped by ΔTS. The solar hot water is directed to the bottom of the storage tank, from where it rises and creates a temperature gradient in the tank, which rises with elevation and is the highest at the top of the tank, from where the hot water is sent to the users. Pressure safety valves (PSVs), drain (V3), and bypass (V6) valves are provided for maintenance and safety reasons.

The operation of the "indirect" system, shown at the bottom of Figure 2.127, is similar to the previously described direct system, except that it has two circulating loops and a heat exchanger between the two. This way, the solar collector loop can be filled with any heat transfer fluid, including antifreeze fluids for winter operation, while the collected heat is transferred to the water circulating in the storage tank loop. This configuration is safer and more convenient than the direct one, but is also more expensive. Its initial cost is higher because of the added equipment, and its operation is more expensive because of the energy required to operate the second pump and because of the heat loss in the heat exchanger.

2.18.2 Combination Hot Water and Electricity Controls

It is possible to combine either one of the solar hot water systems shown in Figure 2.127 with a small organic Rankine cycle unit to also produce electrical power (Figure 2.128). The working fluid in the heat pump can be R-134a;

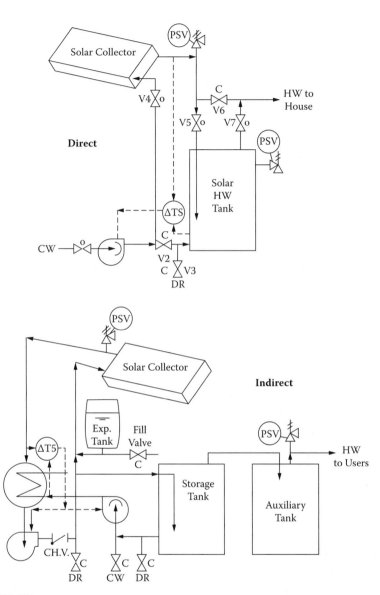

FIGURE 2.127
Direct (top) and indirect (bottom) automatic solar hot water collector systems.

the heat source can be the solar hot water; and the heat sink can be a lake, swamp, or groundwater. A copper coil immersed in the solar hot water tank can serve as the evaporator of the heat pump, whereas its condenser can be a brazed plate heat exchanger or a copper coil immersed in a lake, well, or swamp. A converted automotive scroll compressor can be converted to serve as the expander by removing the compressor's check valve (design by Larry Bingham). The turbine can be shaft-coupled to an alternator that produces the electricity, and the pump can be a DC-motor-driven gear pump.

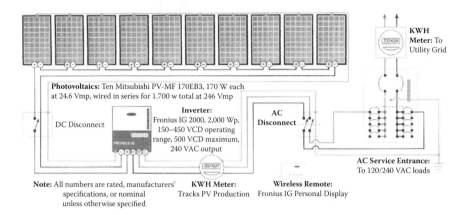

Photovoltaics: Ten Mitsubishi PV-MF 170EB3, 170 W each at 24.6 Vmp, wired in series for 1.700 w total at 246 Vmp

Note: All numbers are rated, manufacturers' specifications, or nominal unless otherwise specified

KWH Meter: Tracks PV Production

Wireless Remote: Fronius IG Personal Display

FIGURE 2.128

Example of PV solar collectors that supply the home with electricity and by being connected to the utility grid, allows the kWh meter to run "backwards" when more electricity is generated than needed. (Reprinted with permission. Copyright 2007 by Home Power, Inc. http://www.homepower.com.)

The simple control scheme in Figure 2.103 measures the temperature difference (ΔTC-1) between the solar collector temperature and the solar water tank temperature, starts the water pump when ΔT > 20°C, and stops it when this temperature difference drops to 10°C. The variable-speed refrigerant pump circulation is similarly controlled. Because the condenser (lake, well, or swamp) temperature is more or less constant, ΔT control might not be essential and measuring only the evaporator (solar water tank temperature) by a standard thermostat (TC-2) can be sufficient. In this configuration, the refrigerant pump is started whenever the solar tank temperature is above 60°C (140°F), and the pump speed is increased to its maximum, as the temperature in the solar hot water tank reaches, say, 70°C (160°F).

2.18.3 Photovoltaic Collector Controls

Figure 2.128 describes the main components of a photovoltaic solar collector system that is connected to the grid. These include the AC service entrance, the AC and DC disconnect switches, the inverter, and the two kWh meters, one tracking the total solar production and the other tracking the electricity flow to (and from) the grid.

Inverters are required to make the PV cell generators compatible with the installed base of electrical consumers, because cells produce direct current and the power distribution grid uses alternating current. A simple electronic inverter is similar in function to the converters used in variable-speed drives to drive the load. One of the many inverters on the market is SMA Technologie's "Sunny Boy," which has been used in nearly one million installations and is available in sizes from 400 Wp up to mWp capacities.

In order to provide a grid-connected inverter, such an inverter must maintain the same frequency as the incoming grid and must detect the loss of the grid and respond to such an event either by isolating itself from the grid or tripping offline. The usual means of detecting a grid disconnect or island situation is to have the output frequency periodically drift up or (more usually) down, because if the grid is in operation, the phase shift will be readily detectable. At 2 Hz away from the nominal grid frequency, the inverter will trip to island mode and either shut down (pending return of the main supply) or disconnect itself from the grid.

The alternative method of detecting grid failure is by detecting the voltage deviation (either high or low) and set the inverter to trip, if the voltage deviation moves outside a +10% or −6% deviation from nominal supply values. The purpose of doing this is to protect linesmen from possible backfeeds from generation systems, which they may not be aware of, when they are isolating the main supply. This system must detect a grid link failure within 2 minutes. With multiple inverters driving the same load, one of the inverters has to be the frequency master for this test to function correctly, whereas the others are operated in load-share mode.

Grid-connected inverters can either drive into the grid as a power supply or can supply part of a large load. The third option is to drive the main's current draw to zero, which requires a more delicate balance. The choice among these options depends upon the specifics of the local electricity supply and upon the receptiveness of the supply authority to generators, which are not owned by them.

2.18.4 Solar Plant Optimization Algorithms

One important aspect of solar farm optimization is the tracking of the sun's trajectory while concentrating the sunlight, so that the mirror reflectors or troughs will be correctly rotated around both axes while concentrating the solar radiation. In order to maximize the collector efficiency, solar and position detectors (Sections 3.15 and 3.17 [Chapter 3]) are used. Some of the tools of positioning include the use of machine vision (Section 3.12) and a variety of positioning devices (Section 3.15).

In addition to the sensors and final control elements, these optimizers are provided with software algorithms to interpret the measurements and to arrive at the optimum responses to them. The traditional algorithms used solar detectors for establishing the angles and direction of the solar radiation and for arriving at the optimum mirror positions for the most efficient focusing and concentration. I have also developed more advanced (shadow-based algorithm) techniques, the details of which are beyond the scope of this book.

Once the desired optimum position is determined for each collector component (mirror or reflector element), the next task is to "herd" all the many thousands of these elements to their different optimum positions. The old algorithms operated by sequentially evaluating and correcting the position of each element. My algorithm follows the strategy of the herding dog (hence its name), by always selecting the one collector out of the thousands of elements that is furthest from the optimum. (The herding dog also directs the whole herd by sequentially going after the animals that are furthest from the desired direction or deviate the most from the desired speed.)

2.18.5 Thermal Plant Optimization

Once the optimized positioning of the collectors has been achieved, the remaining goals include (1) optimizing the pumping rate and distribution of the heat transfer fluid flow among the rows of collectors, (2) monitoring the loads and efficiency of the equipment blocks, and (3) optimizing the equipment reconfiguration and energy use to maximize profitability.

Figure 2.129 illustrates the control of the heat transfer fluid flow rate through each row of parabolic collectors. The control valve is throttled by a temperature difference controller, which keeps the temperature rise constant by matching the flow rate to a rise or drop in solar radiation intensity and closing the valve when no solar energy is received. The set point of the ΔTC is received from the optimizing computer that—on the basis of loads and heat storage—determines it. The opening of the control valve is sent to the valve position controller (VPC), which selects the most open distribution valve on the tank farm and keeps that at 90% opening to minimize the pumping energy consumption (Figure 2.124).

Optimizing the pumping station and the distribution of the heat transfer fluid flow among the rows of collectors is achieved, as was already discussed in connection with Figure 2.124. The optimum pump discharge pressure is selected to keep the most open control valve supplying each row of collectors nearly open. This keeps the pump discharge pressure and the consumption of pumping energy at a minimum. As the valve position controller (VPC-02 in Figure 2.124) lowers the set point of PC-01 and thereby opens all the user valves, it not only minimizes the valve pressure drops but also reduces valve cycling and maintenance.

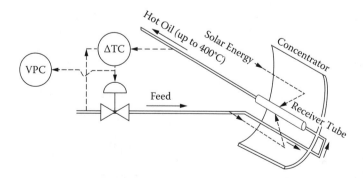

FIGURE 2.129
Differential temperature controls flow through each row of collectors.

If the pump station consists of two variable-speed pumps (Figure 2.124), when only one pump is in operation and the output of PC-01 approaches 100%, PSH-03 is actuated and the second pump is started. When both pumps are in operation and the flow drops to 90% of the capacity of a single pump, the second pump is stopped. This way, the pumping station operating cost is minimized. Once both pumps are operating smoothly and the load drops back to the point, where it can be met by a single pump, the low flow switch FSL-05 stops the second pump. Regardless of the number of pumps in a pumping station, the new pumps are always started on pressure and stopped on flow.

2.18.6 Monitoring and Reconfiguration

The thermal solar farm operation will be optimum if it efficiently responds to both the variation in solar energy availability (insolation) and to both equipment limitations and market conditions. In order to do that, one of the first requirements is to determine the efficiency curves for each combination of equipment blocks as a function of load and then operate the system at that point.

By multiplying the combined oil flow through all the collector rows with the difference between the entering and leaving temperatures, we obtain the rate at which the solar generated heat is being collected (Qs/h). Dividing that rate with the insolation rate (Qins) gives the momentary efficiency of the thermal solar farm (ηs). If Qs is sent directly to a boiler and the heat rate of steam generation (Qstm/h) is measured, the ratio of Qstm/Qs is the efficiency of the boiler (ηstm). If all the generated steam is sent to the turbine generator and the rate of electricity generated is (Qmwh), the turbine/generator efficiency ($\eta\mu$tg) is Qmw/Qstm. Therefore, the overall efficiency of converting the oil heat into electricity is ηe = (ηstm)(ηtg).

By knowing the efficiency curves (as a function of load), one can select the load that will cause the system to operate at maximum efficiency. The loading of the boiler/turbine system can be changed by sending the excess hot oil to storage or by supplementing it from storage, as a function of insolation. In addition to operating equipment at their maximum efficiency, the optimization system also considers the value of the electricity being generated. Naturally, it makes good sense to load the heat into storage when the value of electricity is low, and to supplement the solar heat being generated with hot oil from storage. (In Chapter 4, I will also discuss the added optimization potentials provided by geothermal heat coupling, hydrogen storage, methanol storage, and the use of reversible electrolyzer-fuel cell units.)

In addition to considering market conditions and equipment block efficiency, one should also make the system flexible enough to automatically respond to variations in the supply and demand of solar energy. Knowing the points of maximum efficiency, size of heat storage, and market conditions, the optimizer can automatically select one of at least five operating modes. These are as follows:

1. All solar energy is sent to the boiler/turbine (B/T) to generate electricity (Figure 2.130).

2. Solar energy being collected is insufficient to meet the demand at optimum loading and therefore is supplemented by heat energy from storage.

3. When there is no demand, or market conditions are unfavorable (or for other reasons), all solar energy that is collected is sent to storage.

4. No solar energy is being collected (at night or in unfavorable weather), and all electricity is being generated from hot oil storage.

5. The rate at which solar energy is being collected exceeds the demand for electricity and the excess is being sent to storage.

2.19 Steam Turbine Optimization

Steam turbines are energy conversion machines. They extract energy from the steam and convert it to work that rotates the shaft of the turbine. The shaft output energy of steam turbines range from a few kilowatts to well over 1,000 mW, and there is no reason why still larger machines could not be built. No other prime mover can achieve the shaft output capability that is easily attained by large steam turbines. The energy source for steam turbines is the pressure difference between the supply and exhaust steam. The higher

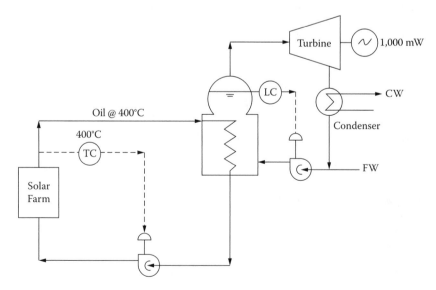

FIGURE 2.130
Thermal solar farm generated heat used for electricity generation by steam turbine.

this pressure difference and the higher the superheat of the steam, the more work the turbine can do.

Rotational speeds vary from approximately 1,800 to 14,000 rpm and can be modulated over a wide range. Such variable speed is an advantage if the turbine is used to drive pumps and compressors or if it is to convert a variable steam flow into electricity, as is the case with thermal solar systems. When provided with the appropriate speed governor, steam turbines can provide excellent speed stability, which is desirable when the turbine serves as the prime mover in electric generators.

The amount of energy that the steam turbine extracts from the steam depends on the enthalpy drop across the machine. The enthalpy of the steam is a function of its temperature and pressure. One can use a Mollier diagram as a graphic tool to determine the amount of energy available under a particular set of conditions. If in Figure 2.131 the inlet conditions correspond to point P_1 and the outlet conditions to point P_2, a line drawn between these two points is called the "expansion line" and represents the operation of the turbine as it is extracting energy from the steam. In an ideal turbine, the steam would expand at a constant entropy (isentropically) and the condition of the exhaust steam, from an ideal machine (which has no losses), would correspond to point P_3.

The efficiency of the turbine, neglecting mechanical losses, is found by

$$\eta = \frac{\Delta h_{1-2}}{\Delta h_{1-3}}$$

as the ratio of the change in enthalpy between P_1 to P_2 (Δh_{1-2}) and the change between P_1 and P_3 as Δh_{1-3}.

The optimization of steam turbine systems is well understood and more widely practiced than the power consumption optimization of alternatives such as electric motors and gas turbines. In addition to the initial cost, the major disadvantage of steam turbines is their low tolerance for wet or contaminated steam. Wet steam can cause rapid erosion, and contaminants can cause fouling. Both will reduce the turbine's efficiency and will shorten its life. Steam quality monitoring is therefore important to maintain the reliability and to reduce the operating cost of steam turbines.

2.19.1 Designs, Applications, and Governors

The two main turbine categories are the condensing and the back-pressure turbines. The exhaust pressure of "condensing" turbines is usually subatmospheric. Condensing turbines are most often used for electric power generation, whereas back-pressure turbines are utilized in cogeneration power plants (Section 2.16), which simultaneously supply steam and electricity for the users. Steam turbine installations can also be configured with a second,

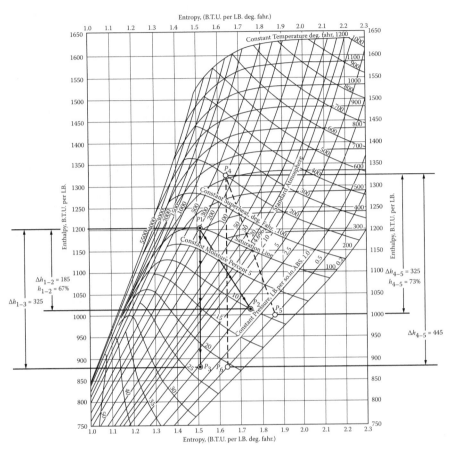

FIGURE 2.131
Mollier diagram showing performance of a steam turbine.

and sometimes even a third, outlet to allow the extraction of steam at different pressures.

A back-pressure turbine with its inlet connected to the plant's high-pressure header and its outlet supplying steam to an intermediate header is called a *topping turbine*. Similarly, a turbine installed between the noncondensed exhaust of another turbine and the condensate system of the plant is called a *bottoming turbine*. A topping turbine could also be described as a noncondensing, or back-pressure unit, and a bottoming turbine as a condensing turbine. Figure 2.132 describes the main variations of turbine configurations. If a turbine's purpose is to generate electricity, it might be called a *generator drive*, whereas if its purpose is to drive a pump or compressor, it can be called a *mechanical drive*.

From an internal design perspective, the steam turbine is either an impulse- or a reaction-type design. In the United States, almost all turbine designs are of the axial flow variety, and only a small number are of the tangential flow variety. In Europe, a significant number of turbines are of the radial flow

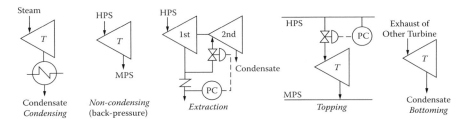

FIGURE 2.132
Steam turbine terminology.

design. The steam turbine can also be single-stage or multistage and, if multistage, one would refer to the turbines as single-flow, double-flow, and so forth. The casing and shaft arrangement is also an important way of categorizing turbines. In a single-casing machine, there is a single casing and one shaft. In a "tandem" design, there are two or more casings connected end to end by a shaft. In "cross-compound" configurations, there are two or more casings connected by multiple shafts.

The governor valve is the valve between the main steam supply and the turbine. This valve is the primary means of controlling the unit. The energy supply and demand is matched, when the turbine speed is constant. Governors can be mechanical, hydraulic, and electrical. They all include a pilot valve or a more sophisticated controller that modulates the turbine's inlet valve in order to keep the shaft speed on set point. Two types of governors are distinguished: (1) in the "isochronous" design, the objective is to maintain the speed of shaft rotation constant regardless of load, whereas in the (2) "droop" design, the speed of the machine is deliberately decreased as load rises.

2.19.2 Steam Turbine Optimization

The simplest application is when the turbine is used to operate a mechanical load at constant speed. In this case, the speed controller (governor) senses the shaft speed and manipulates the steam supply valve to keep the speed on set point. A noncondensing turbine is generally less expensive to buy and to operate than a condensing one, because the energy is extracted from the steam while it has a higher enthalpy and, hence, it has a smaller volume per unit of energy. This has the desirable effect of reducing the size of the turbine and, frequently also increasing its efficiency. Because plants usually have a requirement for low-pressure steam for various loads, such as heating, the low-pressure steam generated by noncondensing turbines can often be used to advantage.

If the plant's demand for low-pressure steam is variable, it is desirable to send that variable amount of steam through a let-down turbine and to recover its energy content in the form of electricity (Figure 2.133). On the other hand, one should not send more high-pressure steam to the turbine than the amount of low-pressure steam demanded by the process. These two goals can be converted into two control loops: a pressure controller

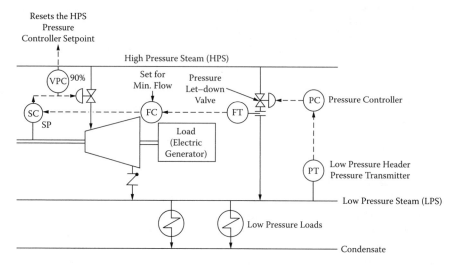

FIGURE 2.133
Back-pressure turbine control system for the generation of LP steam, provided with valve position-based optimizer.

(PC), serving to make sure that all low pressure steam users in the plant are always satisfied, and a flow controller (FC), which keeps the flow in the turbine bypass at a minimum and increases the HP flow to the turbine as soon as the bypass flow starts to increase.

This control configuration can only be used when the excess steam energy is utilized for the cogeneration of electricity, which can vary. Although energy conservation dictates that the flow through pressure let-down line be minimized, control dynamics require its existence. This is because the speed of response of a let-down valve is much faster than that of a turbine. Therefore, the sensitive control of the LP steam pressure is provided by the let-down pressure controller, while the bulk of the steam passes through the turbine and is used to make electricity.

Figure 2.133 illustrates the control system that makes sure that in the short range, the demand for LP steam is continuously met (PC supplements the LP as needed through the turbine bypass), but in the longer range the flow controller (FC) adjusts the turbine's speed and brings the flow in the bypass to a minimum, while the speed controller (SC) throttles the governor valve. The governor valve opening is detected by the valve position controller (VPC), which keeps the governor valve always nearly fully open by modulating the pressure of the HP steam. This way the energy waste resulting from the pressure drop through the governor is minimized.

2.19.3 Extraction (Two-Stage) Turbine Optimization

Extraction turbines are two-stage turbines that can provide steam at a pressure between that of the supply and that of the exhaust, while meeting their

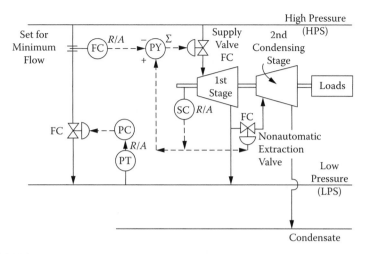

FIGURE 2.134
The addition of a pressure-controlled let-down line increases the speed of response, while the flow controller minimizes the energy waste through that line.

load (mechanical or electrical). As the load changes, the availability of extraction steam also changes. As work load increases, the second stage of the turbine has to contribute more and more energy to meet it and therefore less and less extraction steam is available for the LPS header.

In extraction turbines, in addition to the governor valve, a second "valve" is required (Figure 2.134), which controls the steam flow rate that is extracted from the first stage of the turbine and is sent to the second stage. The extraction rate can be controlled either to keep the shaft speed or the pressure of the LP header constant, or a combination of the two. If the turbine incorporates the controls as a built-in feature, the turbine is referred to as an "automatic-extraction" type. Such turbines are generally designed to deliver 100% shaft power and to provide extraction steam only if the load requirements permit. This is the most common type of extraction machine.

The demand for LP steam can be more or less than the availability of the extraction steam. When the available extraction steam is insufficient, a pressure-controlled bypass valve is used to supplement the LP and maintain the pressure in the low-pressure header. If the interstage steam supply is in excess of the low-pressure header's requirements, it would be wasteful if the excess steam was condensed or vented, and therefore, the pressure set point of the HP pressure controller is reduced.

Figure 2.134 illustrates how the dynamic response of the control system is maintained. Here, the pressure controller (PC) provides the sensitivity required for quick response, while the flow controller (FC) keeps the flow in the bypass line at a minimum, as it slowly opens the HPS supply valve to the turbine whenever the flow in the bypass line exceeds its set point.

Figure 2.135 describes the control configuration required if the LP load of the plant can either exceed the full capacity of the turbine or can drop

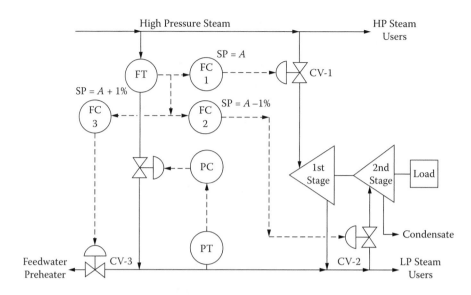

FIGURE 2.135

This control system is both flexible and optimized: FC-1 keeps the bypass flow to a minimum, while FC-3 reduces the LPS demand if it exceeds the work load on the turbine, and FC-3 makes more steam energy available to the turbine if the LPS demand is below the workload.

below the steam flow from the let-down section of the turbine. The figure illustrates a control strategy that utilizes controller set point sequencing to allow optimized and stable operation under any combination of relative load sizes.

If the demand for LPS exceeds the amount of exhaust steam available, the PC opens the pressure let-down bypass. FC-1 serves to minimize the bypass flow. If the pressure controller (PC) opens the bypass and the let-down flow rate exceeds the set point of FC-1, the previously inactive (saturated) FC-3 becomes active and starts cutting back the LPS flow to the boiler feedwater preheater and thereby reduces the plant's demand for LPS. This is an energy-efficient response, because the energy recovered from the LPS supplied to the feedwater preheater is less than the energy content of the HP steam that is needed to produce that LPS.

If the relative loads are reversed and the LPS availability exceeds the demand for LPS, this will cause the pressure in the LPS header to rise and the PC to reduce the bypass flow. When this let-down flow drops below the set point of FC-1, the previously inactive (saturated) FC-2 becomes active and admits that part of the LP steam which is not needed in the LPS header into the second condensing stage of the turbine. Any number of such bypass flow controllers can be used to sequentially respond to changes in the relative sizes of the work and LP steam loads. These controllers should be provided by integral action only, so that they will be saturated (and their control valves closed) until their set points are reached.

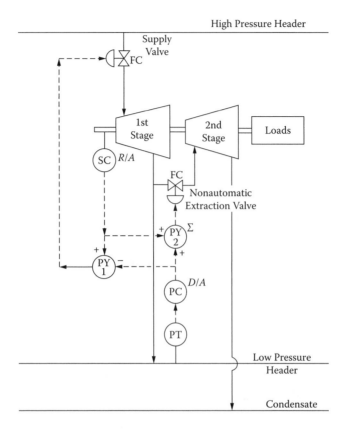

FIGURE 2.136
One way to eliminate interaction between flow and pressure loops is to allow the pressure controller to throttle both the supply and the extraction valves.

2.19.4 Interaction, Decoupling and Safety

The interaction between the pressure and turbine speed controllers (Figures 2.133 and 2.134) can be decoupled. Figure 2.136 illustrates how a drop in the speed of the shaft can open both the inlet and the extraction valves, and an increase in shaft speed can close them both. Therefore, one way to eliminate interaction between flow and pressure loops is to allow the pressure controller to throttle both the supply and the extraction valves.

In this control system, when the LPS header pressure rises, the PC output rises, and therefore, the PY-1 output drops and the supply valve closes. At the same time, the increase in the PC output increases the output of PY-2, which opens the extraction valve. When the LPS header pressure drops, the opposite is the response: the supply valve opens and the extraction valve closes. If the control model is properly tuned and the gains of the summers PY-1 and PY-2 are properly set, there will be no interaction between speed and pressure control, and the speed control will not be adversely affected by the responses to pressure disturbances and vice versa.

An intriguing aspect of this control configuration is the possibility of completely eliminating the need to throttle steam if the turbine capacity is sufficient to meet the demand for LPS. In that case the supply valve is kept fully open, as was the case in Figure 2.133, and the pressure controls of the LPS header determine the distribution of the extracted steam between the turbine's second stage and the LPS header.

The turbine has to be protected against overspeeding, its critical operating parameters have to be monitored, and if a condition exists that could cause equipment damage, the turbine has to be stopped by closing the steam supply valve.

The safety interlocks usually include lube oil failure, high bearing temperature, overspeeding, and vibration. The sudden loss of load will cause the turbine to overspeed. This can happen in mechanical drive applications, but it is a more common occurrence in electrical generator drives. A generator that may have been supplying megawatts of power, and consequently, megawatts of load to its prime mover, can suddenly disappear.

This condition can be detected by a variety of devices, such as centrifugal switches, electronic tachometers, or strain-detecting devices installed on or near such components of the machine that are affected by the overspeed condition.

Continuous information on rotor thermal stress (acceleration and load rates, and maximum allowable initial load pickup) is also needed by the operator. Bearing temperature is an additional indicator that can be used to check if lubrication is functioning properly. It is also a way of detecting the deterioration in bearings before complete mechanical failure occurs. In either case, the bearing begins to dissipate abnormally large amounts of energy, which in turn results in heating. Consequently, a sudden rise in bearing temperature is generally an indication of incipient failure. It is important to quickly stop the machine, because some turbine designs maintain very small clearances between stationary and rotating parts. If the bearing deforms, it may mean the total destruction of the machine.

Especially on larger machines, a stationary vibration monitoring system, consisting of accelerometers or proximity sensors located radially in each bearing and axially on the end of the shaft or the thrust collar, is also provided. Two sensors positioned at right angles are typically used on bearings. The monitoring equipment generally incorporates an "alarm" setting and a "danger" setting which are intended to shut down the machine. In many instances, a vibration-initiated shutdown can prevent major damage.

In power-generating installations, the monitoring and sequential controls serve to automatically bring the turbine from turning gear to generator grid synchronization. These controls evaluate such parameters as bearing temperatures and vibration, water detection, and differential expansion.

Steam turbines provide opportunities for process improvements and energy savings. There usually are four separate and redundant process controller packages that perform operator automatic control (OAC), thermal stress monitoring (TSM), monitoring and sequential control (MSC), and turbine protection (TP). OAC control includes speed, load and steam pressure modulating control, as well as valve testing and remote control operation (auto dispatch, auto synchronizer, and process interface).

3

Sensors and Analyzers for Renewable Energy Processes

3.1 Introduction

The control systems serving the automation and optimization of solar collectors, wind turbines, fuel cells, and geothermal, hydrogen, and other renewable energy processes require reliable information on the operational states of these processes. This feedback is also needed when the goal is energy conservation through the optimization of conventional processes. Obtaining information on the operation of traditional processes or data on solar radiation, wind speed, rate of electricity production, heat and material balances of renewable energy processes requires sensors and analyzers. This chapter describes the various sensors and analyzers that are needed to measure these and many other operating conditions of the alternative energy processes.

The renewable energy processes differ from traditional industrial ones; because some operate at very high temperatures (concentrating and thermal solar collectors), some fuel cells require very small flows, and hydrogen processes can operate at very low temperatures (liquid hydrogen [LH_2]) or at very high pressures (hydrogen gas [GH_2]). These processes require reliable, compact, and accurate means of precisely regulating these variables. Therefore, the descriptions of sensors and analyzers in this chapter will concentrate on the devices that are required for controlling alternative energy processes. Some of the detectors and analyzers that are used in conventional industries will also be mentioned. It is necessary to describe these more conventional sensors not only because some of the renewable energy processes use them, but also because the first step on the road to energy independence is the increase of the efficiency of the already existing conventional processes. For example, the thermal solar collector power plants are integrated with conventional boilers, turbines, and generators, and the geothermal systems depend on the optimization of conventional water-pumping and heat pump systems.

In this book, the first chapter described the nature and operation of the various renewable energy processes, their states of development, costs, efficiencies, and potentials. The second chapter described the methods for controlling and optimizing these processes. In this chapter, the sensors and analyzers that are needed to provide information for optimizing such processes and that can serve in the solar–hydrogen demonstration plants are described.

3.2 Analyzers

In renewable energy processes, our main interest is in determining the composition of stack emissions, monitoring the composition of the atmosphere, and safety. This last concern usually means monitoring the presence of combustibles. As all three of these concerns involve the analysis of gases, the emphasis of this section is on gas and vapor analyzers. In the alternative energy processes, the composition and properties of liquids (water, oil, and other fuels) are of less interest; consequently, these sensors are covered to a lesser depth. Still, because of the geothermal processes, water quality analysis is discussed in this section. If the reader needs more information on liquid and solids analyzers than is provided here, refer to Chapter 8 in Volume 1 of the 4th edition of the *Instrument Engineers' Handbook*.

The composition of process streams can be of interest to guarantee product quality by keeping impurities below specified limits, because of safety and pollution concerns or to make sure that the heating values and other properties of intermediate streams are as they should be. Other reasons for installing process analyzers include reduction of by-products, decrease in analysis time, tightening of specifications, and monitoring of contaminants, toxicants, or pollutants.

The analytical methods used in determining composition include separation, spectroscopic and radiant energy techniques, the measurement of electrochemical properties, and many others. The analyzers used can be intermittent or continuous, can depend on samples, and can be inserted into the flowing streams (probes) or be noninvasive.

Although grab samples taken to a modern laboratory will remain the gold standard of quality control, the effort and time it takes to get a sample, transport it to the laboratory, and wait for the result can seldom be acceptable in modern plants. In selecting the right analyzer, one must also be aware of cost and safety considerations.

3.2.1 Analyzer Selection

In renewable energy processes, the two most important considerations in selecting the right analyzer for a particular analytical task are its specificity (selectivity) and accuracy. Specificity is the characteristic of responding only to the property or component of interest. The more selective a measurement, the less we need to be concerned with interferences from other constituents of the sample matrix.

Frequently, absolute accuracy cannot be established owing to the lack of a suitable calibration standard. For this reason, precision or repeatability can be more important than accuracy. Because the measurement is being made continuously or repeatedly, we are more interested in changes in the reading than its absolute value. *Precision* is defined as the ability of an analyzer to produce the same output each time the same quantity of the component or property is being measured. The terms *stability*, *reliability*, and *reproducibility* are sometimes used synonymously with *repeatability*. However, the term *reliability* is also used to describe the instrument's "up time."

As to the types of analyzers, only the continuous ones will be discussed here. This is because the intermittent information provided by sampling analyzers, such as chromatographs, do not meet the safety requirements of most renewable energy processes. One of the most important families of analyzers utilize the various spectroscopic and radiant energy techniques.

This family of analyzers operates either on absorption or reflection principles, but fluorescence and scattering are also used. If radiation at different wavelengths is passed through a process material, the amount of absorption may be an indication of the sample identity or composition. The region of the spectrum used for the measurement varies with the kind of compound and information desired (Table 3.1). Instruments to make these measurements can vary from single-wavelength photometers to grating and multiplex designs, depending on the spectral region and the needs of the measurement.

The electrochemical analyzers are another important family of liquid analyzers. They include potentiometric, wherein an electric potential is measured and the solution remains unchanged; conductive, in which a minute current is measured but the system is essentially unchanged; and amperometric, in which a chemical reaction occurs during the course of the measurement. Potentiometric analyzers can measure the presence of dissolved ionized solids in a solution. These measurements include pH, oxidation-reduction potential (ORP), and ion-selective electrodes (ISEs) or probes.

TABLE 3.1

Absorption/Emission of Electromagnetic Energy Used for Measurements

Type of Radiation	Wavelength Range	Characteristic Process Probed
Gamma rays	$<10^{-12}$ m	Nuclear transitions
X-rays	1 nm–1 pm	Inner-shell electron transitions in atoms
UV	400 nm–1 nm	n, π (valence) electron transitions in molecules
Vis	750 nm–400 nm	n, π (valence) electron transitions in molecules
NIR	2.5 μm–750 nm	Molecular vibrations
IR	25 μm–2.5 μm	Molecular vibrations
Microwaves	1 mm–25 μm	Rotations in molecules
Radio waves	>1 mm	Rotations in molecules, electron spin flips[a]

[a] Nuclear magnetic resonance (NMR) uses a magnetic field to split the energy levels and the radiofrequency (RF) energy to probe the spin state.

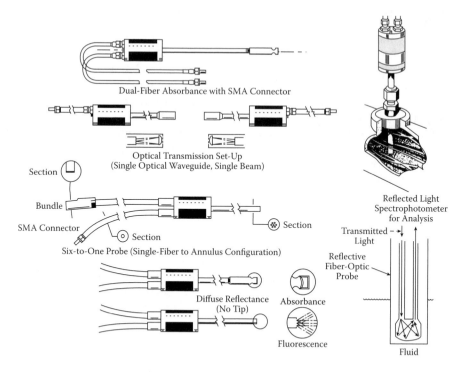

Dual-Fiber Absorbance with SMA Connector

Optical Transmission Set-Up
(Single Optical Waveguide, Single Beam)

Section

Bundle

SMA Connector

Section

Section

Six-to-One Probe (Single-Fiber to Annulus Configuration)

Diffuse Reflectance
(No Tip)

Absorbance

Fluorescence

Reflected Light
Spectrophotometer
for Analysis

Transmitted –
Light

Reflective
Fiber-Optic
Probe

Fluid

FIGURE 3.2
Fiber-optic probes can provide data on absorbance, diffuse reflectance, fluorescence, and scattering. (Courtesy of Guided Wave, Inc.)

Fiber-optic probes (FOPs) use waveguides made of glass, quartz, or other more esoteric material to deliver and return the process-modified light from the probe to the detector located some distance away (Figure 3.2). FOPs can acquire data on spectral absorbance, diffuse reflectance, fluorescence, or scattering. Multiple measurements can be made by multiplexing several FOPs to the same computer-controlled analyzer. Performing spectroscopic measurements over FOPs can be done in the UV-visible, near-infrared (NIR), and IR regions, and insertion probe type critical angle refractometers are also available.

Acousto-optic tunable filters (AOTFs) are used in connection with stack gas analysis. Their main advantage is speed and the elimination of maintenance-prone mechanical elements such as filter wheels, moving gas cells and mirrors, diffraction gratings, and mechanical light choppers. The AOTF acts as an electronically controllable narrow-band filter that can be tuned to any desired frequency in milliseconds. As shown in Figure 3.3, the beam at the selected wavelength is directed across the stack and can be used to simultaneously measure the concentrations of CO, CO_2, C_2H_6, CH_4, NO, NO_2, H_2O, etc. This is achieved by the computer selecting the frequencies required for the measurement of the concentration of each component. This is achieved by the AOTF being tuned at a high speed to each of the

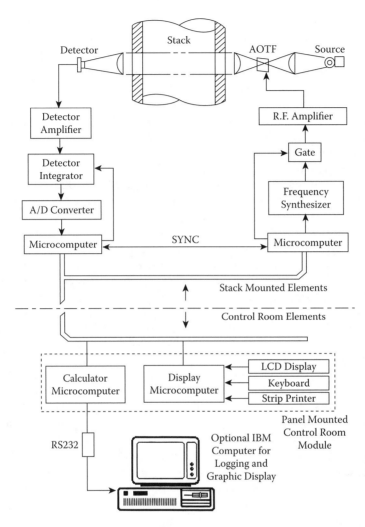

FIGURE 3.3
AOTFs can enhance the speed and increase the reliability of multicomponent analysis in stack genes.

selected frequencies. This results in a "scanning" spectrometer without moving parts.

3.2.2 Analyzer Sampling

In alternative energy processes, in most cases the goal is to eliminate all sampling systems and place the analyzer directly into the process. The "in-line" analyzer designs are widely available, and the various radiant energy and probe-type sensors operate without any sampling.

When the use of a sampling system is unavoidable, it becomes an integral and key part of an analyzer system. It should be designed to obtain a representative sample, transport the sample to the analyzer, condition the sample, accomplish sample stream switching (if necessary), provide facilities for return and disposal of the sample, and provide not only calibration facilities but also preventive maintenance features and alarm functions for online reliability and operator alerts.

Whenever a sample has to be brought to an analyzer, a transportation delay and a potential for interference with the integrity of the sample are inevitable. If an automatic controller maintains the measured composition, the transportation lag can seriously deteriorate the closed-loop control stability of the loop. An even more serious consequence of the use of sampling systems is the potential for interference with the integrity of the sample. This can occur due to filtration, condensation, leakage, evaporation, and so on, and these operations cannot only delay, but also change information and measurement.

Ideally, even if the sample requires little or no conditioning, it is good practice to install a sampling probe (Figure 3.4) for most applications as a precautionary measure to prevent particulates from entering the sample transport system. Sampling of processes that are still reacting chemically may require reaction quenching, or fractionation, at the sample takeoff. This is done by cooling or backflushing with an inert gas or liquid to keep the sample takeoff clean.

The sample is normally transported in three ways: Single-line transport is used when the sample line volume is small in relation to the analyzer sample consumption so that the transport time lag is reasonably short. It is usually used when the analyzer is field-mounted close to the sample point and sample exhaust facilities are available.

Bypass-stream transport is a method for maintaining high sample transport velocity to minimize transportation lag. This method is used when samples are vaporized at the sampling tap and no facilities exist for returning the vapor to the process. If the sample bypass is piped to a drain or vent, this will not only waste the process material but might also pollute the environment. Therefore, the use of a fast bypass-return loop is preferred. After selecting the appropriate sample transport method, the sample time lag should be calculated and used in the tuning of the analyzer controller.

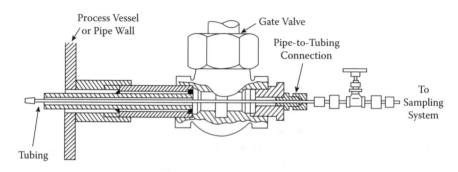

FIGURE 3.4
Sample probe assembly with process shutoff valve.

Automatic liquid samplers are also available to collect intermittent samples from pressurized pipelines and deposit them in sample containers. The sample can be collected on a time-proportional or on a flow-proportional basis. The sampler can withdraw a predetermined volume of sample every time the actuator piston is stroked. In the time-proportional mode, the sampling frequency is constant, whereas in the flow-proportional mode, it is a function of the flow.

3.2.2.1 Filtering and Separation

In the renewable energy processes, the removal of entrainment from liquid or gas samples is normally done by filtering. Filtration can remove both liquid and particulate entrainment from gases, as well as the particulate matter from liquids.

Most analyzer sampling systems require a filter with at least one wire mesh strainer (100 mesh or finer) to remove larger particles that might cause plugging. Available filter materials include cellulose, which should only be considered if it does not absorb the components of interest. Sintered metallic filters can remove particles as fine as 2 μm, cellulose filters can remove down to 3 μm, and ceramic or porous metallic elements can trap particles of 13 μm or larger. When the solids content is high, two filters can be installed in parallel with isolation valves on each. Motorized self-cleaning filters are also available for such services.

If the material to be removed is dust, the self-cleaning bypass filter with automatic blowback constitutes a potential solution, whereas in other instances, cyclone separators should be considered. In the former device (Figure 3.5), the process stream enters tangentially to provide a swirling action, and the cleaned sample is taken near the center. Transportation lag can be kept to less than 1 minute, and the unit is applicable to both gas and liquid samples. This type of centrifuge can also separate streams by gravity into their aqueous and organic constituents.

When liquid droplets are present in a gas stream, glass microfiber filter tubes can efficiently separate suspended liquids from gases. The filter tubes capture the fine droplets suspended in the gas and cause them to run together to form large drops within the depths of the filter tube. The large droplets are then forced by the gas flow to the downstream surface of the filter tube, from where the liquid drains by gravity. This process is called *coalescing*.

If liquid droplets are intermixed in a liquid stream, glass microfiber filter tubes can separate the suspended droplets that are immiscible in the other liquid, using the same process by which they separate droplets of liquid from a gas. The liquid droplets suspended in the continuous liquid phase are trapped on the fibers and run together to form large drops, which are then forced through the filter to the downstream surface. The large drops separate from the continuous liquid phase by gravity difference, settling if heavier than the continuous phase, and rising if lighter. The coalescing action

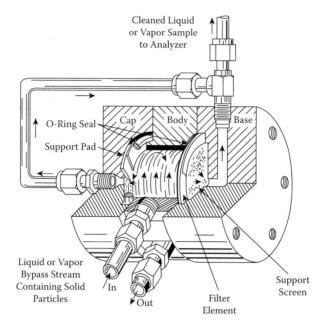

Cleaned Liquid
or Vapor Sample
to Analyzer

O-Ring Seal

Cap

Body

Base

Support Pad

Liquid or Vapor
Bypass Stream
Containing Solid
Particles

In

Out

Support
Screen

Filter
Element

FIGURE 3.5

Bypass filter with its cleaning action amplified by the swirling of the tangentially entering sample.

of glass microfiber filters is effective with aqueous droplets suspended in oil or other hydrocarbons and also with oil-in-water suspensions.

If gas bubbles are to be removed from a liquid stream, the glass microfiber filter tubes can also be used, eliminating the need for deaeration tanks, baffles, or other separation devices. In this case, the flow direction through the filter is from the outside to the inside, and the separated gas bubbles rise to the top of the housing and are vented. If slipstream sampling is used, the separated bubbles are swept out of the housing with the bypassed liquid. Filter tubes rated at 25 μm are a good choice for gas bubble separation. Columns with glass wool packing can also be used.

3.2.2.2 Probe Cleaners and Homogenizers

Although probe-type in-line analyzers can eliminate the transportation lag and sample deterioration problems associated with sampling systems, they are not without problems of their own, the biggest of which is fouling. It is recommended that when an automatic probe cleaner is used, it be placed inside a sight glass so that the operator can continuously observe the performance of the cleaner (Figure 3.6). The cleaners can be of the brush-type, scraper, chemical, hydrodynamic (self-cleaning), and ultrasonic designs.

Homogenizers can be used instead of filters if there is a potential for plugging the sampling system. Also, if the materials that filters remove contain components of interest, the filters are replaced by homogenizers.

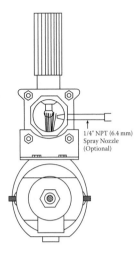

FIGURE 3.6
Probe cleaners should be mounted in sight-flow glasses for good visibility. (Courtesy of Aimco Instruments, Inc.)

3.2.2.3 Stack Gas Sampling

In renewable energy processes, no fossil fuels are burned. The only combustion that takes place is that of H_2 that burns into distilled water. However, for a number of decades, we will not have a completely renewable energy economy but a combination one. Therefore, it is necessary to also discuss the monitoring of fossil fuel combustion processes.

When taking gas samples, the goal is to obtain representative samples with minimum time lag, using short, small-volume sampling lines. Whenever possible, it is preferred to draw dry and clean samples to minimize the need for filters, dryers, knockout traps, or steam tracing.

When sampling hot, wet stack gas, a filter capable of withstanding the gas temperature should be installed in the stack at the tip of the sample line to prevent solids from entering the gas sample line. After the sample is cooled, a coalescing filter is used to remove suspended liquids before the sample goes to the analyzer (Figure 3.7).

A complete U.S. Environmental Protection Agency (EPA) particulate sampling system comprises four major subsystems: (1) a Pitot tube probe assembly for temperature and velocity measurements and for sampling, (2) a two-module sampling unit that consists of a separate heated compartment with provision for a filter assembly and a separate ice-bath compartment for the impinger train and bubblers, (3) an operating control unit with a vacuum pump and a standard dry gas meter, and (4) an integrated, modular umbilical cord that connects the sample unit and Pitot tube to the control unit.

Isokinetic sampling requires the precise adjustment of the sampling rate with the aid of the Pitot tube manometer readings and nomographs. If the pressure drop across the filter in the sampling unit becomes too high, mak-

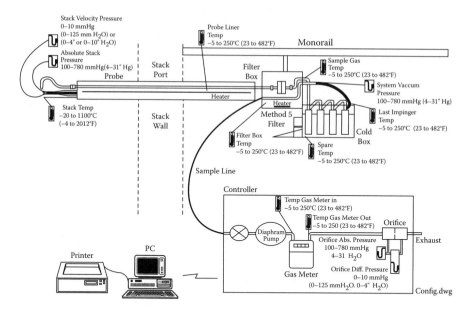

FIGURE 3.7

The components of an automatic stack train. (Courtesy of ThermoAndersen.)

ing isokinetic sampling difficult to maintain, the filter can be replaced in the midst of a sample run.

When the IR analyzer is used for in situ stack gas analysis, two different approaches can be used: (1) the detectors and source on the same side of the stack, and (2) the detectors and source on opposite sides of the stack. For installation purposes, the first is much easier, but it requires an internal filter to remove any solids from the stream. In the second system, shown in Figure 3.8, both the reference and measuring wavelengths are affected by the same scattering bodies in the stream; therefore, there is no need for a filter. This system is frequently used in pollution alarm applications.

When the pollutant of interest is present in the stack, the energy content of the reference path is unaffected (because the absorption is already complete at the selected wavelength). On the other hand, the IR energy reaching the detector through the neutral filter is reduced (due to the absorption of the pollutant gas), and the ratio between the beams reflects the pollutant concentration at the level of concern in the stack.

3.2.3 Air-Quality Monitoring

Even after the conversion to an alternative energy economy, it will be necessary to monitor the air quality. To enforce regulations, air quality is monitored near single sources of pollution and also in urban areas. Regardless of the type of instruments used to measure air quality, the collected data will only be representative if the sampling site is correctly selected. Table 3.9

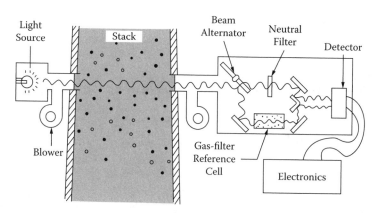

FIGURE 3.8
Dual-beam IR analyzer for stack gas monitoring.

lists the compounds that are of interest when monitoring the quality of air. This table also identifies the types of analyzers used and their measurement ranges and precision.

The simplest air quality monitors are static sensors, which are left in the area being monitored for some length of time and are later analyzed in a laboratory. More commonly, automatic instruments are used that measure several air quality parameters and either retain the collected data on magnetic tape or transmit it by wireless transmission.

The main purpose of air-quality monitoring is to enforce government regulations. In the United States, standards have been promulgated by the federal government and by many of the states to protect human health.

3.2.3.1 Single-Source Sensors

The purpose of some air-quality monitoring systems is to determine the impact of a single source or sources of emission on the surrounding area. In this case, the background level of pollution, the maximum ground-level concentration, and the geographical extent of the air pollutant impact of the source has to be determined. When the source is isolated, such as a single industrial plant in a rural area, the design is straightforward. Utilizing meteorological records, first a wind rose is prepared to estimate the direction of the principal drift of the air pollutant from the source. Next, dispersion calculations are performed to estimate the location of the expected point of maximum ground-level concentration.

As a rule of thumb, with stacks between about 15 and 100 m (50 and 350 ft) tall, this point of maximum concentration will be approximately 10 stack heights downwind. Therefore, the air-quality monitoring system should include at least one sensor at the point of expected maximum ground-level concentration. Additional sensors should be placed not less than 100 stack heights upwind (prevailing) to provide a background reading, and at least

TABLE 3.9

Ambient Air-Monitoring and Meteorological Measurement Parameters

Compound	Range	Accuracy	Technique Employed
Oxides of nitrogen	0–20 ppm	0.5 ppb	Chemiluminescence or DOAS open path
Sulfur dioxide	0–20 ppm	0.5 ppb	Fluorescence dual-channel ratiometric phase detection or DOAS open path
Ozone	0–20 ppm	0.5 ppb	Ultraviolet photometrics or DOAS open path
Carbon monoxide	0–200 ppm	50 ppb	Gas filter correlation or DOAS open path
Carbon dioxide	0–100%	50 ppb	Gas filter correlation or DOAS open path
Benzene, toluene, xylene	0–5 ppm	0.5 ppb	Gas chromatography or DOAS open path
Nonmethane hydrocarbons	0–1,000 ppm	0.01 ppm	FID or DOAS open path
Methane	0–1,000 ppm	0.01 ppm	FID or DOAS open path
Particulates (PM10, TSP, PM2.5)	0–5 g/m³	0.1 μg/m³	Tapered element oscillating microbalance
Carbon particulates	0–5 g/m³	0.25 μg/m³	Thermal CO_2 method
Local visual distance	0–16 km	±10%	Nephelometer
Wind speed	0–70 m/s	0.22 m/s	Anemometer
Wind direction	0–540°C	±3°C	Airfoil vane
Ambient temperature	−50 to 100°C	±0.1°C	Solid-state thermistor
Relative humidity	0–100%	±2%	Thin-film capacitor
Barometric pressure	800–1,200 mbar	±1.3 mbar	Solid-state transducer
Precipitation	NA	0.1 mm	Net radiometer
Solar radiation	250–2,800 nm	9 MV/kWm²	Pyranometer
Net radiation	250–60,000 nm	8 MV/kWm²	Net radiometer

Note: NA = not applicable; FID = flame ionization detector; DOAS = differential optical absorption spectroscopy.
Source: Courtesy of Ecotech.

two or three sensors should be placed between 100 and 200 stack heights downwind to determine the extent of the travel of the pollutants from the source in question.

With such a system for an isolated source, adequate data can be obtained in 1 year to determine the impact of the sources on the air quality of the area. There are very few instances where less than 1 year of data collection will provide adequate information because of variability in climatic conditions on an annual basis.

Some air-quality monitoring is designed for the specific purpose of investigating complaints concerning an unidentified source. This usually happens in urban situations for odor complaints. In these cases, a triangulation technique is used. By the use of this technique, human observers over a period of days can correlate the location of the observed odor and the direction of the wind. Plotting on a map can pinpoint the offending source in most cases.

The most sophisticated and expensive air-quality monitoring systems are installed in large cities, where data collection and analysis are centralized at a single location using wireless data transmission and telemetry. Online computer facilities provide data reduction. The monitoring system can be designed by (1) installing sensors on a uniform area basis (rectilinear grid), (2) installing sensors in areas where pollutant concentrations are expected to be high, and (3) installing sensors in proportion to population distribution.

Adequate coverage of an urban area frequently requires at least 100 sensors. The location of maximum concentration of air pollutants will typically be in the central business district and in the industrial areas on the periphery of the community. One or two sensors are usually placed in clean or background locations, so the average concentration of air pollutants over the entire area can be estimated.

3.2.3.2 Static and Automatic Air Monitoring

The static methods of air monitoring can use dust-fall jars, lead peroxide candles, and sodium carbonate. The manual air-sampling instrument, which is in widest use, is the high-volume sampler shown in Figure 3.10. With this method, ambient air is drawn through a preweighed filter for 24 h. The filter is then removed from the sampler, returned to the laboratory, and weighed. The gain in weight, in combination with the measured air volume through the sampler, allows the particulate mass concentration to be determined in micrograms per cubic meter.

The reference methods usually require wet samplers. Sampling trains have been developed that allow the sampling of five or more gases simultaneously in separate bubblers. These static methods of sampling can be accomplished with a modest initial investment, but the manpower required to distribute and pick up the samples and to analyze them in the laboratory raises the total cost to a point where automated systems may be more economical for long-term studies.

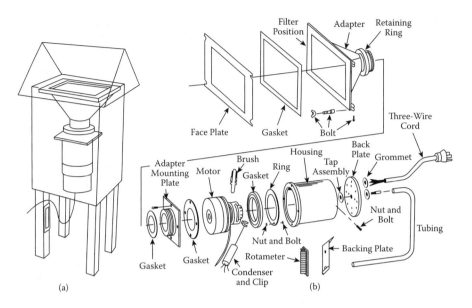

FIGURE 3.10

Description of the high-volume column air sampler: (a) illustrates the assembled sampler and its shelter; (b) shows the components of a typical high-volume air sampler.

As the need for accurate data that can be statistically reduced develops, automated sampling systems are used. The elements of an automated system include the airflow-handling system, sensors, data transmission storage, display apparatus, and data processing facility. The overall system is no more valuable than the weakest link of this chain.

Periodic performance audits are required to validate the accuracy of the air-monitoring system. The Code of Federal Regulations (CFR) requires that performance audits be conducted at least once a year for criteria pollutant analyzers operated at state and local air monitoring stations (SLAMS). The EPA recommends that each analyzer be disconnected from the monitoring station manifold and be individually connected to the audit, from which it will receive the audit gas of known concentration. The audit gas concentrations are usually generated in a van, using a gas calibrator to dilute multi-blend gases with zero air.

3.2.4 Calorimeters

Calorimeters can determine the heat content of their samples by (1) direct burning, (2) calculation from composition, and (3) special designs. Their operation can be continuous, cyclic, or portable. Their errors range from 0.5 to 2% full scale (FS).

In renewable energy processes, the gaseous fuels include GH_2, biodegradation-generated methane, and other gases. Calorimeters are analyzers that measure the heat value or energy content of gaseous fuels. There are two broad categories of this type of instrument: those considered true calorimeters, because they are actually burning the gas and directly measuring its heating value and inferential calorimeters, which analyze the composition of the gas or measure a physical parameter to determine the heating value. A summary of calorimeter features is provided in Table 3.11.

Some of the terms used in connection with calorimetry are listed as follows:

Saturated and Dry Btu: Saturated British thermal unit (Btu) is the heating value that is detected when gas is saturated with water vapor. This state is defined as the condition when the gas contains the maximum amount of water vapor at base ambient conditions. Dry Btu is the heating value when the gas is dry.

Combustion air requirement index (CARI): This is a dimensionless number indicating the amount of air required to support the combustion of a fuel gas.

Gross calorific value: This is the heat value of energy per unit volume at standard conditions, expressed in terms of kilocalorie per cubic Newton meter ($kcal/Nm^3$) or other equivalent units.

Net calorific value: This is the measurement of the actual available energy per unit volume at standard conditions, which is always less than the gross calorific value by an amount equal to the latent heat of vaporization of the water formed during combustion.

Wobbe index: This is a numerical value that is calculated by dividing the square root of the relative density into the heat content ($kcal/m^3$ or Btu/scf) of the gas.

In Europe and in other countries where the metric system is used, natural gas calorimeters are calibrated in megajoule units, and in the United States in Btu units. The output of the calorimeter may represent the gross calorific value (sometimes referred to as upper heating value or gross heating value), the net calorific value (sometimes referred to as lower heating value or net heating value), or the Wobbe index.

Gaseous fuel, including GH_2, is a costly commodity that is being consumed with much more care and efficiency than in the past. For closed-loop control applications, the fast calorimeters (speed of response of less than a minute) are recommended, as shown in Table 3.11. The table also lists calorimeters that are specifically designed for custody transfer applications and offer improved accuracy (at the expense of response time).

3.2.5 Carbon Dioxide

The carbon dioxide (CO_2) analyzer types include (1) nondispersive infrared (NDIR), (2) gas filter correlation (GFC), and (3) Orsat, having measurement accuracies from 0.2 ppm to 1–2% FS for NDIR and 1–2% FS for gas filter cor-

TABLE 3.11

Summary of Calorimeter Features and Specifications

Type	Type (Area Class)				Application[a]					Operation								Performance		
	Direct	Inferential	General Purpose	Ex-Proof	1	2	3	4	5	Continuously	Cyclic	Standard Sample	Empirical	Calibration	Ambient Limits °F (°C)	Local Readout	Remote Transmitters	Range in Btu Full Scale	Accuracy ±% of Full Scale	Speed of Response (90%)
GROSS CALORIFIC VALUE																				
Water ΔT	✓		✓		✓				✓		✓			✓	72–77 (22–25)	✓	✓	130–3300	0.5	3 min
Air ΔT	✓		✓		✓				✓	✓		✓			72–77 (22–25)		✓	120–3600	0.5	15 min
Gas Chromatograph		✓	✓	✓	✓	✓		✓	✓	✓		✓			0–100 (−18–38)		✓	Any	0.5	10 min
Adiabatic Flame Temperature	✓		✓	✓	✓	✓			✓	✓		✓			N/A	✓		N/A	0.5	N/A
NET CALORIFIC VALUE																				
Airflow Calorimeter	✓		✓	✓		✓	✓	✓	✓	✓	✓	✓			50–90 (10–32)	✓	✓	130–3300	1.0	8 sec
Gas Chromatograph		✓	✓	✓			✓	✓	✓		✓	✓			0–128 (−18–53)		✓	Any	0.5	10 min
Expansion Tube Calorimeter	✓		✓				✓	✓		✓		✓			N/A	✓	✓	120–3300	1.0	3.5 min
Specific Gravity		✓		✓			✓	✓		✓					0–128 (−18–53)	✓	✓	Varies	2.0	N/A
Process Chromatograph	✓		✓				✓	✓				✓			60–90 (16–32)	✓	✓	150–3600	2.0	4.5 min
Thermopile Calorimeter	✓	✓	✓			✓		✓		✓		✓			N/A		✓	150–3300	2.0	55 sec
WOBBE INDEX																				
Airflow Calorimeter	✓		✓			✓	✓	✓	✓	✓	✓	✓			50–110 (10–43)	✓	✓	130–3300	0.75	8 sec
Gas Chromatograph		✓	✓	✓	✓		✓	✓	✓		✓	✓			0–120 (−18–49)		✓	Any	0.5	10 min
Expansion Tube Calorimeter	✓		✓			✓	✓	✓		✓		✓			N/A	✓	✓	120–3300	1.0	3.5 min
Thermopile Calorimeter	✓	✓	✓			✓		✓		✓		✓			N/A		✓	150–3300	2.0	55 sec

[a] See feature summary at beginning of section.

relation type (see Table 3.11 for their ranges). The response times are under 30 seconds for NDIR and 90 seconds for gas filter correlation type.

During the industrial age, CO_2 concentration in the ambient air increased from 280 to 360 ppm, and some predict that it could rise to 550 ppm if the use of fossil fuels continues. In addition to monitoring the atmosphere, air-quality-related measurements can also be used in heating, ventilation, and air conditioning (HVAC) systems to monitor the return air quality from occupied spaces. CO_2 is also measured at emission points because some combustion equipment regulations limit, or probably will limit, allowable discharges.

An increase in global CO_2 concentration of only 1% (or about 3 ppm) has significant consequences on the weather. The instruments used to measure atmospheric CO_2 concentrations must be highly precise. With NDIR detectors, because the absorption bands of water and CO_2 somewhat overlap, a freeze-out trap (−80°C or −112°F) is used in the sample preparation system to remove water prior to the measurement.

NDIR-type CO_2 monitors are used in HVAC systems and for industrial applications. The portable units are usually battery operated, and their ambient sample is received by a combination of diffusion and convection effects, without using any pumps or filters. These units are provided with digital displays, one or two alarm settings, and digital output signals.

The more expensive, permanently installed NDIR units often include data loggers, which can store thousands of periodic readings along with their times and dates. Some of these units can also detect other gases, such as CO, H_2S, or O_2.

When very accurate low-level measurements are needed, the gas filter correlation (GFC) analyzer is used. In these designs, the measuring and reference filters are replaced by gas-filled cuvettes. The reference cuvette is filled with CO_2 and the measuring cuvette is usually filled with nitrogen. In addition to being unaffected by the presence of background gases, both the accuracy and the response time of these instruments are better than those using filters. If GFC is used in combination with single-beam dual-wavelength technology, it is virtually immune to obstruction of the optics.

Measurement of the CO_2, carbon monoxide (CO), and oxygen (O_2) concentrations in flue gases from boilers (see Figure 3.7) allows the determination of CO_2 emissions and the precise setting of boiler operating variables for maximum fuel economy. In the past, the CO_2 content of the flue gas was determined only during the few hours of testing. The usual procedure was to slowly withdraw a sample into a plastic bag and analyze it manually using an Orsat analyzer or instrumentally using an NDIR. The operation of the Orsat analyzer was based on the reduction in gas volume resulting from the absorption of CO_2 in a strong alkaline solution.

3.2.6 Coal Analyzers

Coal can be analyzed by thermogravimetry (TG), oxygen combustion bomb, total sulfur analysis, x-ray fluorescence (XRF), atomic absorption (AA) spec-

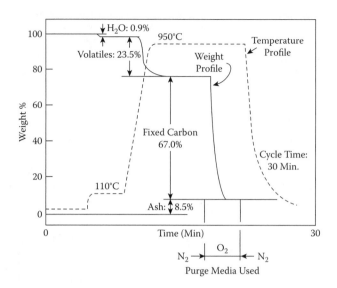

FIGURE 3.12

A typical proximate analysis of coal by microcomputer-controlled thermogravimetry.

trophotometry, coal slurry analyzers, prompt gamma neutron activation analyzers (PGNAAs), and pulsed neutron analyzers. The precisions of these measurements are as follows: heating value, about ±1000 kJ/kg (±250 Btu/lb); ash content, ±0.05 wt%; and moisture content, ±0.05 wt% moisture.

Our goal is to gradually replace all coal and fossil fuel power plants with renewable energy ones, but while they exist (and probably will for most of this century), it is very important to reduce the damage they cause by optimizing their operation. In the United States, there are about 1000 coal preparation plants and coal-fired power plants. One key consideration in operating coal-burning facilities is the control of CO_2 and sulfur dioxide (SO_2) emissions to the atmosphere. The characteristics of coal are monitored for environmental protection, quality assurance, and process control purposes.

The various methods for the online monitoring of coal composition include the TG technique, which automatically performs a multistep analytical sequence by sequentially drying, burning, and weighing the residue. Figure 3.12 shows the analysis results from an automated TG system. The elapsed time of the proximate analysis program and cooling of the tube back to load temperature totals 30 minutes.

Gross calorific value is determined by burning a weighed sample of coal in a water-jacketed bucket. A microprocessor control system controls the jacket temperature and uses the sample weight and temperature data—applying correction for acid, sulfur, fuse, and any added combustion aids—to calculate the gross calorific value. An American Society for Testing and Materials (ASTM) method for determining sulfur in coal uses the washings from the oxygen bomb calorimeter where the sulfur is precipitated as barium sulfate from the washings.

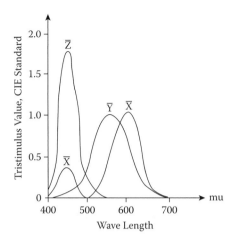

FIGURE 3.13
Spectral response of CIE standard values.

Other procedures include high-temperature tube furnace combustion methods for rapid determination of sulfur in coal and coke, using automated equipment. The instrumental analysis provides a reliable and rapid method for determining sulfur contents of coal or coke. By this method, total sulfur as sulfur dioxide is determined on a continuous basis.

The major or minor elements in coal ash can be determined using XRF techniques. Detector output is converted into concentration by computerized data-handling equipment. All elements are determined, including Fe, Ca, K, Al, Si, P, Mg, Ti, and Na. These elements can also be determined by atomic absorption (AA) spectrophotometry.

Figure 3.13 shows a recording of a continuous monitor of moisture, ash, and the heating value of coal. In this system, a microwave analyzer measures the moisture content without requiring physical contact with the solids. The "fingerprint" of a given type of coal is its distinctive gamma spectrogram, which is displayed using the Commission Internationale de l'Eclairage (CIE). This is produced by the detection and counting of photons released from atomic nuclei in the coal as it passes over a small source of neutron emissions.

Some coal analyzers use several gamma-ray detectors and operate by the use of neutron pulses. Such analyzers can measure the density and sulfur content of coal along with its heating value, moisture, and volatile matter content. This pulsed fast/thermal neutron analyzer can be self-calibrating and can determine such elements as carbon, oxygen, and sodium.

3.2.7 Colorimeters and Shade Detectors

In biological alternative energy processes, in the measurement of color, and in solar energy processes, the measurement and location of shaded areas can

TABLE 3.14

Color and Wavelength Association

Approximate Wavelength (μm)	Associated Color
400–450	Violet
450–500	Blue
500–570	Green
570–590	Yellow
590–610	Orange
610–700	Red

help in optimizing these processes. Visible color can also be measured as an indication of concentration or as an indication of pH in titration using dyes.

Color measurements use the part of the electromagnetic spectrum that is sensed by the human eye and brain. This region is approximately 400–700 nm. The colors of the rainbow are associated with specific wavelengths of visible light, as listed in Table 3.14. If colored filters such as RGB (red, green, and blue) are used, the device is called a colorimeter. If a grating or prism is used, the device is called a spectrophotometer.

In spectrophotometric analyzers, interference filters are selected for desired wavelengths, as determined from the spectral relationship curves. Photodetectors are least sensitive in the blue end of the spectrum. This can be dealt with by using prefilters or narrow spectral ranges, which are calibrated for more sensitivity. Improvements in spectrophotometers include a flashed xenon light source with dual-beam measurement. Dual-beam machines measure the spectrum of both the light source and the reflected light for each measurement.

Recent advances in fiber optics have allowed the light source and the photosensitive device to be remotely located from the actual sensor windows. Various cells can be installed in flow lines measuring from 0.5 to 6 in. in diameter, or in situ probes can be used.

Online shade monitors can be combined with robotics for scanning various surfaces while keeping the light source and spectrophotometer stationary. Fiber optics–based spectrophotometers can be interfaced to host computers that also control stepper motors for positioning the components that can adjust the degree of shading. Spectral information is transmitted serially by RS232, RS422, and USB (IEEE 1394).

3.2.8 Combustibles

In all processes, including the renewable energy ones, a critical safety concern is monitoring the plants for the presence of combustibles. The vaporization rates of the various liquids are a function of their vapor pressures, and vaporization rate increases with increased temperature. Flammable liquids

TABLE 3.15

Properties of Some Flammable Liquids and Gases

Material	Chemical Formula	Specific Gravity Air = 1	Ignition Temperature in Air		Flammability Limits in Air (% vol.)	
			(°F)	(°C)	Lower	Upper
Methane	CH_4	0.55	1193	645	5.3	15.0
Natural gas	Blend	0.65	1163	628	4.5	14.5
Ethane	C_2H_6	1.04	993–1101	534–596	3.0	12.5
Propane	C_3H_8	1.56	957–1090	514–588	2.2	9.5
Butane	C_4H_{10}	2.01	912–1056	489–569	1.9	8.5
Toluene	C_7H_8	3.14	1026–1031	552–555	1.3	6.7
Gasoline	A blend	3–4.00	632	333	1.4	7.6
Acetone	C_3HO	2.00	1042	561	2.6	12.8
Benzene	C_6H_6	2.77	968	520	1.4	6.7
Carbon monoxide	CO	0.97	1191–1216	644–658	12.5	74.0
Hydrogen	H_2	0.07	1076–1094	580–590	4.0	75.0
Hydrogen sulfide	H_2S	1.18	655–714	346–379	4.3	45.0

are therefore more combustible at higher temperatures. As can be seen from Table 3.15, the ranges of air percentages within which some liquids and gases are flammable are extremely wide.

The methods of detecting the presence of combustible gases and vapors can utilize the phenomena of catalytic combustion, electrical resistance, luminosity, thermal conductivity, IR absorption, or gas ionization. Of the foregoing methods, the most widely used is catalytic combustion, in which a change in the resistance or temperature of the sensing filament is caused by the catalytic combustion of the flammable gases, and this change is measured to detect the concentration of the combustibles. Thermal-conductivity-type detectors are used at higher concentrations. Electrochemical and semiconductor sensors can be used when H_2 and other known gases are to be detected.

One of the common limitations of catalytic-combustion-type analyzers is the poisoning of the filament by silicon, sulfur, chlorinated compounds, or lead compounds. A variety of filament protection means have been added to increase the poison resistance of the sensors. Life expectancies are usually defined in terms of exposure concentration hours. One high-concentration exposure of a poison has been known to knock out a sensor; therefore, non-poisoning techniques should be considered when poisoning is an issue.

The installation of the sensors can be (1) remote head system (continuous measurement, continuous readout), (2) multiple head system (continuous measurement, sequential readout), and (3) tube sampling system (sequential measurement, continuous readout). Multiple head systems are used where at

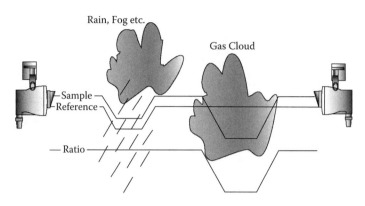

FIGURE 3.16
Open-path IR signal response. (Courtesy of Zellweger Analytics Inc.)

least four or more areas are monitored and a cyclic readout with the accompanying time delay can be tolerated.

Flame ionization detectors (FIDs) and photoionization detectors (PIDs) can be used for the detection of hydrocarbons. Both detectors have been utilized for combustibles monitoring in portable and fixed installation designs. The FID actually burns the sample in an H_2 flame. A charged electrical field is positioned across the flame, and utilizing the ions in the flame can conduct a current. When most combustible materials are introduced into the flame, they produce ions in their combustion products, and these are detected by the increased flow of current across the electric field (flame).

A widely used technique is IR monitoring, which is utilized for both point and area (open-path) measurement applications but cannot detect H_2. The previously discussed detectors were point sensors. To monitor a large area, one would have to locate many monitors (points). In contrast, open-path IR combustibles monitors project their beams in a path that is typically 10–200 m in length and monitor all of the combustibles in that path.

Figure 3.16 depicts how an open-path instrument can compensate for partial blockages of its beam by light-obscuring interference, such as rain, fog, dust, etc. Essentially, it calculates the ratio of the measurement and the reference radiation signal. Most partially obscuring interferences will reduce both signals to the same extent. Therefore, the ratio of the signals is relatively unaffected by the interfering obstruction. On the other hand, the presence of a combustible gas will reduce only the measurement signal and therefore will result in a change in the ratio of the two radiation signals.

It should be kept in mind that the measurement is along a beam or path and, therefore, it is not detecting an area or a point. Percent or parts per million (ppm) readings are obtained by multiplying the concentration of the gas (along the length of the IR beam) by the length of the cloud (along the optical path). It simply measures the total amount of target gas in its optical path.

3.2.9 Fiber-Optic Probes

The renewable energy processes will be monitored and controlled by small, fast, and economical detectors, and fiber optics will play a major role in their designs. The operation and applications of fiber-optic probes have already been discussed in connection with Figure 3.2, so only a brief summary is provided here. Fiber-optic probes can be installed in situ, whereas their readout instruments can be several hundred meters from the probe. The probe can be located in toxic, corrosive, radioactive, explosive, high- or low-temperature/pressure, and noisy environments. Because the measurement signal is optical, the cables are immune to microwave or electromagnetic interference.

Most present-day fiber-optic sensors use linear diode arrays combined with optical gratings and measure the absorption, transmission, fluorescence, and reflection in UV, visible, and NIR regions (see Table 3.1). Light travels to the sampling probe via one fiber-optic cable and returns to the instrument via a second. Laser excitation permits long-distance transmission of excitation radiation to get a useful signal from the sample.

Wand- or dip-type probes are single-sided probes for optical transmission, absorbance, reflectance, and fluorescence measurements. Spectra-caliper probes are used for transmission measurements through solid samples such as plastic films, fibers, biological tissues, and intact plant leaves. Transmission probes are always used in pairs for optical transmission and absorption measurements.

The long-path-length gas probe (see Figure 3.2) is used for the measurement of gases or vapors in stacks, in process pipes, or through the atmosphere. GEM probes use gemstone tips such as sapphire. The six-to-one probe has been developed for fluorescence, reflectance, and Raman-type precious stone measurements. With this probe, it is possible to illuminate the sample with light from an appropriate source (xenon arc, laser, etc.) and then measure the reflected, scattered, or emitted light using six fibers mounted around the light-conducting central fiber.

Four approaches can be used for qualitative and quantitative chemical analysis. These techniques include absorption, fluorescence, scattering, and refractive index change. Absorption sensors can be used in the UV, visible, and IR regions. Remote fiber fluorescence (RFF) systems send a high-intensity light through a large-core quartz fiber to impinge on the sample, which gives off a characteristic fluorescent emission.

As discussed in more detail in Section 3.2.18, Raman laser scattering detects low concentrations of various gases (Figure 3.17). Finally, the refractive index (RI) fiber-optic probe compares the RI of the process material with that of its prism and measures the reflected light as an indication of process RI.

3.2.10 Hydrocarbon Analyzers

Hydrocarbons play an important role in alternative energy processes because many biological degradation processes generate methane, which is often used as the fuel for fuel cells. Ethanol is also important as a gasoline addi-

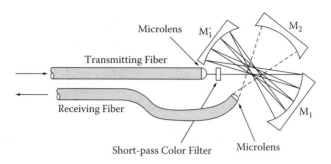

FIGURE 3.17
Multipass reflection sensor used for Raman scattering.

tive. The hydrocarbons comprise a large class of individual chemicals, and Table 3.18 provides a summary of their methods of analysis and the limitations of these methods.

FIDs are used to measure total hydrocarbons at low concentrations such as for pollution, leakage, or safety monitoring. In the FID analyzer, the H_2 flame ionizes the carbon atoms, and the resulting ionic current is an indication of the hydrocarbon concentration. In microprocessor-controlled hydrocarbon analyzers, up to eight channels can be obtained for simultaneous analysis of eight samples. These units also have the capability for remote calibration and automatic self-checking.

The only practical method for the measurement of specific hydrocarbons is by gas chromatography (GC), which can be automated, but is expensive and often used as a laboratory analyzer. Certain specially prepared columns can be used to adsorb specific hydrocarbon classes. For atmospheric hydrocarbons, a common method is to pass a sample of air through a small freeze-out trap, sweep out the air with helium, and then warm the trap and introduce the condensables into the GC column in one concentrated slug.

Either IR or mass spectrometry may be used for individual hydrocarbon determination. Laser-induced Doppler absorption radar (LIDAR) can be used to remotely measure chemical concentrations in the atmosphere. Two different laser wavelengths are selected so that the molecule of interest absorbs one of the wavelengths, whereas the other wavelength is selected to be in a region of minimal interference. The difference in intensity of the two returned laser signals is then used to determine the concentration (see Figure 3.16, which illustrates the concept for IR signals).

A hydrocarbon dew point meter uses a chilled-mirror instrument by which almost invisible films, having sensitivities on the order of 1 ppm, become detectable. Optical fibers are used to detect the reduction of light intensity, and miniature thermocouples measure the surface temperature of the mirror. For total hydrocarbon measurement, the flame ionization analyzer is reliable and accurate, but it requires the attention of operators and also consumes compressed gases.

TABLE 3.18

Atmospheric Hydrocarbon Analyzers

	Hydrocarbon	
Type	**Method**	**Limitations and Interferences**
Total (as carbon)	FID	Some response to carbon-containing nonhydrocarbons
Methane	GC	Expensive equipment (can also be used for carbon monoxide)
Methane-only subtractive	Column preparation fussy column and FID	
	Mass spectrometry	Freeze-out required, expensive
Aromatics, olefins, and paraffins	Subtractive columns and FID	Column preparation fussy
	Mass spectrometry	Freeze-out required, expensive, data reduction requirements large
	Infrared spectrometry	Freeze-out required, expensive, not total class coverage
	Ion mobility spectrometry	Clean sample required, limited knowledge in industry
	Laser-induced absorption	Expensive, specialized support required
	Perimeter monitoring	Concentration/unit length rather than point value
Individuals	GC	Expensive, data reduction requirements large

NDIR sensors can only identify hydrocarbon classes, whereas two-wavelength IR detectors are able to identify individual hydrocarbon species. The GC analyzers that are capable of identifying specific classes of hydrocarbons are complex and are less precise or reliable.

3.2.11 Infrared and Near-Infrared Analyzers

Applications of IR analyzers include the measurement of ammonia, CO, CO_2, ethylene, hexane, methane, moisture, nitrous oxide, propane, and sulfur dioxide. NIR analyzers can detect the concentrations of benzene, caustic, cetane, gasoline boiling point, heating value, molecular weight, octanes, protein, and p-xylene. The measurement errors of these analyzers are: IR—2% FS, NIR—1% FS.

In renewable energy processes, the various types of IR analyzers (including the H_2-sensing Raman version) will play an important role. IR instruments were one of the first analyzers to be moved from the laboratory to the pipeline, and the technology is available for use with gas, liquid, or solid

TABLE 3.19

IR-Analyzer Applications Summary

Analyzer	Organic Vapors					Comments
	Carbon Monoxide	Carbon Dioxide	Simple Molecules	Complex Molecules	Organic Liquids	
Nondispersive infrared (NDIR)	✓	✓	✓			Single-component analysis: methane, ethylene, CO, CO_2, etc.
Mid-IR filter	✓	✓	✓	✓	✓	Same as above, including NH_3, vinyl chloride, methylethyl ketone, etc.
FTIR	✓	✓	✓	✓	✓	The advantage of the FTIR is that it can look at multiple species
Correlation spectrometer	✓					Stack analysis, single-component gas analysis

samples. In the absorption mode of operation on liquid samples, the path length has to be very short, though not as short as in UV. Noncontacting backscatter designs readily measure moisture in solids or composition with FOPs in liquids. NIR is widely used for moisture measurement.

The IR region of the electromagnetic spectrum is generally considered to cover wavelengths of 0.8–20,000 µm. NIR normally covers 0.8–2,500 µm, and classic IR covers the rest. For IR analysis, these limits are normally put in terms of frequency (cm^{-1}, wave numbers, or the number of waves per cm): 4,000–5,000 cm^{-1}, which corresponds to wavelengths of 2,500–20,000 µm. For a summary of IR applications, refer to Table 3.19, and for NDIR applications, refer to Table 3.20.

The detector response of an IR analyzer is not linearly related to concentration. It follows the Beer–Lambert law, and a logarithmic amplifier provides an acceptable linear output. The Beer–Lambert law relates the amount of light absorbed to the sample's concentration and path length.

$$A = abc = \log_{10} I_0 / I$$

where

     A = absorbance

     I = IR power-reaching detector with sample in the beam path

     I_0 = IR power-reaching detector with no sample in the beam path

     a = absorption coefficient of pure component of interest at analytical wavelength; the units depend on those chosen for b and c b = sample path length; sometimes 1 is used

c = concentration of sample component

The advances in IR technology involve several areas of development. The microprocessor has contributed to the development of self-diagnostic and self-calibrating designs. The use of multiple reference cells when stable reference gases are available has also helped auto-calibration. Modular design in conjunction with self-diagnostics has simplified maintenance. The growth of fiber-optic technology has made the probe-type IR analyzer practical and has extended the spectrum by allowing the same probe to use UV, visible, NIR, and IR forms of radiation.

Another area of development is in minimizing the number of moving parts, choppers, shutters, or beam alternators, besides eliminating the need for multiple paths in the IR analyzer. The goal can be met by using AOTFs, which are made of thallium–arsenic–selenide (TAS). This crystal can be tuned by a radio-frequency oscillator, an ultrasonic transducer, over a spectrum of 2–5.5 μm. The TAS AOTF is an electronically controllable narrow-band filter that can be tuned to any desired IR frequency (see Figure 3.3). It can provide a series of chopped and tuned IR beams to simultaneously measure a variety of stack gases. The required frequencies can be selected in milliseconds, and the solid-state AOTF is small and rugged. Therefore, it is insensitive to vibration but must be maintained at a constant temperature. Table 3.20 lists the minimum and maximum ranges of some of the gases and vapors that are commonly detected by NDIR analyzers, and Figure 3.21 shows the absorption bands of NIR-absorbing substances.

As discussed in more detail in Section 3.2.18, the Raman spectrometer obtains its information from the scattering of light from molecules. It is ame-

TABLE 3.20

Typical Applications for NDIR Analyzers

Gas	Minimum Range (ppm)	Maximum Range (%)
Ammonia (NH_3)	0–300	0–10
Butane (C_4H_{10})	0–300	0–100
Carbon dioxide (CO_2)	0–10	0–100
Carbon monoxide (CO)	0–50	0–100
Ethane (C_2H_6)	0–20,000	0–10
Ethylene (C_2H_4)	0–500	0–100
Hexane (C_6H_{14})	0–200	0–5
Methane (CH_4)	0–2,000	0–100
Nitrogen oxide (NO)	0–500	0–10
Propane (C_3H_8)	0–300	0–100
Sulfur dioxide (SO_2)	0–500	0–30
Water vapor (H_2O)	0–3,000	0–5

nable to direct, noninvasive sampling of solids and liquids. Water and glass are weak Raman scatters, providing an opportunity to investigate samples in aqueous media through glass windows.

3.2.12 Mass Spectrometers

Mass spectrometers (MSs) are classified by the technology they use to separate ionic masses into clusters. Their examples include the quadrupole, magnetic sector, time-of-flight, and ion-trap mass spectrometers.

The MS is expected to perform a specific analytical task on a continuous basis for extended periods. It must exhibit long-term stability and accuracy while operating over its environmental range. An advantage of quadrupole instruments (Figure 3.22) over electromagnetic focusing ones is that the fields required to focus a particular mass can be changed very rapidly ($<10^{-3}$ s), which is especially valuable for computer-controlled measurements that require multiple-ion monitoring.

In time-of-flight MS, the ionized sample is subjected to a negative-polarity-accelerating field, and then the ion beam is directed into a time-of-flight drift region. The lighter ions travel faster through this region than the heavier ones, producing mass separation by the amount of time it takes for the ions to traverse this drift tube. To accomplish this type of mass separation, the ionized sample is introduced in discrete pulses of 20,000–35,000 pulses per second. Figure 3.23 illustrates the operation of the time-of-flight spectrometer.

All MSs require a gaseous sample, so if used on a liquid, the sample has to be vaporized. The ionized molecules—and the charged fragments of their decomposition—are then sorted and measured by their mass-to-charge ratio.

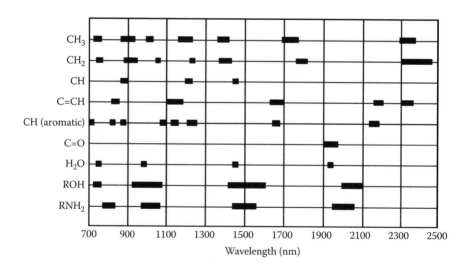

FIGURE 3.21
Absorption bands of some NIR-absorbing chromophores.

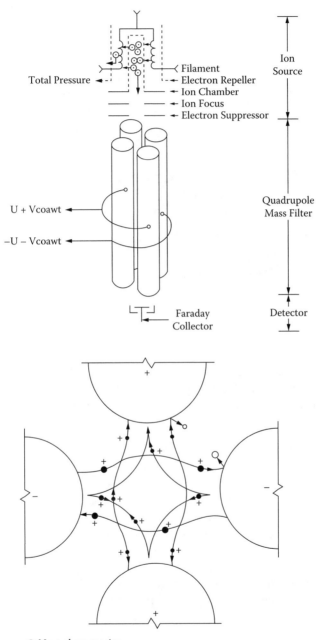

o,O Neutral gas species
• Small masses neutralize on positive rods
• Tuned masses pass through filter
● Large masses neutralize on negative rods

FIGURE 3.22
Top: quadrupole mass spectrometer. Bottom: quadrupole mass filter.

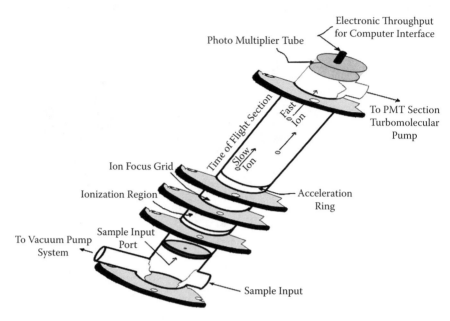

Photo Multiplier Tube

Electronic Throughput for Computer Interface

Time of Flight Section

Fast Ion

Slow Ion

To PMT Section
Turbomolecular Pump

Ion Focus Grid

Ionization Region

Acceleration Ring

To Vacuum Pump System

Sample Input Port

Sample Input

FIGURE 3.23
Time-of-flight mass spectrometer.

This last process can be accomplished by the time of flight of the ions down an evacuated tube. Detection is usually by electron multipliers. The most common application of the MS process is in gas plants, where many gases can be detected simultaneously, or multiple streams can be examined sequentially.

3.2.13 Moisture, Humidity, Dew Point

The sensing elements of these designs include (a) wet–dry bulb psychrometers, (b) hair or fiber element, (c) cellulose element, (d) thin-film capacitance, (e) Dunmore type, (f) lithium chloride for dew point, (g) surface resistivity sensors (Pope cell), and (h) condensation on chilled surface. The inaccuracies of these sensors are: (a) 2% relative humidity (RH), (b) 3–5% RH, (c) 3–5% RH, (d) 2–3% RH standard, (e) Dunmore cell under 1% RH, (f) 1°C (2°F), (g) 3–5% RH, and (h) 0.2–0.4°C (0.3–0.8°F).

In the control and optimization of renewable energy processes, ambient humidity is one of the important parameters. Humidity also plays an important role in building controls and in minimizing the energy use of HVAC systems. Leonardo da Vinci was the first to attempt the measurement of humidity by weighing a ball of wool. Today, as shown in Figure 3.24, a wide range of humidity sensors is available.

The most frequently used unit in expressing the amount of water vapor in air is relative humidity (0–100% RH). It is the ratio of the mol fraction (or amount) of moisture in a gas mixture to the mol fraction (or amount) of moisture in a saturated mixture at the same temperature and pressure.

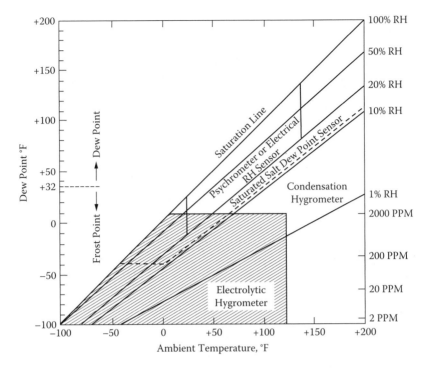

FIGURE 3.24
The operating ranges of different humidity sensors. Most humidity/dew point detectors can make measurements at higher values of humidity and temperature, but are limited at low temperatures and at low concentrations. Most are also limited to a maximum operatng temperature of about 200°F (95°C).

Another frequently used unit is dew point (saturation) or wet-bulb temperature. Dew point is the saturation temperature of a gas–water vapor mixture (the temperature at which water condenses as the gas is cooled). Finally, moisture is also expressed in volume or mass ratio as parts per million (ppm) volume or weight.

Wet- and dry-bulb temperatures are measured by exposing two temperature-sensitive elements to the atmosphere whose moisture level is to be measured. The wet bulb is wrapped with a wick soaked in water; the other element, the dry bulb, is left bare. Water evaporating from the wick lowers its temperature, which is read as the wet-bulb temperature, whereas the other reads the dry-bulb temperature. The relative humidity can be read from a psychometric chart such as the one shown in Figure 3.25.

The operation of RH detectors can be based on the dry- and wet-bulb reading, temperature difference, elongation of cellulose, resistance of lithium chloride (Dunmore), surface resistivity of polystyrene (Pope), and change in capacitance. The capacitors can be formed from aluminum or hygroscopic polymers.

Dew point temperature is the temperature at which water starts to condense on a surface. It is measured from −80 to 100°C (−112 to 212°F). The

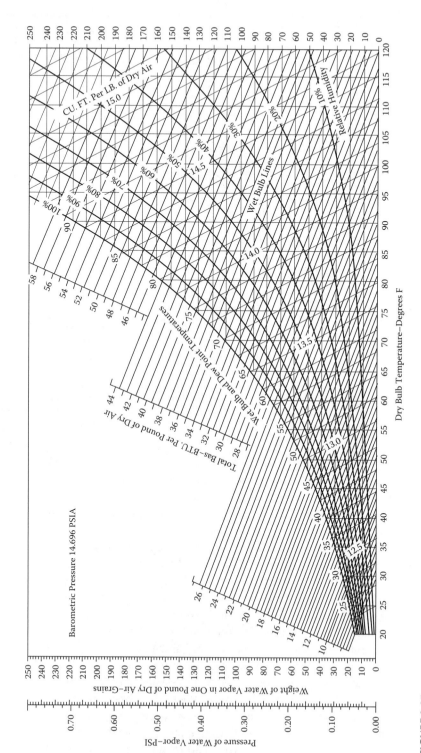

FIGURE 3.25
Psychrometric chart.

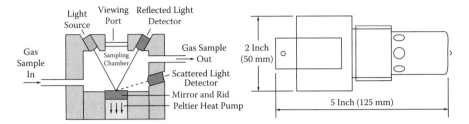

FIGURE 3.26
The cooled-mirror-type sensing element and the probe protector that houses it. (Courtesy of Michell Instruments Ltd.)

dew point reading of an air sample is affected by the air pressure and by the amount of water vapor in the sample. An increase in air pressure increases the dew point reading, whereas the temperature of the air does not affect it. The three most widely used sensors for the detection of dew point are: (1) aluminum oxide or polymer sensors, (2) lithium chloride–saturated salt sensors, and (3) surface-condensation-type sensors.

The temperature of a surface reaches the dew point when the first molecular layer of condensation appears on it. Dew point sensors of this type can be grouped according to the method used to detect the appearance of this condensate into conductivity and optical (mirror) types.

The optical chilled-mirror dew point technique is a fundamental measurement, because the saturation temperature determines the saturation partial pressure of the water vapor. A temperature element is then placed in thermal contact with the mirror, and the mirror temperature is utilized directly as the dew point or saturation temperature. The cooling of the mirror surface in the past was done by acetone and dry ice, liquid CO_2, mechanical refrigeration, and later by thermoelectric heat pumps. The thermoelectrically cooled, optically observed dew point hygrometer probe is illustrated in Figure 3.26.

An improvement on the chilled mirror is the cycled chilled-mirror-type dew point detector. The advantages of cycling the mirror temperature include simplicity, which also reduces the cost and lowers maintenance, because, by making the periods of dew formation short (5% of the time), the probability of contaminant condensation on the mirror surface is also reduced. In this design, the mirror surface is visually monitored by a fiber-optic link.

3.2.14 Moisture in Industrial Materials

Moisture analyzers include a large variety of designs listed here. The list includes their inaccuracies: (1) electrolytic hygrometer (2–5% FS), (2) capacitance (3% FS), (3) impedance (3% FS), (4) piezoelectric (10% AR or 2 ppm by volume), (5) heat of adsorption, (6) infrared (0.5–1% FS), (7) microwave (for a 1–15% moisture range, error is within 0.5%, less if corrected for density), (8) Karl Fischer titrator (0.5–1% FS), (9) drying oven (0.5–1% FS), (10) dipole, (11) cavity ring down, (12) fast neutron (0.2% in solid's density corrected), and (13) radio-frequency absorption (5 ppm).

TABLE 3.27

Summary of Moisture Analyzer Features

Type	Range	Sample Phase	Sample System Required	Remarks
Electrolytic hygrometer	0–20 to 0–2,000 ppm	Clean gas Special sampling for liquids	Yes	Sample flow must be constant
Change of capacitance	0–10 to 0–1,000 ppm	Clean gas or liquids	a	Sample temperature must be constant
Impedance type	0–20,000 ppm	Clean gas or liquids	No, only for liquids	Sample temperature of liquids must be constant
Piezoelectric type	0–5 to 0–25,000 ppm	Clean gas only	Yes	
Heat of absorption type	0–10 to 0–5,000 ppm	Clean gas or liquid; special sampling for liquids	Yes	Sample flow must be constant
Infrared absorption	Gas: 10–100% Liquid: 6–100%	Gas, liquids, and slurries	b	
Microwave absorption	0–1 to 0–70%	Liquids, slurries, and pastes only	No	

a   Available in probe form but can be direct pipeline-mounted only if flow velocity is under 1.6 fps (0.5 m/s).

b   Fiber-optic probe (FOP) designs can be direct pipeline-mounted without sampling.

Some of the industrial process analyzers are of the probe type (capacitance, fiber-optic IR probes), others can look through the process stream in the pipe (microwave), and the majority require some form of a sampling system. Table 3.27 provides a summary of the ranges and other features of many of the process moisture analyzers.

The *electrolytic hygrometer* operates on the basis of the electrolysis of water into O_2 and H_2. Because two electrons are required for the electrolysis of each water molecule, the electrolysis current is a measure of the water present in the sample. If the volumetric flow rate of the sample gas into the electrolysis cell is fixed, then the electrolysis current is an indication of the water concentration in the sample.

The *capacitance hygrometer* detects the change of capacitance, which is a function of plate area, plate spacing, and the dielectric constant of the material between the plates. The dielectric constant of a material has a unique value for each substance. Water, having a dielectric constant of 80, is a good candidate for this measurement. When used to measure the moisture content of solids, compensation is required for particle size, packing, and material density because they all affect capacitance. The same principle can also

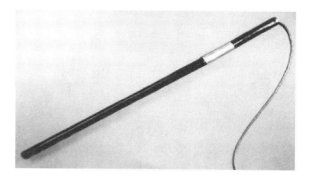

FIGURE 3.28
Dielectric constant monitoring probe for the remote measurement of soil moisture. (Courtesy of Automatika Inc.)

be utilized to detect the moisture content of soil by the insertion of a probe. Such a probe might consist of two electrodes with an insulating gap between them, and the capacitance circuit is used to measure the soil moisture as the dielectric constant of the insulator varies with the soil's moisture content (Figure 3.28). Several probes can be connected to a central computer, which can sound alarms or automatically operate irrigation systems.

The impedance hygrometer measures the water content of a sample by means of a probe whose electrical impedance is a function of the vapor pressure of moisture in the fluid. The probe consists of an aluminum strip that is anodized to form a porous layer of aluminum oxide. A thin coat of gold is applied over the aluminum oxide. Water vapor penetrates the gold layer and equilibrates on the aluminum oxide. Leads from the gold and aluminum electrodes of the probes connect the sensing element to the measuring circuitry. The moisture content of solids can also be measured indirectly by detecting the moisture in the atmosphere above or near the process solids because the atmosphere near the solids is in equilibrium with the moisture content of the process materials.

Piezoelectric crystals are coated with a hygroscopic material and exposed to the sample. Water from the sample is absorbed by the coating, which increases the total mass and decreases the oscillating frequency of the crystals. To measure changes of decreasing moisture concentration and to simplify the frequency measurement, two crystals are used. One crystal is exposed to a wet sample and the other to a dry reference gas for a short period. Then, the sample and reference gas flows are switched so that the moisture is absorbed by one crystal while being desorbed by the other.

The heat-of-adsorption hygrometer uses a column that selectively adsorbs moisture, and the temperature rise due to heat liberation is in proportion to the amount of moisture being adsorbed.

Infrared absorption hygrometers take advantage of the fact that water absorbs electromagnetic radiation in the IR region of the spectrum (see Figure 3.21). By measuring the attenuation of a beam of this wavelength as it passes through a sample, the moisture content of the sample can be determined. A reference wavelength that is not absorbed by moisture but

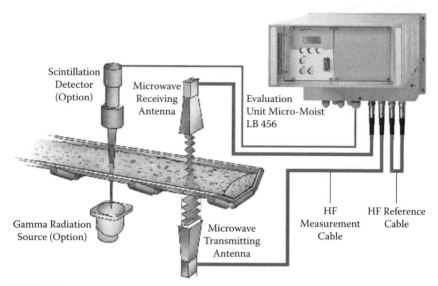

Scintillation Detector (Option)

Microwave Receiving Antenna

Evaluation Unit Micro-Moist LB 456

Gamma Radiation Source (Option)

Microwave Transmitting Antenna

HF Measurement Cable

HF Reference Cable

FIGURE 3.29

The noncontacting moisture analyzer can be provided with gamma radiation densitometer to compensate for variations in process density. (Courtesy of Berthold Technologies.)

is affected by all other factors is also used to compensate for other (non-moisture) effects. When IR analyzers are used for moisture measurement in solids, the measuring and reference beams are not transmitted through the sample but are reflected off the surface, and therefore, only the surface moisture is measured.

The microwave absorption hygrometer operates similarly to the IR absorption hygrometer, but in the microwave frequency band of 20–22 GHz (K band), the wavelength is about 13–15 mm. The unit senses the mass of moisture in the beam path, so the readout is normally in terms of the mass of moisture per unit volume. When using microwave moisture analyzers to measure the moisture content of moving webs or other materials that are thinner than 0.5 mm, multiple transmission is required. Because the absorption of microwaves is affected by both bulk density and solids temperature, both of these variables must be measured and compensated for if precise readings are desired (Figure 3.29).

Dipole polarization-effect moisture sensors utilize the electrical asymmetry around the water molecule (two slightly positively charged H_2 atoms and a slightly negatively charged O_2 atom). When these molecules are subjected to an electric field, they tend to align with the electric field because of their dipole nature. At a frequency of 20 kHz, moisture can be detected by this method.

The neutron backscatter uses a neutron radiation source that is focused on a sample, and the neutron backscatter is measured and correlated to the moisture content. The energy of the neutron after the impact is a function of the mass of the impacted atom. If the mass of the impacted nucleus is equal to the mass of the neutron, all of the neutron's kinetic energy is transferred

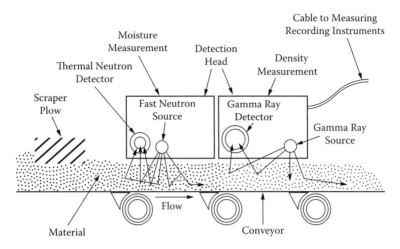

FIGURE 3.30
Nuclear moisture measurement.

to that nucleus. The H_2 atom, because it is most nearly equal in its mass to the neutron, is the most efficient energy absorber (or moderator) of neutrons. However, density also affects the amount of neutrons reflected; therefore, density compensation is needed, as shown in Figure 3.30.

3.2.15 Odor Detection

In some of the alternative energy processes, such as in fuel cell control and optimization, where the fuel gas is generated by wastewater digesters or garbage dumps, odor can indicate both the type and the concentration of these gases. These concentrations can be in the parts per billion (ppb) and parts per trillion (ppt) ranges, so extremely sensitive detectors are needed to measure them.

Odor is a sensation associated with smell, which can be hard to quantify. The same quantities of different materials cause different odor intensities. The unit of odor intensity (W) is based on the odor of tertiary butyl mercaptan or TBM (W = 1.0). Using that reference, H_2S, for example, has an odor intensity of W = 0.08, or 8% of TBM. Most odorant substances contain sulfur. Table 3.31 lists a number of odorant substances and their relative odor intensities (W).

Recent improvements in technology and increased research in the area of instrumentation have made dramatic improvements in the creation of instruments that are capable of surpassing the human olfactory system. In fact, the instruments of today are approaching the sensitivity of the canine olfactory system, which is thought to be as much as a million times more sensitive than that of humans.

Electronic nose technology relies on the use of solid-state sensors, which can either be chemoresistors, chemodiodes, or electrodes. Primary odors are

TABLE 3.31

The Relative Odor Intensity of Different Chemicals

Abbreviation	Name	Formula	(W)
EM	Ethyl mercaptan ethanethiol	CH_3H	1.08
DMS	Dimethyl sulfide methylsulfide	$(CH_3)_2S$	1.0
IPM	Isopropyl mercaptan 20-propanethiol	$CH_3CHSHCH_3$	0.88
MES	Methyl ethyl sulfide methyl thioethane	$(CH_3)_2CH_2S$	0.66
NPM	Normal propyl mercaptan 1-propanethiol	$(CH_3)CH_2CH_2SH$	0.85
TBM	Tertiary butyl mercaptan 2-methyl 2-propanethiol	$(CH_3)_3CSH$	1.00 (ref)
SBM	Secondary butyl mercaptan 1-methyl 1-propanethiol	$(CH_3)_2CH_2CHSH$	1.99
DES	Diethyl sulfide ethyl sulfide	$(C_2H_5)_2S$	0.22
Thiophene	Thiophene tetrahydrothiophene	C_4H_8S	1.63
EIS	Ethyl isopropyl sulfide		0.07

composed of polar molecules, organic vapors, and phthalocyanines. New sensors such as chemosensors are devices that convert chemical composition into a quantifiable electrical format. Other new types include conductometric, optical, gravimetric, amperometric, calorimetric, potentiometric, and chemocapacitor sensors.

This technology allows the analyzers to be small, credit card–size devices. A sensor can be designed to be as specific or as broad in application as desired. There are many possible applications in industrial operation and safety. Wherever an odor can be used to detect the presence or even absence of a substance, these electronic noses can be used. From a water pollution perspective, the analyzer that detects the odor of hydrocarbons in water is valuable.

3.2.16 Open Path Spectrophotometry

Long path monitoring of the atmosphere can be relevant in the control of renewable energy processes for the determination of CO_2 emissions, and for toxic and combustible cloud monitoring. The open path (OP) methods of monitoring use UV; IR and FTIR (Fourier transmission IR); NDIR; TDLAS (tunable diode-laser adsorption spectroscopy); light detection and ranging (LIDAR); and backscatter absorption–based gas cloud imaging technologies.

The use of open path monitoring is driven by either the need for probing an area without physical intrusion or by the need to monitor an area that is larger than one that can be cost-effectively monitored with a requisite num-

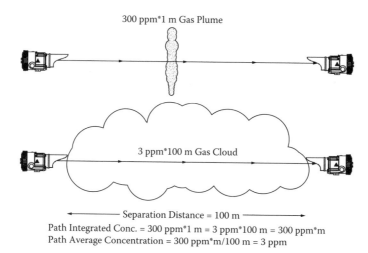

300 ppm*1 m Gas Plume

3 ppm*100 m Gas Cloud

Separation Distance = 100 m
Path Integrated Conc. = 300 ppm*1 m = 3 ppm*100 m = 300 ppm*m
Path Average Concentration = 300 ppm*m/100 m = 3 ppm

FIGURE 3.32

Illustration of path-integrated concentrations and relationship to path average concentration for narrow plumes and widely dispersed vapor clouds.

ber of point detectors. As was discussed in connection with Figure 3.16, combustible detection requires much less sensitivity than does toxic monitoring.

In applications in which FTIR does not have sufficient sensitivity, open path ultraviolet (OP-UV) spectroscopy is frequently employed. This methodology can be used for the detection of homonuclear diatomic molecules (chlorine, bromine, etc.), which have no infrared absorption, or molecules that absorb only weakly in the IR region, such as benzene, sulfur dioxide, and nitrogen oxides.

The concentration profile integrated over the path length is termed path-integrated concentration (PIC). It is simply the path average concentration (PAC) multiplied by the measurement path length. These units can be converted into the more familiar volumetric units of ppm, or for combustible gas determination, the more appropriate, lower explosive limit (LEL). An open path instrument operating over a 100 m path length will produce the same PIC reading for a narrow, concentrated 1-m thick plume having a vapor concentration of 300 ppm as it will for a 3 ppm vapor cloud dispersed across the entire 100 m path length. In each of these situations, the PAC would be 3 ppm (Figure 3.32).

OP-UV spectrometry can be used to measure vapors or gases that have weak absorption characteristics, and therefore, low sensitivities in the IR spectrum. These include such compounds as nitrogen oxides, formaldehyde, ozone, sulfur dioxide, benzene, toluene, and xylenes, and also homonuclear diatomic molecules, such as chlorine. The compounds that can be determined by UV are much fewer (see Table 3.43) than those that are absorbing in the IR spectra.

OP-TDLAS has been applied commercially only in air monitoring. Instruments based on these lasers are capable of making very sensitive mea-

TABLE 3.33

Some Representative Gases and Approximate
PIC Detection Limits for OP-TDLAS, Assuming
Ability to Measure Absorbance to 1 Part in 10^5

Species	Detection Limit (ppm[a])
HF	0.2
H_2S	20
NH_3	5.0
CH_4	1.0
HCl	0.15
HCN	1.0
CO	40
NO	30
NO_2	0.2

[a] From Frish, M. B., White, M. A., and Allen, M. G.,
SPIE Paper No. 4199–05, 2000.

surements, sometimes in the ppt range. They are also very fast, having measurement times as low as 1/10 seconds. On the other hand, complexity and cost limit their acceptance. The need for liquid nitrogen cooling and the effects of pressure broadening limit their use to point monitoring. Substances that can be detected by diode lasers include NO, NO_2, HF, HCl, HCN, HI, NH_3, C_2H_2, CO, CO_2, H_2S, and CH_4. In some cases, diode lasers can measure multiple gases that have closely spaced absorption features. Table 3.33 lists some representative gases of interest for air monitoring that can be measured using TDLAS, along with their approximate detection limits.

Open path detection of combustibles has already been discussed in Section 3.2.8. Figure 3.34 shows a simplified schematic representation of an OP-HC

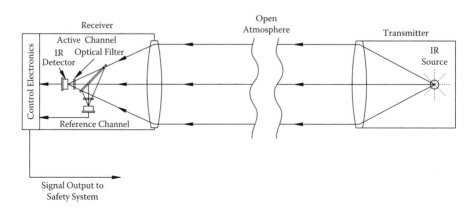

FIGURE 3.34
Schematic representation of an NDIR-based OP-HC detector.

detector. The detection principle relies on a two-channel nondispersive photometer. The optical filters are carefully specified to minimize false gas signals from differential absorption resulting from moisture as well as changes in the spectral output characteristics of the source over time. The source is modulated to mitigate against solar interference. Source modulation is performed at frequencies where there is minimal solar modulation. Distances between the transmitter and receiver vary typically between 10 and >120 m. Most vendors supply at least two models: a short-range unit for operation from 10 to 60 m, and a longer-range unit operating from 60 to >120 m distances.

3.2.17 Oxygen in Gases

The concentration of O_2 can be measured by the following types of analyzers (a) deflection-type paramagnetic; (b) thermal-type paramagnetic; (c) dual-gas paramagnetic; (d) catalytic combustion; (e) low-temperature electrochemical (galvanic, coulometric, and polarographic); (f) zirconium oxide, voltage mode; (g) zirconium oxide, current mode; and (h) NIR spectroscopy. The measurement ranges for types a, b, and d, are in percentages, whereas those for types f and g are from ppm to percentages and type e, from ppb to percentages.

In renewable energy processes, the measurement of O_2 can be used to optimize the efficiency of combustion processes or to guarantee the safety of H_2 production and distribution. Oxygen analyzers can depend on the paramagnetic and electrochemical properties of O_2, or can utilize the catalytic combustion and spectroscopic techniques.

Paramagnetic detectors take advantage of O_2's strong affinity for the magnetic field. This uncommon property is shared by a few other gases such as nitric oxide (Table 3.35). These gases are not normally encountered in processes where oxygen analysis is required. Some gases, such as H_2, are diamagnetic (repelled by the magnetic field). The paramagnetic property of gaseous O_2 has been utilized in three oxygen analyzer designs. These are the deflection, thermal, and reference gas types.

In the deflection-type analyzer, the magnetic force acts on a test body that is free to rotate about an axis. In the thermal design, the O_2-containing sample is attracted by the magnetic field, causing a flow that is measured. In the dual-gas cell, two legs of a flowing reference gas contact the sample, one of which attracts it with a magnetic field. This produces a flow that is related to the O_2 content of the sample.

The advantage of the magnetic wind and the dual gas analyzers is that both are rugged. The disadvantages of the magnetic wind design are the need for temperature and thermal conductivity compensation and the fact that high temperatures can degrade stability and reliability. The dual gas or differential-pressure-type oxygen analyzer is sensitive to vibration, which some manufacturers can compensate for.

In the catalytic combustion design, the sensor consists of a measuring and a reference cell, with a filament in each. The measuring filament is provided

TABLE 3.35

The Magnetic Susceptibility of Different
Gases, with Oxygen Given as 100

Acetylene (C_2H_2)	−0.24
Ammonia (NH_3)	−0.26
Argon (Ar)	−0.22
Carbon dioxide (CO_2)	−0.27
Carbon monoxide (CO)	+0.01
Ethylene (C_2H_4)	−0.26
Hexane–normal (C_6H_{14})	−1.7
Hydrogen (H_2)	+0.24
Methane (CH_4)	−0.2
Nitric oxide (NO)	+43.0
Nitrogen (N_2)	0.0
Nitrogen dioxide (NO_2)	+28.0
Oxygen (O_2)	+100.0

Source: *Guide to the Selection of Oxygen Analyzers,*
Delta F. Corp.

with a catalytic surface to oxidize the fuel, whereas the reference filament serves only to compensate for variations of temperature and thermal conductivity in the sample. The resulting temperature difference is related to O_2 concentration in the sample.

There are three design variations in electrochemical oxygen detectors: (1) the high-temperature fuel cell detectors that utilize the conduction of O_2 ions from one electrode to another through a solid zirconium oxide electrolyte (Figure 3.36); (2) ambient-temperature galvanic detectors that operate by O_2 reduction at the cathode and dissolution of an active anode such as cadmium or lead in an electrolyte; and (3) the polarographic detectors that consist of three electrodes (cathode, anode, and a reference) and an electrolyte where an external potential is applied to the cathode to drive the O_2 reduction reaction. All three types of electrochemical oxygen detectors measure the partial pressure of O_2 and require either temperature control or temperature compensation.

Galvanic oxygen detectors are capable of sensitivities down to the ppb range. In the *coulometric* sensor, the sample gas diffuses to the cathode, where it is reduced to hydroxyl ions, which migrate to the anode. There the ions are oxidized back to O_2, and the resulting cell current is a proportional indication of the O_2 concentration in the gas sample. The *polarographic* cell consists of a sensing electrode (cathode), a reference electrode (anode), and an electrolyte, which is usually potassium chloride. A voltage is applied between the two electrodes, causing the O_2 to be electrochemically reduced, and producing an ionic current that is linear with the O_2 content of the sample.

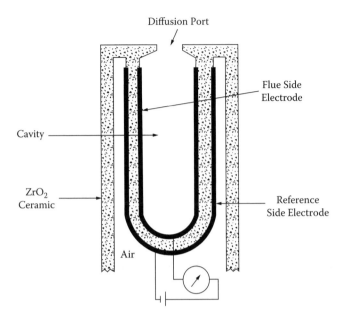

FIGURE 3.36
High-temperature diffusion-limited current mode cell.

The spectroscopic measurement of O_2 is more expensive and more complicated than the previously described methods. On the other hand, those methods allow the monitoring of multiple components simultaneously. As was discussed in some detail in Section 3.2.12, *mass spectroscopy* can be used to monitor O_2 by ionizing the sample and separating the ions according to their mass. As was covered in Section 3.2.11, *NIR spectroscopy* is widely used for stack monitoring, which includes O_2 measurement.

The detection of molecular O_2 by the NIR spectroscopic technique involves a measurement that is often described as *a forbidden transition*. Many molecules can absorb radiation (energy), enter a temporary excited or radiative state, and release energy (reradiation), which is called *fluorescence*. In the case of molecular O_2, there is such a transition that can be monitored in the NIR region of the spectrum near 760 nm. Therefore, O_2 concentration can be measured in this "forbidden region" by NIR methods, such as using the TDLAS.

In the TDLAS detector, a diode laser is scanned across a chosen absorption line. The selection and application of the O_2 spectral line is proprietary to most vendors, but it is usually one of the O_2 lines around 760 nm (Figure 3.37). The laser source wavelength is modulated as the absorption line is scanned, which makes it possible to use the spectroscopic oxygen technique, which previously was not considered possible.

These analyzers are being used in incinerator stacks and other applications where samples are typically dirty or corrosive and high detection speeds are required. The incorporation of fiber-optic technology has provided a high degree of installation flexibility (Figure 3.38).

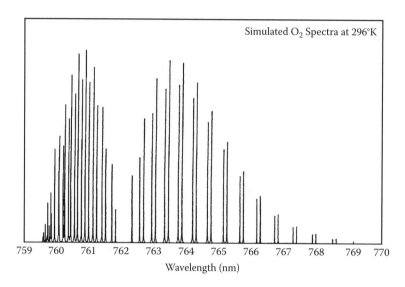

FIGURE 3.37
Simulated NIR O_2 spectra. (Courtesy of Unisearch Associates, Inc.)

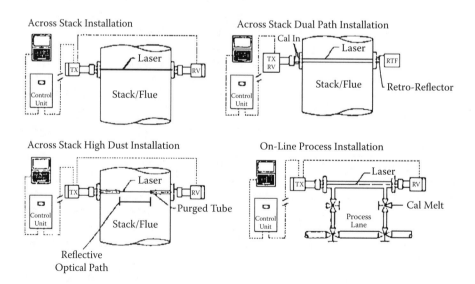

FIGURE 3.38
NIR tunable diode laser instrument layouts. (Courtesy of Analytical Specialties for Norsk Elektro Optikk AS.)

Although neither NIR nor mass spectroscopic techniques are likely to be used to measure O_2 only, they can be justified to simultaneously measure O_2 and a multiplicity of other components in a sample.

3.2.18 Raman Analyzers

In the renewable energy processes, the applications of Raman analyzers are in the same areas where fiber optics (Section 3.2.9) and IR analyzers (Section 3.2.11) are applicable. In addition, Raman analyzers can also detect H_2 and O_2, which are very important measurements in renewable energy processes. The relative capabilities of IR, NIR, and Raman analyzers are summarized in Table 3.39.

Raman analyzers can be considered for applications that IR or NIR detectors are suited for, but the cost of a Raman process analyzer exceeds that of other analyzers. Therefore, as yet, their applications are few and can be found in areas where continuous and fast readings are more important than cost and precision. In the future, the availability of inexpensive, rugged process probes that can work reliably within a high-pressure, high-temperature, and corrosive process environment will increase their use.

When monochromatic light interacts with molecules, most of the photons are scattered without any change in energy. A small number of photons are inelastically scattered and undergo a change in energy. This phenomenon is called *Raman scattering*. The difference in energy between the incident photons and inelastically scattered photons is called *Raman shift*. Homonuclear diatomic molecules (N_2, O_2, H_2, etc.) show Raman bands but not mid-IR absorption bands, because their only vibrational mode does not change the dipole moment of the molecule.

Figure 3.40 shows the layout of a typical Raman analyzer that uses fiber optics for process application. In a Raman process system, light is filtered and delivered to the sample via excitation fiber. Raman-scattered light is collected by collection fibers in the fiber-optic probe, filtered, and sent to the spectrometer via return fiber-optical cables. A charge-coupled device (CCD) camera detects the signal and provides the Raman spectrum. To take advantage of low-noise CCD cameras and to minimize fluorescence interference, NIR diode lasers are used in process instruments.

Optical fibers are used to deliver light and collect Raman signals remotely. Silica fibers with core diameters ranging from 50 to 500 μm are used. The larger the core diameter, the less the bending radius (flexibility). Raman spectroscopy uses low-hydroxy fibers to minimize fiber background interference in Raman data and also to minimize light attenuation caused by fiber absorption.

Invasive and noninvasive Raman probes are used. The requirements for an invasive or insertion probe are that it should be industrially hardened, have a high degree of chemical/corrosion resistance, and be able to withstand very high process temperatures and pressures. Raman probes can withstand up to 350°C and 200 bar pressures, and operate under highly corrosive condi-

TABLE 3.39

Advantages and Disadvantages of Mid-IR, Near-IR, and Raman Methods

	NIR Absorption	Mid-IR Absorption	Raman Scattering
Selection rule	Change in dipole moment	Change in dipole moment	Change in polarizability
Spectral band profile	Overtone/combination bands, broad (nonspecific)	Fundamental band, narrow (high specificity)	Fundamental bands, narrow (high specificity)
Sensitivity	Moderate (ppm, low percentage)	Very high (ppb, ppm)	Low (low percentage, ppm in favorable cases)
Qualitative identification	Difficult	Yes. Extensive spectral library available	Yes. Comprehensive library not available
Sample	Transparent, bubble and particulate-free liquids; Powder samples require diffuse reflectance probe	Transparent liquid, thin solid pellets and ideal for gases; Special ATR probes needed	Same probe is used for all samples: liquids, slurries, emulsions, powders, solids, samples with particulates, and bubbles
Signal-to-concentration relationship	Logarithmic	Logarithmic	Linear
Quantitative analysis	Mathematical/statistical modeling	Simple arithmetic possible	Simple arithmetic possible
Transfer of calibrations	Difficult	Difficult	Relatively easy
Sample handling	Yes	Yes	No. Noninvasive or in situ
Water interference	Good sensitivity to and interference by water	Very high sensitivity to and strong interference by water	No or minimal water interference
Temperature dependence	Sensitive to process temperature variation	Sensitive to process temperature	Wide temperature variation tolerated
Sampling probe	Double-sided, single-sided probes when used with reflectance	Double-sided or special ATR probes	Single-ended
Probe fouling	More likely	More likely	Less likely or not likely with noninvasive
Remote via fiber optics	Long fibers, low-hydroxy silica fibers	Short fibers, expensive chalcogenide fibers	Long fibers, low-hydroxy silica fibers
Wetted optical window	Inert (sapphire, silica, quartz)	Fragile (ZnSe, CaF$_2$, KBr)	Inert (sapphire, silica, quartz)

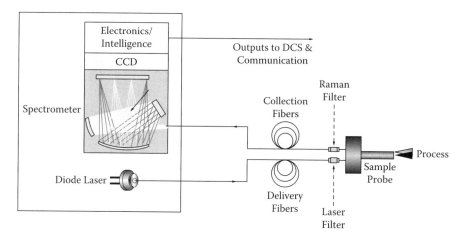

FIGURE 3.40
Schematic of fiber-optic Raman analyzer for process measurements.

tions. The invasive probes can be provided with a retraction mechanism for safely retracting the probe from the process for maintenance. In applications where the process integrity cannot be breached for hazardous or sterility reasons and where the processes are operating at extreme pressure, temperatures, or corrosive conditions, a noninvasive probe should be considered (Figure 3.41). Typically, in noninvasive probe implementation, the probe is placed outside the process, and the laser excitation and Raman collection are done optically through quartz or sapphire windows.

The decision to use a Raman analyzer depends on the availability and practicality of alternative analyzer technologies. Similar to mid-IR, Raman bands represent fundamental modes of vibration. Bands are narrow and molecule specific, and they provide quantitative chemical analysis. The technology is equally amenable to the analysis of gases, liquids, slurries, emulsions, powders, and solids, including samples with particulates and bubbles. Noninvasive probes can be used and are a key feature in many applications where the process must not be breached.

The limitations of Raman spectroscopy are its low sensitivity compared to IR absorption and fluorescence interference from impurities in the sample. Raman spectroscopy is a developing technology, and a good amount of research and planning is necessary before deciding whether or not to employ it. The cost of a Raman process analyzer exceeds that of other analyzers. To reduce cost, Raman analyzers often include multichannel capability. Up to four process streams can be analyzed with a single CCD camera by splitting the lasers.

The availability of inexpensive, rugged process probes that can work reliably within a high-pressure, high-temperature, and corrosive process environment will enable the Raman analyzers to be used in a wider range of process applications. Continuing efforts to improve detection limits, fluorescence elimination, traceability of calibration standards, calibration capabil-

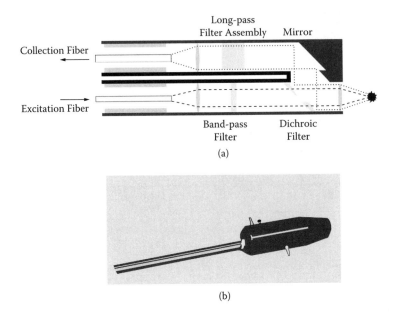

(a)

(b)

FIGURE 3.41

Schematic of focused probe with integrated laser and Raman filters. The beam path (a) and the process probe (b) can be used for both invasive and noninvasive applications. (Courtesy of InPhotonics Inc.)

ity, and mechanical reliability, and lower costs will gradually increase the use of Raman technology.

3.2.19 Sulfur Oxide Analyzers

Renewable energy processes do not generate sulfur dioxide, but coal-burning power plants do; therefore, sulfur oxides (just as CO_2) are present in the atmosphere, contributing to acid rain and other hazards. The predominant form of sulfur oxide in the atmosphere is sulfur dioxide (SO_2) itself. Some sulfur trioxide (SO_3) is also formed in combustion processes, but it rapidly hydrolyzes to sulfuric acid, which is considered to be a particulate matter. In the United States, the ultimate air quality goals (secondary standards) for sulfur dioxide are 60 $\mu g/m^3$ (0.02 ppm) annual arithmetic average and 260 $\mu g/m^3$ (0.1 ppm) maximum 24 h concentration, which are not to be exceeded more than once a year.

The oxides of sulfur are measured both in ambient air, where their concentration is usually a small fraction of one ppm, and in stacks and other industrial emissions, where their concentrations are in hundreds of ppm. As already discussed in Section 3.2.11, sulfur dioxide absorbs radiation over a broad range of wavelengths, which includes both the IR (Table 3.20) and UV regions.

The absorbance of SO_2 in the UV range is shown in Figure 3.42. Because in most stack-monitoring applications there is an interest in measuring the con-

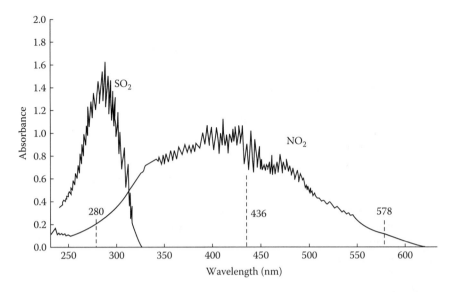

FIGURE 3.42
The absorbance of SO_2 and NO_2 in the ultraviolet range.

centrations of both NO_x and SO_2, some suppliers of UV analyzers offer a single analyzer for the simultaneous monitoring of both. These can be of the probe design and be provided with self-cleaning and self-calibrating features.

The earliest detectors of SO_2 in the atmosphere were automated calorimetric wet chemical devices. The reference technique for determining the SO_2 content of atmospheric air is based on the absorption of SO_2 from an air sample by a solution of sodium tetrachloromercurate, which, upon the addition of formaldehyde and a pararosaniline dye, forms a strong purple dye complex. Several manufacturers have automated this procedure. Properly maintained, these units provide an excellent record of SO_2 concentration in the air, because they operate on the basis of a chemical reaction that is specific for SO_2.

Conductivity measurement can also be used in SO_2 detection applications by measuring the conductance of an absorbent, which changes as a result of the variation in SO_2 concentration in the ambient air. Gases such as hydrogen chloride will cause a positive interference, and basic gases such as ammonia will introduce a negative interference with the readings of this instrument.

In coulometric analyzers, SO_2 serves to reduce a solution of bromine or iodine. These elements can then be detected to give an indication of the SO_2 content of the sample air stream. The required supply of reagents for coulometric analyzers is usually less than the reagent requirements of either colorimetric or conductiometric analyzers.

Flame photometric analyzers use an H_2-rich flame that generates radiation at a wavelength of 394 µm when sulfur is burned. Flame photometric detectors can measure sulfur concentrations in air down to levels less than 0.01 ppm. This instrument measures total sulfur in the sample stream.

In electrochemical analyzers, the ambient air migrates selectively through the membrane and generates a signal in the electrolyte. The electrochemical principle is used as the sensor in pocket-sized portable indicators and alarms.

3.2.20 Ultraviolet Analyzers

The measurements by ultraviolet analyzers can utilize (1) single-beam; (2) split-beam; (3) dual-beam, single-detector; (4) dual-beam, dual-detector; (5) flicker photometer; (6) photodiode; and (7) retroreflector designs. The standard errors of these measurements are 2% FS, whereas it is 1% FS for the fiber-optic diode-array designs. These analyzers can handle process pressures up to 50 barg (750 psig) and temperatures up to 450°C (800°F).

Ultraviolet analyzers have been discussed in previous subsections in connection with the measurement of the concentrations of specific materials. Therefore, only a summary of their features is provided here. Table 3.43 provides a list of UV-absorbing substances.

Fewer compounds absorb in the UV region than in the IR region, and the UV absorption pattern of a compound is not as distinctive (not as narrow) as is its IR "fingerprint." On the other hand, UV analyzers provide better selectivity if the sample contains air and humidity because these materials do not absorb in the UV region. UV analyzers are also more sensitive than IR detectors. On an equal-path-length basis, the UV absorbance of liquids is stronger than that of vapors in proportion to their densities.

TABLE 3.43

Partial List of UV-Absorbing Compounds

Acetic acid	1.3-Cyclopentadiene	Nitrogen dioxide
Acetone	Dimethylformamide	Ozone
Aldehydes	Elemental halogens	Perchloroethane
Ammonia	Ethylbenzene	Phenol
Aromatic compounds	Ferric chloride	Phosgene
	Fluorine	Potassium permanganate
Benzene	Formaldehyde	Proteins
Bromine	Formic acid	Salts of transition metals
Butadiene (1,3)	Furfural	Sodium hypochlorite
Caffeine	Hydrogen peroxide	Styrene
Carbon disulfide	Hydrogen sulfide	Sulfur
Carbon tetrachloride	Iodine	Sulfur dioxide
Chlorine	Isoprene	Sulfuric acid
Chlorine dioxide	Ketones	Toluene
Chlorobenzene	Mercury	Trichlorobenzene
Crotonaldehyde	Naphthalene	Uranium hexafluoride
Cumene	Nitric acid	Xylene (*ortho, meta,* and *para*)
Cyclohexanol	Nitrobenzene	

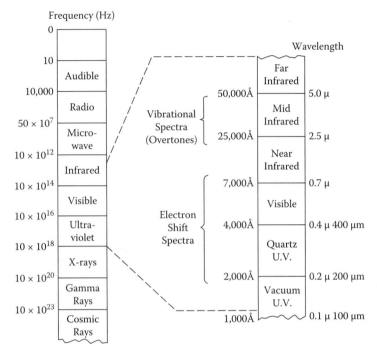

FIGURE 3.44
Electromagnetic radiation spectrum from the audible frequency up to that of the cosmic rays.

The UV and visible regions of the spectrum fall in the electron shift spectra and are generally grouped together. The UV, visible, and IR regions are shown in Figure 3.44. The UV region covers the wavelengths from 200 to 400 nm, the visible region extends from 400 to 800 nm, and the NIR region covers from 0.8 to 2.50 μm. The UV-visible NIR region is a relatively small part of the electromagnetic radiation spectrum, and the shorter the wavelength (the higher the frequency), the more penetrating the radiation becomes.

There are two groups of UV analyzer designs: photometers and spectrophotometers. Photometers are nondispersive analyzers in which the source radiates over its full UV spectrum, and discrete wavelengths are separated by narrow (2–10 nm) band-pass filters. Most process analyzers are nondispersive. Spectrophotometers are dispersive analyzers in which a prism is used to separate the spectral components of the UV spectrum. Most laboratory analyzers are dispersive.

The analyzer optics can be configured as single-beam, split-beam, dual-beam, or flicker photometers. In the flowing-sample-type design, the process stream passes through a measurement chamber, whereas probe-type analyzers can be inserted directly into the process using a retroreflector configuration. Most detectors are capable of measuring the intensity of one wavelength at a time, whereas photodiode arrays can detect all wavelengths simultaneously.

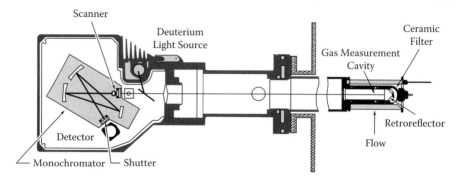

Top View (Horizontal Flow)

FIGURE 3.45

Self-calibrating, self-cleaning, and self-drying probe-type analyzer which simultaneously detects the concentrations of SO_2 and NO at temperatures up to 800°F.

Single-beam designs are limited to easy applications, having only a source and two phototube detectors. *Split-beam* UV analyzers use a single sample cell and two paths for the radiation energy between the source and the two detectors. The *dual-beam, single-detector* UV analyzer uses two optical paths: a single sample cell and a single photomultiplier detector. One path includes the sample cell, and the other is used for reference. The *dual-beam, dual-detector* design isolates the wavelength used for the measurement and uses separate phototubes for the measuring and reference wavelengths.

The *flicker photometer* uses a single cell and a rotating disk with two interference filters that are selected to produce the measuring and reference wavelengths. *Photodiode array (PDA)* spectrophotometers monitor all wavelengths in the spectrum simultaneously. *Scanning spectrophotometers* are dispersive devices that normally utilize diffraction gratings to scan across a spectral region. Scanning spectrophotometers can be used for multiple-component applications in the UV, visible, and NIR regions.

Figure 3.45 illustrates a microprocessor-based UV retroreflector probe, which can be used for monitoring both sulfur dioxide and nitrogen oxide in stack gas. In this design, UV light from a deuterium source is projected through the gas measurement cavity inside the filter at the end of the probe. The retroreflector at the tip of the measurement cell returns the measurement beam to a point where a monochromator with two exits separates the light into discrete spectral bands. The exits, one for sulfur dioxide and the other for nitrogen oxide, allow their corresponding wavelengths to impinge on the detector.

In UV and visible vapor, bubbles cannot be tolerated in the sample, because they will generate noise in the optical reading, and if a pressurized cell is used, the cell pressure must be kept constant. Nitrogen or water is used to purge the cell, and interference or broadband filters are used for checking calibration. Photometers are normally used for single-component measurements, and spectrophotometers are often used for multiple-component measurements. The sample must be relatively free of dirt and must be in a single phase. Trace measurements are also possible. Fiber-optic diode array designs can provide multicomponent liquid composition analysis, and retroreflector probes can provide self-cleaning and self-calibrating stack gas composition measurement.

3.2.21 Water Quality Monitoring

The feed to electrolysis-type hydrogen generators must be distilled water. The quality of water is important in many renewable energy processes, including the geothermal and the solar thermal ones. Water quality parameters of interest include biochemical oxygen demand (BOD), chemical oxygen demand (COD), color, conductivity, dissolved oxygen (DO), odor, oxidation-reduction potential (ORP), pH, total carbon (TC), total inorganic carbon (TIC), total organic carbon (TOC), total oxygen demand (TOD), turbidity, and volatile organic compounds (VOCs). In addition, Table 3.46 lists the various materials that are usually measured in water, together with their methods of detection and ranges.

Water quality monitoring can consist of single-parameter measurements or be complex systems integrating multiple sensors with data manipulation and recording capabilities, and in many cases, with means for remote communications and data transfer. When samples are to be collected at remote locations from below tanks, sewers, channels, sumps, lakes, or rivers, automatic samplers are used. The design of one such "duckbill" sampler is illustrated in Figure 3.47. The sample inlet is at the center of the bottom of the sampler, and the sample enters by gravity. The sampler is installed below the monitored water or process liquid (or sludge) level, and as it fills, it traps some air at the top of the chamber. When a sample is required, compressed air is introduced into the air space, which closes the bottom inlet of the duckbill and discharges the sample. When a new sample is to be drawn, the compressed air is vented, and a new fill cycle is initiated.

A water quality analyzer may be as simple as a probe inserted into the water and connected to a conveniently located recorder or as complex as a data collection system with multiple sensors distributed in several locations and provided with wired or wireless connection to a centralized data-compiling system. Trace concentrations of components are monitored in drinking, industrial, and ultrapure water.

When the purpose of an installation is to monitor water quality of lakes and rivers, the monitoring package must be designed to operate unattended for 30–60 d at a time. The unattended monitoring station (Figure 3.48) usually consists of sensors for both weather-related data and water-quality monitoring. The measurements are transmitted by telephone lines or through radio telemetry to centralized data collection systems, where the collected data are analyzed and stored. When the water-quality monitoring station is located in a remote area, it is common

TABLE 3.46

Partial List of Measured Components in Water

	Typical Instrument Measuring Range	Analysis Method
Acidity	1.0–3.0 pH	Titration
Alkalinity	0–300 ppm	Titration
Aluminum	0–50 ppb	Colorimetry
Ammonia	0–10 ppm	Colorimetry, ISE, chromatography
Boron fluoride	0–500 ppm	ISE
Bromide	0–2.0 ppm	ISE
Cadmium	0–20 ppm	ISE, chromatography
Calcium	0–300 ppm	Titration, ISE
Chloride	0.1–200 ppm	ISE, colorimetry
Chromate	0–100 ppb	Colorimetry
Copper	0–20 ppb	Colorimetry, ISE, chromatography
Cyanide	0–100 ppb	Colorimetry, titration, ISE, chromatography
Fluoride	0–2.0 ppm	ISE
Hardness (Ca + Mg)	0–2.5 ppm	Titration
Hydrazine	0–50 ppb	Colorimetry
Hydrogen sulfide	0–500 ppb	Colorimetry/ISE, chromatography
Iron	0–20 ppb	Colorimetry
Lead	0–1 ppm	ISE, chromatography
Manganese	0–200 pm	Colorimetry
Nitrate/nitrite	0–1 ppm	Colorimetry
Phenols	0–5 ppm	Colorimetry
Phosphate	0–1 ppm	Colorimetry
Silica	0–10 ppb	Colorimetry
Sodium	0–10 ppb	ISE
Urea	0–100 ppm	Colorimetry

Note: ISE = Ion-selective electrode.

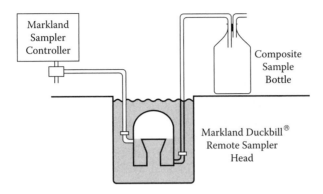

FIGURE 3.47
Automatic wastewater sampler—Markland Duckbill® Sampler.

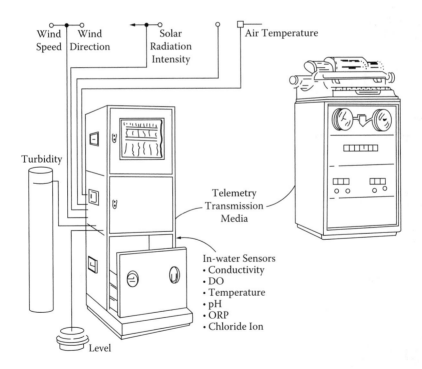

FIGURE 3.48
A complex package of automatic water quality data measurement, collection, and transmission system.

practice to place the system inside a shelter near the open water body being monitored.

3.3 Anemometers

The methods of wind velocity measurement include (a) Pitot, (b) vane, (c) cup, (d) propeller or turbine, (e) thermal, (f) Doppler, (g) acoustic, and (h) laser-type designs. These devices can detect velocities in nautical miles per hour units (1.85 km/h = 101 fpm) or in knots up to 600 knots (60,000 fpm) at the following elevations: designs a, b, and c—tower height; f—60–600 m thickness at up to 1,500 m.

Among the renewable energy processes, it is the wind turbines that benefit the most from the measurement of wind direction and velocity. Doppler-type sensors are used to determine the wind velocity and to obtain three-dimensional air motion profiles and also in the balancing of HVAC systems and measuring of the velocity of wet and dirty gases in industry. For a more detailed discussion of Pitot tubes and thermal flowmeters, also refer to the Sections 3.9.7.2 and 3.9.10. Here, the focus is on mechanical- and Doppler-type anemometers.

Vane-type anemometers rotate with an angular velocity that is proportional to the wind speed. The *three-cup anemometer* is insensitive to wind direction. The tail of the *impeller design* (Figure 3.49) always points the impellers into the wind and therefore can detect both wind speed and wind direction. The response speed of an anemometer is expressed in terms of the *length of*

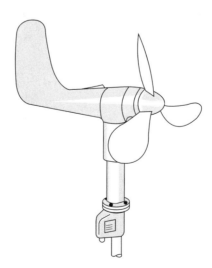

FIGURE 3.49
Impeller anemometer.

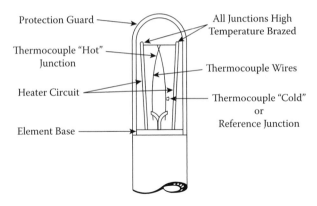

Protection Guard

All Junctions High Temperature Brazed

Thermocouple "Hot" Junction

Thermocouple Wires

Heater Circuit

Thermocouple "Cold" or Reference Junction

Element Base

FIGURE 3.50
Hot-wire anemometer.

wind that has to pass through the meter before the velocity sensor response amounts to 63% of a step change in velocity. This is known as the *distance constant* and is generally expressed in meters (feet). A typical distance constant for commercially available units is 1.8 m (6 ft).

The hot-wire anemometer (Figure 3.50) operates as a heated thermopile that is cooled at a rate proportional to the air (or gas) velocity at the probe tip. It is available in ranges of 0.5–10.0, 0.25–5.0, and 0.1–2.5 m/s (100–2000, 50–1000, and 20–500 ft/min, respectively).

Doppler anemometers take advantage of the fact that when sound or light is beamed into the atmosphere, any nonhomogeneities in the air will reflect these beams. The resulting Doppler shift in the returning frequencies can be interpreted as an indication of wind velocity. The Doppler devices are particularly useful in air-pollution-monitoring applications. When laser-based Doppler anemometers are used, the intensity of the light scattered by the reflecting particles in the air is a function of their refractive index and size (up to 5 μm). At particle sizes under 5 μm, it is safe to assume that the particle velocity is the same as that of the air, but to obtain "perfect" measurements, seeding is recommended.

The operation of laser Doppler anemometers (LDAs) utilizes the Doppler effect or Doppler shift of frequency (color), which occurs as light is dispersed from the surface of moving particles. This shift in the frequency (color) of the light source (laser beam) is proportional to the velocity of the dispersing particles. Relative to the frequency of the light, this frequency shift is very small (from 1 kHz up to 0.1 MHz) and thus cannot be directly measured. Therefore, an arrangement using the interference between the original and the refracted lights is used. This configuration is called the differential mode of the LDA.

In addition to the anemometers using the laser Doppler principle, there are also laser anemometers utilizing two-focus and transmit-time principles. The laser two-focus anemometer (L2F) measures the time needed for particles to pass the known distance between two focused beams. The signal consists of two pulses that are scanned by a photomultiplier. The processing is provided by an autocorrelator. The resulting measurement signal is stronger and has less background noise.

Both types of noncontact Doppler measurements are suitable for nearly all hydrodynamic and aerodynamic velocity measurement applications. When locating other anemometers, the structures on which they are mounted are likely to disturb the airflow. A rectangular building will disturb airflow up to an elevation of about twice its height above grade, six times its height leeward, and twice its height in the windward direction.

3.4 Btu and Heat Rate Measurement

The errors of the various Btu and heat flow detection sensors are as follows: mechanical Btu—2–5% FS; electronic Btu—0.5% FS; and heat flow—1–2% FS.

In all energy systems, including renewable, it is important to measure both the rate at which heat is flowing and also the energy flow that a fuel provides. The detection of the calorific values by calorimeters have already been discussed in Section 3.2.4. Here, first the heat flow sensors will be discussed.

The first step toward an energy-efficient plant design is a reliable energy audit throughout the plant. Such overall heat balance can be prepared only if the individual loads are accurately and separately measured. Figure 3.51 illustrates how the efficiency of a boiler is measured by calculating the ratio of the total energy flows at points 1 and 2. The total useful energy output is obtained by multiplying the totalized steam flow with the difference between the enthalpy of the steam and the enthalpy of the feedwater to the boiler. The ratio of these energy inputs and outputs is the boiler's actual efficiency. Figure 3.51 also illustrates how the performance coefficient of a chiller is measured.

The efficiency of the overall utility distribution system is an indication of the losses that occur as a result of insufficient thermal insulation, leaking steam traps, etc. When considering an alternative optimization strategy, it is recommended to measure the energy consumption both before and after a particular strategy is implemented so that an accurate cost–benefit analysis can be made and the payback periods of the various strategies can be compared.

In *mechanical Btu meters*, the volumetric flow is detected by positive-displacement or propeller-type sensors (Section 3.9.12) and is mechanically multiplied by the temperature difference across the exchanger. Transmitting attachments can also be provided wherever remote readouts are needed. These units are mostly used in HVAC applications and are not recommended if the temperature difference is under 3°C (5°F).

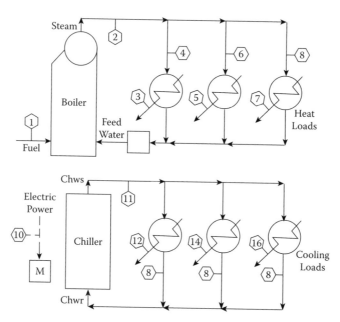

FIGURE 3.51
Plantwide energy audit.

In *electronic Btu sensors*, the *flow* sensor is usually a high-accuracy turbine flowmeter, and the temperature difference is usually detected by RTD transmitters which provide high repeatability and wide turndown. Their total error usually does not exceed ±0.5% of full scale.

The heat flow rate is measured by detecting the mass flow rate of the fuel gas and multiplying it by its heating value, which can be detected by Wobbe index sensors or calorimeters. Continuous and explosion-proof calorimeters are available for measurement of the heating value of any fuel gas (Section 3.2.4). Table 3.52 lists the compositions of various fuel gases.

The heat flow rate (Q) of a gaseous fuel is calculated as the product of its volumetric flow rate at standard conditions (V_0) and its calorific value (CV). The Wobbe index (WI) measures the ratio between the net CV and the square root of specific gravity (SG). With orifice-type flow sensors, the advantage of detecting the WI is that it eliminates the need to separately measure the specific gravity; this is because the product of the WI and orifice pressure drop results in a constant times the heat flow rate ($K \times Q$), without requiring a separate measurement of SG.

Figure 3.53 shows how the heat flow rate represented by the fuel flow into a combustion control process is determined. When optimizing a combustion process, the combustion airflow is ratioed to this Btu flow rate in a feed-forward manner, while responding to the firing rate demand signal in a feedback manner. In alternative energy systems, where waste fuels are also burned, this configuration makes control much easier, as it allows for fixing the Btu flow rate of the total mixture of fuels.

TABLE 3.52

Combustion Constants and Composition of Representative Manufactured and Natural Gases

	Blast Furnace Gas	Coal Gas	Coke Oven Gas	Natural Gas Residual Follansbee, W.Va.	Natural Gas Sandusky, Ohio	SNG Green Springs, Ohio	LNG Columbia Gulf Coast	NG Columbia Gulf Coast	Refinery Gas	Producer Gas
% Methane, CH_4	—	34.0	28.5	—	83.5	98.914	85.136	97.528	27.0	2.6
% Ethane, C_2H_6	—	—	—	79.4	12.5	0.01	10.199	1.238	—	—
% Propane, C_3H_8	—	—	—	20.0	—	—	3.06	0.241	—	—
% Ethylene, C_2H_4	—	6.6	2.9	—	—	—	0.0016	—	2.7	0.4
% Carbon monoxide, CO	26.2	9.0	5.1	—	—	0.025	—	—	10.6	22.0
% Carbon dioxide, CO_2	13.0	1.1	1.4	—	0.2	0.439	0.018	0.487	2.8	5.7
% Hydrogen, H_2	3.2	47.0	57.4	—	—	0.61	—	—	53.5	10.5
% Nitrogen, N_2	57.6	2.3	4.2	0.6	3.8	0.002	0.201	0.224	3.4	58.8
% Oxygen, O_2	—	—	0.5	—	—	—	0.007	—	—	—
% Other[a]	—	—	—	—	—	—	1.37	0.192	—	—
Btu/ft³, high (gross) 60°F, 30 in. Hg, saturated H_2O	93	634	536	1868	1047	—	—	—	516	136
Btu/ft³, low (net) 60°F, 30 in. Hg, saturated H_2O	91.6	560	476	1711	946	—	—	—	461	128
Flame temperature °F	2660	3910	3430	3830	3740	—	—	—	3970	3050

a   Heavier hydrocarbons and traces of compounds, including sulfurs.

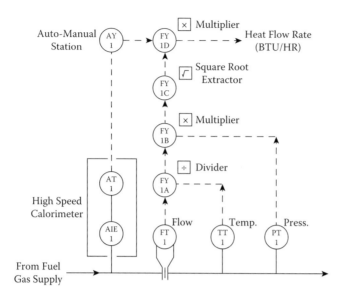

FIGURE 3.53
Heat flow rate detection loop.

3.5 Electric Meters and Peak Shedding

In solar and wind energy systems that are used in residences or in power plants, the electric power generation systems require several electric sensors and other components. Many electric power companies already accept the excess electricity generated by renewable energy sources into their grids so that, for example, a home's electric meter can "run backward" when the solar electricity generated is in excess of the needs of the residence and the excess energy is being sent to the grid.

Figure 2.128 describes the components of such an installation, including the inverter needed to convert the direct current (DC) generated by solar power to the alternating current (AC) of the grid. One of the many inverters on the market is SMA Technologie's "sunny boy," which has been used in nearly one million installations and is available in capacities from 400 Wp up to mWp. The configuration shown in Figure 2.128 allows the kWh meter to run "backward" when more electricity is generated than needed.

In the renewable energy processes, electric meters are required to determine the rate of power generation, the flow of electricity to and from the grid, and the efficiencies of solar collectors, wind turbines, electrolyzers, and fuel cells. Peak shedding is used to increase the cost-effectiveness of the total operation.

Table 3.54 gives a summary of the features and applications of the current, voltage, and wattage instruments. Some of these electric meters have been

TABLE 3.54

Orientation Table for Ammeters, Voltmeters, and Wattmeters

Meters	Type of Meter	Accessories Required	Full-Scale Meter Range (A = amperes; V = volts; W = watts)	Permissible Overload in Multiples of Full Scale and for Noted Time Duration	Recommended Applications
Ammeters					
AC	Rectifier	None	$0.5–20 \times 10^{-3}$ A	*For meters:* 1.2×: 8 h	Low range, high frequency
	Moving iron vane	None	1–50 A	100×: 1 second	General use up to 750 V
	Moving iron vane	Transformer	10–8,000 A	*For transformers:* 50×: 2 seconds	High range, over 750 V, long meter leads
	Digital	None	200×10^{-6} A–10 A	Varies, often protected	General use, medium frequency
	Digital	Transformer/Hall effect probe	2–1,000 A		High range, medium frequency
DC	Permanent magnet moving coil	None	0.02×10^{-3} 50 A	1.2×: 8 h	General use
	Permanent magnet moving coil	Shunt	20–20,000 A	100×: 1 second	High range
	Digital	None	200×10^{-6} A–10 A	Varies, often protected	General use
	Digital	Current probe	2–1,000 A		High range
Voltmeters					
AC	Rectifier	None	3–800 V	*For meters:* 1.2×: continuous	Low range, high frequency
	Moving iron vane	None	3–600 V	100×: 1 second	General use
	Moving iron vane	Transformer	150–18,000 V	*For transformers:* 1.1×: continuous 1.25×: 1 minutes	High range, circuit isolation

	Electrostatic	None	10–1,000 V		High range
	Digital	None	100×10^{-3} V–750 V	Varies, often protected	General use, medium frequency
DC	Permanent magnet moving coil	None	1–600 V	1.2×: continuous 100×: 1 second	General use
	Permanent magnet moving coil	Resistor	250–30,000 V		High range, high sensitivity
	Digital	None	100×10^{-3}–750 V	Varies, often protected	General use
Wattmeters					
	1-element electrodynamic	None	125–1,000 W	*For current:* 1.2×: continuous 100×, 1 second	Low power, single-phase two-wire circuits
	Digital	None	2 mW–15,000 W		General use, single-phase two-wire circuits
Single-phase AC	1-element electrodynamic	Transformer	$1,000$–100×10^{6} W	*For current:* 1.2×: continuous 10×: 1 second	General use, single-phase circuits
Three-phase AC	2-element electrodynamic	Transformer	$1,000$–100×10^{6} W		General use, three-phase, three-wire circuits
	2½-element electrodynamic	Transformer	1×10^{4}–1×10^{8} W		General use, three-phase, four-wire circuits
	Digital	None	2 mW–15,000 W		General use, three-phase, three- or four-wire circuits
DC	1-element electrodynamic	None	100–2,000 W		General use, low-power circuits
	Digital	None	2 mW–15,000 W		General use

in use for nearly two centuries, thanks to the discoveries of Oersted (1819), Ampere (1821), Faraday (1821), Kelvin (1867), D'Arsonval (1881), and Weston (1889). There are four basic types of analog measuring instruments; these are the permanent magnet moving coil (PMMC), moving iron vane, electrodynamic instrument, and electrostatic instrument.

As cost-effective analog-to-digital converters (ADCs) became available, electric metering shifted from analog to digital meters. Whereas instantaneous power is the product of the instantaneous values of the current and voltage, active power is the time-averaged value of instantaneous power. The value presented in the display is calculated from the digitized values of current and voltage. Because the instrument includes a processing unit, several other electric and nonelectric quantities can be derived and displayed, such as RMS values of current and voltage, reactive power (sine wave input), phase angle, energy, and period.

Alternating current indicators are almost exclusively of the moving iron vane type because of their wide range and low sensitivity to frequency variations (less than the electrodynamic type). For current measurement above the range of moving iron vane and digital meters, a current transformer is the simplest and cheapest solution. When inserted in the circuit, it provides the proper ratio between the meter range and the measured current.

Indicators for AC voltage measurement can be digital or analog. Designs include the moving iron vane, electronic PMMC, and electrostatic types. For measurements on circuits above 600 V or on lower-voltage circuits in which isolation is desirable, a moving iron vane meter is used with a potential transformer to provide the proper ratio between circuit voltage and meter movement.

The unit of electric power is the watt, which equals 1 J of energy per second (W = J/s = 3.4 Btu/h). The power in a DC electric circuit is the product of the current flowing through it and the voltage existing across it. In AC circuits, the instantaneous power (p) is the product of the instantaneous values of current and voltage. The active power P is the time average value of the instantaneous power: $P = (iu)_{av}$. If the circuit is under sine wave regime, $P = IU\cos(\mu)$, where I = RMS value of AC current, U = RMS value of AC voltage, and μ = phase angle between U and I. The power factor PF = $\cos(\mu)$ can assume values between 1 and 0. It is 1 in a purely resistive circuit and 0 in an ideal (total reactive) capacitor or inductor. In other words, the power factor is the ratio between the true power and the apparent power ($P_{app} = IU$).

Analog indicators for AC power measurement are usually constructed with electrodynamic movements because the separate fixed and moving coils permit two different types of input signals for the same movement. The fixed coils are usually connected to measure currents; the moving ones are connected to monitor voltage so that the deflection of the moving coil will be proportional to the instantaneous product of the circuit current and voltage. Inertia prevents the moving coil from responding quickly to current and voltage variations in AC circuits above 25 Hz, and the pointer will indicate the average AC power regardless of wave shape.

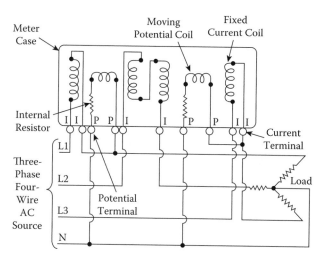

FIGURE 3.55
Two-element wattmeter wired for three-phase, three-wire loads.

In a single-element wattmeter, the current in the current-carrying coils is the load current. However, the voltage across the potential coil is higher than the load voltage by an amount equal to the voltage drop across the current coils. Therefore, there is a positive error in this measurement. In multielement wattmeters, the Aron method is used. It consists of two single-element movements, with the moving coils attached to a common shaft. Each of the elements measures a portion of the power drawn by the load and adds a proportional torque to the common moving coil shaft so that the pointer will indicate total power.

Power measurements can be made in three-phase, four-wire circuits by a three-element wattmeter (Figure 3.55). Such a design is seldom utilized because the two-element wattmeter movement can be modified to permit measurement on three-phase, four-wire systems by reconnecting one of the fixed coils for each element. Meters of this type are known as *21/2-element wattmeters*. They will correctly indicate power for three-phase, four-wire loads as long as the line-to-neutral voltages are balanced for all three phases.

Digital wattmeters are available for both AC and DC applications for both single and multiphase measurements. Without using external components, the ranges can extend from a few milliwatts to tens of thousands of watts. Direct current wattmeters are the same as single-element AC wattmeters, but it is advisable to calibrate them for use on DC circuits.

3.6 Energy Management Systems

When a renewable energy process can operate in several modes, cost-effectiveness will be served by energy management. This includes the cost-effective timing of the users in exporting the generated electricity to the grid when it is the most valuable, and using it for other purposes or sending it to storage (such as H_2) when its value is low. One of the goals of energy management is to shed (reduce) the peaks in electricity demand.

The utility company has to turn on idle equipment to meet peak demand, and that is expensive. Therefore, the cost of operation of a renewable energy plant can be reduced if nonessential electricity is turned off when a peak is approaching, and turned back on when the peak has passed. Load shedding is the simplest, and usually the first, method employed in managing energy use and controlling energy costs.

Energy management systems monitor the energy consumption of the plant, and when it appears that the electrical demand would exceed some preselected limit, they turn off some loads temporarily. More sophisticated systems also consider internal and external temperatures, plant operating level, occupancy in the case of buildings, and other predictable factors or events. Control actions can regulate HVAC equipment, energy storage, and cogeneration systems, as well as the switching of loads on and off.

Electrical demand is defined as the average load measured over a short period of time, usually 15 or 30 minutes. The electrical demand is measured in kilowatts (kWD [kilowatt demand]) and is recorded by the utility meter for each of the measurement periods during the billing period. The highest recorded demand during the billing period is used to determine the cost of each kilowatt hour (kWh) consumed. Energy management devices, particularly load shedders, reduce the demand (i.e., average load) during the critical demand periods by interrupting the operation of connected loads for short periods (Figure 3.56). If, without shedding, the current demand period would have produced the highest average load during the billing period, considerable savings will result.

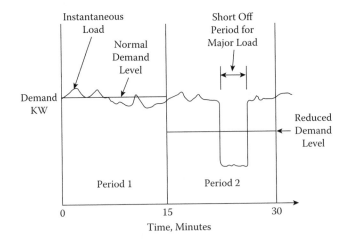

FIGURE 3.56
Effect on demand of cycling a major load off for a brief period.

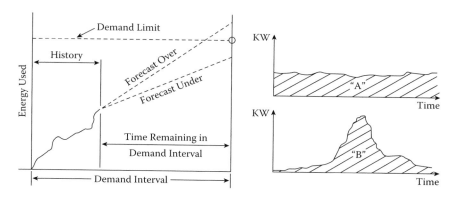

FIGURE 3.57
Electric demand shedding to change electric power use in plant toward a more uniform rate of usage, which reduces its cost.

The simplest load-shedding device is the time clock having a series of cams and switches to turn the electrical loads on and off at preset times. The use of digital systems has made a wide range of control options possible. The digital controls receive inputs from the electrical meter and the sensors that monitor critical parameters within a plant or building. Loads can be regulated based on demand, time of day, day of week, rate of energy consumption, internal and external temperatures, machinery parameters, energy rates, weather predictions, wind direction and velocity, combinations of all of these factors, and more.

The maximum control duration can be adjusted automatically according to these factors. A second demand target can also be used so that if the primary target is exceeded, additional control actions can be taken. Simple demand targets can be made to float according to time of day or other factors, forcing control actions. Multiple plants or buildings can easily be accommodated. Applications include operation of cogeneration and energy storage systems and selection of least-cost energy sources to match current and anticipated requirements.

As shown in Figure 3.57, although users A and B are using the same total amounts of electric energy, user B's bill is going to be higher owing to a higher demand charge; it is more expensive to serve B, as larger generating and distribution equipment is required, but B does not use enough energy to justify operating the equipment all the time. On the left side of the figure, the forecasting strategy is illustrated. If the forecast predicts that without correction the plant will exceed its demand limit, electricity users at the top of the shedding list are turned off. If the forecast is that by the end of the demand period the demand limit will not be violated, some of the loads will be automatically restored.

The shedding list includes all electrical devices in the plant that can be turned off for some limited time period. At the top of the list are the low-priority users (sump pumps, HVAC equipment, etc.), whereas higher up on the list, there can be secondary users (pumps and other equipment that can be operated at partial load levels or cycled on and off).

3.7 Explosion Suppression

The explosion suppression system can be regular or ultra-high-speed deluge (UHSD) systems, and their suppressant velocities of 60–90 m/s (200–300 ft/s) must exceed the radial flame velocities, which range from 0.6 to 24 m/s (2 to 80 ft/s). The response times of explosion suppression systems for the detection are about 25 ms, and the suppressant becomes effective in about 50 ms. In the case of deluge systems, water is applied in about 100 ms from the time of activation.

Renewable energy processes include the operation of combustion and H_2-handling systems and therefore have to be protected against fires (see Section 3.8) and explosions. Because chemicals display different explosive characteristics and processes differ in physical dimensions, an explosion suppression system is usually a design package. In many instances, approval for insurance must be obtained from fire underwriters with evidence of design capability demonstrated in a test.

Explosion suppression and ultra-high-speed deluge systems (UHSD) act within milliseconds to extinguish an explosion or fire almost at its inception. The two techniques are quite different. Explosion suppression systems are designed to (1) confine and inhibit a primary explosion, (2) prevent a secondary and more serious deflagration or a detonation, and (3) keep equipment damage at a minimum.

There are three categories of flame behavior:

1. Burning—the flame does not spread or diffuse but remains at an interface where fuel and oxidant are supplied.
2. Deflagration or explosion—the flame front advances through a gaseous mixture at subsonic speeds.
3. Detonation—advancement of the flame front occurs at supersonic speeds.

The first step in the design of an explosion suppression system is to establish the propagation characteristics of the material in question. First, a sample of the fuel–air mixture is introduced into a cylindrical or spherical vessel, oxidation is initiated by the application of a spark, and the test data are recorded. For example, the radial flame velocity of H_2 in air is 9 m/s (30 ft/s).

The operation of an explosion suppression system is a race against time. On the one hand, there is the buildup in pressure due to the explosion, and on the other, the counterplay is the detection of the explosion, application of the suppressants to extinguish the deflagration, and corrective action to limit the extent of damage. The operation of a typical system is illustrated in Figure 3.58.

In designing the system, the following has to be considered: The explosion pressure front advances at the speed of sound, or 330 m/s (1100 ft/s), whereas the flame front propagates at about 3 m/s (10 ft/s). The detector signal trav-

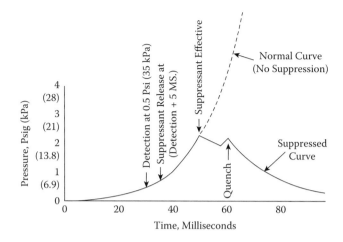

FIGURE 3.58
Explosion suppression sequence.

els at the speed of an electrical impulse. Release of the suppressant can be promoted by the explosive opening of a suppressant bottle or a high-speed hydraulically balanced system. The time period required for triggering the suppressing explosion is much shorter than that of the process explosion. The suppressant is propelled into the explosive zone at a velocity of 60–90 m/s (200–300 ft/s). The quench time of the explosion illustrated in Figure 3.58 is 60 m/s.

Explosion characteristics include the values of average and maximum rate of pressure rise and maximum pressure produced by the explosion (Table 3.59). Effective explosion suppression requires getting sufficient amounts of chemical to the trouble area in a very short time. Water and CO_2 are not generally utilized for explosions. Halogenated compounds, mostly methane derivatives, are popular suppressants. The hardware for explosion suppression falls into three categories: (1) detectors; (2) control units, which initiate the corrective action; and (3) the actuated devices, which blanket the protected area with the suppressant.

The distinction between suppressors and extinguishers lies basically in the method of mounting. Suppressors are mounted internally to the equipment being protected and are actuated by the detonation of an explosive charge within the container. Extinguishers are much larger in volume, mounted outside of the equipment, and pressurized and fitted with a diaphragm that is opened by an explosive charge. Detonator-actuated deluge disks have been developed to reduce the time needed to rupture the disk.

Explosion suppression is used for the protection of extremely hazardous systems in industry. Explosions that develop very high radial flame speeds (such as hydrogen–oxygen) are too fast for existing equipment. Many detonations (ultrasonic) also develop from an initial deflagration. It is possible

TABLE 3.59

Explosion Characteristics of Various Materials

Material	Maximum Pressure, psig[a]	Rate of Rise, psi/s	
		Maximum	Average
Acetaldehyde	94	2,100	1,900
Acetone	83	2,000	1,200
Acetylene	150	12,000	8,800
Acrylonitrile	109	2,800	2,600
Butane	97	2,300	1,700
Benzene	97	2,300	1,600
Butyl alcohol	104	2,700	1,600
Ethyl alcohol	99	2,300	1,550
Hydrogen	101	11,000	10,000
Methyl alcohol	99	3,030	1,500
Cyclohexane	99	3,030	1,500
Ethane	104	2,200	2,000
Ethylene	98	2,500	200
Hexane	119	8,500	6,600
Propane	92	2,500	1,500
Toluene	96	2,500	1,700

[a] 1 psig = 6.9 kPa.

to arrest the flame if detection and extinguishment can be affected before detonation develops, but there is no means of dealing with the detonation once it has developed.

UHSD has a good deal of similarity to explosion suppression, but they differ in where they are applied and how they work. UHSD was developed for extinguishing fires at their inception. Its point of application generally is an open area or room instead of a vessel or container. For UHSD, the speed and sensitivity advantages of UV and IR detectors have been used successfully. Deluge systems must apply a lot of water on the source of ignition within a very short time. There are two basic deluge system designs: the high-speed deluge valve and the pressure-balanced nozzle.

In the high-speed design, an explosion-actuated deluge valve is used to initiate flow. In order to prime the system, a bypass is provided around the valve and the nozzles are sealed with a protective cap (Figure 3.60).

In the pressure-balanced design, the pressures of the main water and pilot lines are balanced at the nozzle to keep it closed. Upon activation of the system, one or more solenoid valves vent the pilot line and the pressure in the main riser opens the nozzle to cause flow.

A typical deluge system specification might have the following requirements: (1) the system must activate within 200 ms of ignition; (2) there must be a discharge at a rate of 305 l min^{-1} m^{-2} (7.5 gal min^{-1} ft^{-2}) of chamber floor

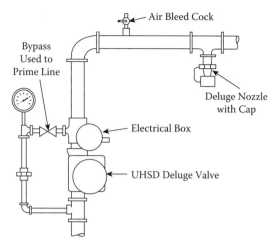

FIGURE 3.60
UHSD system with high-speed deluge valve.

area; (3) water flow must stabilize within half a second; and (4) the system must shut down in 20 seconds and must be reset within 5 seconds.

3.8 Fire, Flame, and Smoke Detectors

A fire occurs in four distinct phases. In the incipient phase, warming causes the emission of invisible but detectable gases. In the second phase, smoldering, smoke is formed; so smoke detectors can be used. In the third phase, when the ignition temperature has been reached, flames and their emitted radiation (IR and UV) can be detected. In the fourth and last stage of the fire, heat is released; the temperature of the space starts to rise, and the use of thermal sensors becomes feasible.

To detect the incipient stage of fire, ionization-chamber-type sensors are used that analyze the composition of the atmosphere by measurement of its conductance. The ionization rises as the invisible combustion gas concentration rises.

Once the fire starts to smolder and smoke is present, photoelectric sensors can be used to activate alarms. As the smoke density rises, less light passes from the source to the receiver, and an alarm is activated.

There are two types of thermal sensors: the rate-of-rise sensor and the absolute temperature sensor. Fixed temperature sensors are usually either bimetallic or low-melting-point fusible link devices that activate sprinklers or other extinguishing devices.

Once there is a flame, it emits a flickering radiation, which is mostly in the IR wavelength. Therefore, IR sensors can be used to detect the presence of flames. The flickering frequency of open flames (5–25 cps) allows the discrimination of flames from the IR radiation generated by light bulbs (at 120 cps) or unmodulated, ambient light sources. Flame detectors should be used when combustible gases or flammable liquids are present, where the ignition is almost instantaneous and has practically no incipient or smoldering stage. Because these are optical devices, it is important to ensure that the detector has an unobstructed view of the hazard and that the selected detector automatically verifies the cleanliness of its lens.

UV detectors are good for sensing hydrogen- and methanol-fueled fires (fires that emit predominantly in the UV spectrum) because these materials burn with a blue flame (i.e., strong UV source). These detectors are generally the fastest responding, typically within 30 ms. The drawback is that they are prone to false alarms from strong UV sources such as arc welding and lightning. Even if not directly in the field of view, UV reflections will trigger an alarm. Because of the sensitivity of UV detectors to lightning, they are generally not recommended for outdoor applications.

In general, IR detectors are good for detecting hydrocarbon-based fires (i.e., fires that have strong IR emissions). The combined UV/IR detectors result in fewer false alarms than UV or IR detectors alone. One disadvantage is that the loss of one (either UV or IR) will prevent alarming, except in the "OR" design. The dual IR sensors seem to be falling behind the more favored multispectrum IR detectors. These detectors have three IR detectors arranged either in voting or median selector configurations. The advantage of closed-circuit TV (CCTV) detectors is that the user is able to verify the presence of a fire before taking any action.

Flame safeguards: Table 3.61 gives a summary of the relative features of the various flame sensors that are used in burner controls. The presence of flame can be established by measuring the (1) heat generated, (2) ability to conduct electricity (ionization), and (3) radiation at various wavelengths, such as visible, IR, and UV.

The response time of thermal flame guard is insufficient for industrial applications. Photocells utilizing the visible range are limited by their inability to reject radiation from the refractory. The IR detector eliminates these shortcomings, but shimmering of hot air could still fool the detecting circuit. UV detectors can be used in most applications, but their cost is high. All flame sensors except the flame rods share the advantage of not being in contact with the fuel and the flame. All detectors must have an unobstructed view of the flame.

3.9 Flowmeters

The following text will emphasize the flow detection devices used in connection with the operation of solar collectors; wind turbines; fuel cells; and

TABLE 3.61

Comparison of Flame Safeguards

Principle of Flame Detection:	Rectification		Infrared	Visible Light	Ultraviolet
Type of Detector	Rectifying Flame Rod	Visible Light Rectifying Photo-tube	Lead-Sulfide Photocell	Cadmium-Sulfide Photocell	Ultraviolet Detector Tube
Advantages					
Same detector for gas or oil flame			✓	✓	✓
Can pinpoint flame in three dimensions	✓				
Viewing angle can be orificed to pinpoint flame in two dimensions		✓	✓	✓	✓
Not affected by hot refractory	✓				✓
Checks own components prior to each start	✓	✓	✓	✓	
Can use ordinary thermoplastic-covered wire for general applications, no shielding needed	✓	✓	✓	✓	
No installation problem because of size			✓	✓	
Disadvantages					
Difficult to sight at best ignition point			✓		
Exposure to hot refractory may reduce sensitivity to flame flicker and require orificing			✓		
Flame rod subject to rapid deterioration and warping under high temperatures		✓			
Not sensitive to extremely hot premixed gas flame	✓		✓		
Temperature limit too low for some applications		✓	✓	✓	
Shimmering of hot gases in front of hot refractory may simulate flame		✓	✓	✓	
Hot refractory background may cause flame simulation		✓	✓		
Electric ignition spark may simulate flame					✓

geothermal, hydrogen, or other alternative energy processes. This is because alternative energy processes are somewhat different from the traditional industrial ones: some solar collectors operate at very high temperatures, some fuel cells require very small flows, and hydrogen processes can operate at very low temperatures (LH_2) or at very high pressures (GH_2).

Advances in designs of processes such as fuel cells and many others require reliable, stable, accurate, repeatable, and compact means of precisely regulating H_2 flow rates. For these reasons, the descriptions of flowmeters will concentrate on the devices that are required for alternative energy processes, while also describing some of the others.

The detection of flow, in particular, requires special consideration. The two-phase flowmeter technology, developed by Kennedy Space Center, was created because no existing flowmeter without moving parts could measure cryogenic fuels such as LH_2 and liquid oxygen (LO_2) when they were transferred to the space shuttle's external tank. The flow of LH_2 can be detected by cryogenic turbine meters, noninvasive microwave, and ultrasonic or Doppler sensors. Doppler sensors use the bubbles to determine velocity and can provide accurate and immediate determination of mass fluid transfer rates for applications ranging from rocket fuel loading to solar–hydrogen measurement.

It has been reported that in the case of orifice plates and nozzles, the correlation of the discharge coefficient with Reynolds number is the same for LH_2 as for water. Similarly, tests of several commercially available turbine-type meters seem to show that the calibration constant (pulses per unit volume) for LH_2 will differ by only about 1% from the value for water.

When H_2 is transported as a gas, its pressure can be very high. Therefore, this type of flow measurement is suited for sensors that operate at high velocities, because the sensor's pressure drop is not a major concern. This consideration is similar to the one that spacecraft designers faced in the 1960s when engineering small rocket engines and thrusters. The flow rates of propellants that are sometimes very low had to be controlled with accurate, robust, and highly reliable flow-regulating equipment. The device selected for that measurement was the sonic nozzle. Other potential sensors for these applications include the head type, thermal, turbine, or vortex meters, all of which are described in the following subsections.

Other alternative energy applications such as the concentrating parabolic mirror reflector-type solar collectors, where the temperature of the circulated oil can be high, around 900°F (500°C), require high-temperature flow sensors. For these applications, in addition to the head-type flowmeters, non-contacting (e.g., ultrasonic) and metallic tubes, variable-area flowmeters can also be considered.

In the alternative energy processes, there still is need for conventional flowmeters, when the flows of air, water, and of various fuels are measured. In this section, all the applicable flow detectors are described, indicating both their capabilities and limitations. Before starting this discussion, first some guidance will be provided on how to select the right flowmeter for a specific application.

3.9.1 Application and Selection

The first step in the flowmeter selection process is to identify the meters that are technically capable of performing the required measurement and have the required size, pressure and temperature ratings, and materials of construction. Once such a list has been developed, one should consider cost, delivery, performance, and other factors to arrive at the best selection.

After this first pass, the designer should concentrate on the performance requirements, such as the maximum error that can be tolerated (defined as either a percentage of actual reading or full scale) and the required metering range. Based on the error limits and range requirements, we can next determine the rangeability required for the particular application (the ratio of maximum and minimum flow limits within which the specified error limit must not be exceeded) and identify the flow sensor categories that can provide it. We should ask if high accuracy is more important or if the emphasis should be on long-term repeatability, low installation cost, or ease of maintenance.

Head-type flowmeters include orifice plates, venturi tubes, weirs, flumes, and many others. They change the velocity or direction of the flow, creating a measurable differential pressure, or "pressure head," in the fluid. Head metering is one of the most ancient of flow detection techniques. There is evidence that the Egyptians used weirs for measurement of irrigation water flows in the days of the Pharaohs and that the Romans used orifices to meter water to households in Caesar's time. In the 18th century, Bernoulli established the basic relationship between the pressure head and velocity head, and Venturi published on the flow tube bearing his name.

The detection of pressure drop across a restriction is undoubtedly the most widely used method of industrial flow measurement. If the density is constant, the pressure drop can be interpreted as a reading of the flow. In larger pipes or ducts, the yearly energy operating cost of differential-pressure (d/p)-type flowmeters can exceed the purchase price of the meter. The permanent pressure loss through a flowmeter is usually expressed in units of velocity heads, $v^2/2\,g$, where v is the flowing velocity, and g is the gravitational acceleration (9.819 m/s^2, or 32.215 ft/s^2, at 60° latitude).

As shown in Table 3.62, different flowmeter designs require different pressure drops for their operation. If the cost of electricity is \$0.1/kWh and the pumping efficiency is 60%, the yearly cost of overcoming a continuously present pressure drop in any water-pumping system is \$0.635/yr (gpm) (psid).

3.9.1.1 Accuracy and Rangeability

If the Reynolds number (Re) and flow rate are both constant, the output signal of a head-type flowmeter will also be constant. However, if the Reynolds

TABLE 3.62

Velocity Head Requirements of the
Different Flowmeter Designs

Flowmeter Type	Permanent Pressure Loss (in velocity heads)
Orifice plates	Over 4
Vortex shedding	Approximately 2
Positive displacement	1–1.5
Turbine flowmeter	0.5–1.5
Flow tubes	Under 0.5

number changes, that will also change the meter reading, even at constant flow. Therefore, it is recommended to calculate the Reynolds numbers at both maximum and minimum flows and check whether the corresponding changes in flow coefficients are within the acceptable error limit. Figure 3.63 shows that as the pipeline Reynolds number drops (the flow is less turbulent), the error (the change in discharge coefficient) of various head-type flow elements rises. Therefore, if at minimum flow, the Reynolds number drops below 20,000, head-type flowmeters should not be used.

The required metering accuracy should be specified precisely and at a realistic value. In some instances, absolute accuracy is less important than long-term repeatability. When the error limit is given as a percentage, it

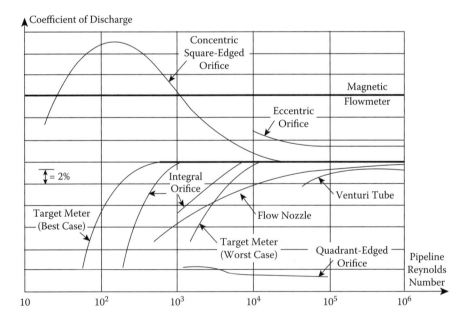

FIGURE 3.63
Discharge coefficients as a function of sensor type and Reynolds number. (Courtesy of The Foxboro Co.)

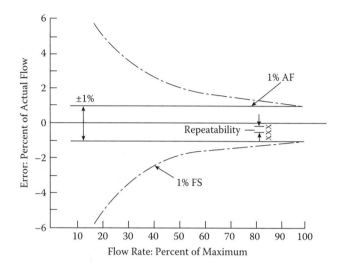

FIGURE 3.64
Comparison of 1% FS inaccuracy with 1% of actual flow inaccuracy.

should be clearly stated whether it is based on full scale (FS) or on actual reading (AR). In case the error limit is in %FS, the absolute error increases as the flow rate drops (Figure 3.64). If the error limit is in AR, the specification should also define the flow range (down to what minimum flow) over which accuracy must be maintained. Therefore, a properly prepared specification might read, "1% actual flow (AF) from 10 to 100% flow," or "0.5% FS from 5% to 100% flow." The metering accuracy is also a function of installation (minimum straight upstream pipe run, for example) and variations in process properties (such as change in fluid viscosity).

3.9.1.2 Safety and Cost

In selecting the flowmeter, both electrical safety (intrinsically safe, flame-proof, or explosion-proof) and leakage safety must be considered. Every joint increases the probability of leakage. Therefore, when metering dangerous or noxious materials, nonpenetrating flowmeter designs are preferred. Also, some organic and inorganic substances, including ordinary lubricants such as oil, grease, and wax, can cause explosions in the presence of O_2 or Cl.

If, in order to do maintenance on the meter, the process flow cannot be stopped and the measurement point cannot be bypassed, the selection choice is limited to clamp-on meters. If the available straight upstream pipe run is insufficient, consider using meters that require less (such as magnetic, which requires five) diameters. Even if the meter installation requirements can be met, their effect on the overall system cost must still be quantified because the cost should also include installation, operation, and maintenance expenses.

Hardware costs, in general, should always be balanced against the potential benefits of increased plant efficiency or product quality. These benefits are usually by-products of increased sensor accuracy, repeatability, and rangeability. In determining the total cost, the costs of flow-conditioning and filtering equipment, electric power supplies, and so on should also be considered. Operating costs are affected by the amount of the routine service required and the level of maintenance personnel needed. These costs also increase if special tools such as flow simulator equipment are required and if the secondary units required for the particular device can also be used on other meters.

Other factors to be considered include the estimated total life of the meter and pressure loss through the meter because the head loss increases the pumping costs throughout the life of the installation.

3.9.1.3 Low-Flow Applications

Measurement and control of low-flow rates are a requirement in such applications as fuel cells, purging, bioreactors, leak testing, and controlling the reference gas flow in chromatographs or in plasma-emission spectrometers. The most traditional and least expensive low-flow sensor is the variable-area flowmeter. It has a high rangeability (10:1) and requires little pressure drop. Due to its relatively low accuracy, it is limited to purge and leak-detection applications.

A much more accurate low-flow detector and controller in gas-metering applications is the sonic flow nozzle. This nozzle accurately maintains constant flow as long as sonic velocity is maintained, which is guaranteed by keeping the inlet pressure at about 50% over the outlet pressure. The disadvantages of the sonic nozzle include its high cost and high pressure drop, and the difficulty in modulating the flow rate.

In laminar flow elements, the pressure drop and flow are in a linear relationship. The laminar flow element can be used in combination with either a differential-pressure- or a thermal-type flow detector. These flowmeters provide better rangeability at about the same cost as sonic nozzles.

Thermal flowmeters also can directly detect low-mass flows without any laminar elements. In that case, they are installed directly into the pipeline as either thermal flowmeters or anemometers. They have a 100:1 rangeability, and can be provided with integral controllers.

3.9.2 Cross-Correlation Flow Metering (CCFM)

The oldest and simplest methods of flow measurement are the various tagging techniques where a portion of the flow stream is tagged at some upstream point and the flow rate is determined by transit time flow measurement (TTFM) over some distance. As illustrated in Figure 3.65, any measurable process variable can be used to build a correlation flowmeter. Flow velocity is obtained by dividing the distance (between the identical pair of detectors) by the transit time. In recent years, the required electronic comput-

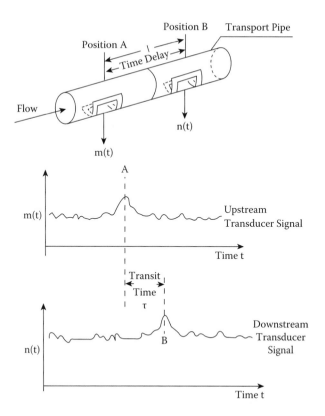

FIGURE 3.65
Cross-correlation flow metering.

ing hardware with fast pattern-recognition capability has become available. Consequently, it is feasible to build online flowmeters using this technique.

The cross-correlation technique of flow metering has been used successfully in several applications, including nuclear power plants, where the thermal hydraulic fluctuations within the reactor coolant system are detectable by temperature, pressure, and radiation sensors. Another flow-metering scheme in pressurized water reactor (PWR) plants is to cross-correlate the signals from a pair of nitrogen-16 radiation detectors that are installed on the reactor coolant pipes. The application is limited only by the imagination and the budget of the design engineers.

3.9.3 Elbow Flowmeters

Elbow taps measure the flow rate by detecting the differential pressure between pipe taps located on the inner and outer radii of an existing elbow in the process pipe. In larger pipes, the installation cost is very low because the pipe size does not affect cost. This is a crude, inaccurate measurement, requiring high flow velocities and long, upstream, straight pipe lengths. The flow rate, size, operating pressure, or temperature of these flow sensors is determined by the size and ratings of the process pipe.

Elbow taps develop relatively low differential pressures. For this reason, they cannot be used for measurement of low-velocity streams. Typically, water flowing at an average velocity of 1.5 m/s (5 ft/s) through a short-radius elbow with a centerline radius equal to the pipe diameter develops about 2.5 kPa (10 in. H_2O) water differential pressure. This is approximately the minimum full-scale pressure drop that is needed for reliable measurement. If the elbow is installed with 25 diameter upstream and 10 diameter downstream straight pipe runs, the measurement error will be under 10% FS over a 3:1 range.

3.9.4 Jet Deflection Flowmeters

Jet deflection flowmeters are used in gas and vapor services at 0–30 m/s velocity in up to 2 m (6.6 ft) diameter ducts. Their error is about 2% FS, their rangeability is about 20:1, and they are suitable for operating at temperatures up to 650°C (1200°F) and pressures up to 0.7 bar (10 psi).

The operation of this sensor requires the blowing of an air (or some other gas) jet through a nozzle and detecting the deflection of this jet caused by the velocity of the process gas in the duct. This deflection causes an increase in pressure at the downstream receiver port and a decrease at the upstream one. Therefore, the process flow is related to the pressure difference between the two receiver ports (Figure 3.66). When there is no flow in the process pipe or duct, the jet is centered between these receiver ports, and the differential pressure is zero. As the process gases start flowing, the jet is deflected, and this deflection is converted into flow.

Jet deflection flow detectors can be considered for flow measurement in large, low-pressure, circular, or rectangular ducts. If they are used in stacks

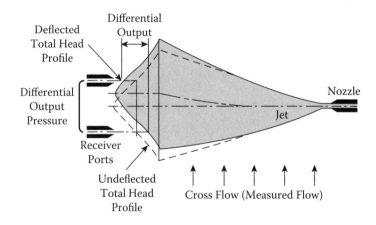

FIGURE 3.66
Pressure profile generated by a jet deflection type flow detector.

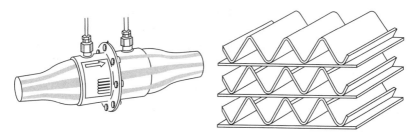

FIGURE 3.67

The laminar flowmeter and its matrix element with miniature triangular duct passage having under 0.1 mm effective diameters. (Courtesy of Meriam Instrument Div. of Scott Fetzer Co.)

or in other dirty services, they can be periodically purged or flushed and can be removed for inspection and cleaning. They are suitable for services that are dirty, abrasive, corrosive, and for plugging in flare headers, stacks, and air ducts. The actual installed accuracy is dependent on inserting the element to a depth at which the velocity corresponds to the average of the velocity profile across the duct.

3.9.5 Laminar Flowmeters

Laminar flowmeters are used in both gas and viscous liquid services. When measuring gas flows, they can detect flows from a few cm^3/min to 50 m^3/min. These meters are available in sizes from 12.5 to 300 mm (0.5–16 in.) in diameter and operate with errors of 0.5–1.0% AF over a range of 10:1. Their operating temperature is up to 150°C (300°F) and pressure range is from atmospheric pressure to 340 bar (5,000 psig).

Most laminar flowmeters (also called capillary flowmeters) measure very low flow rates of gases in applications where other types of meters either give marginal performance or cannot be used at all. Commercial units use either capillary elements or matrix shapes (Figure 3.67). The pressure drop generated by these elements is in linear relationship with the process flow. When higher accuracy and rangeability is desired, thermal instead of d/p detectors are used with the laminar flow element.

Laminar flowmeters are used in the testing of internal combustion equipment, in semiconductor manufacturing, leak testing, fan or blower calibration, and lately in alternative energy processes. The recommended installation practice is to provide 10 to 15 diameters of straight pipe upstream of the flow element. It is also highly recommended that the system be calibrated using the same gas as the process gas to be measured.

3.9.6 Magnetic Flowmeters

Magnetic flowmeters can measure liquid flow rates from 0.05 to 400,000 L/min (0.01–100,000 gpm), and are available in sizes from 2.5 mm to 2.5 m (0.1–96 in.). Their measurement error is 1% AF in the pulsed DC design if the range is < 10.1 and the operating velocity is > 0.15 m/s. The error rises to 1–2% FS if AC excitation is used and the velocity is from 0.1 to 0–10 m/s. These flowmeters have a rangeability of 10:1, and can operate at temperatures from –20 to 180°C (–4 to 350°F) and at pressures from atmospheric to 170 bar (2500 psig).

Magnetic flowmeters measure the velocity of electrically conductive liquids as they cut the magnetic fields that are maintained across these metering tubes. The main advantages of magnetic flowmeters include their completely unobstructed bore and lack of moving parts. Because of these features, they introduce no pressure loss and experience no wear and tear of their components. Other advantages include their chemical compatibility with virtually all liquids; indifference to viscosity, pressure, temperature, and density variations; ability to provide linear analog outputs and to measure bidirectional flows; availability in a wide range of sizes; and ease and speed of on-site re-ranging.

Their major limitation is that they can be used only on electrically conductive fluids. This requirement eliminates their use on all gases and on most hydrocarbon fluids. If the liquid conductivity is 20 μs/cm or greater (Table 3.68), most of the conventional magnetic flowmeters can be used. Special designs are available to measure the flow of liquids with threshold conductivities as low as 0.1 μs/cm.

Another disadvantage of magnetic flowmeters is their somewhat high purchase price and the cost of maintaining the magnetic field. If the location is

TABLE 3.68

Conductivities of Some Common Liquids

Liquid (at 25°C except where noted)	Conductivity (μs/cm)
Acetic acid (up to 70% by weight)	250 or greater
Ammonium nitrate (up to 50% by weight)	360,000 or greater
Corn syrup	16
Ethyl alcohol	0.0013
Formic acid (all concentrations)	280 or greater
Glycol	0.3
Hydrochloric acid (up to 40% by weight)	400,000 or greater
Kerosene	0.017
Magnesium sulfate (up to 25% by weight)	26,000 or greater
Molasses (at 50°C)	5,000
Phenol	0.017
Phosphoric acid (up to 87% by weight)	50,000 or greater
Sodium hydroxide (up to 50% by weight)	40,000 or greater
Sulfuric acid (up to 99.4% by weight)	8,500 or greater
Vodka (100 proof)	4
Water (potable)	70

explosion-proof, the converter and power supply must be remotely located, and intrinsic safety barriers must be installed between them and the tube.

Electromagnetic flowmeters can also be used in corrosive and slurry services. The use of ceramic flow tubes has reduced their costs, while eliminating electrode leakage because the sintered electrodes cannot leak. When selecting the meter size, the 10:1 operating range should be centered on the total range shown in Figure 3.69.

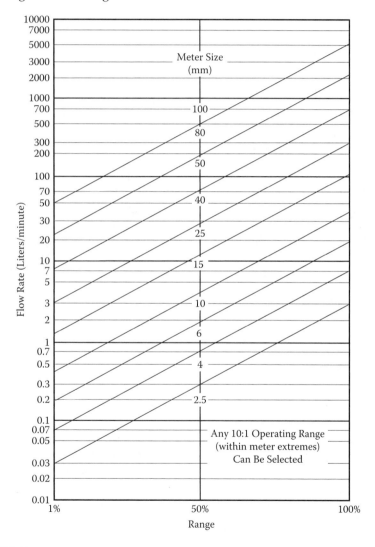

FIGURE 3.69

Selection of flowmeters: In the selection of a suitable flowmeter for a particular application, care must be exercised in handling the anticipated liquid velocities. The velocity of liquid must be within the linear range of the device. For example, a flowmeter with 100 mm internal diameter can handle flows between 50 to 5,000 L/min. An optimum operation will be achieved at a flow rate of 500 L/min.

Magnetic flowmeters are also available in the probe form that are provided with electrode cleaners. In these designs, the magnetic field is cycled to conserve electric energy and to allow automatic rezeroing, which guarantees better accuracy. The addition of intelligence through digital chips has allowed double-range operation, increased turndown, guaranteed the detection of empty pipes, and reduced measurement error to within 0.5% of AF over a 10:1 range. Table 3.70 provides a summary of flowmeter communication capabilities.

3.9.7 Mass Flowmeters

The measurement of mass flow can be obtained by multiplying the volumetric flow with density or by the direct measurement of Coriolis, thermal, impact, and angular momentum effects.

3.9.7.1 Coriolis Mass Flowmeters

Coriolis flowmeters can be used in both gas and liquid services, but are not suited for the measurement of their mixtures. The flow detection capability can range from 0 to 30,000 kg/min (0 to 65,000 lb/min), and these meters are available in sizes from 1 to 250 mm (0.05–10 in.). Their measurement error is 0.1% AF +/– zero offset which is about 0.01% within a 10:1 range. This error rises to 1% FS within a 100:1 range. This flowmeter can operate at temperatures from cryogenic to 450°C (850°F), and at pressures from atmospheric to 345 bar (5,000 psig).

Coriolis mass flowmeters (CMFs) detect the twisting of an oscillating, usually stainless steel, flow tube. This twist is a function of the mass flow through the tube. Figure 3.71 shows a number of tube assemblies in each of which the flow is split into two tubes. Sensors are mounted at the inlet and outlet sections of the tubes, measuring the phase difference between these two points. The tubes are forced into oscillation by the driver that is mounted between the two tubes. Thus, the tubes are automatically driven in a counterphase. The tubes are vibrated at their natural frequency by a magnet-and-coil driving mechanism, with the coil mounted on one tube, and the magnet on the opposite tube. The natural frequency depends mainly on the mass of the system and the elastic properties of the measuring tubes. Because the material properties remain constant, a change in the natural frequency indicates a change in the density of the fluid.

Coriolis meters measure the mass flow directly; they do not have moving parts. They can operate at process flow velocities from 0.06 to 6.0 m/s (0.2 to 20 ft/s) and therefore can provide a rangeability of 100:1. As shown in Figure 3.72, their accuracy is high (0.25% AF). Other advantages include the fact that they do not require flow conditioning or up- or downstream straight runs, and can be used on pulsating or reversing flows and on corrosive, viscous, slurry, or gas streams.

TABLE 3.70

Communication Capabilities of Modern Magnetic Flowmeters

Company	Type			Excitation		Communication			
	Smart	Conventional	Multivariable	AC	DC	HART	FF	PROFIBUS	Serial
ABB	✓			✓	✓	✓	✓	✓	✓
Advanced Flow Solutions	✓		✓	✓	✓	✓		✓	✓
Bopp & Reuther		✓			✓			✓	
Brooks Instrument		✓			✓	✓			
Brunata		✓			✓				✓
Danfoss		✓			✓	✓		✓	
Diessel		✓			✓				✓
Elis Plzen		✓			✓				✓
Endress+Hauser	✓	✓			✓	✓	✓	✓	✓
Foxboro	✓			✓	✓	✓			✓
Isoil	✓			✓	✓	✓	✓		✓
Krohne	✓			✓	✓	✓			✓
Liquid Controls	✓	✓		✓	✓	✓			✓
McCrometer		✓			✓	✓			✓
Oval Corp.	✓	✓		✓	✓	✓	✓		✓
Rosemount		✓			✓	✓			
Siemens		✓		✓	✓	✓			
Sparling Instruments	✓	✓		✓	✓	✓			
Toshiba Intl.	✓	✓			✓	✓			
Venture	✓	✓	✓	✓	✓	✓			
Yamatake	✓				✓	✓	✓		✓
Yokogawa	✓				✓	✓	✓		

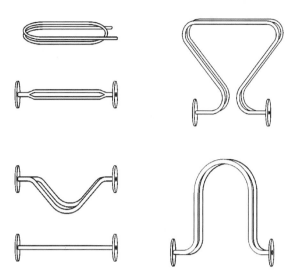

FIGURE 3.71
Selection of geometries of various Coriolis flowmeters.

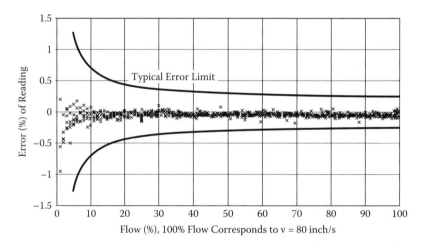

FIGURE 3.72
The measuring uncertainty for a 1 in. (DN25) Coriolis flowmeter. The maximum flow speed is 80 in./s (2 m/s), which is 20% of the maximum specified flow speed of the flowmeter. The curves show the specified error limits.

The oscillation amplitude is too small (typically, 100 μm) to cause damage to the meter. Because the excitation current is also very small, intrinsically safe CMF versions are available for use in hazardous areas. The electronics can be mounted on the flowmeter directly, forming one compact unit, or the flowmeter can be interfaced to the electronics via a cable. The remote assembly may be necessary for high-temperature meters, or it may be convenient if the sensor is installed in a place that is not easily accessible.

The mass flow rate is calculated directly by multiplying the time difference or the phase shift with the calibration constant of the flowmeter; thermal effects on the mass flow and density reading have to be included as well. This is commonly done with a microprocessor. The primary output from a CMF is mass flow. However, most electronic designs are also capable of providing temperature, density, and volumetric flow data. Further, totalizers provide mass or volume totals. Analog (4 to 20 mA) and digital output protocols are supported (e.g., PROFIBUS, FOUNDATION Fieldbus, HART, Modbus, scaled pulse, and others).

The meter is not suited to measure gas–liquid mixtures or low-pressure gases and develops high pressure drops when the gas velocity is high. To keep the tubes full, they should be installed in vertical pipes with an upward flow direction. Besides their relatively high cost, CMF limitations also include their relatively small sizes (up to 150 mm, or 6 in.), their vibration sensitivity, and the limitation of standard designs to 205°C (400°F) and special ones to 425°C (800°F). The Coriolis-based mass flowmeters are popular in fuel cell, fuel flow, reactor feed, and in other applications where mass flowmeters are needed.

3.9.7.2 Thermal Mass Flowmeters

Thermal flowmeters are mostly used in gas services, but can also detect liquid flows. They can measure flows from 0 to 10,000 kg/h (22,000 lb/h) with 1 to 2% FS errors. Their rangeability is from 10:1 to 100:1, their operating temperature range is up to 500°C (950°F), and they can operate at pressures from atmospheric to 83 bar (1,200 psig).

The mass flow of homogeneous gases, including H_2, is most frequently measured by thermal flowmeters. The main advantage of these detectors is their good accuracy and very high rangeability. The main disadvantage is their sensitivity to specific heat variations in the process fluid due to composition or temperature changes. Therefore, compensation is needed to eliminate the effect of changes in these variables. Thermal devices such as hot wire anemometers can also detect volumetric flow rates and the flow velocities of the process streams.

Thermal flowmeters can be divided into the following two categories: (1) flowmeters that measure the rise in temperature of the fluid after a known amount of heat has been added to it, which can be called *heat transfer flowmeters;* and (2) flowmeters that measure the effect of the flowing fluid on a hot body, which are sometimes called *hot-wire probes,* or *heated-thermopile flowmeters.*

In the heat-transfer-type flowmeters, heat is added to the fluid stream with an electric heater, and the resulting temperature rise is detected (Figure 3.73). These types of flowmeters are best suited for the measurement of homogeneous gases (such as H_2) and are not recommended for applications in which composition or moisture content is variable, because, for these sensors to work, both the thermal conductivity and the specific heat of the process fluid must be constant.

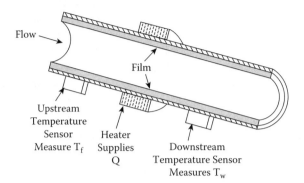

FIGURE 3.73
Thermal flowmeter with external temperature sensors and heater.

To facilitate measurement and control of larger flow rates, the heat-transfer-type flowmeters can also be placed in a small bypass around a capillary element in the process pipe. The laminar flow element serves to ensure laminar flow and also acts as a restriction forcing a portion of the flow into the sensor tube. The small-size bypass serves to minimize the electric power requirement and to increase the speed of response, but it requires upstream filters to protect it against plugging.

In hot-wire probe designs, the cooling effect of the flowing process gas stream is detected, as was already discussed in connection with anemometers (see Figure 3.50). A major limitation of the hot-wire-type mass flowmeters is that they do not detect the mass flow across the full cross section of the pipe, but do so only at the tip of the sensor. Therefore, if the sensor is installed in a nonrepresentative location, the resulting reading will be in error.

Accuracy is improved if the flowing gas stream is directed at the probe by a venturi nozzle, or by placing the sensor in the throat of the venturi (Figure 3.74). The venturi ensures a smooth velocity profile and eliminates boundary layer effects while concentrating the flow onto the sensor. These units are available for both liquid and gas services. Other designs are of the insertion probe type. Their flow ranges are a function only of the size of the pipe into which they are inserted, and their performance is a function of the correctness of the insertion depth (as are all Pitot tubes).

3.9.7.3 Indirect Mass Flowmeters

One of the earliest methods of mass flow determination was to install two separate sensors: one to measure the volumetric flow, and the other to detect the density of the flowing stream. On the basis of these two inputs, a microprocessor-based transmitter can measure mass flow. A further improvement occurred when the density and volumetric flow sensors were combined in a single package (Figure 3.75). These units are composed of either a Doppler ultrasonic flowmeter or a magnetic flowmeter and a gamma radiation–

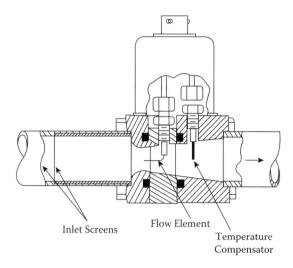

Inlet Screens   Flow Element   Temperature Compensator

FIGURE 3.74
Venturi-type thermal mass flowmeter. (Courtesy of TSI Inc.)

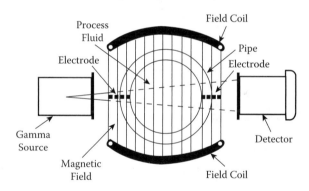

FIGURE 3.75
Mass flowmeter combining a magnetic flowmeter and a radiation-type densitometer in a single unit.

based densitometer. These mass flow units do not require compensation for changes in process variables and are obstructionless.

3.9.7.4 Angular Momentum Mass Flowmeters

The mass flow of fluid can be determined if an angular momentum is introduced into the fluid stream and both the fluid velocity and the produced torque are measured. The impeller-turbine-type mass flowmeter implements this principle by two rotating elements: The impeller is driven at a constant speed by a synchronous motor through a magnetic coupling and imparts an angular velocity to the fluid as it flows through the meter. The turbine downstream of the impeller removes this angular momentum, and

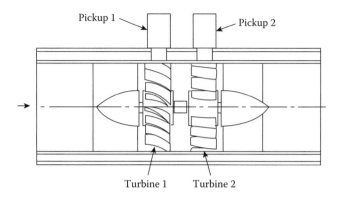

FIGURE 3.76
Twin-turbine mass flowmeter.

its restraining spring deflection is proportional to the torque, thus measuring mass flow.

Another angular-momentum-type device is the twin-turbine mass flowmeter. In this instrument, two turbines are mounted on a common shaft (Figure 3.76) connected with a calibrated torsion member. A reluctance-type pickup coil is mounted over each turbine, and a strong magnet is located in each turbine within the twin-turbine assembly. Each turbine is designed with a different blade angle; therefore, there is a tendency for the turbines to turn at different angular velocities. In the twin-turbine assembly, the turbines are not restrained by a spring, but the torsion member that holds them together is twisted. The angle enables calculation of torque, which is proportional to the mass flow rate.

Another angular momentum mass flowmeter was attempted using the principle of a gyroscope. It consisted of a pipe shaped in the form of a circle formed in a plane perpendicular to the direction of the process flow. If this pipe is oscillated around one axis, a precession-type moment is produced about the axis perpendicular to it, which is proportional to mass flow. The gyroscopic mass flowmeter can handle slurries in medium pressure and temperature ranges, but its industrial use is very limited because of its high cost and inability to handle high flow rates.

3.9.8 Metering Pumps

Metering pumps serve the multiple purposes of pumping, metering, and also controlling (accurately charging) relatively small quantities of fluids. As shown in Figure 3.77, they generate a pulsating flow. Their three basic design variations are the peristaltic, plunger, and diaphragm versions.

Peristaltic pumps can handle very low flows in laboratory applications. The plunger pump provides better accuracy than the diaphragm version, and much higher discharge pressure (up to 3,500 bar, or 50,000 psig). The diaphragm-type designs are preferred for dangerous or contaminated fluid

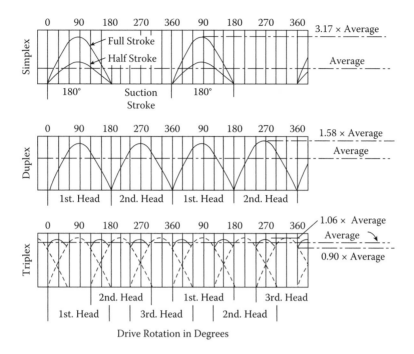

FIGURE 3.77
Pressure fluctuations are dampened by the use of multiple pistons.

services. Their design variations include the double-diaphragm and pulsator versions.

Pulsator pumps can be valuable when very difficult fluids are being metered, such as boiling sulfuric acid, pyrophoric fluids, fluorinated hydrocarbons, and slurries suspended in Freon®. In this design, hydraulic oil is pumped into the cavity within the pulsator piston, which expands and thereby displaces the process fluid from the pressure chamber. When the hydraulic oil is removed from the pulsator, fresh process fluid is drawn into the pressure chamber through the suction port (Figure 3.78). Although the illustration shows only one pulsator, normally two are used to smooth out the flow.

The pulsator pump produces an infinitely adjustable flow rate. In hazardous areas, hydraulic fluid can be pumped from a remote location, thereby removing all electrical components from the process area. The pump can be easily sterilized because only the outside of the pulsator is exposed to contamination.

Actuators can adjust either the speed or stroke of metering pumps. The controls used for stroke adjustment include micrometers, positioners, and reversing motors with slide-wire feedback. Metering pump advantages are as follows: they are self-contained, easy to install, and generally provide good accuracy (0.25–1% FS over a 10:1 range). Metering pump performance is a function of both the process fluid, which must be clean and must contain no bubbles, and the process pressure and viscosity, which must be constant

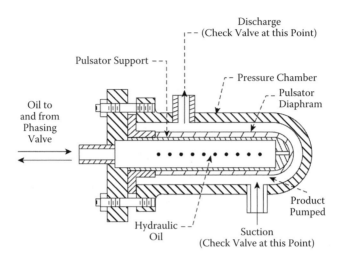

FIGURE 3.78
The pulsator-type metering pump.

to keep the leakage flow constant. Disadvantages include high cost, the need for periodic recalibration, and the requirement for such accessory equipment such as filters and air releases.

As far as maintenance is concerned, the pump motors must be lubricated periodically, and the pump must not be operated without liquid in it. The inlet piping must be designed to prevent cavitation. A metering pump should be calibrated not only before it is first used, but also periodically during its operation. The calibration should duplicate fluid properties, suction and discharge pressures, and inlet and outlet piping configuration.

3.9.9 Orifice Plates

The size of orifice-type flow sensors is limited only by the pipe size they are installed in, and their pressure and temperature ratings depend only on the limitations of the d/p detector used. Their measurement error is the combined orifice and d/p readout error, which in a standard installation over a 3:1 range is about 2% FS, and with an intelligent and multirange transmitter it can be reduced to 1% AF over a 10:1 range.

Head-type flow measurement derives from Bernoulli's theorem. Work on the conventional orifice plate for gas flow measurement was commenced by Weymouth in the United States in 1903.

Velocity head is defined as the vertical distance through which a liquid would fall to attain a given velocity. *Pressure head* is the vertical distance that a column of the flowing liquid would rise as a result of the static pressure if placed in an open-ended tube. Typically, the velocity at the throat of an orifice is increased relative to the velocity in the pipe, and there is a corresponding increase in velocity and decrease in pressure head. The difference

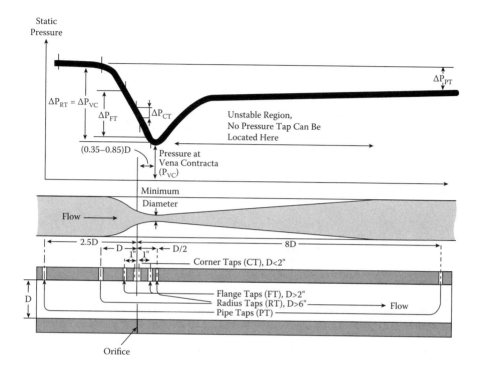

FIGURE 3.79

Pressure profile through an orifice plate and the different methods of detecting the pressure drop.

between the pressure in the pipe just upstream of the restriction and the pressure at the throat is measured. Flow rate derives from velocity and area. Figure 3.79 describes the locations on an orifice-generated pressure profile, where pressure differentials can be detected.

Orifice plates are the simplest and least expensive flow elements within the head-type and d/p-type sensor family. The total installed cost is relatively independent of pipe diameter because only the cost of the plate is a function of pipe size—the d/p manifold and the d/p readout or transmitter are relatively constant. Consequently, the orifice-type installations become more and more cost effective as the pipe size increases.

Orifices can be used in a wide range of applications because these plates are available in a variety of materials and in many designs, such as concentric, segmental, or eccentric. Another advantage is that the orifice plate can be badly worn out or damaged, and yet will still provide a reasonably repeatable output, albeit significantly inaccurate. Yet another very convenient feature of the orifice-type installation is the possibility to service or replace the readout or transmitter without the need to remove the orifice or to interrupt the process flow.

The main disadvantages are the low accuracy (Figure 3.80) and low rangeability of standard orifices, although substantial improvements have been

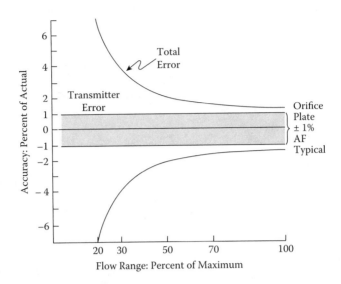

FIGURE 3.80
Total error of an orifice-type flow measurement, using a ±½% full-scale d/p cell, as a function of actual flow.

reported (error under 1% AF over a 10:1 range) when intelligent and multi-range d/p cells are used. Other disadvantages of orifice-type installations include the high irrecoverable pressure loss (40 to 80% of the generated head), and the deterioration in both measurement accuracy and long-term repeatability as the edge wears out or as deposits build up. High maintenance is another disadvantage in installations where manifold leakage and pressure tap plugging are likely.

The orifice-type flow measurement has been modified, and new, special-purpose devices have been introduced to meet particular process requirements. One such unique design is the annular orifice used to measure the hot and dirty gases in the steel industry. Here, the process flow passes through an annular opening between the pipe and a disk-shaped, concentrically located plate; the pressure difference is detected between the upstream and downstream faces of that disk. (Refer to Table 3.81 for recommendations on selecting the right orifice plate design for various applications.)

For paper pulp or slurry flow detection, the segmental and eccentric orifices, venturi cones, and the segmental wedge elements have been developed. The venturi cone is shaped as a restriction in the center of the flow path, forcing the flowing stream into an annular space between the cone and the pipe. The segmental wedge element restricts the flow passage because the top of the pipe is blocked. Segmental and eccentric orifices are all used on dirty fluids or fluids at higher temperatures.

Concentric orifice plates should not be used in slurry or other services, where solids may accumulate behind them. On erosive or corrosive fluids, the orifice plates should be made of resistant materials so that the sharp

TABLE 3.81

Selecting the Right Orifice Plate for a Particular Application

Orifice Type	Appropriate Process Fluid	Reynolds Number Range	Normal Pipe Size, in. (mm)
Concentric, square edge	Clean gas and liquid	Over 2000	0.5–60 (13–1500)
Concentric, quadrant, or conical edge	Viscous clean liquid	200–10,000	1–6 (25–150)
Eccentric or segmental square edge	Dirty gas or liquid	Over 10,000	4–14 (100–350)

edge is maintained. For flows with less than 10,000 Reynolds number, the quadrant-edged orifice plate is recommended. When measuring compressible fluids, the orifice differential should not exceed 25% of the upstream pressure, in order to minimize the errors caused by density changes through the orifice.

For liquids with entrained gas or vapor, a "vent hole" should be provided, and for gases with entrained liquid, a "drain hole." Meters for liquid with entrained gas or gas with entrained liquid services should be installed vertically. Normally, the flow direction would be upward for liquids and downward for gases. The use of vent and drain holes is discouraged, if in order to keep them from plugging, the holes would need to be large and would adversely affect accuracy. On severe entrainment applications, eccentric or segmental orifice plates should be used.

For concentric orifice plates, the β ratio (bore diameter divided by pipe diameter) should be limited to a range of 0.2–0.65 for best accuracy. The lower the β ratio, the higher the pressure drop (Figure 3.82). In case of large flows, the orifice pressure loss can result in significant energy costs and should be replaced with higher recovery sensors, such as the various venturi and proprietary flow tube designs.

The accuracy of orifice-based flow measurements can be increased by the use of several orifices in parallel and by opening or closing the pipes to some of them to keep the flow in the active paths within 80–90% of full flow. Another less expensive choice is to use two (or more) transmitters, one for high (10–100%) pressure drop and the other for low (1–10%), and to switch their outputs depending on the actual flow.

As smart d/p transmitters became available, the orifice rangeability could be increased at a lower cost by using dual-span transmitters. Some smart d/p transmitters are currently available with 0.1% span accuracy, and their spans can be automatically switched based on the value of measurement. Therefore, a 100:1 pressure differential range (10:1 flow range) can theoretically be obtained by automatically switching between a high (10–100%) and a low (1–10%) pressure differential span (in practice, some overlap is recommended). The overall result can be a 1% of AF accuracy over a 10:1 flow range.

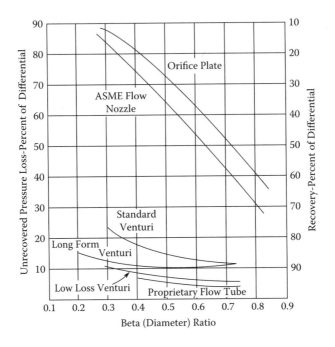

FIGURE 3.82
Pressure loss curves.

3.9.10 Pitot Tubes

Pitot tubes can be installed in any pipe size. Their pressure and temperature ratings depend only on the limitations of the d/p detectors used. Their measurement error for the single-port design is 2 to 5% FS over a 3:1 range. This error is the combined orifice and d/p readout error, which in a standard installation, over a 3:1 range, if the Reynolds number exceeds 50,000.

Henri de Pitot invented the Pitot tube in 1732. It is a small, open-ended tube that is inserted into the process pipe with its open end facing into the flow. The differential between the total pressure on this open impact port and the static pipeline pressure is measured as an indication of the flow. The Pitot tubes provide a low-cost measurement with negligible pressure loss and can also be inserted into the process pipes while the system is under pressure (wet- or hot-tapping). They are also used for temporary measurements and for the determination of velocity profiles by traversing pipes and ducts.

Their disadvantages include their low accuracy, low rangeability, and the fact that they are suitable only for clean liquid, gas, or vapor services unless purged. Their principal limitation is that they measure the flowing velocity only at one point and, therefore, even after calibration, their measurement will be in error every time the velocity profile changes. Therefore, they are used only when low-accuracy volumetric readings are acceptable, such as in HVAC applications. To reduce the effect of velocity profile changes and

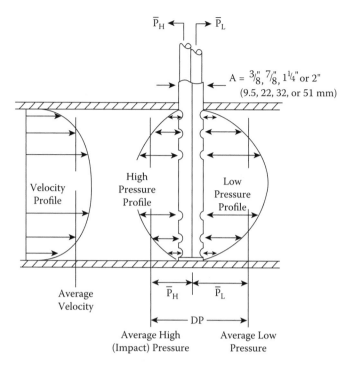

FIGURE 3.83
The design of a particular averaging Pitot tube. (Courtesy of Dietrich Standard.)

thereby improve measurement accuracy, multiple-opening Pitot tubes (Figure 3.83) and area-averaging Pitot traverse stations (Figure 3.84) have also been developed.

Another limitation of Pitot tubes is that they do not generate strong output signals. Pitot venturi and double venturi elements have been developed to amplify these signals. Also, the conventional d/p detectors have been replaced by narrow differential-membrane-type d/p cells or, if the Pitot tube is purged, by flowmeters. Pitot installations also require long, straight runs, but this requirement can be reduced by the installation of straightening vanes.

3.9.11 Polyphase Flowmeters

These applications are for the measurement of the flows of one or more liquids intermixed with gases. (For details on this somewhat complicated measurement, refer to Section 2.17 in Volume 1 of the 4th edition of the *Instrument Engineers' Handbook*.)

There are many circumstances in which the ability to distinguish the components of a liquid–vapor–gas mixture (or, multiple immiscible liquids, vapors, and gases) in the presence of solids is highly desirable. The liquids and gases do not normally travel at the same velocity. The liquids are being

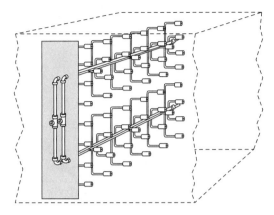

FIGURE 3.84
Installation of area-averaging Pitot tube ensembles in rectangular duct for metering the flow rate of gases. (Courtesy of Air Monitor Corp.)

dragged to the surface by the expanding gas flow. The traditional method of determining the flow of the phases (such as gas, oil, and water) was to use a *test separator.* The accuracy of this technique was about 10%.

To improve the precision and reduce the cost of measurement, new techniques have been developed. These techniques differ depending on the liquid-to-gas ratio (known as LGR, or the inverse, GLR). The extreme end of the LGR (say, <10% by mass) is classified as *wet gas.*

The standard approach is to treat the wet gas as if it was a gas and measure its flow by d/p flowmeters (preferably venturi or venturi cone [V-cone]). Differential devices used in wet gas service will generally overestimate the dry gas flow rate, and various algorithms have been developed to compensate for this. It has long been known that a gas stream carrying a well-dispersed liquid content through a differential flow element (orifice, nozzle, venturi, or V-cone) will develop a higher differential than the corresponding gas flow. If the properties of the gas and liquid are known, it is possible to calculate the liquid content in the gas. If two dissimilar pressure differential devices are installed in series (Figure 3.85), the combination can determine changes in the GLR. This enables a continuous estimation of the LGR.

In multiphase flow metering, it is usually required to distinguish hydrocarbon from water. If the liquid phase is "oil continuous," the water fraction can be determined by dielectric constant measurement at microwave frequencies because the dielectric constant of dry hydrocarbon is on the order of 2 to 4 and that of water is 82. Naturally, density measurement can also distinguish water from oil. The next requirement is to distinguish the flow of liquid from the flow of gas in a system where the two will try to separate and travel at different velocities. Cross-correlation by nuclear techniques can measure the density of the stream twice (a short vertical distance apart) and correlate the fluctuations in density with time to determine velocity. Multiphase flow metering is a new and evolving technology,

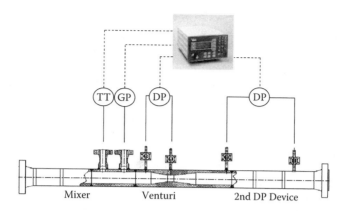

FIGURE 3.85
Solartron's ISA Dualstream II wet gas flowmeter.

and its advancement will also contribute to the optimization of alternative energy processes.

3.9.12 Positive-Displacement Flowmeters

Positive-displacement (PD) flowmeters are used when the total quantity of the flowing process stream is of interest or when a recipe is being formulated in a batch process. These meters operate by trapping a fixed volume of fluid and transferring that volume from the inlet to the outlet side of the meter. The number of such calibrated "packages" of fluid is counted as a measure of total volumetric flow. These measuring devices are used in both gas and liquid services.

3.9.12.1 PD Gas Meters

The flow rate in household units is 0–7 scmh (0–250 scfh), and that of industrial units is up to 4,000 scmh (100,000 scfh). The errors of the standard designs of these sensors are 0.5–1% FS over a rangeability of 200:1, and their high-precision designs are available at 0.5% AF over a 100:1 range.

PD gas meters measure volumetric flow by internally passing isolated volumes of gas in a process of successively filling and emptying compartments. This filling and emptying process is controlled by valves and is translated into rotary motion to operate a calibrated register or index that indicates the total volume of gas passed through the meter.

The liquid-sealed drum meter is the oldest commercial PD gas meter, which is no longer used in industry because of difficulties such as changes in liquid level and freezing. These limitations were overcome by the development of the two-diaphragm and the sliding vane meters. Their principle of operation, however, has remained the same for almost 150 years.

The diaphragm meter consists of four chambers. Differential pressure across the diaphragms extends one diaphragm and contracts the other, alternately filling and emptying the four compartments. The inaccuracy of the diaphragm-type PD meters is typically ±1% of registration over a range in excess of 200:1. This accuracy is maintained over many years of service.

The rotary design can be temperature-compensated and can operate at higher flows than the diaphragm design. There are (1) lobed impeller, (2) sliding vane, and (3) rotating vane designs. The lobed impeller design is used in high-flow applications. The close clearance of moving parts requires the use of upstream filters to prevent deterioration of accuracy, which is ±1% over a 10:1 flow range.

The sliding vane meter has four radial vanes in a single rotating drum that is eccentrically mounted, and when the drum revolves a single time, four volumes of the gas are passed. The rotating vane meter is an improvement on the lobed impeller meter, having four compartments formed by the vanes. Typical inaccuracy for the rotating vane meter is ±1% over a 25:1 range.

Higher precision and rangeability are achieved by eliminating pressure drop and thereby eliminating slip or leakage flows by using a motor drive. This flowmeter is claimed to limit the error to 0.5% over a 100:1 range.

Dirt and moisture are the worst enemies of the performance of all PD gas meters, so inlet filtering should be used when indicated. Pressure and temperature should either be controlled or compensated. The testing (or *proving*, as it is called in the gas utility industry) of gas meters is usually done by an accurately calibrated "bell" of cylindrical shape that is sealed in a tank by a suitable liquid. The lowering of the bell discharges a known volume of air through the meter under test. Other standards used to calibrate gas meters are calibrated orifices and critical flow nozzles. These devices compare rates of flow rather than fixed volumes.

The chief advantages of gas PD meters are their high accuracy and wide rangeability. Their chief disadvantages are maintenance costs and the fact that wear and tear can degrade their performance.

3.9.12.2 Liquid PD Meters

The flow capacities of liquid PD meters range from 0.04 L/h to 100,000 L/min (0.01 gal/h to 20,000 gal/min), and their sizes range from 6 mm to 0.4 m (0.25 to 16 in.). Their average error is 0.5% AF over a rangeability of 15:1, and their precision designs are good for 0.25% AF over a 50:1 range. Units are available for operating temperatures from −270°C to 300°C (−460°F to 550°F) and for pressures from atmospheric to 210 bar (3,000 psig).

PD meters split the flow of liquids into separate known volumes and act as counters or totalizers of these volumes. The energy to drive the moving parts is extracted from the flowing stream itself, resulting in a pressure loss through the meter (Figure 3.86). The error of these meters is a function of the clearance, and the longer the length of the leakage path, the better the precision of the meter. Meter accuracy tends to increase with larger meter sizes.

When high precision and high rangeability (0.25% AF over 50:1) are required, the leakage flow must be eliminated. This is achieved by pro-

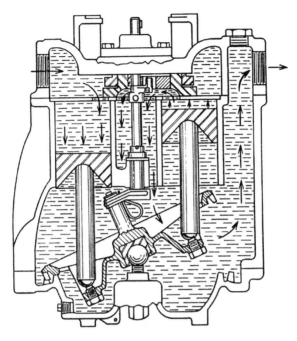

FIGURE 3.86
Cutaway of reciprocating piston meter with two opposing pistons.

viding a motor drive that introduces as much pumping energy as needed to keep the pressure differential at zero across the meter. Periodic recalibration is a requirement and can be provided with permanently installed inline provers.

Design variations include the oval gear, rotating vane, reciprocating piston, viscous helix, rotating disk, and impeller models. Liquid PD meters offer good accuracy and rangeability (>10:1), and are particularly suited to measuring the flow of viscous fluids. Their performance is virtually unaffected by upstream piping configuration. With local readouts, they do not require a power supply. When operated as a transmitter, the PD meter's output signal is linear with flow.

The PD meter applications are limited to clean fluids because their operation depends on close meshing surfaces. Their leakage error increases with dropping fluid viscosity. Both density and viscosity can change with temperature, and therefore, compensation may be required. Contaminant particle sizes must be kept below 100 μm, and most of these meters are not adaptable to the metering of slurries and dirty, nonlubricating, or abrasive fluids.

These meters also require regular recalibration and maintenance, particularly when used to measure the flow of nonlubricating liquids. Another disadvantage is that they are bulky and heavy. Their installed cost is high, because, in addition to the need for block and bypass valves, they also require filters and air releases for proper operation.

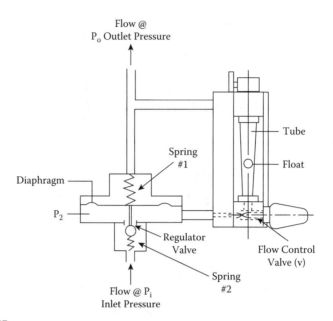

FIGURE 3.87
Purge flow regulator consisting of glass tube rotameter, an inlet needle valve, and differential pressure regulator.

3.9.13 Purge Flowmeters

Purge flow regulators serve the regulation of low flow rates of air, nitrogen, oil, or water. They are most often used in purging air bubblers or electrical housings (in explosion-proof areas) and optical windows of smokestack analyzers. Water and liquid purge meters are most often applied to protect process connections from plugging. The low flow rates of purge media can be detected and controlled by a variety of devices including rotameters (variable area), chromatographic flow controllers, spring-loaded and flexible orifice flow regulators, and thermal flow controllers.

The rotameter-type purge meter is the least expensive and most widely used purge flowmeter design. When a variable-area flowmeter is combined with a d/p regulator (Figure 3.87), it becomes a self-contained flow controller. The purge flow is fixed by adjusting two springs, #1 and #2, for a particular pressure difference, usually in the range of about 150 to 200 cm (60 to 80 in.) H_2O.

The purpose of the flow regulator shown in Figure 3.88 is not purge flow regulation, but the balancing of the water flow to the various heating/cooling coils in large HVAC systems where the water pressure varies on different floors. In order to keep the flow constant, although the inlet pressure varies, it is necessary to vary the orifice openings. As water pressure rises, the element is pushed in against the spring, and therefore, fewer orifices are available for the water to pass through. This increases the pressure drop across the regulator while keeping the flow relatively constant.

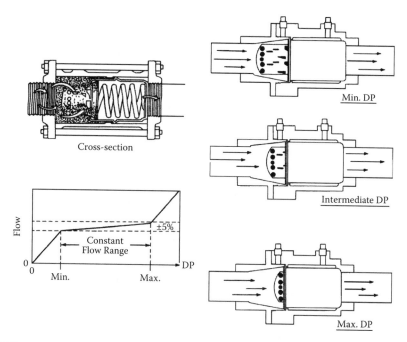

FIGURE 3.88

As inlet pressure rises, the orifice element is pushed in against the spring, thereby exposing fewer openings to the flow, and as a result, increasing the pressure drop so as to maintain the flow relatively constant. (Courtesy of Griswold Controls.)

3.9.14 Solids Mass Flowmeters

Solids flow measurement is more important in the control and optimization of coal-fired power plants than in alternative energy processes. The mass flow of solids can be detected by impact flowmeters, which are relatively low-accuracy devices (1–2% FS). Better accuracy and rangeability are provided by belt-type gravimetric feeders (0.5% AF over a 10:1 range), which measure both the speed and loading on the moving belt, as shown in Figure 3.89.

Other, less accurate methods (1–3% FS) of solids flow measurement include the impulse and the accelerator flowmeters. In the loss-in-weight-type measurement, the total weight of the supply tank is measured and that signal is differentiated by time. The rate at which the total weight is dropping is the mass flow from the tank. These systems do not provide high precision (1% AF over a 10:1 range), but are suited for the measurement of hard-to-handle process flows because they do not need physical contact with the process stream.

Cross-correlation flowmeters in combination with concentration detectors are available for the measurement of the mass flow of solids in pneumatic conveying systems or for volumetric flow measurements. The cross-correlation flowmeter uses a microwave (or gamma ray, ultrasonic, or photometric detectors) as the densitometer and a measurement of the time it takes for particles to travel a known distance to determine velocity.

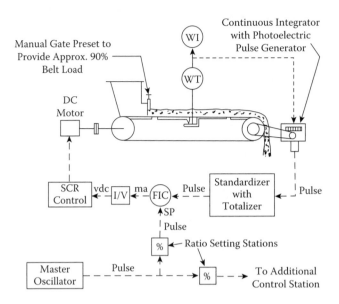

FIGURE 3.89
Belt-type gravimetric feeder with digital controls.

In the case of pulverized coal flow measurement, the concentration of the pulverized coal is measured by low-power, low-frequency microwave sensors. The variation in the microwave transmission characteristic (dielectric load) is caused by the changing coal concentration, which produces shifts in measurement frequency. The resulting quantifiable values indicate the coal density. This concentration measurement is performed by a microwave transmitter and a microwave receiver, as shown in Figure 3.90. The velocity of the pulverized coal is measured by two identical microwave devices by cross-correlation. Here, the pair of sensors detect the stochastic signals resulting from the charged coal particles, which are nearly identical but shifted by the time the pulverized coal gets from one sensor to the other.

Such noninvasive designs are suitable for the measurement of the flow of solids and two-phase flows, including heavy slurries, and very corrosive and difficult measurement applications. Their disadvantages include high cost, a

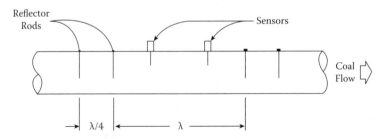

FIGURE 3.90
Standard sensor and rod arrangement. (Courtesy of Air Monitor Corp.)

fairly high minimum requirement on the operating Reynolds number, and poor accuracy (0.5–1% FS).

3.9.15 Target Flowmeters

The size of the probe-type target meters is unlimited, and the size of inline designs range from 12.5 mm to 0.2 m (0.5 to 8 in.). The measurement error of inline units is about 0.5% FS, and that of the probe designs, about 5% FS at a rangeability of 3:1. They can be used at temperatures up to 650°C (1,200°F) and at pressures from atmospheric to 200 bar (3,000 psig).

In a target flowmeter, a target or impact plate is inserted into the flowing stream, and the resulting impact force is detected. The target meters are more expensive than orifices, but because they have no pressure taps to plug or freeze, they are better suited for applications where the process fluid is "sticky" or contains suspended solids.

Material buildup in front of orifice plates can cause both measurement errors and plugging when the process stream is a liquid slurry or a gas carrying wet solids. The drag body target flowmeter detects the impact forces produced by the flowing fluid by means of a strain gauge circuitry. This unit is available in standard full-flow configurations and also in retractable probe designs (Figure 3.91), which are used in larger pipe sizes, if it is desirable to withdraw the sensor periodically for cleaning without opening the process line. The target meter allows unimpeded flow of condensates and extraneous material along the bottom of a pipe, while allowing unimpeded flow of gas or vapor along the top of the pipe.

The flow range through a particular-sized meter can be varied by changing the target size and by replacing or readjusting the transducer. Repeatability of output is good. Because the transducer and the primary element are calibrated as a unit, overall accuracy of calibrated target meters is better than that of orifice-type systems.

Fixed

Retractable

FIGURE 3.91
The insertion-type target flowmeter.

3.9.16 Turbine Flowmeters

Turbine flowmeter designs include (a) single rotor, (b) dual rotor, (c) impeller, and (d) probe versions. They are available in the following sizes: (a) 5 mm to 0.6 m (3/16 to 24 in.), (b) 6 mm to 0.3 m (1/4 to 12 in.), (c) 75 mm to 1.8 m (3 to 72 in.), and (d) unlimited. The measurement errors are as follows: (a) 0.25% AF over 10:1, (b) 0.25% AF over 100:1, (c) 2% AF, and (d) 2 to 5% FS over 10:1. The temperature ratings range as follows: (a) −200 to 450°C (−330 to 850°F), (b) −270 to 450°C (−454 to 850°F), and (c) up to 71°C (160°F). The operating pressure ranges are (a) atmospheric to 345 bar (5,000 psig), (b) up to 100 bar (1,500 psig), and (c) up to 10 bar (150 psig).

Turbine meters are available for liquid and gas, for very low or high flow rates in both full-bore and insertion designs. They are widely used when high-accuracy measurements are required. The principal limitations of single- and dual-rotor turbine meters are high cost, incompatibility with viscous or dirty liquids, and the potential for being damaged by overspeeding if slugs of gas or vapor are sent through the liquid meter. The installation of upstream filters is often recommended. They consist of a multibladed rotor suspended in the fluid stream on a free-running bearing. The fluid impinging on the rotor blades causes the rotor to revolve. Within the linear flow range of the meter, the angular speed of rotation is directly proportional to the volumetric flow rate. The speed of rotation is monitored by an electromagnetic pickup coil. Each blade produces a separate and distinct voltage pulse. Because each blade sweeps a discrete volume of fluid, each electrical impulse represents the same discrete volume of fluid.

The liquid turbine meter is one of the most accurate meters having the highest rangeability, and can handle practically any pressure and extremely high and low temperatures. Their capacities are listed in Table 3.92, and their installation is illustrated in Figure 3.93. Turbine flowmeters in gas service are similar but designed for higher velocity and lower torque.

Dual-turbine or twin-turbine designs were developed for the aerospace industry, where much higher rangeabilities (up to 1,000:1) are needed than in industry.

Impeller and shunt flowmeters are widely used in steam and gas flow metering. They comprise an orifice plate in the main flow line and a self-operating rotor assembly in the bypass. As gas flows through the meter body, a portion of the flow is diverted to drive the fan shaft assembly. The rotational speed of the shaft is proportional to the rate of flow at all rates within the normal range of the meter.

Impeller- and propeller-type flowmeters are widely used in wastewater and irrigation application for large flows and large pipes, and cost is more important than accuracy. In this meter, a corrosion-resistant plastic impeller is connected to a flexible and self-lubricating cable, which through a magnetic coupling, drives an external mechanical register.

The insertion or Pitot turbine meter, a small turbine meter, is mounted on a probe in large (over 100 mm to 4 in.)-diameter pipe. The insertion meter

TABLE 3.92

Typical Flow Capacities of a Range of Turbine Meters for Liquid Service

Nominal Diameter		Minimum Linear Flow		Maximum Linear Flow	
in.	mm	gpm	m³/h	gpm	m³/h
0.75	20	2.5	0.68	25	6.8
1	25	3.3	0.90	50	13.6
1.5	40	7.2	1.96	108	29.5
2	50	20	5.45	160	43.6
3	75	60	16.3	400	109
4	100	180	27.2	1,000	272
6	150	250	68.1	2,000	545
8	200	415	113	4,150	1,130
10	250	715	195	6,400	1,750
12	300	1,025	280	9,160	2,500
14	350	1,210	330	10,800	2,950
16	400	1,830	500	14,650	4,000
18	450	2,310	630	18,500	5,050
20	500	2,930	800	24,000	6,540

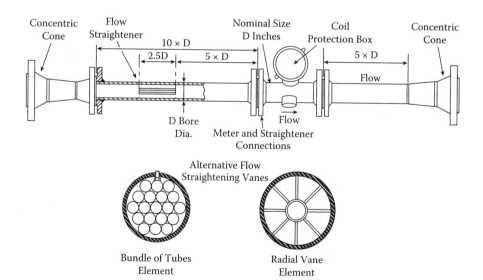

FIGURE 3.93

Recommended turbine flowmeter installation pipework.

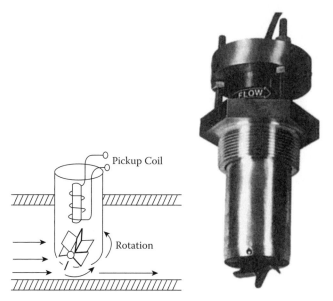

FIGURE 3.94
Paddlewheel flowmeter. (Courtesy of Data Industrial Corp.)

is inaccurate as it is measuring the velocity only at one point on the cross-sectional area, but it can provide low-cost metering in large-diameter gas or liquid pipelines. Other advantages of this design include its optical sensor and high rangeability (100:1). These probes can be hot-tapped into existing pipelines through a valving system without shutting down the pipeline.

Paddlewheel flowmeters are one of the least expensive ways of measuring liquid flow in larger pipes (Figure 3.94). The rotation of the paddlewheel can be detected magnetically or optically. Accuracies, pressure ratings, and temperature ratings are low, but rangeability is reasonable, as these units can handle both low and almost any maximum velocity.

3.9.17 Ultrasonic Flowmeters

The ultrasonic flowmeter designs include the following types: (a) transmission, (b) reflection—Doppler, and (c) open channel. Their size ranges are as follows: (a) 3 mm to 3 m (1/8 to 120 in.), (b) 13 mm to 1.8 m (0.5 to 72 in.), and (c) unlimited. Their measurement errors range from (a) 1 to 2% AF with calibrated units, and (b) and (c) 2 to 5% FS. The temperature ratings range from (a) and (b) −180 to 260°C (−300 to 500°F), and (c) unlimited. The operating pressures are as follows: (a) 200 bar (3,000 psig), (b) unlimited, and (c) atmospheric.

Ultrasonic flowmeters were first introduced in Japan, in 1963, by Tokyo Keiki (now Tokimec). Today, they are used on all types of process fluids (Table 3.95). Transit-time designs are used on clean fluids, and Doppler reflection types are used on dirty, slurry-type streams. The clamp-on designs can

TABLE 3.95

Models and Types of Ultrasonic Flowmeters by Supplier

Company	Type			Operating Principle			Fluid		
	SP	CL	IN	TT	D	H	G	L	S
American Sigma		x			x			x	
Automated Sonix	x	x						x	
Caldon	x	x		x				x	
Controlotron	x	x		x	x		x	x	x
Danfoss	x			x				x	
Daniel	x			x			x		
Datam Flutec			x	x				x	
D-Flow				x				x	
Durag			x	x			x		
Dynasonics		x		x	x			x	
Eastech Badger	x	x		x				x	
EES		x		x				x	
Elis Plzen	x			x				x	
EMCO		x		x				x	
Endress+Hauser		x		x				x	
Flexim		x		x				x	
Flotex UK		x		x				x	
Fluenta			x	x			x		
FMC Smith Meter	x			x			x		
Fuji Electric		x		x	x			x	
GE Panametrics		x		x		x	x	x	x
Greyline		x			x			x	
Honda		x		x				x	
Instromet	x		x	x			x		
Kaijo	x	x		x			x	x	
Kamstrup	x			x				x	
Krohne	x	x		x			x	x	x
Laaser		x			x			x	
Matelco	x	x	x	x				x	
Mesa Laboratories	x	x		x	x			x	
Micronics		x		x				x	
Monitor Labs				x			x		
Oval Corp.	x			x			x		
Polysonics		x		x	x			x	
Quality Control		x			x			x	
Rittmeyer		x	x	x				x	
SICK MAIHAK			x	x			x		
Siemens	x			x				x	
Solartron Mobrey		x				x		x	
Sparling	x			x				x	
Teksco USA		x			x			x	

Continued

TABLE 3.95 (*Continued*)

Models and Types of Ultrasonic Flowmeters by Supplier

	Type			Operating Principle			Fluid		
Company	**SP**	**CL**	**IN**	**TT**	**D**	**H**	**G**	**L**	**S**
Thermo MeasureTech		x		x				x	
Tokimec		x		x	x		x	x	
Tokyo Keiso	x	x		x				x	
Ultraflex		x	x	x				x	
Ultrasount Tes. Ctr.		x	x	x				x	
Yokogawa		x		x				x	

SP = spoolpiece, CL = clamp-on, IN = insertion, TT = transit time, D = Doppler, H = hybrid, G = gas, L = liquid, S = steam.

be installed without process shutdown, and do not generate any pressure drop. Their prices are unaffected by pipe size.

The clamp-on designs (Figure 3.96) are well suited for difficult fluid services, including corrosive ones. The limitations of clamp-on designs include that it can be difficult to maintain a good acoustic coupling that would not fail from thermal expansion or drying out of the coupling material (at higher temperatures for prolonged periods) and would not short-circuit by "ringing around the pipe." For wetted designs, the changes in velocity profiles, turbulence, refractive index, and the velocity of sound with temperature all can reduce accuracy. Piping configurations (e.g., 20 diameter upstream, 5 diameter downstream, of straight pipe) are also important.

Transit-time flowmeters measure the time taken for an ultrasonic energy pulse to traverse a pipe section both with and against the flow of the liquid within the pipe (Figure 3.97). The flow rate is the difference in transit times. Transit-time flowmeters are widely used in water treatment and chemical plant applications. This type of ultrasonic meter is considerably more expensive than the Doppler version, but it offers better accuracy. Unlike the Doppler meter, it is usable only on relatively clean fluid applications. Its advantages

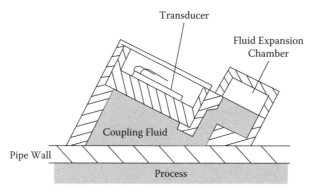

FIGURE 3.96
Liquid-filled ultrasonic coupling assembly.

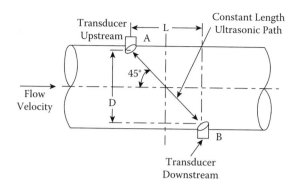

FIGURE 3.97
Wetted transducers communicate over a path that is fixed and independent of fluid speed, unless sound speed is nonuniform. (Courtesy of Panametrics, Inc.)

include that it introduces no restriction or obstruction to the flow, so it causes no pressure drop.

To improve performance and accuracy, some suppliers offer flowmeters with two, four, or more pairs of transducers arranged to interrogate multiple acoustic paths. The accuracy of multipath flowmeters is improved, and for that reason they are widely used in custody transfer of natural gas.

In 1842, Christian Doppler discovered that the wavelength of sound is a function of the receiver's movement. The transmitter of a Doppler flowmeter projects an ultrasonic beam into the flowing stream and detects the reflected frequency, which is shifted in proportion to stream velocity. The difference between the transmitted and reflected velocities is called the *beat frequency*, and its value relates to the velocity of the reflecting surfaces (solid particles and gas bubbles) in the process stream. For accurate readings it is important that the ultrasonic radiation be reflected from a representative portion of the flow stream. The main advantage of Doppler meters is their low cost, which does not increase with pipe size, whereas their main limitation is that they are not suitable for the measurement of clean fluids or gases.

3.9.18 Variable-Area Flowmeters (Rotameters)

Variable-area flowmeters in liquid service can measure flow rates up to 15 m³/min (4,000 gal/min) and in gas service up to 40 m³/min (1,200 scfm). The measurement error of laboratory units is 0.5% AF and the errors of industrial ones are 1 to 2% FS over a 10:1 range. In special designs their operating temperatures and pressures are up to 540°C (1,000°F) and 400 bar (6,000 psig), respectively. (Glass tubes are limited to 20 bar, or 300 psig.)

The variable-area flowmeter is a head-type flow sensor, but it does not measure the pressure drop across a fixed orifice; instead, the pressure drop is held relatively constant, and the orifice area is varied to match the flow (Figure 3.98). In gravity-type variable-area flowmeters, increase in flow lifts the float, piston, or vane, and it is the weight of these flow elements that has

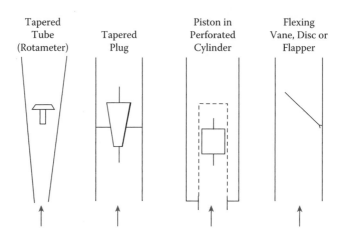

| Tapered Tube (Rotameter) | Tapered Plug | Piston in Perforated Cylinder | Flexing Vane, Disc or Flapper |

FIGURE 3.98

In a variable-area flowmeter the area open to flow is changed by the flow itself. Either gravity or spring action can be used to return the float or vane as flow drops.

to be balanced by the kinetic energy of the flowing stream. These units can operate only in a vertical position.

Variable-area meters are well suited for the local indication of low flow rates, but are not limited to those applications. They are available in both glass and metal tube constructions (Figure 3.99). In the glass tube design, the position of the float can be visually observed as an indication of flow rate. Their advantages include their low cost, low pressure loss, direct flow indication, and the ability to detect very low flow rates of both gases and liquids, including viscous fluids.

The limitations of most variable-area meters include the need for vertical mounting (exceptions are the rotary vane and tapered plug designs), and the relatively low pressure ratings of the glass tube designs. The metallic tube units are readily available as transmitters and can be obtained in larger sizes, with higher pressure ratings. They provide good rangeability (10:1) and a linear output, but their use, too, is limited to clean fluids, and they must be mounted vertically.

In some variable-area flowmeters, gravity has been replaced by spring loading. In these units, an increase in flow results in a compression or deflection of a spring, and this motion is used as an indication of flow. These units can be mounted in any position, including horizontally, as flow-through pipeline devices.

3.9.19 V-Cone Flowmeter

V-cone flowmeters can be used in liquid, gas, or steam services and are available in 12 mm to 1.8 m (0.5 to 72 in.) sizes. Their measurement error is 0.5% FS over a range of 10:1 if a dual-range transmitter is used. Their operating temperatures range from cryogenic to 370°C (700°F), and their operating pressures range from atmospheric to 40 bar (600 psig).

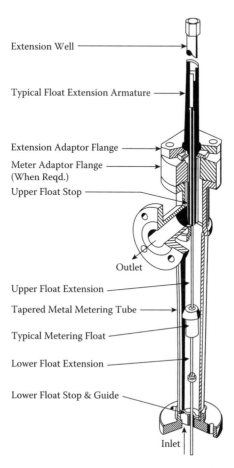

Extension Well

Typical Float Extension Armature

Extension Adaptor Flange

Meter Adaptor Flange
(When Reqd.)

Upper Float Stop

Outlet

Upper Float Extension

Tapered Metal Metering Tube

Typical Metering Float

Lower Float Extension

Lower Float Stop & Guide

Inlet

FIGURE 3.99
Metallic tube rotameter.

In a V-cone meter, a cone is positioned in the center of a metering tube (Figure 3.100). This cone reduces the cross-sectional area available for the process flow and, much similar to an orifice restriction, generates a low-pressure region downstream of the flow element. The square root of the pressure difference is related to the flow through the meter.

The main difference between an orifice plate and a V-cone element is that at lower Reynolds numbers (where the velocity profile is no longer flat, as in the highly turbulent region, but starts to take on the shape of an elongated parabola, with the maximum velocity in the center of the pipe), the cone element tends to flatten the velocity profile. This is caused by the cone, which interacts with most of the flowing stream and tends to slow the flow velocity in the center while increasing it near the wall. For this reason, the square root relationship is maintained longer, down to about a Reynolds number of 10,000. Below that, the relationship between the flow and pressure drop gradually changes from square root to linear.

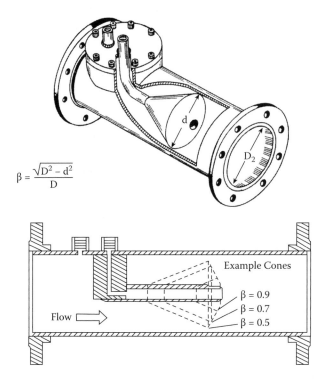

$$\beta = \frac{\sqrt{D^2 - d^2}}{D}$$

FIGURE 3.100

The V-cone flowmeter requires less upstream straight pipe and maintains the square root relationship between flow and pressure drop at lower Reynolds numbers than does an orifice plate. (Courtesy of McCrometer Div. of Danaher Corporation.)

The basic rangeability of this meter is the same as that of an orifice plate (3:1), but if two (a high span and a low span) transmitters are used, and the flow element is accurately calibrated over the complete flow range, it can be increased to 10:1. This performance can be obtained from all properly calibrated d/p flow elements, not just from the V-cone design. The V-cone flowmeter should be installed horizontally so that the two pressure taps are at the same elevation. This guarantees that the d/p cell will detect a zero pressure differential when there is no flow.

The V-cone design requires less maintenance than the sharp-edged orifice because the flow is directed away from the cone edge. Therefore, the edge is not likely to wear. The cone geometry also provides a sweeping action that eliminates stagnant areas and prevents gas accumulation or solids entrapment that can occur in front of sharp-edged orifices. The straight-pipe run requirements are also substantially below those required by orifices.

3.9.20 Venturi Tubes and Nozzles

The sizes of venturi tubes are a function of only the pipe size. Standard sizes for sonic venturi tubes are 25 to 200 mm (1 to 8 in.). These units can be used at

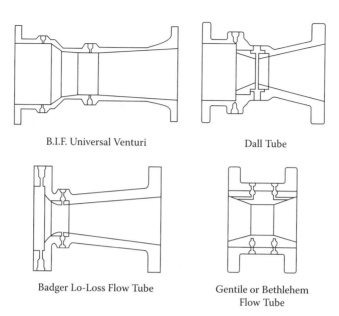

B.I.F. Universal Venturi

Dall Tube

Badger Lo-Loss Flow Tube

Gentile or Bethlehem
Flow Tube

FIGURE 3.101
Proprietary of flow tubes.

operating temperatures and pressures from the cryogenic to 650°C (1,200°F) and up to 690 bar (10,000 psig), respectively. The standard error for calibrated units (combined error of sensor and d/p readout) is 0.25% AF over a 3:1 range, while for uncalibrated ones it is 1% AF. The rangeability can be extended by using intelligent and multirange transmitters to over 10:1 with standard and to over 100:1 with sonic venturi.

Venturi tubes, flow nozzles, and flow tubes, similar to all differential pressure producers, are based on Bernoulli's theorem. Meter coefficients for venturi tubes and flow nozzles are approximately 0.98–0.99, whereas for orifice plates it averages about 0.62. Therefore, almost 60% (98/62) more flow can be obtained through these elements for the same differential pressure (see Figure 3.82).

The various shapes of these tubes (see Figure 3.101) and nozzles have been obtained with the goal of minimizing the pressure drop across them. These tubes are often installed to reduce the size of (and therefore capital expenditures on) pumping equipment and to save on pumping energy costs. In contrast with the sharp-edged orifice, these tubes and nozzles are resistant to abrasion and can also be used to measure the flow of dirty fluids and slurries. They are, however, considerably larger, heavier, and more expensive than the orifice plate. Their installation is also more difficult. Table 3.102 provides a summary of their performance.

Flow tubes are broadly defined by the American Society of Mechanical Engineers (ASME) as any d/p-producing primary whose design differs from the classic venturi design. Flow tubes fall into three main classes, depending

TABLE 3.102

Venturi, Flow Tube, and Flow Nozzle Inaccuracies (Errors) in Percentage of Actual Flow for Various Ranges of Beta Ratios and Reynolds Numbers

Flow Sensor	Line Size, Inches (1 in. = 25.4 mm)		Beta Ratio	Pipe Reynolds Number Range for Stated Accuracy	Inaccuracy, Percentage of Actual Flow
Herschel standard	Cast[a]	4–32	0.30–0.75	2×10^5–1×10^6	±0.75
	Welded	8–48	0.40–0.70	2×10^5–2×10^6	±1.5
Proprietary true venturi	Cast[b]	2–96	0.30–0.75	8×10^4–8×10^6	±0.5
	Welded	1–120	0.25–0.80	8×10^4–8×10^6	±1.0
Proprietary flow tube	Cast[c]	3–48	0.35–0.85	8×10^4–1×10^6	±1.0
ASME flow nozzles[d]		1–48	0.20–0.80	7×10^6–4×10^7	±1.0

[a] No longer manufactured because of long laying length and high cost.
[b] Badger Meter Inc.; BIF Products; Fluidic Techniques Inc.; Primary Flow Signal Inc.; Tri Flo Tech., Inc.
[c] ABB Instrumentation; Badger Meter Inc.; BIF Products; Preso Industries.
[d] BIF Products; Daniel Measurement and Control.

on the hydraulic position of the inlet and throat pressure taps. Type 1 has static pressure taps at both the inlet and the outlet, type 2 has a corner tap in the inlet and a static tap in the throat, and type 3 has a corner tap each at both the inlet and the outlet.

There are two types of flow nozzles. The common nozzle used in the United States is the so-called long-radius, or ASME flow nozzle. This nozzle comes in two versions, known as *low-beta-ratio* and *high-beta-ratio* designs. Flow nozzles represent a transition between orifices and flow tubes. At the same pressure drop, they pass about 60% more flow than an orifice. They are more compact and less expensive, but they also produce more head loss than do flow tubes.

Flow nozzles are particularly suited for measurement of steam flow and other high-velocity fluids, fluids with some solids, wet gases, and similar materials. Because an exact contour is not critical, the flow nozzle can be expected to retain good calibration for a long time even under erosion or other hostile conditions.

Sonic venturi flowmeters provide very high rangeability. They can be obtained by inserting venturi tubes into the ports of multiport digital control valves (Figure 3.103). The on–off ports (each twice the area of the previous) are opened through binary manipulation, and therefore, the meter rangeability is a function of the number of ports used. With eight ports, the rangeability is 255:1, and so on. A sonic velocity venturi element passes a known and constant flow rate when the flow velocity at its throat reaches sonic velocity.

Therefore, for the proper operation of this flowmeter, the meter pressure drop should exceed 40% of the absolute upstream pressure. This will guar-

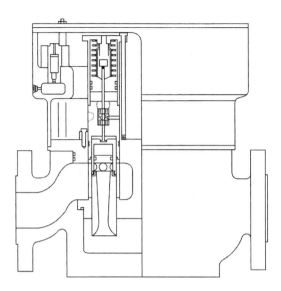

FIGURE 3.103
Sonic venturi digital flowmeter featuring extremely wide rangeability.

antee the continuous presence of sonic velocity at the throat of the venturi tubes. Because of the inherent requirement for this high pressure drop, this meter is ideal for applications in which it is desirable to lower the pressure as well as to measure the flow. The accuracy of the sonic venturi is 1/2 to 1% AF throughout the meter range. With the addition of inlet gas pressure, temperature, and density sensors, it can be converted for mass flow measurement.

The sonic venturi can also meter the flow of liquids. This flowmeter is available in sizes from 1 to 8 in. (25 to 200 mm). Units have been built for up to 10,000 psig (690 bar) pressure services and for temperatures from cryogenic to 1,200°F (650°C).

3.9.21 Vortex Flowmeters

The basic design categories of vortex flowmeters include (1) vortex shedding, (2) vortex precession (swirl), and (3) fluidic oscillation or Coanda. The flow range of (1) and (2) on water is 8 L/min to 40 m³/h (2 to 10,000 gal/min), on steam at 10 bar is 0.2 to 2,000 kg/min (0.5 to 4,000 lb/min). The flow range for (3) is 0.1 to 4 m³/h (2 to 1,000 gal/min). The available sizes of vortex flowmeters for (1) and (2) are 12 mm to 0.3 m (0.5 to 12 in.), and that of (3) are 12 to 100 mm (0.5 to 4 in.). Error ranges are 0.5 to 1% AF for (1) and (2), and 1 to 2% AF for (3). Their rangeability is a function of the Reynolds number. Rangeability of (1) and (2) is Reynolds number at maximum flow divided by 20,000, and that of (3) is Reynolds number at maximum flow divided by 3,000. Temperature ratings for (1) and (2) are −200°C to 400°C (−330 to 750°F), and for (3) are −20°C to 120°C (−4 to 250°F). Pressure ratings for (1) and (2) are up to 140 bar (2,000 psig) and for (3), they are up to 40 bar (600 psig).

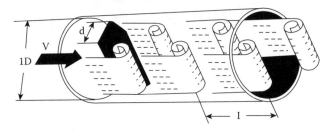

FIGURE 3.104
The distance between the Kármán vortices (1) is only a function of the width of the obstruction (d), and therefore, the number of vortices per unit of time gives flow velocity.

While fishing in Transylvania, Theodore von Kármán noticed that downstream of the rocks the distance between the shed vortices was constant, regardless of flow velocity (Figure 3.104). From that observation evolved the three types of vortex meters: the vortex shedding, the vortex precession, and the fluidic oscillation (Coanda) versions. All three types detect fluid oscillation. They have no moving components and can measure the flow of gas, steam, or liquid.

The fluid oscillations can be detected by shuttle balls, ultrasonic and piezoelectric detectors. In noisy installations, dual piezoelectric elements provide noise compensation. The various detectors can measure one of the following: (1) the oscillating flow across the face of the bluff body, (2) the oscillating pressure difference across the sides of the bluff body, (3) the flow through a passage drilled through the bluff body, (4) the oscillating flow or pressure at the rear of the bluff body, or (5) the presence of free vortices downstream to the bluff body.

Vortex-shedding meters require a fully developed flow profile. Typical upstream and downstream pipe work requirements for a variety of disturbances are shown in Figure 3.105. Where there is a severe upstream disturbance, the resulting long, straight-length requirement can be reduced by fitting a radial vane or bundle of flow-straightening tube elements in the upstream pipe. The unit can be insulated for cryogenic or high-temperature services and can be provided with extension bonnets. It should be installed in self-draining low points in the piping or in vertical upward flows to keep the meter flooded and to avoid air bubbles and standing liquid pools. Block and bypass valves should be provided if the meter is to be serviced while the process is in operation.

Construction of a typical vortex-precession (swirl) meter and its operating principles are illustrated in Figure 3.106. The fixed helical vanes at the entrance to the meter introduce a spinning or swirling motion to the fluid. The swirling fluid then enters an enlarged section in the meter housing, which causes the axis of fluid rotation to change from a straight to a helical path. The resulting spiraling vortex is known as *vortex precession*. The frequency of precession is proportional to velocity, and hence, to volumetric flow rate above a given Reynolds number.

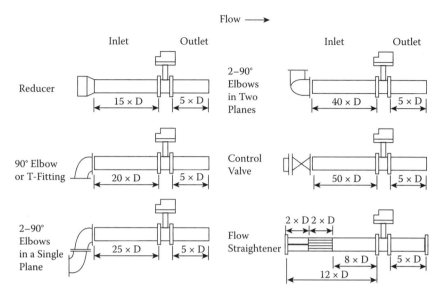

FIGURE 3.105
Straight-pipe run requirements as a function of upstream disturbance. (Courtesy of Endress+Hauser, Inc.)

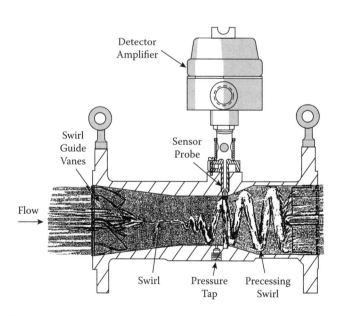

FIGURE 3.106
Construction of a typical vortex precession (swirl meter).

The swirl meter operates in most of the same applications as the vortex-shedding flowmeter but has the advantage that, because flow conditioning is done at the inlet and outlet of the meter body, virtually no upstream or downstream straight run is required for optimal installation.

In fluidic (Coanda) meters, fluid entering the meter is entrained into a turbulent jet from its surroundings, causing a reduction in pressure. The jet can be made to oscillate by one of two methods. In the relaxation oscillator, the two ports are connected, and fluid is sucked from the high-pressure side to the low-pressure side, causing the jet to switch to the other wall. The more commonly used system is the feedback oscillator, where the deflected jet causes a low-pressure area at the control port. The feedback flow intersects the main flow and diverts it to the opposite side wall, resulting in a continuous, self-induced oscillation of flow between the side walls of the meter body. The frequency of oscillation is linearly related to the volumetric flow rate. One advantage over vortex meters is that fluidic meters can operate down to a Reynolds number of 3,000. The maximum flow range (dependent on size and viscosity) is 30:1. Fluidic meters have been used almost exclusively in liquid applications.

The advantages of vortex-shedding flowmeters include their suitability for liquid, gas, and steam services; independence from viscosity, density, pressure, and temperature effects; low installation cost in smaller sizes; good accuracy and linearity without requiring calibration; wide rangeability; low maintenance using simple, easily accessible, and interchangeable spare parts; simple installation; and direct pulse output capability. These meters can be general-purpose, economically competitive alternatives to the orifice plate; they are also used in many more demanding applications because of their superior accuracy and rangeability.

The disadvantages are that (1) they are not suitable for services that are dirty, abrasive, viscous, or mixed-flow (gas with liquid droplets, liquid with vapor bubbles), or that have low Reynolds numbers (below 20,000); (2) the available choices in materials of construction are limited; (3) the pulse resolution (number of pulses per gallon or liter) drops off in larger sizes; (4) the pressure drop is high (two velocity heads); and (5) substantial straight runs are required, both upstream and downstream.

3.10 Leak Detectors

Some of the renewable energy processes operate under high pressures (H_2), and others such as thermal solar systems circulate oils under high temperatures. In these and many other applications, leakage in tanks (under- or aboveground), valves, steam traps, or pipes must be detected. Table 3.107 lists the detectable leak rates of some common gases, including H_2.

The most common type of combustible gas detectors utilizes the catalytic combustion principle. A platinum wire filament (in a Wheatstone bridge cir-

TABLE 3.107

Minimum Detectable Leak Rates of Some Common
Gases Using Thermal-Conductivity-Type Leak
Detectors

	Minimum Detectable Leak Rate	
Gas	cc/s	ft³/yr
Helium	0.00001	0.11
Hydrogen	0.000005	0.06
R-12	0.000006	0.07
R-1301	0.000006	0.07
R-134(A)	0.000004	0.04
SF_6	0.00001	0.11
CO_2	0.00003	0.34
CH_4	0.00002	0.23
Argon	0.00002	0.23

cuit) is usually mounted in a diffusion-sensing head. Other designs that utilize the phenomenon that fluids escaping from openings generate sonic and ultrasonic waves, which can be detected from a distance. The method can be considered for locating leakage of buried pipes and tanks. One application of the ultrasonic leak detectors is to monitor the leaking of steam traps from a distance. These portable probe-type units can also detect in-leakage of atmospheric air into vacuum equipment.

Other leak detectors utilize thermal or halogen technologies. Electron-capture-type halogen leak detectors are the most sensitive leak detectors available. Leak testing can be done by first evacuating the vessel and then filling it with a halogen gas, such as Freon-12, and scanning the welds and other joints by a halogen sniffer probe. When refrigeration systems are being leak-tested, halogen is the working fluid inside the equipment.

Underground leaks can be detected by tank-level or soil monitoring. Aspirated underground leak sensors pull a vacuum inside an underground probe that is provided with slots to allow in-leakage of vapors. The slotted portion must never be fully covered by groundwater, and the slots must be protected from plugging. With multiple probes in the ground, continuous concentration profile measurements are possible.

3.11 Level Detectors

The most important level detection tasks in the renewable energy processes include cryogenic (H_2) and boiling applications, where vapor bubbles are present in the liquid (steam boilers). About two dozen different types of level-measuring devices are available for use in the renewable energy processes, and some guidance is given here concerning their selection.

In order to make the right level detector selection, one should ask the following questions: Is it acceptable if the sensor's operation depends on motion (float, paddle, slip-tube, and tape types)? Can the detector have dead-ended cavities that might plug (some diaphragms, d/p types, and sight gauges)? Can a sensor be considered that will not operate properly when coated (some capacitance, conductivity, displacer, float, all optical, and thermal types)? Is it acceptable if for its operation it requires purging (bubbler type)?

If any of these features are objectionable from the reliability or maintenance point of view, either the noncontacting designs (proximity capacitance, radar, laser, sonic, and ultrasonic) or the nonpenetrating designs (time-domain reflectometry [TDR] and microwave for fiberglass tanks, nuclear gauges, and load cells) should be considered. For guidance in making the right selection, refer to Table 3.108. In addition, level detectors that measure the weight of a column will be affected by density variations caused by bubbles, or by composition and temperature changes.

When a pressurized vessel contains a clean liquid and local indication only is needed, the traditional choice is the armored level gauge. When a transmitter is needed, either the external cage-type displacer or a d/p transmitter is the most frequent choice. The most common problem with the d/p transmitter arises when the low-pressure connection cannot directly be connected to the vapor space, and compensation is needed.

Accounting-grade measurements for storage tank gauging systems require good reliability, high accuracy, and high resolution. Differential pressure transmitters (Figure 3.109) for hydrostatic tank gauging (HTG), which detect level while compensating for density and temperature, are popular for these high-accuracy measurements. Radar is a favored technology to provide the 0.125 in. (3 mm) accuracy, which is usually required for storage tank applications. In that method, the actual level is measured directly and entered into a calibrated strapping table to obtain volume, but measuring to this accuracy is made difficult by roof deflection and thermal tank expansion.

On *cryogenic applications,* such as LH_2 storage, the storage tank is usually located inside a "cold box" or in a double-walled high-vacuum dewar tank (Figure 3.110). As the d/p cell-sensing lines approach the dewar tank, at some point in the sensing line, the LH_2 boils, causing a liquid–gas interface. From that point to the d/p cell, the sensing line is filled with H_2 vapor. Boiling should occur in a large-diameter, low-slope section of the sensing line as it approaches the penetration of the outer wall of the vessel. The line diameter should be large (1 in. or 25 mm) because the interface between the liquid and vapor can be turbulent during consumption transients. For most cryogenic-level measurements, including H_2, the density of gas in vertical sections of the sensing lines can safely be ignored.

TABLE 3.108

Level Sensor Selection Guide

	Liquids	Liquid/Liquid Interface		Foam		Slurry		Suspended Solids		Powdery Solids		Granular Solids		Chunky Solids		Sticky Moist Solids	
	Continuous	Point	Continuous	Point	Continuous	Point	Continuous	Point	Continuous	Point	Continuous	Point	Continuous	Point	Continuous	Point	Continuous
Beam breaker	—	—	—	2	—	—	—	—	—	1	—	1	—	3	—	1	—
Bubbler	1	—	—	1	2	3	2	—	—	—	2	—	—	—	—	—	—
Capacitance	1	1	1	1	2	1	2	—	—	2	2	1	2	2	2	1	2
Conductive	—	2	—	1	—	1	—	—	—	3	—	3	—	3	—	1	—
Differential pressure	1	2	2	—	—	2	2	—	—	3	3	—	—	—	—	—	—
Electromechanical																	
Diaphragm	1	2	—	2	—	2	2	—	—	1	3	1	—	3	—	2	3
Displacer	2	2	2	3	—	3	2	—	—	—	—	—	—	—	—	—	—
Float	1	2	—	3	—	3	—	—	—	—	—	—	—	—	—	—	—
Float/tape	1	—	—	—	—	—	3	—	—	—	—	—	—	—	—	—	—
Paddle wheel	—	—	—	3	—	3	—	—	—	2	—	1	—	3	—	2	—
Weight/cable	1	—	—	—	—	—	1	—	1	—	1	—	1	—	1	—	1
Gauges																	
Glass	1	2	2	3	3	3	3	—	—	—	—	—	—	—	—	—	—
Magnetic	1	—	—	3	3	3	3	—	—	—	—	—	—	—	—	—	—
Inductive	—	—	—	2	—	2	—	—	—	2	2	2	2	2	2	3	3
Microwave	1	1	—	1	—	1	1	—	—	1	2	1	1	1	1	1	1
Radiation	1	1	—	1	—	1	1	—	—	1	1	1	1	1	1	1	1
Sonic echo																	
Sonar	—	2	2	—	—	—	3	1	1	—	—	—	—	—	—	—	—
Sonic	1	3	3	1	—	1	1	2	2	—	3	1	1	1	1	2	1
Ultrasonic	2	2	2	1	—	2	2	1	1	—	3	2	2	1	2	2	2
Thermal	—	1	—	2	—	2	—	—	—	—	—	—	—	—	—	—	—
Vibration	—	3	—	2	—	2	—	1	—	1	—	1	—	1	—	1	—

1 = Good; 2 = Fair; 3 = Poor or not applicable.

Source: I&CS/Endress+Hauser, Inc.

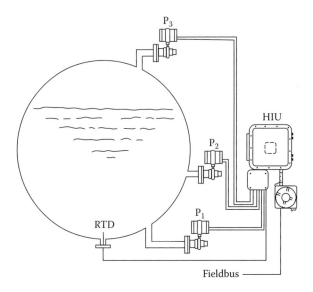

FIGURE 3.109
A hydrostatic tank gauge applied to a pressurized, spherical tank. (Courtesy of The Foxboro Co.)

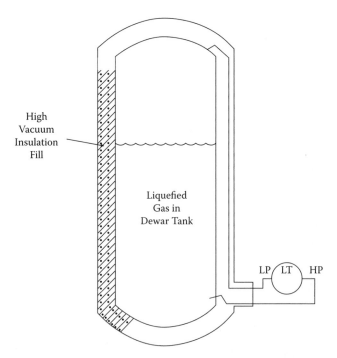

FIGURE 3.110
Cryogenic level measurement in vacuum-insulated tank.

On *hard-to-handle services*, such as the fluidized-bed level measurement in combustion processes, there is little choice but to use radiation gauges. On slurry and sludge services, d/p units with extended diaphragms eliminate the dead-ended cavity and bring the sensing diaphragm flush with the inner surface of the tank. Other level transmitters that can be considered for hard-to-handle services include the capacitance/RF, laser, radar, sonic, and TDR types.

When the application involves *foaming*, one must detect both the liquid–foam interface and foam level. Radiation sensors can detect the liquid–foam interface and TDR transmitters, or conductance and RF switches can detect the foam level if it is conductive. In the case of heavier foams, vibrating or tuning-fork switches and beta-radiation gauges have been used; in some cases, optical or thermal switches have also been successful.

In case of *boiling*, the density rises and the level drops as the rate of boiling is reduced. Therefore, the measurement of hydrostatic head alone can determine neither the level nor the mass of liquid in the tank, and hence, the volumetric percentage of bubbles must be determined by a separate measurement. For a more detailed discussion of the measurement of the level of boiling fluids, refer to Figure 3.116.

When detecting the *interface* between two liquids, electrical conductivity, thermal conductivity, opacity, or sonic transmittance of the liquids can be used. Interface-level switches are usually of the sonic, optical, capacitance, displacer, conductivity, thermal, microwave, or radiation types. Differential pressure transmitters can continuously detect the interface, but, if their density differential is small relative to the span, the error will be high. On clean services, float- and displacer-type sensors can also be used as interface-level detectors. In specialized cases, such as the continuous detection of the interface between the ash and coal layers in fluidized bed combustion chambers, the best choice is to use the nuclear radiation sensors.

Liquid–solid interface measurements are extremely demanding and require nuclear or sonic sensing. The sonic sensor must always be submerged because a gas phase will either disrupt the measurement entirely or appear to be the solid. In special noncoating cases, optical sensors have also worked.

Following this overall review of level measurement sensors and applications, the individual detector designs will be described.

3.11.1 Capacitance Probes

Capacitance probes can detect the levels of both conductive and nonconductive liquids, liquid–liquid interfaces, and levels of solids. Their measurement error is about 1% FS, and their temperature and pressure ratings are up to 1,100°C (2,000°F) and 1,400 bar (20,000 psig).

RF probes operate by applying a constant voltage to a metallic rod and monitoring the current, which is proportional to the admittance or capacitance (if conductivity is absent). The current then flows to a second electrode, such as the tank wall. The difference between conductance and RF probes is that the conductance types use DC or low-frequency AC, whereas the latter operate in the range of 0.1–1.0 MHz. Single-point switches have a substantial performance advantage over the continuous type in terms of accuracy, temperature capability, coating rejection, self-checking, and reliability.

The most basic probe configuration is a metal rod, insulated from the process vessel via threads into a half-coupling. The next step in complexity adds an insulating coating to the rod that isolates it chemically as well as electrically from the process. To inactivate a section of the probe, an *electronic guard* (a tight-fitting metal tube, insulated both from the rod and the mounting element) can be used, having a voltage identical to that on the rod.

When signaling high *interface level* between an insulating and a conducting material, the measurement is completely independent of temperature and density variation. The instrument will indicate a high level as soon as the conductive material contacts the tip of the bare probe. To detect the high level of solids in large silos, usually a flexible cable probe is used with a weight at the bottom.

When used as a continuous level detector on a conductive process fluid, conductive coating rejection should be provided. When measuring insulating liquids, the probe should be provided with its own parallel ground reference. It is possible to obtain a level signal that is independent of dielectric, and even conductivity, if a short inactive section is provided at the bottom of the probe, and there is no stratification in the process fluid. The compensation is accomplished by making two independent measurements. The first measurement is made with a short sensor (composition probe) below the tip of the level probe (Figure 3.111) and detects the electrical character of the process material. The second measurement uses the level probe. By dividing the output of the level probe by the output of the composition probe, the transmitter output becomes independent of the electrical properties.

Early capacitance probe designs were subject to (1) errors resulting from changes in the dielectric constant of insulating process fluids, (2) errors resulting from the tank geometry and fiberglass construction, and (3) conductive coating buildup on the probes. In the newer designs, these problems have been largely solved by increasing the operating frequency, incorporating a phase-detector component in the electronic circuits, and modifying the design of the sensors and their guards. These improvements have made the capacitance/RF admittance-type probes a powerful means of level detection. The only types of measurements that these instruments cannot make are the detection of interfaces between two conductive liquids of the measurement and liquid–solid interface.

3.11.2 Differential Pressure Sensors

Standard d/p transmitter ranges are available from 0–5 cm to 0–125 m H_2O (0–2 in. to 0–400 ft H_2O). Their standard error is 0.5% FS, whereas intelligent units with compensation can reduce the error to 0.1% FS. Their pressure and temperature ratings are up to 700 bar (10,000 psig) and 650°C (1,200°F).

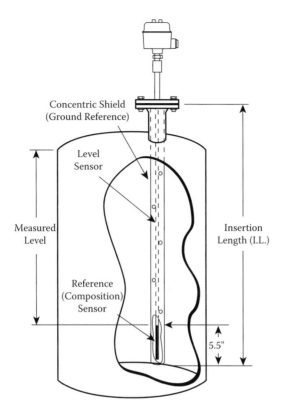

FIGURE 3.111
A probe with a second element for dielectric compensation. (Courtesy of AMETEK-Drexelbrook.)

Liquid level can be detected (inferred) by measuring a d/p. The high-pressure side of the d/p cell is connected to the bottom of the vessel, and the low-pressure side (reference) to the vapor space is connected above the liquid (pressurized tanks) or is vented to the atmosphere (atmospheric tanks). In atmospheric HTG, the low-pressure side reference leg must produce a constant head either by a column of fluid of fixed height (reference leg) or by a gas-filled reference leg. The sensor can be direct-acting (0 to 100%) d/p or reverse-acting (100 to 0%), and the densities (therefore temperatures also) of both process and reference liquids must be constant.

The d/p cell can be modified for use on viscous, slurry, or other plugging applications by providing extended diaphragms on the d/d cells (or extended diaphragm chemical seals). These extensions will completely fill the tank nozzle and place the diaphragm flush with the inside of the vessel wall. This design cannot be serviced without depressurizing and draining the tank. These designs can be used in combination with various purging schemes (Figure 3.112).

When using chemical seals, errors are introduced in addition to the error of the d/p cell itself, because the spring constant of the diaphragms and the

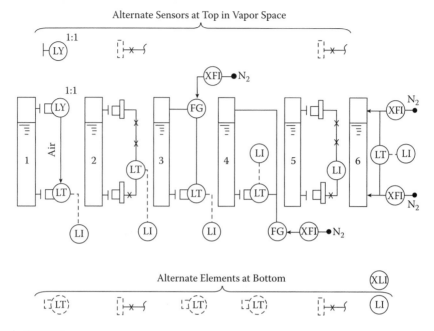

FIGURE 3.112
d/p Level detection choices for "hard to handle" process fluids stored in pressurized tanks.

fill fluid will introduce additional errors, which become more pronounced at very small d/p ranges and at high-static pressures. A larger and less predictable error can result from the temperature-sensitive nature of the seal and capillary system, because temperature differences between the low and high sides will cause differing amounts of thermal expansion.

To improve the accuracy of HTG systems, a third transmitter located at a fixed distance above the bottom transmitter can be used for density compensation (Figure 3.109). These tank expert packages, in addition to level, can also calculate mass, density, and volume on the basis of measurements from three or more d/p cells and one temperature transmitter. Most manufacturers offer optional digital communication and the ability for remote adjustments of suppression and linearization. "Smart" level transmitters can convert the level readings of spherical or cylindrical tanks into actual volume percentage readings (Figure 3.113).

All d/p cells can be provided with zero, span, elevation, and depression adjustments, either mechanical or electronic. Table 3.114 shows some typical d/p cell ranges and the available elevation and suppression setting adjustments for each. Whenever the d/p is at an elevation other than the connecting nozzle on an atmospheric tank, the zero of the d/p cell needs to be elevated or depressed. It is important to realize that two zero-reference points exist. One is the level in the tank that is considered to be zero (lower-range value) when the tank is almost empty. The other zero-reference point is the point at which the d/p cell experiences a zero differential pressure (zero value of the

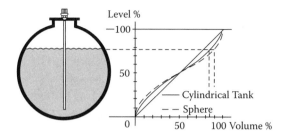

FIGURE 3.113
Intelligent transmitters can automatically convert level readings into volume.

TABLE 3.114

D/P Cell Capsule Capabilities

	Low Range	Medium Range	High Range
Minimum span:			
in. H$_2$O	0–2	0–25	0–30 psid
kPa	0–0.5	0–6.2	0–210
Maximum span:	0–150	0–1,000	0–3,000 psid
in. H$_2$O			
kPa	0–37.5	0–250	0–207 bar
Maximum zero suppression	(maximum span) minus (calibrated span)		
Maximum zero elevation	minimum span		

measured variable). The terms *elevation* and *depression* as used in this discussion refer to the zero experienced by the d/p cell.

The elevation can be set to cancel out any initial pressure exerted on the high-pressure side of the diaphragm capsule. Similarly, the depression spring can be adjusted to compensate for initial forces on the low-pressure side of the d/p cell. The amount of depression setting is limited to the full range of the capsule, whereas the sum of elevation setting and span cannot exceed the full range of the cell.

When measuring the level of the interface between two liquids, the span of the cell is the product of the density difference between the two liquids (SG2–SG1) and the distance between the maximum and minimum interface levels (X). An example of determining the required range depression is given in Figure 3.115.

With boiling applications, reverse-acting d/p level transmitters can be used (Figure 3.116), where the reference leg is the high-pressure side of the d/p cell. The constant height wet leg is filled by installing an uninsulated condensate pot that remains at ambient temperature. The output signal of this d/p cell indicates not the level inside the steam drum, but the mass of water inside it. The density inside the drum will drop when the steaming rate rises (swelling), and it will rise when the steaming rate drops. When cold feedwater (FW) is added to a steam drum, bubbles collapse and the actual

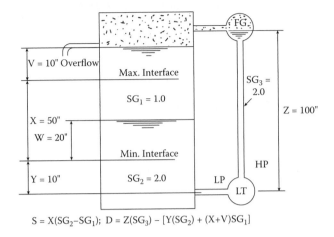

$$S = X(SG_2 - SG_1); \quad D = Z(SG_3) - [Y(SG_2) + (X+V)SG_1]$$

FIGURE 3.115
Span (S) and depression (D) settings for interface detection.

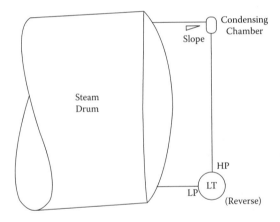

FIGURE 3.116
Level measurement in steam drums or on other boiling liquid applications.

level drops, whereas the mass of water in the drum rises (reverse action). Therefore, the d/p cell output can be converted into a true level reading only if the density and void fraction of the boiling fluid are separately determined. The shrinkage may be caused by an increased FW flow or a sudden reduction in steaming rate. The swelling is caused by an increase in the demand for steam. In either case, there is an inverse relationship between the apparent level and the mass of water in the drum.

3.11.3 Displacement-Type Level Detectors

The standard ranges of displacement level transmitters are from 36 to 150 cm (14 to 60 in.), whereas special units are available up to 18 m (60 ft). Their error is 0.5% FS, whereas their pressure and temperature ratings are up to 410 bar (6,100 psig) and from −212 to 454°C (−350 to 850°F).

Archimedes' (ca. 287 to 212 B.C.) principle states that a body wholly or partially immersed in a fluid is buoyed up by a force equal to the weight of the fluid displaced. If the displacer is heavier than the process fluid, and its cross-sectional area and the density of the liquid are constant, then a unit change in level will result in a reproducible unit change in the apparent weight of the displacer. When the liquid level is below the displacer, the scale shows the full weight of the displacer. As the level rises, the apparent weight of the displacer decreases, thereby yielding a linear and proportional relationship between spring or torque tube tension and level. Displacers detect the weight of the liquid column; float detects their height. The two readings are the same only if the liquid density is constant.

On pressurized tanks, the seal has to be frictionless and useful over a wide range of pressures, temperatures, and corrosion conditions. Displacement detectors can be magnetically coupled, or they can use a torque tube, diaphragm and force bar, spring balance, flexible disk, or the flexible shaft design. All of them can be used to detect a liquid–vapor interface, a liquid–liquid interface, and if the level is constant, they can detect density as well. The external displacers are usually installed with level gauges (Figure 3.117) so that the operator can visually inspect their calibration and performance.

The external-cage-type displacement level transmitters and controllers are popular in power plants. Applications include the control of level in high-temperature and high-pressure vessels where the process cannot be shut down for reasons of sensor replacement or maintenance. They are limited to use on clean fluids because dirt and material buildup on the displacer would change its weight and therefore cannot be tolerated. They are suitable for foaming services, and changes in vapor density can be tolerated. They detect the weight of the liquid column and not the maximum level. Therefore, on steam drums, where the volume of steam bubbles vary with steaming rate, they do not respond to the swelling or collapsing of steam bubbles. If one is also interested in knowing the total level, a float or similar sensor should also be added.

3.11.4 Laser and Optical Detectors

Optical level detectors can be used to measure the levels of solids, slurries, or opaque liquids. Their ranges are from 0.2 to 250 m (0.5 to 800 ft), and their measurement errors are about 5 mm (0.2 in.). These units are not limited in their pressure ratings and can be used from cryogenic temperatures up to those of molten metals at 1,500°C (2,732°F).

Laser-based level sensors are used in noncontact level measurements in difficult applications. Three types of laser technology are commonly used: pulsed, continuous-wave (frequency-modulated), and triangulation. The laser level gauge is a nonintrusive instrument that can look through sight glasses while being mounted externally to the process. This design is ideal

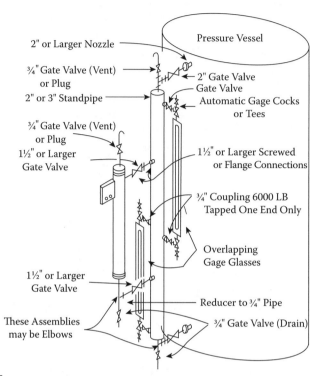

FIGURE 3.117
The installation of an external cage displacer on a standpipe with two level gauge sections. (Courtesy of the American Petroleum Institute API RP 550.)

for corrosive process applications if clean air or process-compatible gas or fluid can be continuously or intermittently injected to keep the glass surfaces clean and laser beam transmittance losses to a minimum. The laser is ideal for vacuum applications because, unlike ultrasonic sensors, the laser does not need a medium for propagation.

Pulsed technology is time-of-flight based; it measures the time for laser pulses to travel from the transmitter of the instrument to the target (where the beam is reflected) and back to the receiver of the instrument. Pulsed lasers are used for most industrial level-monitoring applications because they offer better range and penetration characteristics (through dust and steam) without sacrificing accuracy and repeatability (Figure 3.118).

Continuous-wave laser technology directs a continuous laser beam at the target, and the distance is calculated on the basis of frequency, wavelength, and phase shift in the returning beam. It is used for short-range, extremely high-accuracy, clean-air applications usually for positioning and only infrequently for level.

Triangulation measurement is done at an angle, with a sharply focused laser beam directed toward an object. This method is suitable for only short-range measurements, primarily for the positioning of objects and robotics, not for level sensing.

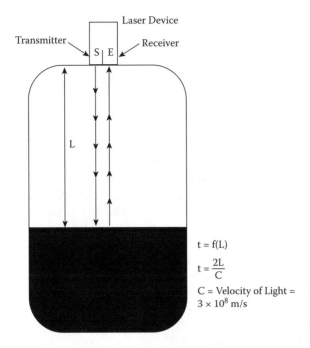

$$t = f(L)$$

$$t = \frac{2L}{C}$$

C = Velocity of Light = 3×10^8 m/s

FIGURE 3.118
Detection of level by measuring the time of reflection. Light travels at a speed of about 0.3 m/ns (10^{-9} s).

In addition to laser technology, conventional optical devices are also used for level detection. Reflected visible (or infrared) light beams can also detect the level of liquids or solids. The sensor can be provided with several light-sensitive detectors to permit multipoint switch actuation. *Transmission-type* optical gauges can detect both the concentration of solids in a liquid or detect the level or interface between sludge and supernatant. The level sensor can be a point-sensing switch (Figure 3.119) or a continuous sludge depth detector.

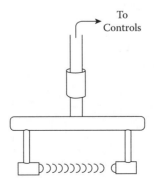

FIGURE 3.119
Optical sludge level detector.

Optical fibers can also be used for liquid level detection. A light beam travels in the fiber and, as long as no process fluid contacts the fiber, the return beam will have the same intensity as the source beam. As the level rises, and liquid covers some of the fiber, the index of refraction increases, allowing some of the light to escape into the liquid. This reduces the strength of the return beam, a change that is an indication of level.

3.11.5 Microwave and Radar Gauges

Microwave switches are beam-breaker-type point sensors with an accuracy of 13 mm (0.5 in.) and with pressure and temperature ratings up to 28 bar (400 psig) and 300°C (600°F). Pulse-type radar gauges have ranges up to 200 m (650 ft) and are accurate to 0.5% FS, whereas frequency-modulated carrier wave (FMCW) units have errors from 1 to 3 mm (0.04–0.125 in.). Their pressure and temperature ratings are up to 80 bar (1,200 psig) and up to 400°C (750°F).

Microwave switches are used in hard-to-handle solid, liquid–solid, and liquid–liquid interface applications, whereas the radar design is used for continuous level measurement. The beam-breaker microwave switches are used in both solids or liquid applications. Microwaves do not pass through metal walls, but they do pass through fiberglass or plastic tank walls and through windows of plastic, ceramic, or glass that are installed in metal vessel walls. Therefore, the beam breaker microwave switches can be placed outside the tanks, located in front of windows that are built into the tank walls. Thick windows can withstand heavy abrasion on solids service and can isolate the sensor from hazardous and toxic liquids on high-pressure service. The presence of dust, mist, or nonmetallic foam has a negligible effect on their performances.

The beam-breaker switch sends a beam across the measurement zone (Figure 3.120), and if process material breaks the beam path, it reduces the strength of the signal received. Separation distance can be up to 100 ft (30 m), which is considerably greater than with ultrasonic or nuclear devices. Micro-

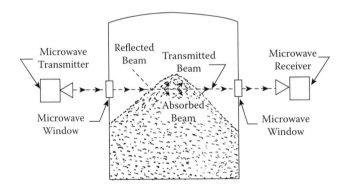

FIGURE 3.120
Beam-breaker detector.

FIGURE 3.121
Frequency modulated carrier wave (FMCW) radar transmitters. (Courtesy of Thermo Measure Tech Inc.)

wave devices can tolerate more coating than ultrasonic or laser units but less than radiation-type level switches.

The beam-breaker technique is useful for detecting large and abrasive materials such as coal, minerals, wood chips, and vegetable pulp. It is also useful for detecting very light materials such as dry sawdust and powdered materials in fluidized beds. This technique can also be considered for use on difficult-to-handle liquids that are viscous, toxic, or hazardous because the detector is isolated from the vessel contents.

Radar level transmitters and gauges use electromagnetic waves, typically in the microwave bands to make a continuous liquid and some solid level measurements. The radar sensor is mounted on the top of the vessel and is aimed down, perpendicular to the liquid surface. Most tank-farm gauges are operated on the FMCW principle (Figure 3.121). Other gauges and transmitters, particularly the lowest-cost units, are operated on the pulse principle. Both principles are fundamentally based on the time of flight from the sensor to the level of the surface to be measured. In the FMCW method, this time of flight is tracked on a carrier wave; in the pulse method, it is the *echo return*.

Generally, conductive products such as water and other water-based liquids (acids, strong bases, and so on) can be measured, regardless of dielectric constant. Nonconductive materials have reflectivity based exclusively on the dielectric constant. Materials with low dielectric constants absorb microwaves and provide much lower reflected signal strength than do materials with high dielectric constants. The velocity of radar wave transmission is equal to the speed of light divided by the square root of the medium's dielectric constant. Fortunately, the dielectric constants of different gases at different pressures and temperatures vary only slightly.

FMCW radar level gauges are the primary tank farm radar devices because of the inherent accuracy. Pulse transmitters are not suitable for tank farm inventory and custody transfer applications. The combination of foam and low-dielectric material can cause errors or provide an insufficient signal strength on the return signal.

3.11.6 Radiation Gauges

Radiation-type level detectors can be point source, multiple source, strip source, motorized tape-supported traveling source, or backscatter types with cesium-137 or cobalt-60 (for thick-walled tanks) source materials. The ranges of external units are up to 7 m (23 ft) and of the traversing backscatter-type units up to 45 m (150 ft). Measurement errors range from 6 mm to 1% FS.

To measure the level of LH_2 inside a double-walled cryogenic storage tank, one of the safest and most reliable level sensors is the radiation gauge. Nuclear radiation can pass through metallic walls, and it is sufficiently penetrating to also pass through the contents of the tank, although radiation gauges are also used in the backscatter mode.

The relative penetrating powers of the three kinds of radiation are approximately in the range of 1,100, and 10,000 for the alpha, beta, and gamma rays, respectively. The penetrating power of alpha rays is less than 0.2 m (8 in.) of atmospheric air. Beta radiation can penetrate approximately 6 mm (0.25 in.) of aluminum. Gamma rays are used for level measurement because of their high penetrating power, and because they cannot be deflected.

The most often used unit to quantify the activity of any radioactive material is the curie (Ci). For most level detection applications, source strengths of 100 millicuries (mCi) or less are satisfactory. A 1 Ci source will produce a dose of 1 roentgen (r) at a receiver placed 1 m (3 ft) away from the source for 1 h. Radiation is attenuated when it penetrates liquids or solids, and the rate of attenuation is a function of the density of the material. The higher the density, the more attenuation the shielding material will provide. Figure 3.122 shows how various thicknesses of different materials will attenuate (reduction factor—NB) the intensity of radiation and result in different degrees of attenuation.

A person receives the dose of 1 rem of radiation when exposed to 1 r of radiation in any time period. As illustrated in Figure 3.123, a person should not receive more than 250 rem over an entire lifetime. The rate at which this exposure is accumulated is also important. It is desirable to keep the yearly dose below 5 rem, and it should definitely not exceed 12 rem/year or 3 rem per quarter.

Radiation-based level detection continues to be very appealing for cryogenic, hard-to-handle, toxic, and corrosive processes, because it does not require vessel wall penetrations. Costs and licensing requirements do limit the number of applications but are not serious impediments to the implementation of carefully designed systems.

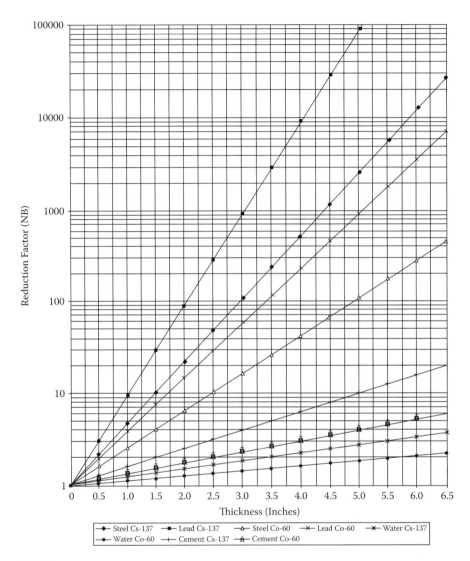

FIGURE 3.122
Radiation reduction (NB) as a function of the source material, the material that the narrow-beam radiation is passing through, and of the thickness of this material. So, for example, if the source is Cs-137, which is passing through a 1-in.-thick steel plate, NB = 4, and therefore, the field intensity will be reduced to 25%.

3.11.7 Tank Farm Level Monitoring

The designs of tank farm level detectors include (1) wire-guided float, (2) servo-operated float, (3) surface detector (plumb-bob), (4) radiation backscatter, (5) radar, (6) HTG, and (7) hybrid gauges. Their ranges go up to 60 m (200 ft).

Hydrogen tank farms will require the same types of inventory controls and level monitoring as do liquefied natural gas (LNG) and oil tank farms

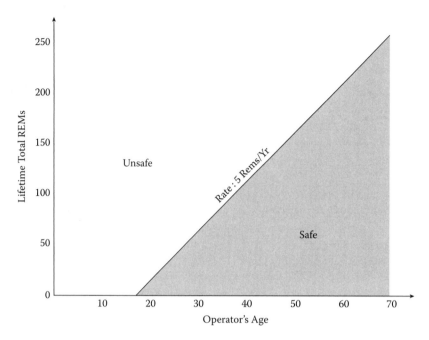

FIGURE 3.123
Radiation exposure as a function of time and safety.

today. In the 19th century, oil could not be measured more accurately than about 5%, so producers agreed on the unit of the barrel (42 gal), thereby making sure that there would be at least 40 gal in every barrel. A hundred years later, the precision of custody transfer improved to about 0.5%, and today, if every error source except nonuniformity in the tank's cross section (Figure 3.113) is carefully eliminated, the error will be about 0.25%.

As shown in Table 3.124, some automatic tank gauges (ATGs) are better suited for "outage" detection and others for "innage" sensing. When measuring outage, the distance from the top of the tank to the liquid surface is detected, and multiplying this distance by the cross-sectional area of the tank gives the volume of the vapor space, called *ullage*. Outage detection is less accurate than innage detection because the outage distance has to be compared to a reference height (the mounting location of the gauge), which, if the ATG is supported from the tank shell or roof, varies with liquid depth, temperature, and age. Therefore, if the ATG error is 3 mm but the gauge mounting can travel 10 mm, the actual error can reach 13 mm.

In custody transfer of full or nearly full tank volumes, manual gauging is still practiced in the United States, whereas ATGs are preferred in Europe. In either case, the inherent tank accuracy is a factor, because the filling of a large tank causes the bottom to sink, the shell to bulge, and the top to sink, and changes in temperature also cause changes in tank dimensions.

Automatic tank-gauging systems are found in almost all tank farms for inventory monitoring. The wire-guided float tape gauge systems are most

TABLE 3.124

Features of Automatic Tank Gauges

Type of Design	Approx. Number in Use in United States	Cost (for 40-ft Floating Roof Tank with Temperature Measurement)	Measures	Accuracy	Maintenance Required
Float	±300,000	$4,000±	Outage	Low	Highest
Servo	±10,000	$6,500±	Outage	Good	High
Radar	±5,000	$8,500±	Outage	Good	Low
HTG	±5,000	$8,500±	Innage	Low	Low
Smart Cable	±5,000	See Footnote a	Innage	Good	Varies
Hybrid	±200	See Footnote b	See Footnote b	See Footnote b	See Footnote b

ᵃ Costs vary for the different types of smart cable systems. A magnetostrictive system for a 40-ft high floating roof tank costs ±$3,500, including average temperature measurement.

ᵇ The cost, measurement method, accuracy, and maintenance of a hybrid ATG depend on the level and temperature measurement system. A pressure transmitter adds ±$1,500 to the cost of the system.

common on the existing tank farms (Figure 3.125) and have been used for the last 50 years on liquid services. They were developed to reduce the need for the operator to climb the tank for manual "dipping." To protect against tank farm accidents such as overfilling, they should be backed up with high-level switches, and computer monitoring should be provided to detect sensor failures resulting from float or tape hang-up. When the distance from the tanks to the remote readout is long, a satellite multiplexer system can be used for a pipeline transmission installation where the various bulk storage facilities are hundreds of miles apart.

3.11.8 Ultrasonic Level Detectors

Ultrasonic level transmitters can be used on liquids or solids and can be of the noncontacting or the submerged designs. These errors are from 0.5 to 2% FS over measurement ranges up to 75 m (250 ft) in silos and up to 600 m (2,000 ft) in wells.

Sonic (up to 9,500 Hz) and ultrasonic (10–70 kHz) level sensors operate either by the absorption (attenuation) of acoustic energy as it travels from source to receiver, or by generating an ultrasonic pulse and measuring the time it takes for the echo to return. If the transmitter is mounted at the top of the tank, the pulse travels in the vapor space above the tank contents, and if it is mounted on the bottom, the time of travel reflects the depth of liquid in the tank. In water, at ambient temperature, the ultrasonic pulse travels at 1,505 m/s (4,936 ft/s).

Temperature compensation is essential in ultrasonic level measurement, because the velocity of sound is proportional to the square root of tempera-

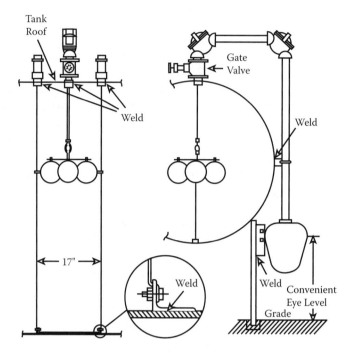

FIGURE 3.125
Wire-guided float detector installation for high-pressure tanks.

ture, and in the case of air, it changes by about 0.6 m/s for each degree Celsius change in temperature. Fluff and loose dirt have poor reflecting characteristics because they tend to absorb the sonic pulse. Irregular surfaces result in diffuse reflection in which only a small portion of the total echo travels vertically back to the source; the echo of sloping surfaces will not be directed back to the source at all. An ultrasonic sound wave traveling in dry and dust-free ambient air is attenuated by 1–3 dB for each meter of travel.

The continuous level detector designs can be point sensors or continuous, can be penetrating or wetted, and can be located below or above the process fluid. Point sensors are either vibrating devices and are damped out when the process material contacts them or absorption devices, where the source and receiver are at the same level and the rising level blocks the passage of the pulse (Figure 3.126).

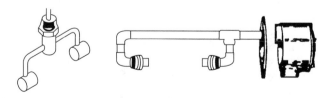

FIGURE 3.126
Top- and side-mounted level switches for sludge or slurry services. (Courtesy of Delavan, Inc.)

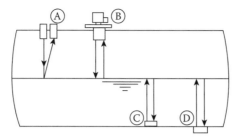

FIGURE 3.127
Continuous ultrasonic level detectors.

The continuous ultrasonic level detector (SONAR) measures the time required for an ultrasonic pulse to travel to the process surface and back. Figure 3.127 illustrates some of the possible installations. Intelligent micro-processor-based ultrasonic transmitters, using FOUNDATION Fieldbus and Profibus PA connections, convert the level in cylindrical, horizontally mounted tanks into actual volume or can automatically recalibrate themselves by correcting for the effects of changing vapor space composition using a fixed reference. They can also be shared among the 24 or 48 storage tanks.

The main advantages of ultrasonic transmitters are the absence of moving parts and the ability to measure the level without making physical contact with the process material. The reading can be unaffected by changes in the composition, density, moisture content, electrical conductivity, and dielectric constant of the process fluid. If temperature compensation and automatic self-calibration are included, the resulting level reading can be accurate to 0.25% of full scale.

In terms of limitations, the echo will be reduced if the vapor space is dusty or contains foam, water vapors, or mists. In addition, one should not use them to detect sound-absorbing (fluffy solids), sloping, or irregular surfaces. Therefore, the best guide for using ultrasonic level instruments is past experience on similar installations.

3.12 Machine Vision

Machine vision will be an important tool in the operation of unattended, fully automated renewable energy processes. Vision is the response of the eye and brain to light. Machine vision is the artificial response of a device to spectral radiation. Machine vision also extends the human range of the visible spectrum into the IR, UV, and x-ray regions. Machine vision can be used to make decisions faster or more accurately and precisely than a human can. Machine vision may combine online defect monitors with shade monitoring, which will be an important tool in optimizing the positioning of solar collectors (Table 3.128).

TABLE 3.128

Machine Vision Cameras

Type of Camera[a]	Cost ($)	Pixels	Output Format	Frame Grabber and Software Notes
Linear PSD	500–1,000	N/A	Analog	N/A
Linear PDA or CCD array (may have 3 arrays for RGB)	1,000–10,000	128–8,192; @8–12 bits/pixel	Serially transmitted pixel data	Custom software
Two-dimensional array monochrome or RGB, digital cameras	1,000–10,000	Up to 4,096 × 4,096	Digital, USB 2.0, IEEE 1394, Ethernet	RS-170, NTSC, PC-based with custom software
Two-dimensional array monochrome	500–5,000	Up to 4,096 × 4,096	Analog or discrete	Camera internally runs algorithms based on pixel counting
VCRs, analog video cameras	100–1,000	N/A	Analog to USB	Desktop publishing

[a] Maybe ultraviolet, infrared, VIS, external lighting, or built-in, such as a laser beam.
PSD, photosensitive device; PDA, photodiode array; CCD, charge-coupled device; RGB, red–green–blue.

Line scan diode arrays can be used to detect leading and trailing edges of passing objects. Here the digital logic is based on which pixel is illuminated by the reflected beam. Another application for a linear array or photosensitive device (PSD) operation is light triangulation depicted in Figure 3.129. Combining lenses, external lighting, and microprocessors yields smaller and faster spectrophotometers.

The use of line scan cameras for a real-time color pattern inspection grew out of pattern recognition systems in the textile industry, where a single camera with ample lighting can image about 1-m-wide webs. In that application, the complete system was controlled by a 32-b Versa Module Europa (VME) bus microcomputer that also recorded an encoder input for web position.

A two-dimensional array of CCD photo-detectors is used to store all the pixels (square dots) within a cycle time of about 1 ms. This is called *pixel counting*. Many simple inspection applications can be handled by this method. Thus, processes should be designed ahead of time to incorporate machine vision rather than adding it later and having to compensate for variation in lighting, vibration, and excessive movement. The Automating Imaging Association provides a good Web site for learning.

If one takes a magnifying lens and looks closely at a color TV monitor, it is seen that the screen is made up of thousands of dots of red, green, and blue

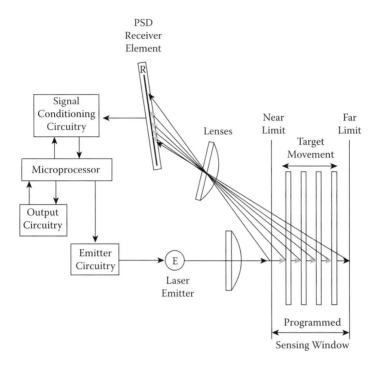

FIGURE 3.129
Optical triangulation. (Courtesy of Banner Engineering.

(RGB) phosphors. These RGB pixels are laid out and controlled by NTSC (National Television Standard Code) formats for displaying an output from an RGB-transmitting device or recording. The video camera must transmit separate RGB signals for the horizontal and vertical sweeps of the display. Display devices will eventually become totally digital with flat screen technology. Analog-to-digital (A/D) conversion is best done internally to the camera onboard so that high-speed serial data transmission is used to get the data into a PC.

For small space and robotics needs, smaller cameras (not needing frame grabbers) are desirable because these devices are external to the host. With the Universal Serial Bus (USB) protocol, the sending of 16-bit data values is done at about 12 Mbps, but higher speeds are available up to 1 Mbps. To use 32-bit data words containing RGB color information takes dedicated graphics boards running at over 1 GHz plus about 10 MB of memory per image.

Today's software graphics design packages offer a resolution finer than the unaided human eye can detect. Such a resolution (or one outside human vision—IR or x-ray) makes sense if one is trying to inspect flaws in materials or to apply advanced optimization strategies in renewable energy processes, such as the colorization of IR-temperature sensor data.

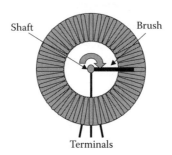

FIGURE 3.130
Wire-wound potentiometer.

3.13 Position Detection

In the automation and optimization of solar, wind, and wave energy systems, linear or angular position measurement plays an important role. There is a difference between absolute position sensors and sensors that detect only displacement. Position measurement requires a sensor and a transmitter. Mounting of position-sensing equipment in most applications must be custom-made. In position sensing, the linearity is only about 1%, but that is acceptable, because repeatability matters more.

Position sensors convert the position into an electrically measurable signal such as resistance, voltage, current, inductance, pulses, or capacitance. The simplest and most widely used position sensor is the potentiometer. The potentiometer has three terminals, one for each end of the resistive element and one for the brush. As the brush moves, the resistance between the center tap and end terminal changes (Figure 3.130).

The position can also be determined in a resistive or voltage mode. In the resistive mode, a current is sent to the center tap and to one of the end terminals, whereas the resulting voltage drop is measured by the transmitter. A potentiometer has an infinite resolution. Linearity for a precision rotary potentiometer can be as good as 0.25%. However, due to linkages and gears, the linearity for the entire assembly usually is about 0.5–1%. A linear variable differential transformer (LVDT) is mostly used in linear motion applications and also inside some pressure transmitters (Figure 3.131).

In the *magnetostrictive* sensor, a coil in the head end starts the measuring cycle by periodically sending a trigger pulse down a wave-guide wire encased in a protective tube. A movable magnet assembly slides along the outside of the tube. The magnet assembly is moved by the part whose position is being detected (Figure 3.132). The wave guide twists when the magnetic field accompanying the pulse down the wave guide intersects the magnetic field from the movable marker.

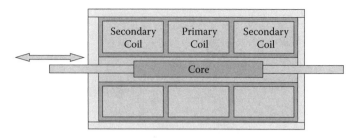

FIGURE 3.131
The operating principle of an (linear variable differential transformer) LVDT-type sensor.

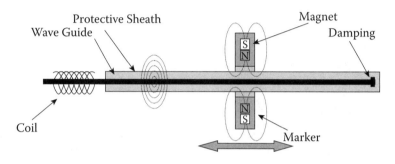

FIGURE 3.132
The operation of the magnetostrictive position sensor.

In the *Hall effect* sensor, a magnet assembly is moved by the part whose position is being detected. The magnet assembly can be constructed for either linear or rotary motion. For rotary quarter-turn motion, a round magnet assembly with the two magnets in line is used. The magnetic flux picked up by the Hall sensor is a sinusoidal function of the angle. Encoders are either absolute or incremental. Incremental encoders detect the distance moved and not the position. Absolute encoders can be linear but are mostly rotary. In the rotary configuration, the encoder shaft is turned by the part whose position is being detected. The shaft turns a disk that has a radial pattern with a unique code provided for each of a finite number of distinct positions (Figure 3.133).

Intelligent devices have a built-in microprocessor for providing digital communication. The leading technologies today are FOUNDATION Fieldbus and highway addressable remote transducer (HART), whereas PROFIBUS DP is widely used in the automation of machinery. FOUNDATION Fieldbus is digital and has a higher speed (31.25 kbps). This makes it possible to connect as many as 16 bus-powered devices per network into a multidrop system, which reduces wiring costs. Digital communication is also used for closed-loop control where control strategies typically are executed in the field devices themselves. Therefore, FOUNDATION Fieldbus makes the control system architecture leaner by virtually eliminating controller hardware.

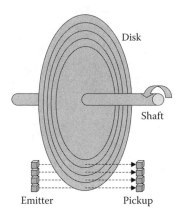

FIGURE 3.133
Encoder principle of operation.

3.14 Pressure Detection

Accurate pressure control is required not only in biofuel and ethanol refining but also in the operation of the steam boilers of thermal solar systems and in the handling and transportation of high-pressure H_2.

3.14.1 Application and Selection

Pressure sensors can measure pressures from ultrahigh vacuums, such as 10^{-13} mmHg, to ultrahigh gauge pressures up to 26,000 bar (400,000 psig). The unit of pressure often used is the bar (bar = 0.981 kg/cm^2 = 14.7 psi). Pressure sensor errors range from 5% for dial gauges to 0.01% FS for fused-quartz helix sensor. Table 3.134 provides a summary of the different designs, their ranges, and accuracies.

Most pressure sensors detect the difference between the measured value and a reference. In the case of absolute pressure sensors, the reference chamber cannot be evacuated to absolute zero, because it can only be approached within a few thousands of a millimeter of mercury (torr). In the case of positive pressure detectors, when the barometric pressure is the reference, atmospheric pressure variations can cause errors up to 25 mmHg. Intelligent transmitters can also operate with multiple references and switch them as required.

For the measurement of near-atmospheric positive or vacuum pressures, bellows-type diaphragm sensors and manometers are the most likely choices. Force or motion balance transmitters can detect vacuums down to 1 mmHg and gauge pressures up to 7,000 bar (100,000 psig). The operating principles of the various electrical designs include the capacitance, frequency change, LVDT, piezoelectric, potentiometric, reluctive, and strain gauge designs.

The mechanical designs utilize bellows, diaphragm, d/p, and Bourdon tube designs. In 1852, Eugene Bourdon patented a curved or twisted tube,

TABLE 3.134

Pressure Detector Errors, Ranges, and Costs

Type	Range	Inaccuracy (%)	Approximate Cost ($)
General-purpose Bourdon-tube indicator	15–10,000 psig (1–690 bar)	2	100
High-accuracy test gauge	Low vacuum to 3,000 psig (207 bar)	0.1–0.01	300–6,000
Bourdon/spiral case-mounted Indicator/recorder	Low vacuum to 50,000 psig (Low vacuum to 3,450 bar)	0.50	1,200
Spring-and-bellow case-mounted recorder	Low vacuum to 50 psig (Low vacuum to 3.5 bar)	0.50	1,600
Nested capsular case-mounted recorder	10–90 psig (0.7–6.2 bar)	0.50	1,600
Low-pressure bell-case-mounted indicator	−0.1 to 0.1 in. H_2O (−3 to 3 mm H_2O)	2	2,200
Beam-mounted strain gauge (sensor only) 4–20-mA DC output	0–1,000 psig (0–69 bar)	0.25	800
Piezoresistive transducer 4–20-mA DC output	0–5,000 psig (0–365 bar)	0.50	500
"Smart" piezoresistive transmitter 4–20 mA DC output	0–6,000 psig (0–414 bar)	0.10	1,200–2,000
"Smart" field communicator for remote calibration and configuring of "smart" transmitter	—	—	1,000–3,000
Capacitive sensor/transmitter	1 in. H_2O–6,000 psig (25 mm H_2O–414 bar)	0.2	1,000

which, if pressurized at its open end, produced a proportional movement at its closed end (tip travel). The Bourdon tube elements are suited for the measurements of medium to high pressures using atmospheric reference. The sensitivity and accuracy is increased by joining C-Bourdon tubes into spiral elements. When pressure is applied, this flat spiral tends to uncoil and produces a greater movement of the free end, requiring no mechanical amplification.

Figure 3.135 shows the construction of a helical Bourdon element. This sensor produces an even greater motion of the free end than the spiral element. Other advantages include the high over-range protection and that it is suitable for continuously fluctuating services. High-pressure elements might have as many as 20 coils. Helical coils can also be used as the element in d/p

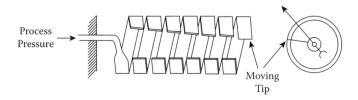

FIGURE 3.135
Helical Bourdon pressure sensor.

sensors if one of the pressures is acting on the outer surface and the other on the inner surface of the coil.

In the fused-quartz design, the helix is mounted above a large precision gear that can rotate concentrically around the pressure sensor (Figure 3.136). A lamp and a pair of photocells are attached to the gear. The light from the lamp is reflected from the mirror and distributed equally on the two photocells. When the process pressure changes, the free end of the helix rotates the mirror, unbalancing the photocells. This signal causes the balancing motor to turn the gear until the light falling on the photocells is equally distributed once again. This laboratory sensor provides high resolution and sensitivity, whereas its hysteresis is negligible.

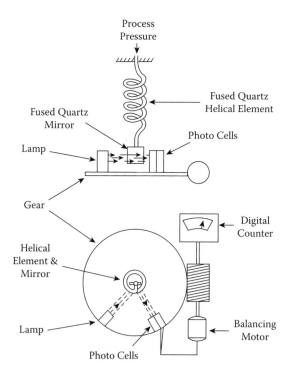

FIGURE 3.136
Fused quartz helical element pressure detector design.

FIGURE 3.137
The variety of diaphragm- and capsule-type pressure and differential pressure detectors and transmitters is great. (Courtesy of Foxboro Invensys.)

Differential pressure-type detectors have already been discussed in connection with both flow (Section 3.9.9) and level (Section 3.11.2) measurements. Therefore, only their ranges and accuracies will be briefly mentioned here. The basic error of d/p transmitters ranges from ±0.1 to ±0.5% of the actual span. Added to this are the errors caused by the temperature and pressure effects on the span and zero of the instrument. For intelligent transmitters, the pressure and temperature corrections are automatic, and the overall error is ±0.1 to ±0.2% of span.

The ranges of membrane-type d/p cells can be as low as 0–0.25 mm H_2O (0.01 in. H_2O). Standard d/p cells are available from 0–15 cm to 0–4 m H_2O (0–5 to 0–850 in. H_2O) ranges. Motion balance bellows indicator ranges go up to 28 bar differential (400 psid). Some intelligent transmitters provide self-selection of multiple ranges, reaching 40:1 overall rangeability. Some of the d/p cell design variations are illustrated in Figure 3.137.

Microprocessor-based pressure and d/p transmitters perform the manipulations of ranging, linearizing, error checking, unit converting, and transmitting the measurement in a digital or analog form. They also check their own calibration and incorporate self-diagnostics. Some of their features include multiple ranges, remote zero, and span settings and reconfiguration capability.

3.14.2 High-Pressure Sensors

High pressures can be measured by

(a) optical
(b) piezoelectric
(c) magnetic
(d) deadweight testers
(e) helical Bourdon
(f) Manganin cells
(g) strain gauges
(h) bulk modulus-type detectors

Their ranges and errors are the following:

Ranges:
(a)   up to 5,000 bar (60,000 psig)
(b, c, d, e)   up to 7,500 bar (100,000 psig)
(g, h)   15,000 bar (200,000 psig)
(f)   25,000 bar (400,000 psig)

Errors:
(d)   0.1% of span
(c)   0.1–0.5% FS
(f, g)   0.2% of span to 0.25% FS
(e)   1% of span
(h)   1–2% of full span (FS)

In the renewable energy processes, particularly in the GH_2 storage and transportation applications, accurate measurement of high pressures is required. The term "high pressure" is relative, but for the purposes of this discussion, pressures exceeding 1,000 bar (15,000 psig) are considered to be high.

Deadweight testers are also used as primary standards in calibrating high-pressure sensors. The tester generates a reference pressure when National Institute of Standards and Technology (NIST)-certified weight is placed on a known piston area, which imposes a corresponding pressure on the filling fluid. NIST has found that at pressures exceeding 3,000 bar (45,000 psig), the precision of the test is about 1.5 parts in 10,000.

One mechanical high-pressure sensor, the *helical Bourdon* element, has already been discussed in connection with Figure 3.135. An improvement on that design detects tip motion optically, without requiring any mechanical linkage. Other electronic sensors (capacitance, potentiometric, inductive, and reluctive) are also capable of detecting high pressures, but none can go as high as the strain gauge.

The operating principle of strain gauges is more than 100 years old, and it was discovered when Lord Kelvin reported that metallic conductors sub-

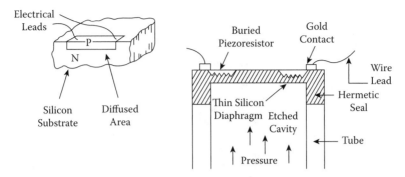

FIGURE 3.138
Diffused semiconductor (piezoresistor) strain gauge.

jected to mechanical strain exhibited a corresponding change in electrical resistance. This design eliminated the mechanical frame by attaching the sensing view directly to the strained surface. Foil gauges have also been used in which the foil thickness can be as low as 0.025 mm. Semiconductors, such as silicon and germanium, also came into use as the gauge elements. The most attractive characteristic of semiconductors is their sensitivity, which is close to 100 times greater than that of metallic wires.

Strain gauge transducers can be built by using *unbonded or bonded wire*. In the latter, the wire filament is attached to a test specimen by plastic cement, and the strain in the wire is measured in terms of its resistance. Constantan, nichrome, platinum, or Karma-type alloy wires, foils, or semiconductor materials can be permanently bonded to the strained surface. Such strain gauge designs are inherently unstable (subject to creep) due to degradation of the bond, temperature sensitivity, and hysteresis caused by thermoelastic strain. *Thin-film* sensors use a molecular bond instead of cement. In the diffused semiconductor strain gauge, which consists of resistance elements diffused into a single silicon chip, the silicon wafer acts as the sensor diaphragm, reducing both the size and the cost of the sensor (Figure 3.138).

For high-pressure measurement, the strain gauge sensors can detect 15,000 bar (200,000 psig) and can provide measurement precision in the range of 0.1% of span to 0.25% of full scale. Temperature compensation and periodic recalibration are desirable because the temperature variation or drift can also produce an additional error. The sensitivity of semiconductor-type strain gauges is 100-fold greater than that of wire strain gauges. The thickness of the wafer is controlled by chemical etching, and it is this thickness that determines the pressure range of the sensor.

For temperature compensation, the most frequently employed method is the use of dummy elements. The dummy gauge is mounted on the same surfaces as the active element. It is exposed to the same temperature but is not subject to the forces applied. If such a dummy is connected in a Wheatstone bridge arm adjacent to the active element, it will automatically compensate for temperature effects.

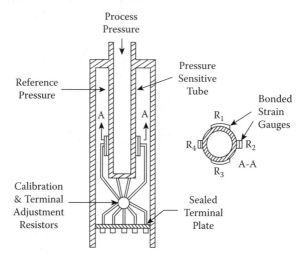

FIGURE 3.139
Strain gauge transductor with elements bonded to tube surface.

There are several ways to bond strain gauge elements. In Figure 3.139, the working element is a tube closed on one end, with the other end open to the process pressure. Four strain gauges are bonded to the outside of this tube. Two of the elements are strained under pressure, and two are not because they are mounted longitudinally and circumferentially.

The resistance change in a monocrystalline semiconductor (a piezoelectric effect) is substantially higher than that in standard strain gauges. Conductivity in a doped semiconductor is influenced by a change (compression or stretching of the crystal grid) that can be produced by small mechanical deformation. Integrated circuits are used for temperature compensation. In the silicon pressure sensor shown in Figure 3.140, the piezoresistors are joined in a bridge configuration and attached to the bond pads for circuit interconnection. The two pairs are placed diagonally in the bridge such that applied pressure produces a bridge imbalance.

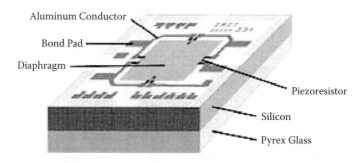

FIGURE 3.140
Silicon pressure detector. (Courtesy of Maxim Integrated Products.)

FIGURE 3.141
Miniature transducer suited to the measurement of both hydrogen pressure and temperature. (Courtesy of Kulite.)

Figure 3.141 shows a piezoresistive transducer with a pressure range of 1,500 to 5,000 bar range and a millivolt output. It provides the internal calibration and 0.3% FS accuracy.

The bulk modulus cell consists of a hollow, cylindrical steel probe with a projecting stem (Figure 3.142). When exposed to a process pressure, the probe is compressed, and the probe tip is moved to the right. This stem motion is then converted into a pressure reading. The hysteresis and temperature sensitivity of this unit are similar to that of other elastic element pressure sensors. The main advantages of this sensor are its fast response and safety; in effect, the unit is not subject to failure. The bulk modulus cell can detect pressures up to 15,000 bar (200,000 psig) with 1–2% full-span error.

In another high-pressure detector, Manganin, gold–chromium, platinum, or lead wire sensors are wound helically on a core. The electrical resistance of these wire materials will change in proportion to the pressure experienced on their surfaces. The pressure–resistance relationship of Manganin

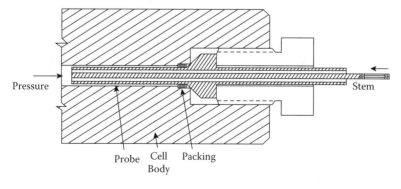

Pressure

Stem

Probe Cell Packing
       Body

FIGURE 3.142
Bulk modulus cell.

is positive, linear, and substantial. Manganin cells can be obtained for pressure ranges up to 25,000 bar (400,000 psig) and can provide 0.1–0.5% of full-scale measurement precision. The main limitation of the Manganin cell is its delicate nature, making it vulnerable to damage from pressure pulsations or viscosity effects.

3.14.3 Pressure Safety Devices

Alternative energy processes, just like any others, need to be protected from excessively high or low pressures. The methods of protection include pressure regulation, alarm, or safety interlock actuation when preset pressure limits are violated, and providing pressure relief devices, which need to be replaced after each operation (rupture disks) or can automatically reclose (relief valves). The features and characteristics of these devices are discussed in the following subsections.

3.14.3.1 Alarm and Interlock Switches

Pressure switches are used to energize and de-energize electrical circuits when the process pressure reaches their set points. The standard switch contact configurations are shown in Figure 3.143. The contacts are designated as normally open (N/O) and normally closed (N/C). In case of a single-pole double-throw (SPDT) switch, as the process pressure reaches the set point, the N/C contact opens and the common (C) side of the switch is closed to the contact designated as N/O.

Some of the standard arrangements include the single-pole single-throw (SPST), single-pole double-throw (SPDT), and the double-pole double-throw (DPDT) designs, but units are available with up to four poles. These switches

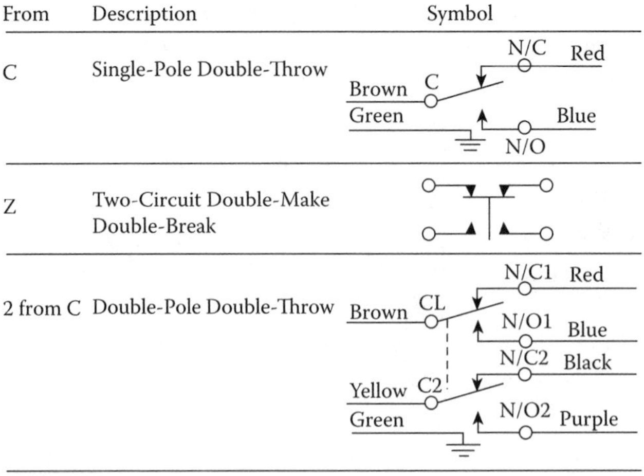

FIGURE 3.143
Standard electric contact configurations used in pressure switches.

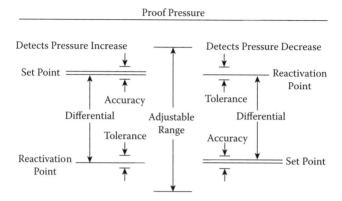

FIGURE 3.144
Pressure switch terminology.

can detect absolute, compound, gauge, and differential pressures with inaccuracies up to ±0.5% of span. The sensing elements can be bellows, Bourdon tubes, diaphragms, or other elastic components. They can be the snap-acting, mechanical, or mercury switches. The latter has no mechanical moving parts and must be mounted level and in vibration-free locations.

Figure 3.144 lists some of the terminology used in connection with pressure switches. The pressure range within which the actuation point can be set is called the *adjustable range*. The switch may actuate at its set point on rising or falling pressure as distinguished in the sketch. The set point pressure is the pressure that actuates the switch to open or close an electric circuit. The set point accuracy defines the band within which repetitive actuations will occur. Differential or dead band is the difference between set point and reactuation point.

Most elastic elements will have a service life of close to a million cycles, if the cycle time is not less than 5 seconds. If longer service life or higher-cycle frequencies are required, metal fatigue tends to limit the usefulness of elastic elements, and special designs, such as sealed piston elements, should be considered.

3.14.3.2 Pressure Regulators

The pressure regulator is a complete pressure control loop, incorporating sensor, controller, and valve. It is called a regulator (and not a controller) because it is mechanical and self-contained, requiring no external energy source. In small sizes, regulators are less expensive to buy, install, and maintain than control loops. Regulators are fail-safe devices, because they are not affected by air-supply failure, which in critical applications or at locations remote from a source of compressed air, is an important consideration. However, diaphragm failure in a regulator usually causes valve opening, which can be unsafe.

The set point of a regulator is provided integrally, and remote control is usually not possible. It is a single-mode (proportional only) controller,

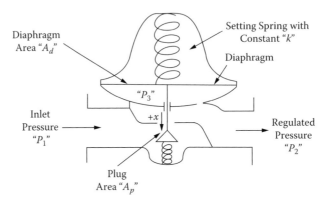

FIGURE 3.145
The basic components and operation of a spring-loaded pressure-reducing valve.

which is subject to set point droop that varies with throughput. Materials of construction are limited, as is interchangeability, and accessories are not available.

In a self-contained pressure regulator, the process pressure acts on the diaphragm, which moves the valve plug (by compressing the spring). The initial spring compression sets the pressure at which the valve begins to open. As shown in Figure 3.145, the diaphragm compares the set point (spring force), to the regulated pressure (force on the diaphragm), and adjusts the valve opening to reduce the difference between the two. Pressure regulators can also be loaded by weights or gas pressure and can control back pressure, d/p, or vacuum.

The regulator has a fixed proportional band of about 5% (gain of about 20), and therefore, the regulated pressure cannot control pressure at a set value but will be offset by changes in the load (flow range of $q_1 - q_2$). This is called the "droop" of the regulator, which is a function of the spring rate, k; valve lift, x; and diaphragm area, A_d. Figure 3.146 illustrates the phenomenon of droop. The minimum flow rate is a function of plug design and is typically 5–10% of maximum capacity, q_2.

In order to provide tight shutoff, extra force is needed, and therefore, the pressure difference on the diaphragm must rise. Consequently, at near-zero flow, the regulated pressure will rise. What the manufacturers call the "set point" of the regulator, in fact, is only the pressure at minimum flow (q_1). Maximum regulator capacity is not at full-valve opening (q_2) but at maximum acceptable droop. Information on droop versus flow is therefore essential to check if regulator performance will be satisfactory.

Pressure regulators have been successfully used as cascade slaves receiving their set points from temperature control masters. Figure 3.147 illustrates such an installation on a steam-heated water heater. This configuration works well, because the air-loaded pressure regulator is extremely fast and corrects for load changes or steam-supply pressure variations instantaneously.

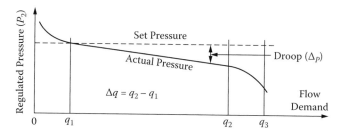

FIGURE 3.146
The actual value of the regulated pressure equals the set pressure at minimum flow (q_1). Once the flow rises above minimum, the actual controlled pressure drops. This "droop" or offset is permanent and increases with flow.

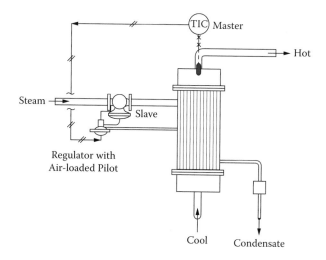

FIGURE 3.147
A pressure regulator is an ideal slave controller in a cascade loop. This is possible due to its speed of response.

Maintaining gas flows at less than sonic velocities at the regulator outlet is recommended to avoid noise. Sonic velocity can be avoided by limiting pressure reduction to less than the critical ratio of 2 to 1. Higher ratios can be achieved with regulators in series. Two to three stages are common, both to reduce noise and improve regulation. On liquid service, high pressure drops can cause temporary vaporization at the vena contracta that results in erosion or corrosion caused by cavitation. To avoid cavitation, one can change the operating temperature, use multiple regulators, or choke fittings in series.

Diaphragm rupture is the most common failure in a regulator. Most regulators fail to open completely on diaphragm failure, which is an unsafe condition. In installations where it is imperative that the process continue to be supplied even on a regulator failure, two regulators can be placed in series.

TABLE 3.148

PRV Nozzle Orifice Areas in Square Inches
($1 \text{ in.}^2 = 645 \text{ mm}^2$)

Orifice Designation	API Area	Actual (ASME) Area
D	0.110	0.1279
E	0.196	0.2279
F	0.307	0.3568
G	0.503	0.5849
H	0.785	0.9127
J	1.287	1.496
K	1.838	2.138
L	2.853	3.317
M	3.60	4.186
N	4.34	5.047
P	6.38	7.417
Q	11.05	12.85
R	16.0	18.60
T	26.0	28.62

3.14.3.3 Relief Valves

Pressure relief valves (PRVs) can be (1) spring-loaded, (2) weight-loaded, (3) balanced by bellows seals, or (4) pilot-operated, and are available in sizes from 12 to 300 mm (0.5–12 in.), with American Petroleum Institute (API)-designated orifices: D, E, F, G, H, J, K, L, M, N, P, Q, R, and T (for orifice areas, see Table 3.148). Their actuation error is about 3% up to 200 barg (300 psig), and it is 1.5–2% above. They can operate at vacuum to 700 bar (10,000 psig) pressures and at –270 to 554°C (–450 to 1,000°F) temperatures.

The purpose of relief valves is to protect tanks and other equipment from overpressure damage and to protect against the spilling of their contents. One of the main causes of overpressure is fire. During fires, it is important that tanks do not rupture and their contents do not feed the fire. This is particularly true in hydrogen processing because of the high pressure of the gas or cryogenic temperature of the liquid, but relief valves are needed in all renewable energy processes.

Determination of the required capacity of the pressure safety valves (PSVs) involves a calculation that considers the approximated amount of heat that a fire would send into the storage tank and the properties of the tank contents as they are being vaporized by that heat and generate the vapors to be relieved through the valve. The details of this calculation are beyond the scope of this discussion, and the reader is advised to review Chapter 7.15 in the fourth edition of Volume 1 of the *Instrument Engineers' Handbook* if that information is needed. Once the required PSV capacity is determined, the orifice area required to handle that capacity can be obtained graphically.

Example 1 in Figure 3.149 illustrates the graphical method of sizing a PSV for the following service: a vapor of 120 molecular wt, at a temperature of 149°C (300°F) is to be relieved at a rate of 9.090 kg/h (20,000 lb/h) when the pressure in the vessel reaches 18.7 barg (275 psig). By following the dotted line on the figure, we find that the orifice area required is 3.75 cm² (0.58 in.²), and therefore, the next larger orifice that should be selected is a size H.

The back pressure can affect the set pressure of the pressure relief valve (PRV; Figure 3.150). If the back pressure builds up as the valve opens, the valve may chatter and reclose. If it is constant and is already present when the valve starts to open, it should be possible (if all code and insurance requirements are fully satisfied) to compensate for it by raising or lowering the spring setting. It is recommended that spring-setting adjustment be limited to the noncritical applications and pilot or bellows seal designs be used to guarantee that the PRV will open on set pressure in all critical applications.

Blowdown is the difference between the set pressure and the reseating (closing) pressure of a PRV, expressed in percent of the set pressure or in bar or pounds per square inch. Figure 3.151 describes both the API values and the recommended values used in connection with PSVs.

The phenomenon of blowdown is caused by the use of springs. The valve spring balances against a pressure that equals the pressure in the protected tank minus the kinetic effects. When the PSV is open, its inlet pressure is less than the pressure in the protected vessel because of the inlet pressure drop. Blowdown is the amount by which the protected tank's pressure has to drop below the PSV's set pressure for the valve to reseat. The normal blowdown of a PRV is between 2 and 7% of set pressure. Pilot-operated PRVs can reduce the blowdown to about 2%.

Section I of the ASME code for fired boilers requires that the PSVs reach their full lift at a pressure not greater than 3% over their set point. Section VIII of the ASME code for unfired vessels does not provide a blowdown requirement, and the industrial practice is about 7%, which means that the normal operating pressure must be under 93% of set pressure. The position of the adjustable ring on the PSV nozzle controls the blowdown. This position establishes a secondary orifice area as the valve opens and closes.

3.14.3.4 Rupture Disks

The available rupture disk (RD) materials and temperature limitations are listed in Tables 3.152 and 3.153. Their rupture tolerance (error) is 2% of set pressure, which can be up to 8,000 barg (100,000 psig). The RD in its simplest form is a metallic or graphite membrane that is held between flanges. It bursts at some predetermined pressure and relieves the pressure in the protected equipment before the system components fail. Another common use of an RD is as single use, fast-acting valves used in fire suppression systems. Table 3.152 lists some RD design variations and their features.

Rupture tolerance is the range on either side of the marked or rated burst pressure. *Operating ratio* is the ratio of the maximum operating pressure to

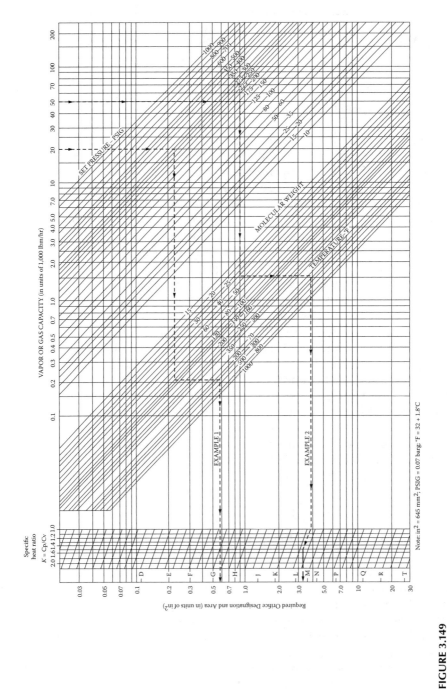

FIGURE 3.149
Graphical method of sizing pressure relief valves for vapor or gas services.

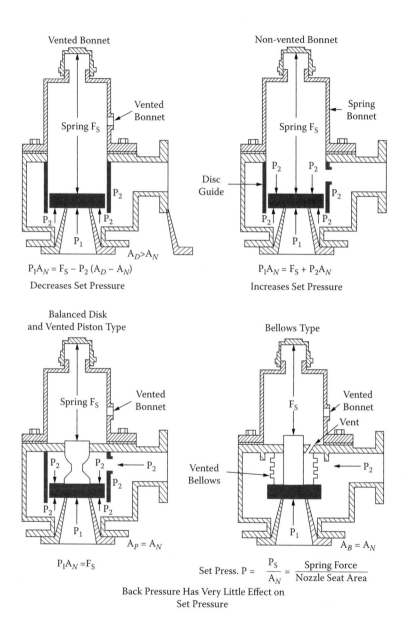

Vented Bonnet

Vented Bonnet

Spring F_S

P_2

P_2

P_2

P_1

$A_D > A_N$

$P_1 A_N = F_S - P_2 (A_D - A_N)$

Decreases Set Pressure

Non-vented Bonnet

Spring Bonnet

Spring F_S

Disc Guide

P_2

P_2

P_2

P_2

P_2

P_1

$P_1 A_N = F_S + P_2 A_N$

Increases Set Pressure

Balanced Disk and Vented Piston Type

Spring F_S

Vented Bonnet

P_2

P_2

P_2

P_2

P_2

P_2

P_1

$A_P = A_N$

$P_1 A_N = F_S$

Bellows Type

F_S

Vented Bonnet

Vent

Vented Bellows

P_2

P_1

$A_B = A_N$

Set Press. $P = \dfrac{P_S}{A_N} = \dfrac{\text{Spring Force}}{\text{Nozzle Seat Area}}$

Back Pressure Has Very Little Effect on Set Pressure

FIGURE 3.150

Effect of back pressure on set pressure (A_N = nozzle area; A_D = disk area; A_P = piston area; A_B = bellows area).

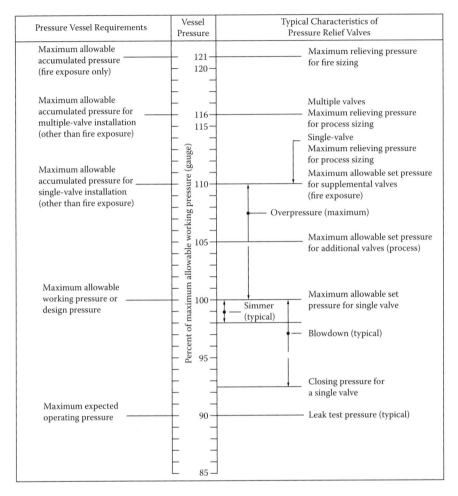

FIGURE 3.151
Definitions of terms used in connection with overpressure relief systems. (Reproduced courtesy of the American Petroleum Institute from API Recommended Practice 520, Sizing, Selection, and Installation of Pressure-Relieving Devices in Refineries, Part I – Sizing and Selection, 7th Edition, January 2000.)

the marked burst pressure expressed as a percentage. The ASME Section VIII Division 1 (ASME Code) is the primary pressure vessel code used in North America. ISO 6718 is used primarily in Europe as well as parts of South America, Asia, and the Middle East. The API RP520 is an industry standard, which goes beyond the ASME Code by providing additional guidance in sizing and applying pressure relief devices.

RDs differ from relief valves, because once burst, they cannot reclose. Therefore, they are only used by themselves as primary relief devices if this characteristic can be tolerated. For a given venting area, they are less expensive than the relief valve. Their main advantage is that they do not leak, as PSVs can. If a tank is protected by both an RD and a PSV (mounted on differ-

TABLE 3.152

Characteristics of Rupture Disk Types

Characteristics	Forward Acting (FA)					Reverse Acting (RA)	
	Prebulged (FAB)	Composite (FAC)	Scored (FAS)	Flat	Graphite	Knife (RAK)	Scored (RAS)
Cost	Low	Medium	High	Medium	Medium	Medium	High
Maximum-operating ratio	70%	80%	90%	50%	80%	90%	90%
Life under cyclic conditions at maximum-operating ratio	High	Low	Medium	Low	Medium	High	High
Fragmenting	Yes	Varies	No	No	Yes	No	No
Vacuum resistant	With support	With support	Yes	With support	With support	Yes	Yes
Suited for low pressure	Yes	Yes	Yes	Yes	Yes	Yes	Yes
Suited for high pressure	Yes	No	Yes	No	No	No	Yes

ent tank nozzles) and the relief valve is the primary-relieving device, the RD is set higher as permitted by the ASME Code. Therefore, the relief valve will open on mild overpressures, relieve a small amount of material, and reclose. The RD will not burst unless a more extreme condition arises.

Often, the RD is mounted under the PSV so that it is sealed tight and protects the relief valve from being contacted by corrosive, plugging, hazardous, freezing, or regulated processes. This way the best characteristics of both devices are utilized. The RD can also be installed after the PSV. This installation can be used when the valve discharges into a vent header that might contain corrosive vapors. Table 3.153 provides data on RD materials, sizes, and minimum rupture pressures.

RDs can be classified in two general categories: forward acting or reverse acting. Forward-acting disks are pressurized on the concave side of the disk such that the material in the dome of the disk is subjected to tensile stresses. Forward-acting RDs can be prebulged, composite, scored, flat, and graphite. Reverse-acting disks are pressurized on the convex side of the disk such that the material in the dome of the disk is subjected to compressive stresses. Reverse-acting types include those that use knife edges to cut the disk membrane and are scored.

3.15 Positioning and Proximity

In a number of renewable energy processes, information on the linear or rotational position of solar collector or wind turbine components is needed to control and optimize the operation of these processes.

The operation of proximity sensors can be based on a wide range of principles, including capacitance, induction, Hall and magnetic effects; variable reluctance, linear variable differential transformer (LVDT), variable resistor; mechanical and electromechanical limit switches; optical, photoelectric, or fiber-optic sensors; laser-based distance, dimension, or thickness sensors; air gap sensors; ultrasonic and displacement transducers. Their detection ranges vary from micrometers to meters, and their applications include the measurement of position, displacement, proximity, or operational limits in controlling moving components of valves and dampers. Either linear or angular position can be measured:

Linear position: Linear motion is typically driven by the use of lead screws, ball screws, or worm drives with ranges from less than a 25 mm to over 6 m. Linear sensors for position feedback in the lower range include LVDTs, magnetic, and optical encoders. For longer strokes, the linear feedback devices include encoders and magnetostrictive position transducers. Optical encoders are practical up to 2 m. Magnetostrictive position transducers can be used up to 20 m.

Angular position: Gearboxes are normally used to transmit rotary motion, but direct drive is also used. The motion, often driven by a

TABLE 3.153
Minimum Rupture Pressures[a] as a Function of Size, Based on 70°F (21°C) Operating Temperature and on Standard Disk Design[b]

Size Inch (mm)	Aluminum	Aluminum (lead lined)	Copper	Copper (lead lined)	Silver	Platinum	Nickel	Monel	Inconel	316 Stainless
¼ (6.2)	310	405	500	650	485	500	950	1085	1550	1600
½ (12.5)	100	160	250	330	250	250	450	530	775	820
1 (25)	55	84	120	175	125	140	230	265	410	435
1½ (38)	40	60	85	120	85	120	150	180	260	280
2 (50)	33	44	50	65	50	65	95	105	150	160
3 (75)	23	31	35	50	35	45	63	74	105	115
4 (100)	15	21	28	40	28	35	51	58	82	90
6 (150)	12	19	25	25	24	26	37	43	61	70
8 (200)	9	17	35	35	27	—	30	34	48	55
10 (250)	7	16	42	42	—	—	47	28	—	45
12 (300)	6	10	55	55	—	—	—	360	—	45
16 (400)	5	8	55	55	—	—	—	270	—	33
20 (500)	3	8	70	70	—	—	—	215	—	27
24 (600)	3	8	60	60	—	—	—	178	—	65
Maximum recommended temperature, °F (°C)	250 (121)	250 (121)	250 (121)	250 (121)	250 (121)	600 (315)	750 (399)	800 (427)	900 (482)	600 (315)

a In psig (14.5 psig = 1.0 barg).
b These data apply to forward-acting prebulged disks only. Reverse-acting disk designs are not limited by the values listed.

stepper or DC motor, can be 0–360° or can be multiturn or continuous rotation. Typical angular feedback sensors have a range of 270–360° rotation and, when using a 360° sensor, the turns can be counted for multiturn applications. A complete control loop for one axis of motion comprises several components including a motor, motor driver, controller, and one or more feedback elements.

Positioning systems can use either an open-loop or a closed-loop control system. In closed-loop motion control, such as the optimized positioning of solar collectors based on measuring their shadows, the positions of both the collector and the shadow are continuously detected. Based on this feedback, the position and velocity of the collector can both be controlled. The reported position is continuously compared to the desired one, and the collector is moved to reduce the error between the two. This is called *servo control* (Figure 3.154).

The complete control system for positioning on an actual control axis includes several levels of control loops, which are acting in a synchronized manner, as shown in Figure 3.155. The intermediate control loops are called *embedded loops*, and together they guarantee that the original command for a specified movement is correctly completed. Any command for a movement, which is executed under closed-loop control, will result in a *motion profile*. This, usually trapezoidal profile, defines the way how the component is intended to be moved.

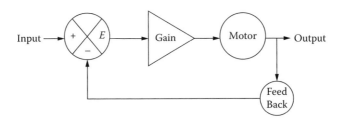

FIGURE 3.154
Basic block diagram of a closed-loop motion control system.

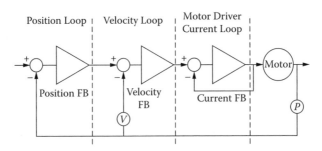

FIGURE 3.155
Motion control system with embedded loops.

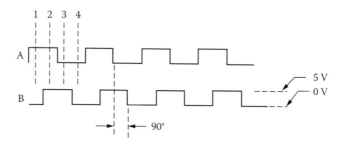

FIGURE 3.156
Incremental encoder pulse trains, showing the output trains A and B, that are separated by 90°.

Position can be detected by optical or other encoders that generate pulses in proportion to the amount of movement detected. Encoders can provide incremental or absolute readings. An incremental encoder outputs a number of pulses corresponding to the amount of motion. There are usually two outputs, called A and B, that are separated in time by 90° to indicate the direction of travel. This is shown in Figure 3.156.

The two outputs being separated by 90° are called *quadrature outputs*. A counter counts up all of the increment and decrement outputs to arrive at the total count that represents the present position. There are four states of the combination of A and B levels per the 360° of one cycle. The phase relationship between outputs A and B during a state change indicates the direction of motion (increment or decrement of the count).

The drawback of an incremental encoder is that the count can be corrupted by power interruption or electrical noise. Then, in order to restart the system, it needs to be homed to re-establish the zero count position. A pictorial representation of an incremental rotary optical encoder is shown in Figure 3.157.

In an optical encoder, a linear pattern or rotor disk is provided with adjacent transparent and opaque sections, allowing pulses of light to reach the detectors as the move progresses. These light pulses produce the electrical output pulses that indicate the linear or angular position change and direction.

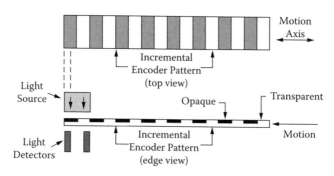

FIGURE 3.157
Incremental rotary optical encoder.

An LVDT is suitable for measuring relatively short travel distances at high accuracy, in the range of microns to several centimeters. It is usually cylindrical in shape and is provided with a core that moves into the center bore of the cylinder. A magnetostrictive position transducer can have as high a resolution as 1 μm and a range from less than 25 mm to over 20 m. Typical error is 0.01% of range.

Positioning systems can use a network communication configuration, where the components operate as nodes on a network. Network communications protocols include ARCnet, CANbus, DeviceNet (a version of CANbus), Ethernet, PROFIBUS, IEEE 1394 (FireWire), IEEE 1451, Interbus-S, SERCOS, and Seriplex, among others. PC bus-type protocols include the normal backplane ISA/EISA (PC-XT/AT) connection for a PC, MAC PCI (Nubus) for MacIntosh computers, Multibus, PC 104, PCI bus, cPCI bus (compact PCI), PCMCIA, VME bus, and VXI.

The best approach to implement a motion control system is to first specify a motor and drive that are capable of meeting the performance requirements. Once they are specified, the remaining parts of the system should be selected to match the motor and drive combination. In addition to the performance of the control axis, networking capability, user interface, and environmental considerations should also be addressed.

3.16 Temperature

In the operation of renewable energy processes, one of the most important measurement variables is temperature. It has to be detected both at very high (for example, in concentrating solar detectors) and very low cryogenic temperature levels (for example, in LH_2-related operations).

The International Practical Temperature Scale is the basis of most present-day temperature measurements. The scale was established by an international commission in 1948 with a text revision in 1960. A third revision of the scale was formally adopted in 1990 and is reproduced in Table 3.158. Reproducible temperature points have been established by physical constants of readily available materials define the scale.

Interpolation between these fixed points is made by platinum resistance thermometers when the temperature is below 1,832°F (1,000°C), and by platinum–platinum and by 10% rhodium thermocouples (TCs) when it is higher. The NIST (formerly the National Bureau of Standards) has the capability for calibrating temperature-measuring devices against these primary temperature points. These devices are secondary standards that are then used by manufacturers and users to calibrate other equipments. NIST's capability for calibrating temperature-measuring devices is illustrated in Figure 3.159. This figure also shows the error (uncertainty). Table 3.160 provides information on the ranges and accuracies of temperture sensors.

TABLE 3.158

Primary Temperature Points Defined by the
International Practical Temperature Scale (IPTS-90)

Equilibrium Point	°K	°C
Triple point of hydrogen	13.81	−259.34
Liquid/vapor phase of hydrogen at 25/76 standard atmosphere	17.042	−256.108
Boiling point of hydrogen	20.28	−252.87
Boiling point of neon	27.102	−246.048
Triple point of oxygen	54.361	−218.789
Boiling point of oxygen	90.188	−182.962
Triple point of water	273.16	0.01
Boiling point of water	373.15	100
Freezing point of zinc	692.73	419.58
Freezing point of silver	1235.08	961.93
Freezing point of gold	1337.58	1064.43

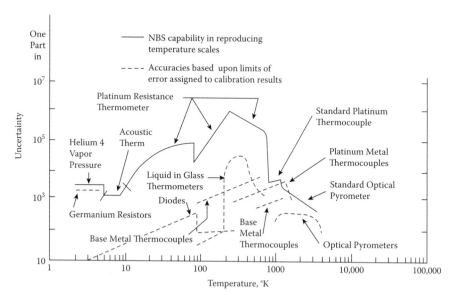

FIGURE 3.159

Uncertainties in calibrating different temperature sensors at various temperatures. (From NBS
Technical Note No. 262.)

TABLE 3.160
Temperature Sensor Accuracy and Range

Sensor Type	Useful Range °F[a]	Maximum Range °F[a]	Accuracy Standard Grade	Accuracy Premium Grade
Thermocouples				
Type J iron vs. constantan	32 to 1,382	−346 to 2,192	±4°F or 0.75%[b]	±2°F or 0.4%
Type K chromal vs. alumel	−238 to 2,282	−454 to 2,502	±4°F or 0.75% >32°F; ±4°F or 2% <32°F[b]	±2°F or 0.4%
Type T copper vs. constantan	−328 to 662	−452 to 752	±2°F or 0.75% >32°F; ±2°F or 1.5% <32°F[b]	±0.9°F or 0.4%
Type E nickel–chromium vs. constantan	−274 to 1,832	−454 to 1,855	±3°F or 0.5% >32°F; ±3°F or 1% <32°F[b]	±1.8°F or 0.4%
Type C tungsten–5% rhenium vs. tungsten–26% rhenium	0 to 4,200	−32 to 4,208	8°F to 767; 1% to 4,200	n/a
Type R platinum–13% rhodium vs. platinum	32 to 2,642	−40 to 3,214	2.7°F or 0.25%[b]	1.1°F or 0.1%
Type S platinum–10% rhodium vs. platinum	32 to 2,642	−40 to 3,214	2.7°F or 0.25%[b]	1.1°F or 0.1%
Type B platinum–30% rhodium vs. platinum–6% rhodium	32 to 3,092	32 to 3,308	0.9°F over 1472°F	n/a
Tungsten vs. tungsten–26% rhenium	32 to 4,200	0 to 4,200	±8°F for 32 to 800°F; ±1% for 800 to 4,200°F	n/a
Tungsten–5% rhenium vs. tungsten 26%–rhenium	32 to 4,200	0 to 4,200	±8°F for 32 to 800°F; ±1% for 800 to 4,200°F	n/a
Tungston–3% rhenium vs. tungsten 25%–rhenium	32 to 4,200	0 to 4,200	±8°F for 32 to 800°F; ±1% for 800 to 4,200°F	n/a

Resistance thermometers

Platinum Class A RTD Alpha = 0.0385	−328 to 1,000	−328 to 1,562	$°F = ±(0.27 + 0.0036^a[t])^c$
Platinum Class B RTD Alpha = 0.0385	−328 to 1,000	−328 to 1,562	$°F = ±(0.54 + 0.009^a[t])^c$
Nickel 672 RTD	−94 to 572		±0.2 to 0.5% FS
Copper 427 RTD	−130 to 500		±0.2 to 0.5% FS
			±0.09°F spans <90°F
			±0.18°F spans < 135°F
Thermistor	−150 to 600		±0.36°F spans < 212°F
Solid state	−67 to 300		±3.6°F for 32 to 250°F
Nonelectric			
Liquid-in-glass thermometers	−40 to 700		±1 scale division
Bimetallic thermometers	−100 to 1,000		±1% FS
Filled-system thermometers	−320 to 1,200		±1% FS
Phase change	100 to 3,000		±1°F

[a] °C = 5/9 (°F − 32) and −°C = 5/9 (°F + 32).
[b] Error is the larger of the two values given.
[c] t = temperature span.

3.16.1 Application and Selection

It is believed that Galileo invented the liquid-in-glass thermometer around 1592. Thomas Seebeck discovered the principle behind the TC—the existence of the thermoelectric current—in 1821. The same year Sir Humphry Davy noted the temperature dependence of metals, but C. H. Meyers did not build the RTD until 1932. Today, some 20 different types of temperature sensors are available, and Table 3.160 lists the temperature ranges and accuracies of a number of them.

The simplest thermometers are *bistate/phase change sensors*. These low-cost nonelectric sensors can be crystalline solids that change from a solid to a liquid at a particular temperature or can be crayons, lacquers, pellets, or labels over a wide range of temperatures from 38 to 1,650°C (100–3,000°F). They offer a very inexpensive method for visual verification within about 1°F.

Liquid-in-glass thermometers used mercury or alcohol as the liquid that expands as it gets warmer. Most countries mandate the removal of any mercury-filled devices due to its extreme toxicity, but alcohol and other fillings are still used. The expansion rate is linear with temperature and can be accurately calibrated. *Bimetallic thermometers* bond two dissimilar metals with different coefficients of expansion to produce the bimetallic element in thermometers, temperature switches, and thermostats. *Filled System Thermometers* can be filled with either liquid or vapor. Liquid-filled units are the most popular although they require compensation for the weight of the liquid head and for capillary length. Vapor-filled elements cannot be used if the operating temperature crosses the vapor/liquid point.

In the category of electronic thermometers, the thermocouples (TCs), resistance temperature detectors (RTDs), thermistors, integrated circuitry (IC), and radiation thermometers will be discussed in separate subsections. The IC and diode detectors will be discussed in connection with cryogenic thermometry. Their characteristics are shown in Figure 3.161.

In the 1980s, Rosemount developed the highway addressable remote transducer (HART) protocol to enable detailed information to be superimposed onto the 4–20-mA signal. In 1979, Modicon introduced the MODBUS protocol that serves to share data among multiple vendor devices and to support temperature multiplexers that can communicate with a host system using MODBUS, MODBUS Plus, or MODBUS TCP/IP protocols. In the 1990s, a trend emerged for more open protocols to enable plug and play of instruments from varying manufacturers to work as part of a Fieldbus structure. Two protocols that have emerged as leaders are Foundation Fieldbus and Profibus. Each has its proponents and support groups that have been promoting their acceptance. There are a variety of products introduced beginning in 2002 that will interface process measurements, including temperature, over high-speed Ethernet links to host systems using object linking and embedding for process control (OPC) servers.

The array of intelligent temperature transmitters on the market seems almost endless. Some common features of the leading models are universal

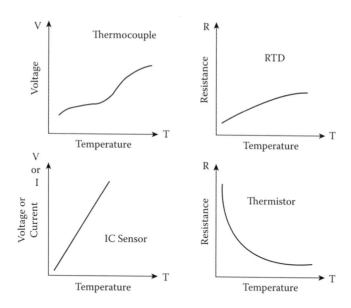

FIGURE 3.161
The characteristics of electronic thermometers.

inputs from any TC, RTD, mV, resistance, or potentiometer; loop powered with 4–20 mA output; digital outputs, and configuration with pushbuttons, PC software, or a handheld configurator. Choices must be made regarding which protocol is required: HART, FOUNDATION Fieldbus, PROFIBUS, vendor proprietary, Ethernet, or just 4–20 mA. Universal transmitters check their own calibration on every measurement cycle, incorporate self-diagnostics, and provide a reconfiguration process that is quick and make the transmitters interchangeable. Some transmitters are capable of handling dual RTD elements. This allows for temperature averaging, temperature difference measurement, or automatic RTD sensor switchover if the primary sensor fails in a redundant installation.

Measuring the surface temperatures of such moving objects as concentrating solar collectors requires special consideration. IR pyrometers offer high accuracy and fast response, although, for stationary surfaces, either a TC or an RTD may be good enough to get continuous information. In some renewable energy processes, the measurement of interest is the average temperature. For a temperature profile in a storage tank that is to be used for volume correction for inventory, an array of high-accuracy RTDs would allow the calculation of the average temperature. This also requires a level measurement so that only those RTDs are averaged which are covered by product.

Noise protection is needed to protect against RFI/EMI problems caused by mobile and stationary radio, television, and handheld walkie-talkies; radio-controlled overhead cranes; radar; induction heating systems; static discharge; high-speed power-switching elements; high-AC current conductors; large solenoids and relays; transformers; AC and DC motors; welders; and even fluorescent lighting. A properly designed temperature transmitter effectively negates the effects of incoming RFI noise by converting a sensor's low-level signal to a high-level analog signal (typically 4–20 mA). This amplified signal is resistant to RFI/EMI and can accurately withstand long-distance transmission from the field, through a noisy plant, back to the control room. When specifying a transmitter, the application engineer should always check for RFI/EMI protection and for input/output/power signal isolation.

3.16.2 Cryogenic Temperature Sensors

In the renewable energy economy, hydrogen will have an important role, and in the processing of LH_2, the measurement and control of cryogenic temperatures is one of the most important tasks.

As was shown in Figure 3.159, cryogenic temperatures can be detected by integrated circuit diodes; types K, T, and E thermocouples (TCs); class A and B resistance temperature detectors (RTDs); acoustic and ultrasonic thermometers; germanium and carbon resistors; and paramagnetic salts. As TCs and RTDs will be discussed in separate subsections, here the focus will be on the other sensors.

Diodes are highly sensitive and linear temperature sensors. Silicon and germanium diode temperature elements are available from −272 to 202°C (−458 to 395°F). They are accurate to 0.2% FS. The main advantages of the diode-type sensors are their high accuracy, particularly at cryogenic temperatures; small size; low cost; and good linearity (Figure 3.162). The diodes are small enough for most applications, but where sensor size is a concern, microdiodes can be used. One disadvantage is that the variations between diodes require calibration, which also increases their costs.

Acoustic time domain reflectometry operates on the principle that the velocity of sound and ultrasound increases in gases and decreases in liq-

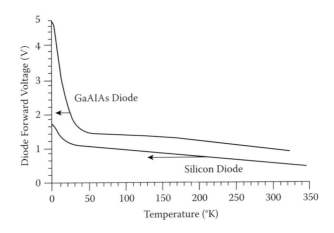

FIGURE 3.162
Characteristics of silicone and gallium–aluminum–arsenide diodes.

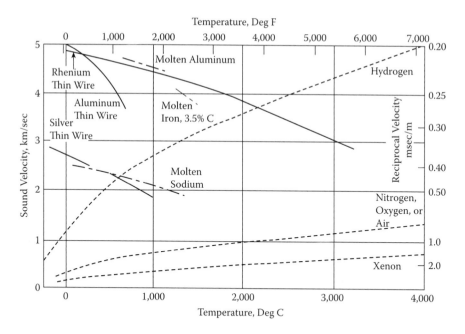

FIGURE 3.163
The velocity of sound increases in gases and decreases in liquids and solids as the temperature rises.

uids and solids as the temperature rises (Figure 3.163). This measurement is made by detecting the time needed for the acoustic pulses to travel from the transducer to the impedance demarcation point (which may be the junction between the wire and wall of the tank) and back to the transducer. These devices can be highly accurate, particularly in the cryogenic and even in the sub-milli-Kelvin range.

Commercially available carbon resistors have been used as temperature sensors in the cryogenic temperature area near absolute zero, from about –253°C to –272°C (–424°F downward to below –458°F). One major benefit of the carbon resistor at low temperature is its lower susceptibility to adverse effects caused by a magnetic field and stray radio interference. They do require individual calibration to keep the measurement error under 1%. Carbon resistors may be incorporated into resistor networks to improve linearity. These sensors exhibit a large increase in resistance below –253°C (–424°F). Reproducibility on the order of 0.2% is obtainable when calibrated individually. Small size, low cost, and general availability make their use attractive in cryogenic work.

Magnetic thermometry has been developed chiefly to measure temperatures near absolute zero (below –458°F, or –272°C). These measurements are obtained by adiabatic demagnetization of a paramagnetic salt. Inductance can be measured with an AC bridge (as shown in Figure 3.164) whose balance is independent of frequency. The relationship between self-inductance and sus-

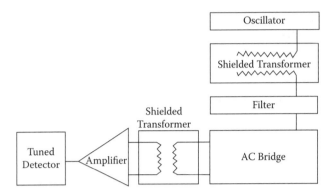

FIGURE 3.164
Circuitry for temperature measurement by paramagnetic salts.

ceptibility of a salt has been found to be linear when the ellipsoidal or spherical salt piece is placed in the homogeneous part of the measuring field.

3.16.3 High-Temperature Sensors

In several renewable energy processes, including the concentrating solar collectors, boilers, and combustion systems, the accurate measurement and control of high temperatures are required. These (over 1,000°C) temperatures are most often detected by thermocouples (types B, C, R, and S) and by optical and IR-radiation pyrometers. These devices are only briefly mentioned here, because they will be discussed in detail later. Here, the emphasis will be on some of the other high-temperature detectors such as sonic and ultrasonic sensors.

One of the most popular high-temperature sensors is the platinum thermocouples, which are usually installed inside protective thermowells or protection tubes. When installed horizontally, wells tend to droop, causing binding of the TC element, making replacement difficult. The latest designs incorporate a sheath with a flexible cable that can easily be inserted into even badly drooping wells. Ceramic wells do not suffer from droop but have other limitations such as low surface strength, brittleness, and low erosion resistance.

IR pyrometers offer a very viable noncontact method to measure temperatures all the way up to 3,600°C (6,500°F) and can be the best choice for most applications. When high temperature is to be detected in hard-to-reach locations, IR radiation pyrometers are combined with the use of optical fibers. These applications will also be discussed later.

Sonic and ultrasonic thermometers have a unique role for high-temperature applications that involve detection of the average temperature in harsh and abrasive process environments. These sensors operate on the basis of detecting the speed of sound, which is proportional to temperature (Figure 3.163). When the temperature is detected by the measurement of transit

TABLE 3.165

Applications for Ultrasonic Thermometers

Measured Object	Temperature (°C)	Sensor or Medium	Sound Conductor	Transducer	Signal
Gas	1,500–15,000	N_2	Quartz	Piezoelectric	One period 1 MHz
Liquid	1,000	Molten Na	Stainless steel	Piezoelectric	One period or period series 3–10 MHz
Solids with holes	3,000	Rewire	Tungsten	Magnetostrictive	Period series 0.1 MHz
Solids without holes	1,500	Steel wire	Steel	Piezoelectric	Period series 1 MHz

time, calibration is not required; because the gas itself is the thermometer element, there are no errors due to leakage, and fast temperature changes can be followed.

Acoustical temperature sensors can theoretically measure temperature from the cryogenic range to plasma levels. Their accuracy can approach that of primary standards. Temperature measurements can be made not only in gases but also in liquids or solids, on the basis of the relationship between the sound velocity and temperature shown in Figure 3.163. The acoustic velocity can be detected by immersing a rod or wire into the fluid or by using the medium itself as an acoustic conductor. The sensor rod can measure the temperature at a point or, by means of a series of constrictions or indents, can profile or average the temperature within the medium.

An acoustic sound source and a sound receiver can be located on the outside of opposing walls of a boiler or furnaces. Ultrasonic thermometry can be used at temperature extremes, in high electrical fields, or when the medium being measured is inaccessible. It is also useful for averaging the temperature of bulk materials or profiling furnace temperatures. Some of the applications are tabulated in Table 3.165.

3.16.4 Infrared and Fiber-Optic Thermometers

Pyrometric thermometers can be of the following types: optical (brightness), ratio (two-color), total (wide-band), and narrow-band, whereas the fiber optic designs can be of the: light-pipe, black body, dual-wavelength, crystal, gap, and fluoroptic variety. Their temperature ranges are from –40 to 4,000°C (–40 to 7,000°F), and their errors range from 0.25% to 2% FS.

It was only a few hundred years ago that physicists abandoned the view that heat is a substance and accepted that it is a form of energy that is transferable from one material to another by conduction, convection, or radiation. Since that time it has been discovered that all materials, at all temperatures, including down to near absolute zero, radiate electromagnetic energy, which

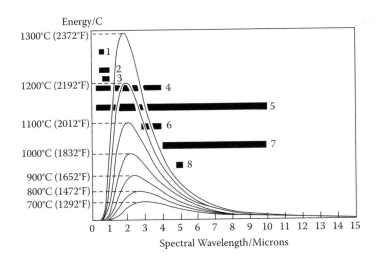

FIGURE 3.166
Pyrometers and their wavelength bands.

travels at the speed of light (186,000 mi/s). It was also learned that the temperature of the material determines both the quantity and the type of energy radiated. As the temperature rises, the wavelength of the radiation drops and its frequency rises.

An IR thermometer is a noncontact radiant energy detector. Every object in the world radiates IR energy. Noncontact thermometers measure the intensity of the radiant energy and produce a signal proportional to the target temperature. The physics behind this broadcasting of energy is called *Planck's Law of Thermal Radiation*. This radiated energy covers a wide spectrum of frequencies, but the IR spectrum (0.3–15 μm) is most commonly used for temperature measurement.

Pyrometers can be portable or permanently installed; the radiated energy can reach them through focusing on lenses or through optical fibers. In addition, they can be categorized according to the number and width of the wavelength bands used. When the full spectrum (0.3–15 μm) is utilized (#5 on Figure 3.166), the device is called a *wideband* or *total radiation pyrometer*. When a single, small segment of the spectrum is used (for example, #8 on Figure 3.166), the design is called a *narrow-band pyrometer*. If that narrow band falls within the visible spectrum (#1 on Figure 3.166), this subgroup of the narrow-band pyrometers is called as *optical, color,* or *brightness pyrometer*. When, instead of measuring the radiation intensity of a single wavelength band, the instrument detects the ratio of the intensities of two wavelengths, the pyrometer is called a *ratio* or *two-color thermometer*.

The amount of thermal radiation leaving an object depends on the temperature and emittance of that object. If the object is a perfect emitter (a blackbody), its emittance is unity. The emissivities of almost all substances are known (Table 3.167), but emissivity is only one component in determining

TABLE 3.167

Total Emissivities of Various Substances

Material	Emissivity	Material	Emissivity
Aluminum		Iron	
Unoxidized	0.06	Oxidized	0.89
Oxidized	0.19	Rusted	0.65
Brass (oxidized)	0.60	Lead (oxidized)	0.63
Calorized copper	0.26	Monel (oxidized)	0.43
Calorized copper		Nickel	
(oxidized)	0.19	Bright	0.12
Calorized steel		Oxidized	0.85
(oxidized)	0.57	Silica brick	0.85
Carbon	0.79	Silver (unoxidized)	0.03
Cast iron		Steel (oxidized)	0.79
Oxidized	0.79	Steel plate (rough)	0.97
Strongly oxidized	0.95	Tungsten (unoxidized)	0.07
Copper (oxidized)	0.60	Wrought iron	
Fire brick	0.75	(dull oxidized)	0.94
Gold	0.03		

the emittance of an object, because other factors, such as shape, oxidation, or surface finish, also affect the actual emissivity.

The uncertainties concerning emittance can be reduced by creating blackbody conditions (target tubes or target holes) or by using short-wavelength or ratio pyrometers. The use of short wavelengths is useful because the signal gain (the amount of energy radiated per unit temperature) is high in this region. This high output response tends to swamp the effects of emittance variations, the absorption effects of steam, dust, or water spray.

Noncontact thermometry has many advantages, particularly in regard to measuring temperatures of objects that are extremely hot or corrosive, moving or fragile, located in strong electromagnetic fields, or experience heat loss by conduction. Until the glass fiber-optic cables became available, noncontact thermometry required a line-of-sight vision between the sensor and the target object. The elimination of this restriction makes it possible for the fiber-optic thermometer to solve many difficult measuring problems. Optical fiber thermometry (OFT) depends on total internal reflection within a thin-fiber element.

Perhaps the most outstanding feature of fiber-optic systems is the ability of the fibers to withstand and function in hostile environments, including intense heat. At extreme high temperatures and under other severe conditions, the fibers can be protected with air or inert gas purging or by water cooling. The available variations in fiber-optic probe sensor designs are shown in Figure 3.168. They fall into four categories: light-pipe, blackbody, dual-wavelength, and gap types. The light-pipe design transmits the radia-

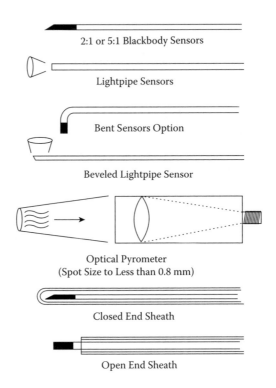

2:1 or 5:1 Blackbody Sensors

Lightpipe Sensors

Bent Sensors Option

Beveled Lightpipe Sensor

Optical Pyrometer
(Spot Size to Less than 0.8 mm)

Closed End Sheath

Open End Sheath

FIGURE 3.168
Configurations of the sensor portion of the OFT probe.

tion from the target to the detector through an open tip. The blackbody-type unit radiates heat from a cup of material, such as a thin coat of precious metal, surrounding the tip.

The advantages of OFT pyrometry include the following:

Small size of the sensor unaffected by electromagnetic radiation fields or by shock and vibration

Wide range

Fast response

Ability to provide temperature profiles or average temperatures

Relatively rugged

Ability to look around opaque objects

Ability to reduce the unit cost through multiplexing

Their disadvantages include the following:

Measurement errors can be caused by variations in the emittance of the target

Aging of radiation source

Background interference

Lack of focusing because the target is too small to fill the view

Radiation absorption by dust, smoke, moisture, atmosphere, or by dirt on the windows

3.16.5 Resistance Temperature Detectors (RTDs)

The temperature ranges detectable by RTDs are listed in Table 3.169. They are available in a variety of materials, including platinum, which is used for high accuracy, and gold/silver, used on cryogenic temperature applications. The measurement errors are according to classes: class A—0.03°C (0.06°F) or 0.01% FS; class B—0.12% FS; total errors: 0.15% of span for platinum, 0.25% of span for nickel.

In 1821, Sir Humphrey Davy discovered that the resistivity of metals depends on their temperature. The application of this property using platinum was first described by Sir William Siemens. When, in renewable energy or any other processes, accurate measurements are needed; RTDs tend to be the first choice.

RTDs are constructed of a resistive material with leads attached and placed into a protective sheath. Platinum resistance thermometers are the international standard for temperature measurements between the triple point of H_2 at 13.81 K (24.86°R) and the freezing point of antimony at 630.75°C (1,167.35°F). The RTD elements include platinum, nickel of various purities, 70% nickel/30% iron (Balco), and copper, listed in order of decreasing temperature range. Their features and relative performance characteristics in comparison with other sensors are tabulated in Table 3.169.

The relationship between the resistance change of an RTD versus temperature is referred to as its alpha curve (see Figure 3.161). Commonly, the RTDs are manufactured with a protective sheath that provides a hermetic seal to protect the sensor from moisture or contamination. The sheathed elements are often installed into a protective well to isolate the sensor from the process. One manufacturer offers a universal model that has only a 1-in.-long sheath but has long leads encased in a spiraled spring. The unit may be cut in the field to fit in any length of thermal well. This greatly reduces the requirements for stocking sensors in varying lengths (Figure 3.170).

To eliminate the errors caused by lead wire resistance changes, it is recommended to use transmitters that accept 4-wire RTD inputs as a standard feature. In order to eliminate errors by self-heating, the measuring current should be minimized. For platinum elements, it should not exceed 250 µA. The best installation practice is to place all electronics directly on top of the thermowell and thereby eliminate lead wire and noise effects. If this cannot be done, the lead wires should be twisted and shielded; the wires should also not be stressed, strained, or made to go through steep gradients. The

TABLE 3.169

Comparison Chart of Various Temperature Sensors

Evaluation Criteria	Platinum RTD 100 Ω Wire Wound and Thin Film	Platinum RTD 1,000 Ω Thin Film	Nickel RTD 1,000 Ω Wire Wound	Balco RTD 2,000 Ω Wire Wound	Thermistor	Thermocouple	Semi-Conductor Devices
	High	Low[a]	Medium	Medium	Low[a]	Low[a]	Low[a]
Temperature range	Wide −400°F to +1200°F (−240°C to +649°C)[a]	Wide −320°F to +1000°F (−196°C to +538°C)[a]	Medium −350°F to +600°F (−212°C to +316°C)	Short −100°F to +400°F (−73°C to +204°C)	Short to medium −100°F to +500°F (−73°C to +260°C)	Very wide −450°F to +4200°F (−268°C to +2316°C)[b]	Short −57°F to +257°F (−49°C to +125°C)
Interchangeability	Excellent[a]	Excellent[a]	Fair	Fair	Poor to fair	Good[a]	Fair
Long-term stability	Good[a]	Good[a]	Fair	Fair	Poor	Poor to fair	Good to fair
Accuracy	High[a]	High[a]	Medium	Low	Medium	Medium	Medium
Repeatability	Excellent[b]	Excellent[b]	Good[a]	Fair	Fair to good	Poor to fair	Good[a]
Sensitivity (output)	Medium	High[a]	High[a]	Very high[a]	Very high[b]	Low	High[a]
Response	Medium	Medium to fast[a]	Medium	Medium	Medium to fast[a]	Medium to fast[a]	Medium to fast[a]
Linearity	Good[a]	Good[a]	Fair	Fair	Poor	Fair	Good[a]
Self-heating	Very low to low[a]	Medium	Medium	Medium	High	N/A	Very low to low[a]
Point (end) sensitive	Fair	Good[a]	Poor	Poor	Good[a]	Excellent[b]	Good[a]
Lead effect	Medium	Low[a]	Low[a]	Low[a]	Very low[b]	High	Low[a]
Physical size/ packaging	Medium to small	Small to large[a]	Large	Large	Small to medium	Small to large[a]	Small to medium

[a] Good rating.
[b] Best rating.

FIGURE 3.170

A flexible WORM sensor in a transmitter system. (Courtesy of Moore Industries-International, Inc.)

typical performance capabilities of different RTD transmitters are summarized in Table 3.171.

Intelligent transmitters are capable of working with eight types of TCs or a number of different types of RTD elements. This increases their flexibility and reduces the need for spare parts. They are also provided with continuous self-diagnostics, with automatic three-point self-calibration, which is performed every 5 seconds and does not interrupt the analog or digital output, remote reconfiguration capability, which can change their zero, span, or

TABLE 3.171

Performance Capabilities of Standard and "Smart" RTD Transmitters

Performance Criteria	Standard		Smart	
	Platinum Element	Nickel Element	Digital Output	Analog Output (4–20 mA DC)
Inaccuracy	±0.15% or ±0.15°F (0.08°C)	±0.25%	±0.035% or ±0.18°F (0.1°C)	±0.05% or ±0.18°F (0.1°C)
Repeatability	±0.05%	±0.05%	±0.015% or ±0.18°F (0.1°C)	±0.025% or ±0.18°C (0.1°C)
Zero shift/6 mo.	±0.1%	±0.2%	±0.06% R or 0.18°F (0.1°C)	±0.1% R or 0.18°F (0.1°C)
Span shift/6 mo.	±0.1%	±0.4%		
Supply voltage variation	±0.2% or 0.02°F (0.01°C)		—	(0.005%)/V
Ambient effect (100°F or 55°C)	±0.75%		Included in the values listed above	±0.1%

Note: When two values are given, the error is the higher of the two. When percentage is given, it refers to percentage of span or percentage of calibrated span, except if %R is shown, which means percentage of actual reading.

many other features without requiring rewiring. The intelligent RTD transmitter can also be furnished with dual-RTD elements that can be used to measure temperature differentials, averages, or high/low sensors, or used as redundant backup elements.

RTDs are among the most accurate, reproducible, stable, and sensitive thermal elements available. Some of the precision platinum RTDs can measure within a few thousandths of a degree. Other advantages include their ability to use conventional copper lead wire (instead of more expensive TC wire), and in the case of copper elements, the TC effect is minimized at their junction.

A concern common to all RTDs is the error produced by self-heating. It can be reduced by improving heat transfer and by minimizing or eliminating (null balance) the current flow through the RTD. Other disadvantages of RTDs include their higher cost, more fragile construction, and larger size, relative to TCs. Because of their size, their thermal response time is also relatively slow. Errors can be introduced if the RTD insulation resistance is affected by moisture being sealed in the sheath or by contact between the element and sheath.

3.16.6 Thermistors

Although the temperature measurement ranges of thermistors are fairly wide, −200 to 300°C (−300 to 600°F), their spans can be rather narrow: 1°C (2°F). These are high-resistance sensors: 5,000–1,000,000 Ω, with 250 Ω/degree sensitivity and about 2% of span error.

It is important to minimize the cost of the optimization of renewable energy processes, and therefore one should always consider the use of such inexpensive temperature sensors as thermistors. Thermistors are thermally sensitive resistors and have either a negative (NTC) or positive (PTC) resistance/temperature coefficient. They have a very large change in resistance per degree change in temperature, allowing very sensitive measurements over narrow spans. Because of their very large resistance, lead wire errors are not significant. They are constructed from ceramic semiconductor materials made from specific mixtures of pure oxides of nickel, manganese, copper, cobalt, tin, uranium, zinc, iron, magnesium, titanium, and other metals. PTCs are manufactured from silicon (silistors), barium, lead, and strontium titanates with the addition of yttrium, manganese, tantalum, and silica (switching PTC thermistors).

These types of bead, disk, and chip thermistors are increasingly used in low-cost temperature probes in the transport, medical, food processing, and more recently in the renewable energy industries. Figure 3.172 illustrates the relative behavior of thermocouples (TCs), RTDs, and NTC thermistors. On the lower part of the figure, one can observe both the sensitivity and the nonlinearity of a particular thermistor.

The most common design is the bead type, which is usually glass-coated. However, thermistors can also be made into washes, disks, or rods. They can also be encapsulated in plastic, cemented, soldered in bolts, encased in glass

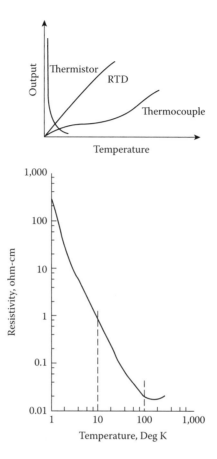

FIGURE 3.172
NTC thermistors are very sensitive and highly nonlinear.

tubes or needles, or a variety of other forms (Figure 3.173). NTCs can be classified in one of two groups: bead type and surface electrode type. PTCs can be divided into thermally sensitive silicon resistors (silistors) and switching PTC thermistors.

For most temperature-measuring applications, self-heating is not a problem, because thermistor currents used are relatively low. The power dissipation constant of a thermistor element can vary from a few microwatts to several watts per degree of resultant temperature rise. Other than temperature measurement, thermistors can be used where physical phenomena produce a temperature change, such as in surge protection and inrush current limiting. PTC applications make general use of either the voltage–current characteristics (resettable fuses, heater and thermostat, liquid and flow sensing) or the current–time characteristics (motor starting, time delaying).

Thermistors can be connected to a microprocessor-based memory element and packaged as a portable element. When the temperature-memory-pack unit is retracted from the process, it can be interrogated by a computer to store the

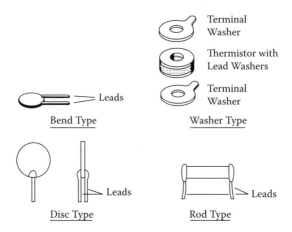

FIGURE 3.173
Variations in thermistor sensor packaging.

temperature history. The micropack weighs less than 100 g, has a battery life of about 500 h, and a temperature range of −40 to 150°C (−40 to 302°F).

Thermistors have the desirable characteristics of small size, narrow spans, fast response (their time constant can be under 1 second), and a very high sensitivity. They do not need a cold-junction compensation, errors due to contact or lead-wire resistance are insignificant, and they are well suited for remote temperature sensing. They are inexpensive, their stability increases with age, and they are the most sensitive differential temperature detectors available.

Their disadvantages are that they are nonlinear, fragile, and not suited for wide spans. Their high resistance necessitates the use of shielded power lines, filters, or DC voltage. One of the most serious limitations of thermistors is their lack of stability (drift and decalibration) at higher temperatures. Thermistors also have a low temperature limit, because as the temperature drops, their resistances rise to such levels that measuring it becomes difficult.

3.16.7 Thermocouples

For data on the available thermocouple spans, ranges, and errors, refer to Table 3.177. For the relative merits of the different TC types, refer to Table 3.178, and for comparison information between TCs and other types of temperature sensors, refer to Table 3.169.

The most often used temperature detectors in renewable energy and most other processes are the thermocouples (TCs). Their operation is based on the principle known as the Seebeck effect. T. J. Seebeck discovered that heating the junction of dissimilar metals generates a small, continuous electromotive force (EMF). The name is a combination of *thermo* and *couple* denoting heat and two junctions, respectively. The dissimilar TC wires are joined at the hot (or measurement) end and also at the cold junction (reference end),

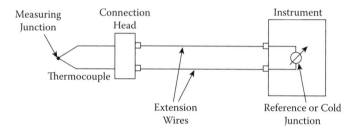

FIGURE 3.174
Thermocouple terminology.

while the open-circuit (Seebeck) voltage is measured (Figure 3.174). This voltage (EMF) is a nonlinear function of the temperature difference between the hot and the cold junctions and of the Seebeck coefficients of the joined two metal wires.

The EMF developed at wire junctions is a manifestation of the Peltier effect. The flow of current causes the liberation or absorption of heat at every junction. The resultant heating or cooling depends on the direction of current flow. Applications of this principle are becoming increasingly useful in electric heating and refrigeration. A second EMF develops along the temperature gradient of a single homogeneous wire. This is the Thomson effect. The circuit EMF depends only on the metals employed and the temperature of their junction; hence, the circuit EMFs are independent of both the length and diameter of wires. Holding temperatures constant at all junctions except the measuring one results in a thermometer. TCs drift because the junction of the two dissimilar metals degrade.

The unknown temperature is determined as ([voltage measured]/[Seebeck coefficient – reference temperature]). The measured temperature is obtained by converting the voltage to temperature using graphs or, for more accuracy, TC tables. Unfortunately, the voltage-to-temperature relationship is not linear, because the Seebeck coefficient changes with temperature (Figure 3.175). Thanks to the microprocessors, what used to be a tedious and time-consuming conversion is no longer a problem.

Thermocouples are relatively inexpensive. They can be of rugged construction and can cover a wide temperature range, from 262 to 2,760°C (440–5,000°F). Approximate errors of a type J thermocouple are described in Figure 3.176. For spans, ranges, and errors of the various TC designs, refer to Tables 3.177 and 3.178. Most temperature transmitter, indicator, controller, or data logger will accept TC inputs. However, TCs produce a very small microvolt output per degree change in temperature, and they are very sensitive to electromagnetic interference (EMI) from motors, especially radio frequency interference (RFI) from walkie-talkies. Installing a top-quality transmitter in the connection head of the TC will minimize many of the problems and concerns described earlier. A top-quality instrument will offer a common-mode noise rejection of 100 db, normal-mode rejection of about 70 db, and RFI immunity of 10–30 V/m.

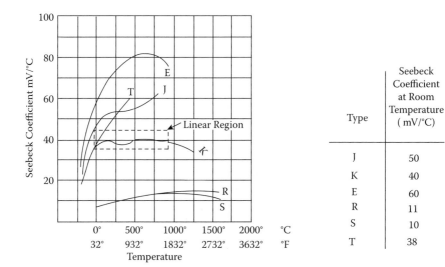

FIGURE 3.175
The Seebeck coefficient gives the amount of voltage generated (in microvolts) by a 1°C change in temperature. The value of the Seeback coefficient varies not only with the thermocouple type but also with temperature.

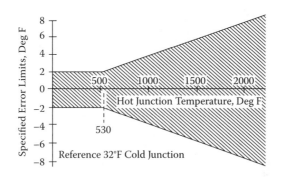

FIGURE 3.176
The error limits of an iron–constantan thermocouple manufactured to meet the "special" limits of ISA.

TABLE 3.177

Thermocouple Errors and Spans

TC Type	Measured Temperature Range in °F[a]	TC Wire Errors For Wires of Different Qualities[a]		Transmitter Error Is Additional and in the Case of "Smart" Transmitters Is ±0.05% of Span or Value Given Below, Whichever Is Larger[a]	Recommended Span Limits[a]	
		Standard	Special		Minimum	Maximum
B	32–3,380	NA	NA	±1.89°F	63°F	2,020°F
E	32–600	±3°F	—	±0.81°F	45°F	2,100°F
	600–1,600	±0.5%	—			
J	32–530	±4°F	±2°F	±0.81°F	45°F	2,500°F
	530–1,400	0.75%	±0.375%			
K	32–530	±4°F	±2°F	±0.81°F	45°F	2,750°F
	530–2,300	±0.75%	±0.375%			
R	32–1,000	±5°F	±2.5°F	±1.53°F	360°F	2,950°F
	1,000–2,700	±0.5%	±0.25%			
S	32–1,000	±5°F	±2.5°F	±1.53°F	360°F	2,900°F
	1,000–2,700	±0.5%	±0.25%			
T	−300 to −75	—	±1%	±0.81°F	45°F	1,025°F
	−150 to −75	±2%	±1%			
	−75 to 200	±1.5°F	±0.75°F			
	200–700	±0.75%	±0.375%			
N	32–530	±4°F	±2°F	NA	NA	NA
	530–2,300	±0.75%	±0.4%			

[a] $°C = \dfrac{°F - 32}{1.8}$

TABLE 3.178
Thermocouple Comparison Table

ISA-Type Designation	Positive Wire (Numbers = Percentages)	Negative Wire	Millivolts per °F	Recommended Range Limits Temperature (°F^a)		Scale Linearity	Atmosphere Environment Recommended	Favorable Points	Less Favorable Points
				Min	Max				
B	Pt70-RH30	Pt94-Rh6	0.0003–0.006	32	3,380	Same as for type R couple	Inert or slow oxidizing	—	—
E	Chromel	Constantan	0.015–0.042	–300	1,800	Good	Oxidizing	Highest EMF/°F	Larger drift than other base metal couples
J	Iron	Constantan	0.014–0.035	32	1,500	Good; nearly linear from 300 to 800°F	Reducing	Most economical	Becomes brittle below 32°F
K	Chromel	Alumel	0.009–0.024	–300	2,300	Good; most linear of all TCs	Oxidizing	Most linear	More expensive than T or J
R	Pt87-Rh13	Platinum	0.003–0.008	32	3,000	Good at high temperatures poor below 1,000°F	Oxidizing	Small size, fast response	More expensive than type K
S	Pt90-Rh10	Platinum	0.003–0.007	32	3,200	Same as R	Oxidizing	Same as R	More expensive than type K

T	Copper	Constantan	0.008–0.035	–300	750	Good but crowded at low end	Oxidizing or reducing	Good resistance to corrosion from moisture	Limited temperature
Y	Iron	Constantan	0.022–0.033	–200	1,800	About same as type J	Reducing	—	Not industrial standard
—	Tungsten	W74-Re26	0.001–0.012	0	4,200	Same as R	Inert or vacuum	High temperature	Brittle, hard to handle, expensive
—	W94-Re6	W74-Re26	0.001–0.010	0	4,200	Same as R	Inert or vacuum	Same as previous	Slightly less brittle than previous
—	Copper	Gold–cobalt	0.0005–0.025	–450	0	Reasonable above 60°K	—	Good output at very low temperature	Expensive laboratory-type TC
—	Ir40-Rh60	Iridium	0.001–0.004	0	3,800	Same as R	Inert	—	Brittle, expensive

^{a} $°C = \dfrac{°F - 32}{1.8}$

The features, advantages, and disadvantages of the various types of thermocouples are tabulated in Table 3.178. For most process applications, the TC is manufactured with a protective outer sheath that uses an insulating material to electrically separate the TC from the sheath and provide mechanical and environmental protection. In some cases, the TC junction is placed in direct contact with the tip of the sheath to increase speed of response. Recommended practice is to always use an instrument with full isolation. The minimum span of even the best TC transmitters is limited to about 35°C (60°F).

3.17 Weather Stations and Solar Detectors

Because of the concern about climate change and global warming, it is important to have reliable data on the trends in various areas of the planet. Weather stations monitor the following variables over the noted typical ranges (for the errors in these measurements, refer to Table 3.179):

Barometric pressure: 95–108 kPa (28–32 in. Hg)
Dew point: −40 to 50°C (−40 to 120°F)
Precipitation: each tip of bucket is 0.25 mm (0.01 in.)
Relative humidity: 0–100% RH
Solar radiation: 0–10 $kWh/m^{-2} d^{-1}$
Other units: 0–100 $(mV/cal)/(cm^2/min)$
Langleys: $ly = g\text{-}cal/cm^{-2} min^{-1} = 221$ $Btu/h^{-1} ft^{-2} = 70$ $mW/cm^2 = 4.18 J/cm^2$
Temperature: −34 to 50°C (−30 to 120°F)
Wind direction: 0–360°
Wind speed: 0.2–56 m/s (0.5–125 mph)
Wind chill: −87 to 28°C (−125 to 85°F)

The sources of renewable energy are natural processes, and weather plays an important role in nature. Because the operation of complex weather stations and weather-modeling software packages is beyond the scope of this book, the systems described here are often used on solar, ocean, and wind-farm-type power plants.

C.S.M. Pouillet of France (1830) and J. Ericsson of the United States (1870) devised pyrheliometers that measured the energy of the direct rays of the sun by the rate of temperature rise of blackened metal masses. C.G. Abbott of the Smithsonian Institution perfected this technique with his silver-disk pyrheliometers, which are still in use as secondary standards. The inventor of the primary standard radiometer used today was K.J. Angström of Sweden. His instrument, invented in 1899, uses two blackened strips of manganin, each of which can be heated either electrically or by the rays of the sun. The measurement is made by exposing one strip to the sun, and then measuring the current required to heat the adjacent strip to exactly the same

TABLE 3.179

The Sensors Used at the Florida Solar Energy Center Field Test Site and Their Inaccuracies[a]

Parameter	Sensor	Inaccuracy
Wind speed	Three-cup anemometer photon-coupled chopper 0.5 mph threshold	±0.15 mph or 1%
Wind direction	Lightweight vane low torque potentiometer 0.5 mph threshold	0.5% linearity
Dry-bulb temperature	Platinum resistance thermometer	±0.1°C
Dew point temperature	Chilled mirror system with platinum resistance thermometer	±0.3°C
Barometric pressure	Piezoresistive element	0.08% of reading
Precipitation	Tipping bucket 0.01 in. resolution	0.5% at 0.5 in./h

[a] Courtesy of FSEC.

temperature. By reversing the operation of the strips several times, an average value of the radiation intensity within 1% of the absolute value can be obtained. The results may be expressed in Langleys or other units.

Although each station is different, they all collect solar radiation–related data (horizontal solar radiation, total radiation, direct normal radiation, total horizontal infrared sky radiation). Table 3.179 provides a summary of the data collected.

The intensity of solar radiation can be measured by (1) absorbing the incoming radiation and determining the rate of heat absorption, and (2) making use of photovoltaic or photoresistive transducers. One of the solar radiometers in use today employs a multijunction thermopile to detect the difference in temperature between whitened and blackened segments of divided or concentric circles, which are generally protected by hemispherical glass domes. Because the response of the thermopile instruments is independent of wavelength, they can be used with filters to determine radiation intensity in selected portions of the solar or long-wave spectrums (Figure 3.180).

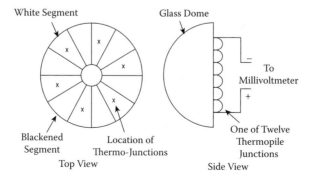

FIGURE 3.180
Thermopile-type pyranometer.

Photovoltaic cells are solar radiometers using silicon photovoltaic cells that produce a short-circuit current that is directly in proportion with the intensity of the solar radiation falling on them. Thermal radiometers that are used to measure solar radiation are called *pyranometers*. They measure the shortwave radiation coming directly from the sun and diffusely from all parts of the sky. They consist of a blackened disk containing temperature-measuring sensors. When exposed to the sun, the disk heats up and glass domes prevent cooling by the wind. The temperature of the disk is a function of the amount of solar radiation hitting it.

Pyrheliometer is an instrument that measures only the direct radiation from the solar disk itself, without bouncing off clouds or the atmosphere. Concentrating solar collectors utilize only this part of the total solar radiation. Its measurement gives an indication of the clearness of the sky. The normal incident pyrheliometer (NIP) measures this form of radiation. A tracker, called the *equatorial mount*, is used to keep the NIP pointed at the sun. The difference between the NIP and the total pyranometer readings is referred to as the diffuse solar radiation.

Wind speed- and direction-measuring anemometers have been described in Section 3.3. Rain gauges measure the rainfall by a tipping bucket. Rain is collected in a funnel and alternately fills two small cups (buckets). These two cups are placed on a pivot, and at each tip, the full cup is emptied and the empty cup placed under the funnel. Each 0.25 mm (0.01 in.) of rain causes the alternate fill and tip, and a magnet causes a momentary switch closure with each tip. Rain gauges should be installed on a level plot of ground near windbreaks (fences, bushes, etc.), which reduce the prevailing wind speed in the vicinity of the gauge.

Barometric pressure is detected by bellows against a reference, which is provided by a sealed-in vacuum. Exposing the instrument to direct sunlight, radiant heaters, or to direct drafts, such as from open windows or doors, should be avoided. Air temperature is measured with a platinum resistance thermometer. It is placed in a long tube that has been painted white on the outside to reflect solar radiation. A small blower pulls ambient air into the tube and across the thermometer.

Dew point is detected by either lithium chloride or, if high precision is required, by the cooled mirror-type sensors, discussed in Section 3.2.13. For relative humidity measurements, the elongation and contraction of hygroscopic elements are used.

A small weather station is illustrated in Figure 3.181. Such stations are usually available in both the cabled and the wireless forms to provide all the information that is needed to control and optimize the various alternative energy processes. These units are usually preprogrammed to provide simple meteorological data, and to perform the complex calculation of ETo (evapotranspiration). The microprocessor-based systems automatically measure the variables and store the data before transmitting it to a user-supplied remote PC over distances of up to about 10 mi. The units are usually provided with a rechargeable battery pack and a spread spectrum radio transceiver.

FIGURE 3.181
WeatherHawk 916 Wireless Weather Station providing measurements of air temperature, wind speed, solar radiation, relative humidity, wind direction, barometric pressure, and rainfall, storing up to 22 days of data and daily ETo (evapotranspiration) information. (Courtesy of WeatherHawk.)

4

The Design of the World's First Full-Size Solar–Hydrogen Demonstration Power Plant

4.1 Introduction

It is debatable how much climate change we can live with. It is debatable whether we are close to the tipping point that leads to catastrophe or not. It is also debatable how much of our economic resources should be devoted to stabilizing or reversing humankind's growing carbon footprint. What is not debatable is that the time for debating is over. We must start dealing with facts. The time for action has arrived. If we care about the future of our children and grandchildren, we must stop debating and start building.

As was shown in Chapter 1, Section 1.7.1, 1% to 2% of the gross world product (GWP) would be sufficient to completely convert our energy economy to renewable energy sources (such as solar–hydrogen) by the end of the century. What is proposed in this book is to build the first such demonstration plant and thereby prove its feasibility, determine its performance, and evaluate its initial and generating costs.

This chapter describes the design details of such a power plant. The proposed solar–hydrogen demonstration plant is sized to generate electricity at a yearly average rate of 1 gW, which, at peak insolation on a summer day, will result in about 5 gW. In contrast, a 1 gW nuclear or fossil power plant cannot generate at a rate more than 1 gW, and on a yearly average generates only at a rate of 0.6–0.8 gW. Because the solar–hydrogen plant generates five to six times that during peak demand periods, and because electricity is more expensive during these peak periods, the plant's profitability greatly exceeds that of the fossil power plants even if one disregards the "cap-and-trade" carbon emission income of renewable energy plants.

4.2 Solar–Hydrogen Demonstration Power Plant

Figure 4.1 (gatefold) describes a 1 gW power plant that can be located in either the equatorial, subtropical, or temperate zones. This power plant, which generates an *average* of 1 gW power during the year, will generate about 5 gW at peak insolation, and the energy not sent to the grid in the form of electricity can be put into storage in the form of hot oil or hydrogen (H_2).

In order to determine the solar plant area requirement for the above described power plant if it is located in southern California, one can assume that Figures 1.28 and 1.29 accurately give the average yearly insolation as about 2,600 kWh/m²/year. If the efficiency of the solar collectors is 20%, the collectors could be arranged on a square with 2.5 mi sides (6.5 mi² or 16.6 km²). Naturally, if the plant is located in the equatorial zone (such as the Sahara or on a floating island), the area requirement would be less. For example, if the insolation in the selected area is 3,000 kWh/m²/yr and the solar plant efficiency is 30%, the radius of a circular plant would be about 1 mile, while in the temperate zone, with 1,500 kWh/m²/yr insolation and a 15% efficient system the required radius would be about 2.0 miles.

If we assume that there is no electric grid in the area and therefore all the solar energy will have to be converted, stored, and transported in the form of chemical energy (hydrogen), we can determine the yearly average hydrogen production of the plant by considering the energy efficiencies of the electrolyzers, compressors, liquefiers, and related equipment. Table 1.46 estimated these efficiencies and also gave the total energy cost of generating a kilogram of liquid hydrogen (LH_2) as 66 kWh². If that estimate is correct, the average LH_2 production rate would be about 15 tons/hour. This being a yearly average, the peak production would be 4–5 times higher during periods of maximum insolation and naturally would drop to zero at night.

The main features of this demonstration power plant are the following: (1) integration of several technologies that include not only solar and hydrogen, but also geothermal, methanol, and dual-function regenerative fuel cell technologies, (2) provision of multiple forms of energy storage through the capability of selling the excess generation to the grid, or storing it in hot oil, H_2, or methanol, and (3) full integration and optimization of the components of the plant to maximize profitability of the operation, while responding to variations in the electricity or hydrogen cost and other market conditions.

4.2.1 The Process

In this design, sun-tracking and concentrating thermal solar collectors are assumed to be used (Chapter 1, Section 1.4.4.1), which generate 400°C (752°F) hot oil. It should be noted that all equipment for this demonstration plant (including the solar collectors) will be purchased on the basis of competitive bidding, and therefore, it is possible that at the time of issuing the purchase orders, the type of equipment in the quotation of the successful bidder will not be the type shown in Figure 4.1 (gatefold). (See, for example, Section 4.2.4.5.)

Assuming that thermal solar collectors are installed, they will be operated by sending the hot oil either to a boiler to generate steam or to storage for use at night or during periods of low insolation. The steam generated by the boiler will drive the steam turbine generators, which produce electricity. If an electric grid exists in the area, and if that is the most profitable option, the generated electricity is sent to the grid during periods when electric energy is most valuable. During off-peak periods, the energy can be stored either as heat (hot oil) or as chemical (H_2 or methanol) energy. The generated H_2 can either be sold as transportation fuel or converted back into electricity during the next peak period, whichever is more profitable.

The heat input into the boilers can be supplemented by geothermal heat. Depending on the temperature of the groundwater in the selected location, the geothermal heat is either used directly or as preheat for the boiler feedwater. If substantial geothermal potential exists in the area, the integration of the solar and geothermal heat sources will increase the continuous energy availability and reduce the need for storage.

When it is more profitable to use the generated electricity to make H_2, the electricity is sent to electrolyzers, which convert it to chemical energy by splitting water into H_2 and O_2. The H_2 will be collected at about 3 bar (45 psig) pressure, and will be either liquefied and sent to storage or compressed to up to 1,000 bar (15,000 psig) and sent to high-pressure gas storage.

When, during peak periods, electricity is more valuable than H_2 or when solar energy is unavailable (such as at night), and it is profitable to reconvert the stored hydrogen back into electricity, this will be done in either conventional fuel cells or if available by that time in my reversible fuel cells (RFCs) that were operating as electrolyzers when hydrogen was generated and now will be switched into their fuel cell mode of operation so that using the stored hydrogen as a fuel they will generate electricity for the grid. Today, my RFCs do not exist (except as my design concept, described in Figure 4.2), but by the time this plant is built, they could be available in the required sizes.

If there is a fossil power plant in the area, or if carbon dioxide (CO_2) is available from other sources, the generated H_2 can also be used to produce methanol from CO_2. Assuming that the CO_2-capturing and methanol conversion technologies have matured by the time of building this demonstration plant, the plant's flexibility will further increase because income can be generated by both the recapturing of CO_2 and also from the sale of methanol. An added benefit to this dual income is that storage and transportation of methanol is less expensive than that of H_2.

4.2.2 Unique Design Features and Inventions

Although the basic technologies utilized in this demonstration plant (such as solar, geothermal, H_2, and fuel cell technologies) are well established, the described solar–hydrogen power plant does include some unique and original ideas and new design features.

Probably the most important feature of this plant is that it is not only fully automated, but also that this automation includes profit-based plantwide optimization. This is different from the operation of traditional power plants, which basically have only one operating mode and only adjust their rate of production to meet the load. In the control system schematically shown in Figure 4.1 (gatefold), the material and heat balances are only constraints, and the key optimized variable is profitability. Depending on the cost and availability of the various forms of energy, the plant operation will be automatically reconfigured to continuously maximize profitability. This optimization is done by considering the amounts and momentary values of the various forms of energy that are being produced or are in storage, and the availability of markets for them.

The optimization of the solar farm operation is assisted by the automatic switching between five equipment configurations. These configurations are a function of the supply and demand for solar heat energy:

1. The solar heat collected on the farm matches the demand and is sent directly to the boiler.
2. The heat demand exceeds the available solar heat from the farm, and therefore it is supplemented by hot oil from storage.
3. There is no demand for solar heat, and therefore all of it (the hot oil) is sent to storage.
4. No solar heat is collected on the solar farm, and therefore the heat demand of the boiler is met from hot oil storage.
5. Solar heat from the farm exceeds the demand of the boiler, and therefore the excess is sent to storage.

In addition to optimized supply–demand matching, the individual solar collectors are optimized to track the sun by my "shadow-minimizing" algorithm. In addition, my "herding" algorithm is also applied. This algorithm looks at the total performance of the solar farm and corrects their orientation (when needed) or performs maintenance on the collectors that are the furthest from the average orientation. (I gave the name *herding*, to this control strategy because the herding dog also goes after one sheep at a time, the one that is furthest away from the desired direction or speed of the herd.)

On the H_2 side of the plant, the main inventions involve the fuel cells. They include the idea of using the "dual-function" electrolyzer and fuel cell combination units, which I call reversible fuel cells (RFCs) (Figure 4.2). This way, in the electrolyzer mode, the RFCs convert electricity into H_2, and in the fuel cell mode, they generate electricity from the H_2 in storage.

Another important, fuel cell–related invention is to reduce the differential pressure (d/p) between the cathode and anode chambers and thereby reduce not only the thickness of the separation diaphragm, but also the overall size, weight, and cost of the whole RFC unit. This goal is achieved by replacing the presently used pressure difference controllers with much more sensitive ones.

If there is a source of CO_2 near the solar power plant, the plant will also include a methanol conversion subsection, which will provide further flexibility and can increase the profitability of the overall operation.

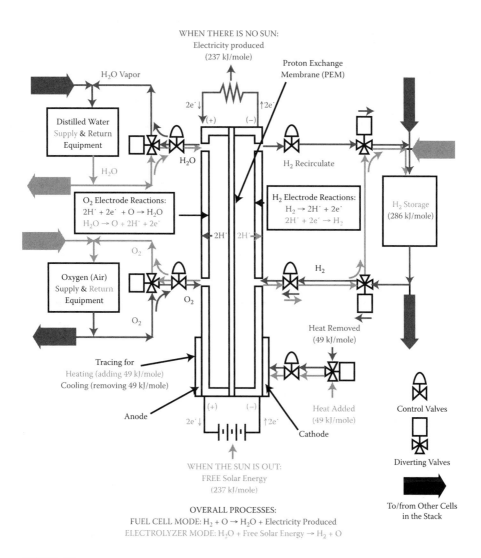

FIGURE 4.2

(See color insert following page 140.) My reversible fuel cell (RFC) design will make the intermittently available solar energy continuously available by storing the excess solar energy when the sun is out and using this stored energy to make electricity when it is not. Compared to the cost of todayís fuel cells, mass production and optimizing control algorithms (not shown) will probably lower the cost of RFCs by an order of magnitude. Small RFCs are most likely to be used in ìzero energyî homes, if electric storage is not available and the electric cars are to be charged at night. Once large RFCs are also available, they can be used in full size power plants.

4.2.3 Profitability Optimization

The operating mode of the plant is selected to always maximize profitability. The optimizing model compares the net profits under the possible modes of operation and selects the most profitable mode. In each possible mode, the actual costs of operation, market conditions, rate of energy generation, and size of energy storage are considered. The main inputs used in this optimized operating mode selection model are the following:

1. Market value of the various possible products (electricity, LH_2, hydrogen gas [GH_2], and methanol), in kWh equivalent or $ units (M1, M2, etc.)
2. Production costs of each of the products in kWh equivalent or $ units (P1, P2, etc.)
3. Energy collection rates (solar, geothermal, and fuel cell) in kW or $ equivalent units (ER1, ER2, etc.)
4. Quantities of the various forms of energy (hot oil, GH_2, and LH_2) in storage (ES1, ES2, etc.)
5. Conversion efficiencies (boiler, turbine generator, electrolyzer, liquefier, compressors, etc.) and coefficients of performance (heat pumps) of the various blocks of equipment (EFF1, EFF2, etc.)
6. Other factors such as equipment availability or capacity limits (F1, F2, etc.)

The foregoing inputs will be continuously updated by the model, and the recommended optimum modes and associated data will be continuously displayed for the operators' approval, prior to implementation. The model inputs are listed as follows:

Market values in units of ¢/kWh

M1—Electricity to the grid (during both peak and std. periods)

M2—LH_2 to market

M3—High-pressure GH_2 to market

M4—Methanol from converter using GH_2, including CO_2 emission income

Costs of production in units of ¢/kWh

P1—Electricity

P2—LH_2

P3—High-pressure GH_2

P4—Methanol, including CO_2 income

Rates at which energy is being collected or generated (kW)

ER1—Insolation [F1(T2 − T1)]

ER2—Geothermal [F7(T11 − T12)]

ER3—Steam (F11, P1, T15)

ER4—Electricity (gW)

ER5—LH$_2$ (F26)

ER6—GH$_2$ (F28)

ER7—Methanol (F8)

Quantities of energy in storage (kWh)

ES1—Total heat content of hot oil (L1 × T6 + L2 × T7)

ES2—Total energy content of LH$_2$ (L5)

ES3—Total energy content of GH$_2$ (V × P18, etc.)

ES4—Total energy content of methanol (L6)

Equipment block efficiencies at present loads (%)

EFF1—Solar farm (insolation) ([F1(T2 − T1)])

EFF2—Boiler [(F11@P1 and T15)/[F6(T8 − T9) + F9(T11 − T12)]]

EFF3—Steam turbine generator [gW/(F11@P1 and T15 + CT)]

EFF4—Electrolyzers (F4/gW)

EFF5—H$_2$ liquefiers ([(FCP + RCP) − ETGs]/F26)

EFF6—H$_2$ compressors [(CP^1 + CP^2)/F28]

EFF7—Geothermal ([F9(T11 − T12)]/(CP1 − ETG1))

Other Factors

F1—Maximum capacity of equipment reached or not (all equipment)

F2—Operating efficiency below maximum or not (all equipment)

F3—Market restricted or not (for electricity, H$_2$, methanol, etc.)

F4—Market unlimited or not (for electricity, H$_2$, methanol, etc.)

F5—Peak period approaching, or near its end

The total plant profitability model is configured in an ANN (artificial neural network) format, which is self-correcting and can be trained on the actual performance data of the plant based on past operation.

4.2.4 Optimizing Control Systems

The dual goals of optimization are to implement the most profitable operating mode based on the economic model and to keep the operating equipment at their maximum efficiency of operation. These suboptimization strategies are briefly described in the following text, referring to Figure 4.1 (gatefold). The abbreviations used are explained at the end of this chapter.

4.2.4.1 Solar Farm Optimization

The solar farm shown in Figure 4.1 (gatefold) consists of parabolic thermal collectors (discussed in detail in Chapter 1, Section 1.4.4.1), which concentrate the sun rays onto pipes located at their focal points in which oil is flowing. (During the bidding phase of the project, quotations will also be obtained on other designs, such as Stirling nanosolar and other collectors, and the design of the successful bidder will be selected.)

If the successful bidder is a PV supplier, some of the DC electricity will be directly used to generate hydrogen by electrolysis. This would not only save the cost of DC-to-AC conversion, but also would lower the cost of energy transportation because it costs less to transport LH_2 than electricity.

The key requirements to obtain maximum efficiency include accurate tracking of the sun and focusing the solar radiation onto the oil pipe, as well as keeping the mirror surfaces clean. The proposed solar tracking controls (STCs) will be provided for each collector unit, and depending on the collector design, will adjust the collector position around one or two axes. It is expected that as the size of the orders increase, the cost of the optimizing controls and sensors will drop until it approaches the unit costs of automobile automation components. Components such as machine vision sensors have been discussed in Section 3.12, position detectors in Section 3.13, and positioning devices in Section 3.15 of Chapter 3.

The methods for finding the optimum collector positions for the individual collector will use the author's minimum shadow area strategy. At the higher level of overall solar farm optimization, the herding optimizer algorithms will also be used to identify individual units that might be defective or to correct their position, starting with the collector that is furthest from the optimum.

The hot oil circulating system is controlled by a temperature rise controller (ΔTC-1), which maintains the discharge temperature of the circulated hot oil from each row of collectors at about 400°C (752°F) by modulating the speed of the variable-speed pumping station (VP-1 and VP-2). This control configuration is used in all modes of operation in which the solar farm is active. In Mode D (nighttime mode), when no solar energy is being collected and the plant operates using the stored hot oil, the controls are switched from ΔTC to constant pump discharge pressure control.

The minimum size of the hot oil storage tanks (T1 and T2) is based on the requirement to provide energy for one night, though preferably it should be for several days of unfavorable weather conditions. The hot oil system is a closed one, and an expansion tank is provided (not shown) to accommodate the thermal expansion as the oil temperature changes. Until a storage tank is full of hot oil, the temperature rises with the level in the tank and, therefore, the cold oil is taken from and returned to the bottom of the active tank, while the hot oil is taken from and returned to the top of the tank.

The diverting 3-way valves (DVs 2, 3, 4, and 5) are used to select the tank that is supplying the hot oil to the boiler (modes B and D) or the tank that is being recharged with hot oil from the solar farm (modes C and E). The plant-monitoring software continuously monitors the heat storage capacity of the oil tanks and selects the operating modes accordingly.

Each row of collectors is provided with a balancing valve (BV-1), which is throttled by its temperature rise (or discharge temperature) controller (ΔTC-

1), thereby adjusting the residence time of the flowing oil to keep the exit temperature on set point. In order to achieve the balancing of the rows while minimizing pumping costs, the most open balancing valve is kept nearly fully open by the valve position controller (VPC-1).

VPC-1 measures the opening of the most open balancing valve (selected by the high-signal selector) and compares that opening with its set point of about 90%. When even the most open valve is less than 90% open, the VPC reduces the set point of the pump discharge pressure controller (PC-2) which in turn lowers the speed of the pumps. As the pump discharge pressure (the inlet pressure to the balancing valves) drops, all the valves open up until the most open one reaches 90% opening. This control configuration is implemented in all modes of operation when the solar farm is active (modes A, B, C, and E).

In the rectangle below the pump symbols in Figure 4.1 (gatefold) 2.17.2 refers to the section in which the applicable optimizing controls have been described in full detail. Depending on the number of pumps in the station and on their mix of constant and variable-speed pumps, the details of the recommended control systems will change (see Figures 2.123 and 2.124). Note that this is the case with all pumping stations used (not only VP-1) in the demonstration plant.

In the "night mode" of operation (Mode D), the set point of the pump discharge pressure controller is switched to the constant pressure that is required to return the oil from the boiler back into storage.

4.2.4.2 Boiler and Heat Storage Optimization

The balance between the steam energy supplied and the electric power generated is indicated by the shaft speed of the turbine generator. In the case of a conventional steam boiler, this speed would be maintained constant by adjusting the firing rate of a fossil-fueled boiler. Here, in the case of the solar boiler, the speed controller (SC-3) adjusts the set point of the superheated steam pressure controller (PC-1), which pressure keeps the pressure at around 200 bar (3000 psig). If the shaft speed is dropping, the steam pressure is increased, and if the speed is rising, the pressure set by the steam pressure controller is lowered.

The output signal of PC-1 controls the heat input into the boiler. When the PC output is 50%, the heat generated by the solar farm is exactly the same as is needed to maintain the desired rate of electric power generation on the set point (Mode A). As solar energy is becoming insufficient and needs to be supplemented by hot oil from storage (Mode B), the PC-1 output rises above 50%, and the hot oil pumping station, VP-3, is started. As the output of PC-1 rises, the pumping speed increases, and the amount of hot oil delivered from storage is increased.

Conversely, when the available solar energy is in excess (Mode E) of the energy needed to maintain the desired rate of power generation, the output signal of PC-1 drops below 50%, and the three-way throttling valves (PCV-

1A and PCV-1B) start to send some of the heat collected by the solar farm to storage. Thereby, as the PC-1 output drops from 50 to 0%, more and more of the solar hot oil is sent to storage.

In this split range operation, when the collected solar energy is insufficient (Mode B), it is supplemented from storage by hot oil energy (PC output 50 to 100%), and when the solar energy is in excess of what is needed (Mode E), the hot oil excess is sent to storage (PC output 0 to 50%). In Mode C, the boiler is off (or is supplied completely by geothermal energy), and all the collected solar energy is sent to the hot oil tanks for storage, whereas in Mode D (night operation), no solar energy is available, and the heat to the boiler is supplied completely from storage.

4.2.4.3 Boiler Feedwater Optimization

The boiler is provided with a three-element feedwater control system, which was described in detail in Chapter 2, Section 2.2, in connection with Figure 2.1. The boiler is also provided with a variable-speed feedwater pump station, having the controls described in Chapter 2, Section 2.17.2 (see Figures 2.123 and 2.124).

The condensing turbine efficiency is maximized by minimizing the turbine exhaust pressure. This is achieved by maximized cooling in the condenser (CO-1) by the condensate temperature controller, TC-10, which controls the speed of the variable-speed cooling tower water-pumping station. The control system details for the pumping station can be found in Section 2.17.2, and the controls for optimizing the steam turbine system are described in detail in Section 2.19.2 (see Figure 2.136).

The optimization of the cooling tower operation is achieved by ΔTC-9, minimizing the energy cost for circulating the cooling water and air in the cooling tower. The details of the required control system that guarantee that the minimum amounts of pumping and fan energy will be used are described in Section 2.4.3 (see Figure 2.16).

The collected condensate is supplemented by fresh feedwater under level control (LC-11) and is sent to the boiler feedwater preheater (HE-1).

4.2.4.4 Geothermal Heat Pump Optimization

In this demonstration plant, the boiler feedwater can be preheated by geothermal heat if available. Depending on the availability and temperature of the geothermal heat in the area, the geothermal plant's design would vary. If the groundwater temperature exceeds 150°C (302°F), a "flash steam" power plant would be used, and if it is between 100 and 150°C (212 and 302°F), a "binary cycle" power plant would be used.

The system shown in Figure 4.1 (gatefold) assumes that the groundwater temperature at the site is less than 100°C, and therefore, only geothermal heat pumps (GHPs) can be used. The GHP operates the same way as domestic refrigerators, but instead of removing the heat from the inside of the refrigerator and pumping it outside, it takes the heat from the groundwater and moves it to the feedwater of the solar boiler.

In this configuration, the groundwater is pumped by a variable-speed pumping station that is optimized in accordance with the design described in Chapter 2, Section 2.17.2 (see Figures 2.123 and 2.124). The pumping rate is maximized under level control (LC-8) to match the maximum capacity of the well. (If a lake is used as the heat source, this control is not needed, and the flow rate can be set by the maximum capacity of the pumping station or the capacity of the evaporators.) The groundwater heat is removed in the evaporator (EV-1), and the cooled water is returned into the ground. This removed heat serves to evaporate the working fluid in the heat pump. These working fluid vapors are compressed in the compressor (CP-1) to a pressure at which they will condense in the boiler feedwater preheater (CO-2). After the working fluid of the heat pump has transferred the heat to the feedwater, it is condensed at the higher pressure of the compressor discharge.

Some of the compressor energy that was introduced by CP-1 is recovered in the expander turbine generator (ETG-1) as the working fluid pressure is reduced from that of the condenser (CO-2) to that of the evaporator (EV-1). The compressor station operation is optimized under differential temperature control (ΔTC-7) by the control system described in detail in Section 2.5.3 (see Figures 2.20 and 2.22). The electricity generated by ETG-1 is used to lower the total power consumption of the plant.

4.2.4.5 Electrolyzer Optimization

Electrolyzers operate at around 66% efficiency. This efficiency is defined as the ratio of the energy content of the generated H_2 divided by the electrical energy required to produce that H_2. The purpose of the electrolyzer station is to store the solar energy in the chemical energy of H_2, which later can be transported to users or reconverted into electricity, if that is more profitable. It takes the same amount of energy to split water into H_2 and oxygen (O_2) as that obtained when oxidizing H_2 into water.

The only difference between the two operations (fuel cells and electrolyzers) is that electrolysis increases the entropy; therefore, not all the energy needs to be supplied in the form of electricity, as the environment contributes 48.7 kJ of thermal energy. Therefore, during electrolysis, waste heat can be provided to furnish the required thermal energy.

When it is more profitable to use the electricity to make H_2, the electrolyzers split the feedwater into H_2 and O_2. The H_2 is collected at about 3 bar (45 psig) of pressure, and is either liquefied and sent to LH_2 storage, or compressed to a high pressure, which can be up to 1,000 bar (15,000 psig), and sent to high-pressure gas storage.

If it is available by the time this demonstration power plant is built, the reversible (dual function) fuel cell (RFC) will be used (Figure 4.2), which in its reverse mode operates as an electrolyzer. This device and its controls are described in detail in Section 2.12.5. When it is more profitable to produce electricity from the stored H_2, the operation of these units will be reversed, and they will be switched into their fuel cell mode. In this mode, H_2 will be used as the fuel to generate electricity for the grid.

In the case of the fuel cell mode of operation, the oxidation of 1 mol of H_2 will generate 237.1 kJ of electrical energy and 48.7 kJ of thermal energy (heat). This waste heat can be used for heating buildings or for preheating the boiler feedwater. (The fuel cell mode of operation is not shown in Figure 4.1 [gatefold].)

If the liquid electrolyte design is selected for the electrolyzer, the optimization controls in Figure 4.1 (gatefold) include the electrolyte balancing controls based on the valve position control (VPC-32) of the variable-speed pumping station (VP-6). These controls are the same as those described for VP-1 and elaborated on in Chapter 2, Section 2.17.2. The power distribution controller (PoC-15) serves to control the electric power sent to the electrodes of the electrolyzer, and the pressure controller PC-14 serves to maintain the H_2 pressure in the distribution header at around 3 bar (45 psig).

A new concept is proposed to be used for balancing the O_2 and H_2 pressures on the two sides of the electrolyzer's separation diaphragm. This control strategy will result in a substantial reduction in the d/p between the cathode and anode chambers, and therefore, will allow a reduction in the thickness of the separation diaphragm and in the overall size, weight, and cost of the RFC unit.

The proposed optimization strategy will replace the traditional method of controlling the release of O_2. Today, the rate of O_2 released is controlled to maintain the d/p between the electrolyte chambers in order to limit the force that the separation diaphragm has to withstand. When the pressure differential is detected and controlled by conventional d/p cells, the measurement cannot be sensitive or accurate; therefore, the diaphragm has to be strong, and the electrolyzer (or fuel cell) must be bulky and heavy. In this optimized design (if a liquid electrolyte design is selected), differential level control (ΔLC-12) will be used, which can control minute differentials.

The solar–hydrogen power plant design shown in Figure 4.1 (gatefold) assumes that the RFCs will use liquid electrolytes. This is just an assumption based on the fact that on this large-scale proton-emitting solid membrane designs are not yet available. If the RFC becomes reality and its development is completed by the time this plant is built, it is possible that the liquid electrolyte will be replaced by a solid polymer.

It should be noted that much progress is being made in the use of catalysts in fuel cells and electrolyzers. In both the cost and supply of fuel cells, the need for platinum catalysts used to be a serious limitation because platinum is expensive and its availability is limited. This limitation is being solved by the development of nanocarbon catalysts. Similarly, in the area of electrolyzer efficiency the new NanoNi catalysts promise a more than doubling of the H_2 output of electrolyzers by drastically increasing the electrode surfaces (QuantumSphere, Inc.).

4.2.4.6 H_2 Liquefier Optimization

The H_2 generated by the electrolyzers can either be sent to gas or liquid storage, or used as a feed stream to a methanol generator station or to any combinations of them. The total rate of H_2 generation is measured by FT-04, and the flow is distributed by flow ratio controllers (FFC-20, 30, and 33) among these destinations. The ratio settings of these controllers are received from the plant optimizer, which assigns these settings on the basis of market conditions and profitability.

The control and operation of H_2 liquefiers is an old and well-established technology, which will not be discussed in detail. Instead, the emphasis will be placed on its optimization through the recovery of compressor energy in expander turbines and through optimized compressor and temperature controls.

The higher heating value (HHV = maximum heat of combustion) of H_2 is 39.3 kWh/kg (141,500 kJ/kg = 61,000 Btu/lb). The energy consumption of liquefier plants ranges from 10 to 20 kWh/kg of LH_2 produced. Some estimates suggest that with optimization of the process, this energy consumption could be reduced to 5 to 10 kWh/kg. (It is hoped that the demonstration plant described here will achieve this goal.)

Theoretically, electrolysis of a kilogram of H_2 requires 32.9 kWh/kg which, at 66% efficiency, corresponds to about 50 kWh/kg. Assuming that liquefaction, transportation, and storage take another 16 kWh/kg, the total energy cost of producing a kilogram of LH_2 is 66 kWh/kg. Therefore, if the average rate of electricity generation of the demonstration plant during the year is 1 gW, and if all of that energy is devoted to the production of LH_2, the plant will produce LH_2 at about 15 tons/h.

The liquefaction process uses a number of heat pumps in series to reduce the H_2 temperature from ambient to the condensation temperature of H_2 (20.3°K, −253°C, −423.2°F, or 36.5°R). In Figure 4.1 (gatefold) the precooling exchanger, which is usually cooled by external refrigeration, is not shown. There are several pipes running through these exchangers, because H_2 is also used as the working fluid (refrigerant) in a number of heat pumps. The actual liquefier plant will use many more heat exchangers in series, but for the purposes of describing the control and optimization system, two is enough. The path of the H_2 stream being cooled through the exchangers, HE-1 and HE-2, is shown by heavy lines, whereas the heat-pumping-related paths are shown by light lines.

The evaporation of liquid nitrogen (LN_2) in HE-1 brings down the H_2 temperature below its inversion point. In the process of liquefaction, one must consider this inversion point temperature (−5°C, or −95°F), because the behavior of H_2 changes (reverses) at this point. Below the inversion temperature, when the pressure is reduced, the temperature of H_2 will drop. Above that temperature the opposite occurs: a drop in pressure causes a rise in temperature.

In Figure 4.1 (gatefold), H_2 is cooled to below its inversion temperature in HE-1 using LN_2. The discharge temperature of the H_2 leaving HE-1 is controlled to be below −6°C (−97°F) by TC-25. This controller adjusts the set point of LC-24 on the LN_2 accumulator, which controls the LN_2 flow to HE-1.

After the H_2 has been cooled to below its inversion temperature, it enters HE-2 for further cooling. HE-2 serves to condense some of the total H_2 feed (the F23 portion of the total F25). F23 is shown by the heavy line. The cooling required to condense F23 is provided by evaporating some of the LH_2 (F27) in the evaporator pipe in HE-2.

The pressure of the LH_2 product is "let down" by PC-29, and the resulting vapor–liquid mixture is returned into HE-2 for recondensation. The temperature of the LH_2 product at the outlet of HE-2 is controlled by a similar cascade loop as was shown for HE-1, where TC-26 keeps the LH_2 temperature under −253°C by a split range control configuration. This loop operates by increasing the LH_2 feed to the accumulator by adjusting the set point of LC-27 until the maximum level is reached, and if that is not sufficient, starts cutting back on the product flow (F26) by throttling TCV-26 and thereby reducing the rate of product flow through HE-2.

In this heat pump configuration, both the refrigerant and the process fluid are H_2. After the flash compressor (FCP-2) compresses the GH_2, part of it (F27) is condensed (as it passes through HE-2) and is sent as the refrigerant liquid into the LH_2 accumulator. The LH_2 liquid from the accumulator (at temperature T33) travels as the working fluid (refrigerant) into the evaporator path of HE-2. When evaporating, it removes its heat of evaporation from the H_2 product stream (F23) that is being cooled and leaves HE-2 as a low-pressure flash vapor stream at temperature T30 to the suction of the flash compressor. The flash vapors are recompressed by FCP-2, and at the higher compressor discharge pressure are recondensed; at a temperature of T34, they return as the LH_2 feed to the accumulator. Thus, the heat pump removes the heat from the lower-temperature H_2 product stream (in the evaporator portion of HE-2) and moves it to a higher temperature in the condenser portion.

If the energy of compression is not recovered in the heat pumps, the liquefaction efficiency will be low (35–60%). If the letdown valves are replaced by turboexpanders (ETG-2 and ETG-3), which will recover some of the compression energy during pressure letdown, and if helium or neon refrigerants are used, the liquefaction efficiency can theoretically reach 80% (see Table 1.46).

Controlling the two compressors (FCP-2 and RPC-3), which are operating in parallel, requires the use of load distribution controls. Section 2.5.4 (Chapter 2) describes how two compressors can be proportionally loaded and unloaded, while keeping their operating points at equal distance from their surge curves (see Figures 2.19 and 2.22).

The load distribution can be computer-optimized by calculating compressor efficiencies (in units of flow per unit power) and loading the compressors in their order of efficiencies. The pressure controller (PC-22) directly sets the set points of SC-21 and SC-23, whereas the balancing controllers (FFC-22 and FFC-24) slowly bias those settings as they follow the total H_2 generation of the electrolyzers (FT-4). The flow ratio controllers (FFCs) are also protected from reset windup, as was explained in Section 2.5.4.

4.2.4.7 Methanol Converter System

When the most economical mode of operation is to send the H_2 to the methanol converter, and when CO_2 is available (being captured from some external fossil source such as a coal-burning power plant), each mol of CO_2 will require 3 mol of H_2 to generate 1 mol of methanol (CH_3OH). Therefore, in that mode of operation, the amount of H_2 sent to the methanol converter (FT-30) is ratio-controlled by FFC-30 proportioning the H_2 flow to match the flow of CO_2 (FT-34).

The details of the methanol conversion process and its control are in development. The process itself is described in Dr. George Oláh's book titled *Beyond Oil and Gas: The Methanol Economy*, published by Wiley-VCH, and I understand that it is being developed by Universal Oil Products (UOP), which is also developing related process technologies and joint ventures for demonstration and de facto plants. Therefore, it is not certain, but it is likely that the methanol process could mature by the time the solar–hydrogen demonstration plant is completed, and in that case, it could be made part of the total power plant.

4.2.4.8 Optimized H₂ Compression

H_2 can be stored as a gas compressed to 350–1,000 bar pressure (about 5,000–15,000 psig) or as a cryogenic liquid. The energy requirement of liquefaction is about 15 kWh/kg. The energy cost of compression varies with the discharge pressure, compressor size, and the number of compression stages. Multistage units with intercoolers are more efficient (about 90%), and they consume only 8–12% of the H_2's higher heating value (HHV), or about 4 kWh/kg (Table 1.46). Therefore, when storage space is available and H_2 can be reused to generate electricity in fuel cells (for example, at night), this method of storage is more economical.

The compressor designs used in H_2 services have been described in Chapter 1, Section 1.5.5, and the controls of high-pressure compressors have been covered in Chapter 2, Section 2.5. Therefore, here the discussion will be limited to the optimization aspects of this operation.

In Figure 4.1 (gatefold), for purposes of simplicity, the controls of only one compressor stage are shown. The surge protection, described in detail in Figure 2.19 in Section 2.5.2, is implemented by recirculation controllers (FC-33 and 35). The majority of high-efficiency installations consist of multistage stations with intercoolers between the stages. For the control and optimization of multistage systems or for systems in series, refer to Figure 2.21. In Figure 4.1 (gatefold) only two compressors are shown, operating in parallel. In the actual compressor station there might be a number of parallel compressor units, and each unit might consist of more than one compressor stage or more than one compressor in series.

The compressor loading in Figure 4.1 (gatefold) is determined by the central plant optimizer, which determines the percentage of the H_2 production that is to be compressed, and sets the flow ratio controllers (FFC-33 and FFC-35). The discharge pressure from the parallel compressors (CP-2 and CP-3) is controlled by PC-40, which adjusts the set points of the speed controllers (SC-36 and SC-37).

When charging gas into several tanks or serving other users, the workload on the compressors is minimized by minimizing the pressure drop across the distribution valves. Therefore, the valve position controller (VPC-41) keeps opening all distribution valves by lowering the discharge pressure (PC-40) until the most open valve reaches nearly full opening (90%).

The individual distribution valves are under pressure control (PC-42 to PC-44). Knowing the volumes of the H_2 storage tanks and the pressures in each (P18, P19, P20, etc.), the central optimizer terminates charging when a tank is full, or advises the operator concerning the additional amount of H_2 a particular vessel can receive.

5

Conclusions

With this book I hope I have convinced the reader that it is time to stop writing articles and holding debates on what to do about the energy crisis. The time has arrived to stop talking and to start building so that we can establish the facts concerning the costs and feasibility of renewable energy plants.

In this book I have described the steps needed to convert our present exhaustible, unsafe, and polluting energy economy into an inexhaustible, clean one and accomplish that by the end of this century.

In Figure 5.1 a "bird's-eye view" is provided of the proposed solar–hydrogen power plant that makes solar energy continuously available both in the form of electricity and in the form of fuel. This transition from todayís exhaustible energy economy to the clean, free and inexhaustible energy economy of the future can take place without causing any interruptions just by not spending our economic resources on drilling and building more nuclear plants but instead building only renewable energy power plants. If this plant is built close to a fossil power plant, it can also convert the carbon dioxide emission into fuel (methanol).

5.1 Comparing Energy Options

In terms of carbon emission and its impact on climate change, the beneficial effect of my solar–hydrogen power plant is the equivalent of planting 10,000 acres of rain forests. Once this plant is built, its fuel is free, its emission is distilled water and its products are both fuel (hydrogen) and electricity.

In Table 5.2 I have compared the features of traditional fossil and nuclear power plants with those of this solar–hydrogen power plant of the future. This comparison shows that BAU (business as usual) technology is inferior even in the short run. This data shows that there is no clean fossil or safe nuclear power and that mankind must fully convert to a renewable energy economy, in the shortest possible time, but definitely not later than the end of this century.

Unfortunately, the advertisements of the BAU interest groups and leaders of both political parties still talk of "clean coal" and "safe nuclear" energy (as if such existed) instead of standing up to the powerful business interests for whom maintaining the "status quo" is profitable. A majority of voters tolerate this, because to them BAU vs. renewable is a left–right issue and renew-

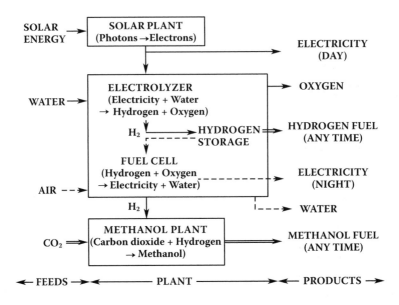

FIGURE 5.1

My proposed solar–hydrogen power plant uses solar energy and water to continuously produce both electricity and hydrogen fuel. If, during the transition period conventional power plants are also used and built close to a fossil power plant (in the future), it will probably be able to convert the CO_2 from the fossil plant to methanol fuel.

able energy is seen as supported only by the political left. BAU corporations pay lip service to having renewable energy in some distant future, while still drilling, polluting and implying that energy wars are both patriotic and unavoidable, because they prevent the interruption of our energy supplies.

Although all fossil fuels are exhaustible, their rate of discovery is already below their rate of consumption (Figure 1.8) and despite the rising cost China is building a dirty coal–fired power plant every week. According to the 2007 UN Intergovernmental Panel on Climate Change, the CO_2 content of the atmosphere has NEVER been as high as it is today (http://en.wikipedia. org/wiki/Carbon_dioxide_in_the_Earth's_atmosphere), and if the present trends continue a "new planet" will result, one that is not only warmer, but has a 20% reduced land area because of the flooding of coastal regions. In short, not only the "left" of the planet is warming!

5.2 Nuclear Option

Some believe that a nuclear energy future is a possible option to follow the age of the fossil fuels. Many believe that because it does not emit carbon, nuclear power can save us from climate disaster. These people do not realize that uranium-235 (the only naturally occurring fissionable material) is exhaustible. The total amount of U_{235} is from 2.8 to 17 million tons, which is a function of its price (at higher prices the mining of low concentrations also becomes economical) and the potential of obtaining more by dismantling nuclear weapons). Knowing that the consumption rate today is 80,000 tons, we can calculate its R/P ratio as about 70 years. In addition to being exhaustible, it is unsafe, expensive and when the useful life of the plant is over, nuclear power plants are also expensive to dismantle.

In the past only two nuclear accidents (Three Mile Island and Chernobyl) were widely reported, while over 100 went unreported. These other accident were not caused by only earthquakes, design errors or terrorist acts, but more recently also by software virus attacks through the Internet. For example, on January 25, 2003 a Slammer worm penetrated the private computer network of Ohio's Davis-Besse nuclear power plant, and stopped its control computer. The only reason a meltdown did not result is because the plant was not in operation.

Another major safety concern is waste disposal. Today, there is no safe nuclear waste depository anywhere on the planet. Therefore, used fuel rods are stored locally near each plant in steel casks. These casks can be destroyed by conventional weapons and if that happens, they can release radioactive gases.

5.3 Breeder Reactors

Shortages of oil and coal will be followed by one of uranium. The nuclear industry knows that the fuel of today's thermal nuclear reactors (U_{235}) is exhaustible and therefore in a few decades they plan to shift to breeder reactors. They say little to the public, except that this conversion would make nuclear power inexhaustible. This is true, because the conventional "slow neutron" thermal reactors are "once through" (in the sense that they consume their uranium fuel), while fast neutron breeder reactors make more fuel than they use.

In breeder reactors, uranium-238 (a non-fissionable material) is bombarded by fast neutrons and, as a result, fissionable plutonium-239 is produced. When Pu_{239} is bombarded by fast neutrons, it produces 2–3 neutrons per fission, hence the name "breeder." Breeder reactors are much more dangerous than regular thermal ones because the fuel of regular reactors is of low concentration and therefore not directly usable in nuclear weapons while Pu_{239} is. Therefore, if breeder reactors are ever used, the "safety margin" between the "peaceful" use of nuclear technology and nuclear weapons will disappear. Today's nuclear reactors have this "safety margin" because terrorists cannot use their fuel to build "dirty bombs," and because the concentration of uranium-235 in their fuel is only 3% while nuclear weapons require 90%. To obtain this 90% concentration, the use of sophisticated centrifuges is needed.

TABLE 5.2

Comparing the Costs and Other Features of 1,000 mW Fossil, Nuclear, and Solar Power Plants

Electric Power Plant Types	Fossil	Nuclear	Solar–Hydrogen[1]
First cost[2] (in 2007 $billions)	Conventional: $2b Advanced: $5b	$5b–$6b[3]	Grid connected: $3b With H$_2$ storage: $4b
Time to build (Years)	Conventional: 5 Advanced: 10	10–20	Grid connected: 3 H$_2$ With storage: 4
Fuel cost per 1,000,000 Btus of energy in 2007	Coal: $6/mBtu Natural Gas: $10/mBtu Oil: $30/mBtu	$3/mBtu[4]	Fuel is free
Fuel cost increase during the last decade	Coal: $25–$140/ton Natural Gas: $2.50–$10 per 1000 ft^3 Oil: $15–over $100/barrel	Uranium: $10–$75/kg	None
Reserve/production[5] (R/P ratio in years)	Oil: ~40 Natural Gas: ~50 U.S. Coal: ~100[6]	70	Unlimited
Electricity[7] (Wholesale[8] cost per kWh in 2007)	Pulverized Coal: 5.7¢ Coal Gas: 6.6¢ Natural Gas: 7.3¢	6.4¢	Grid connected: 12¢[9] H$_2$ storage: 14¢
Other Factors			
Emissions	Carbon dioxide, NOx, SOx	Nuclear radiation	O$_2$ and distilled water
Safety concerns	Unsafe mines	Terrorism, earthquakes, melt-down, waste disposal	None
Causes global warming	Yes	No	No

1 The DOE Hydrogen Program of the United States has made significant progress since its inception is 2003. By 2007 it reduced the cost of producing a "gallon gasoline equivalent (gge)" quantity of hydrogen (which is about 1 kg) from $5.00/gge in 2003 to $3.00/gge in 2007, when made from natural gas and has a goal of reducing it to $2.00/gge by 2015. Considering that this was achieved by using "pay fuel" (natural gas), it is safe to assume that the cost will be further reduced when it is made from water using "free" solar energy.

2 The power plant construction cost index (PCCI) tracks the costs of building coal, gas, wind, and nuclear power plants, indexed to the year 2000 — registered 231 index points in the third quarter of 2007 ending in October — indicating that a power plant that in 2000 cost $1 billion in 2007 would have cost $2.31 billion.

3 On March 12, 2008, testifying before the House Select Committee on Energy, Sharon Squassoni of Carnegie Endowment estimated this cost at between $5,000/kWe to $6,000/kWe.

4 One kilogram of fuel provides 360 mWh of electrical energy (1.22 gBtu). To obtain 1 kilo of fuel requires 9 kilos of uranium. The cost of conversion, enrichment, and fuel fabrication results in a total cost of about $4,000/kg. Therefore, the fuel cost is 1.1¢/kWh or $3.22/mBtu.

5 The time when the demand exceeds supply arrives much sooner than the R/P ratio suggests. The R/P values given are based on the rate of use in 2007. As consumption rises because of increased demand in the third world, the R/P ratio numbers are likely to drop further.

6 Estimated by the American Academy of Science.

7 According to the *New York Times* (11/7/07), the actual average "wholesale" costs of one kWh of electricity in 2007 in the United States was as follows: Pulverized Coal - 5.7¢, Nuclear - 5.7¢, Coal Gas - 6.6¢, Natural Gas - 7.3¢, Wind - 9.6¢, Biomass - 10.7¢, Solar Thermal - 12.0¢. These numbers do not yet reflect the coming carbon charges, nor do they reflect transportation and distribution costs required to deliver the electricity to the residences.

8 The "wholesale" costs of electricity are approximately half of what the average homeowner pays because transportation, distribution, and other charges and profits are added. USA average residential end-user cost in 2007 was 10.7¢/kWh and in peak periods it ranged between 15 and 25¢/kWh.

9 The 12¢/kwh electricity cost is based on $3,000/kW solar collector cost, 25 years of equipment life expectancy and 5% interest on investment. This price does not reflect that most of the electricity, is generated during peak periods, when the peak prices are often twice as high as the yearly average. This price does not consider either the drop in collector prices that will result from design advancements (nanosolar, etc.) and mass production.

It is this requirement that provides the "safety margin" for today's plants, and it is this protection that disappears when breeders are used because breeder technology utilizes weapons-grade fuel.

Mankind is already on the slippery slope of breeder technology. Russia is the leader in breeder nuclear power generation (BN-600). India also has an operating fast breeder test reactor (FBTR), is building a 500 mWe unit at Kalpakkan and is planning four more by 2020. India works in cooperation with the California-based American firm Dauvergne Brothers Inc. (DBI). China is near completing an experimental unit (CEFR) and is planning to build a full-sized one by 2020. Japan has a small unit (Joyo) in operation, is planning to reopen its 300 mWe Manju reactor and is building a 1,500 mWe JSFR unit. France's 233 mWe Phoenix unit will probably end its operation in 2014. In the United States the founding of the Clinch River Breeder Reactor was terminated; and the Idaho National Laboratory operated a unit named EBR-I but that unit suffered a partial meltdown. Its EBR-II test unit is functioning.

I have devoted more space to explaining the dangers of nuclear power than to the consequences of using fossil fuels, because while the consequences of carbon emission are well understood, the inexhaustible nature of thermal power and the implications of terrorists using breeder reactor fuel for military purposes are largely unknown.

5.4 Costs of Converting to Renewable Energy

Understanding the economic impacts of business as usual (BAU) versus the costs of conversion to a clean and renewable energy economy is necessary for intelligent decision making. The cost of transformation to renewable energy was discussed in Section 1.7.1. These cost estimates vary with the time period allotted for completing the transformation. The cost models also change with the assumed discovery rate of new inventions, the cost reductions expected from mass production and also with the population growth and global lifestyle assumed to exist at the end of that period.

The various future models differ because they assume varying stabilized global energy consumptions at the end of this century. These assumptions range anywhere from 650 to over 1,000 Q (Q = 10^{15} btu). What we do know as a fact is that based on today's lifestyle and today's global population of 6.8 billion, energy consumption is nearing 500 Q. We also know that if the population doubles by 2100 and the global lifestyle rises to that of the United States today, mankind would require about 2,500 Q. While in Figure 5.3 I have assumed this energy target to be 1,000 Q, the reader can calculate the costs for any other level of energy consumption by following the logic of the calculations described in Section 1.7.1.

In addition to lifestyle and population, the cost of transition will also be influenced by the likely cost reductions due to mass production and innovations, similarly to the trend that the computer industry experienced. The innovations already on the horizon include my reversible fuel cell (RFC), inexpensive solar collectors using nanosolar technology, solar roof shingles and inexpensive new catalysts being developed for both electrolyzers and fuel cells (FC). One such invention is the Tokyo Institute of Technology's carbon nanosphere catalyst that reduces FC cost (compared to platinum-based FC) by an order of magnitude. This would make 100 to 200 kW automotive and 1 to 2 kW residential RFCs (used to make our homes "energy free" while recharging our electric cars at night) economically feasible.

Probably the most important cost factor is the economic expansion that this transformation will cause as solar shingles are installed on our buildings, solar–hydrogen power plants are built and a hydrogen infrastructure is established. Transformation to a permanent energy economy will create the economic boom of the century, the GWP (gross world product) will rise and, therefore, our cost estimates in Section 1.7.1 will improve. For all these reasons I believe that starting the transformation right now is economically feasible, but I do not suggest that anybody should take my word for this! What I believe makes no difference! What matters is building the world's first solar–hydrogen power plant to obtain hard and undeniable facts so that the debate will finally be ended and the conversion start.

5.4 Costs of Inaction

Sir Nicholas Stern, former vice president and chief economist at the World Bank, estimated that by 2020 the cost of continued reliance on fossil and nuclear fuels will reach 20% of the gross world product. The GWP today is approaching $50 trillion. This estimate of inaction costing $10 trillion/yr does not even include high energy costs and the resulting recessions or "rescue plans" that governments will waste our tax dollars on instead of spending those $ trillions on conversion to renewable energy. In addition to Sir Stern's 20% figure, inaction will also result in social, environmental and "energy war" costs. (Just the cost of the Iraq war is nearly $3 trillion while the U.S. GDP is around $15 trillion.)

There is another way to look at the cost of inaction: In June 2008 the International Energy Agency (IEA) estimated that $45 trillion would have to be invested (between now and 2050) just to maintain the status quo. In other words, this is how much it will cost to build the needed fossil and nuclear power plants to prevent energy shortages and greenhouse gas emissions from slowing global economic growth. IEA also estimates that if we want to cut the carbon emissions in half by 2050, using existing technology, 32 nuclear and 50 carbon-capturing power plants will need to be built annually.

In addition, inaction in converting to a renewable energy economy also means that our economic resources will be wasted on useless projects. For example, in the United States the ban on offshore drilling was lifted. The 11 billion barrel offshore reserves and the 7 billion barrel reserves in the Arctic National Wildlife Reserve will meet American oil consumption for about 2.5 years. To obtain this oil will take 10 years and cost $0.4 billion dollars (not counting environmental costs). If the same $0.4 trillion were invested to build over 100 solar–hydrogen power plants in the Mojave Desert, the collected energy would equal the total energy content of these reserves in 40 years and it would not run out in 2.5 years but be available forever. (One can make the same calculation with other wasted costs including oil imports, wars, bailout packages and other economic incentives, etc.)

Therefore, business as usual is irresponsible not only because it does not provide a permanent solution (it uses exhaustible fossil and nuclear fuels), but also because investments in them block the financing of the conversion to a permanent energy solution.

5.5 The Road to a Sustainable Future

For the first time in human history, the future of mankind is in our own hands. Onto the road leading to a sustainable and peaceful future, we ourselves have placed several "roadside bombs." We must safely pass these. We must avoid energy wars, the collapse of the global economy, creating a drastic change in the climate or bringing a nuclear winter.

As shown in Figure 5.3, the carbon dioxide concentration of the atmosphere, which is already the highest level ever measured, is rising at the staggering rate of 3.5%/year. The population of the planet is increasing at a yearly rate of 2.2%, adding a billion people to our population every dozen years. In addition, people in the less developed areas of the planet justifiably expect an improvement in their living conditions. All this necessitates the generation of more and more energy.

Models predicting the future differ both in the time period assumed for the conversion and the per capita energy need at the end of the transition period. Yet, almost all models agree that fossil and nuclear energy deposits are exhaustible (as shown in Figure 5.3) and must eventually be replaced by renewable energy. Only the advocates of nuclear breeder reactors disagree and suggest that mankind could meet its energy needs by using weapons-grade plutonium as reactor fuel.

The models disagree on the time we still have to convert to renewable energy. They disagree because they represent different views (some self-serving) on how high energy costs can rise before the global economy collapses, how much climate change we can live with and when competition for the remaining fossil resources will trigger a nuclear war?

It seems to me that since renewable energy power plants can be built faster and with less investment than nuclear or fossil ones, from now on we

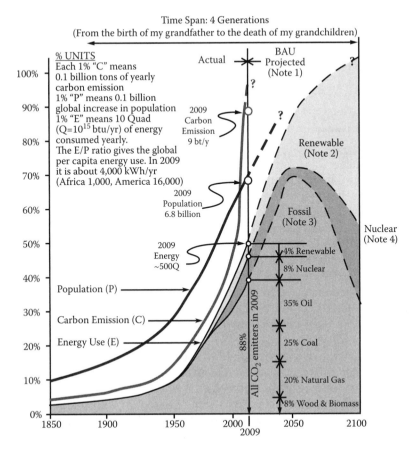

FIGURE 5.3

(See color insert following page 140.) Global energy trends showing actual data up to 2009 and projections after that. The projection represents a "business as usual" case which assumes that the transition to renewable energy will be very slow and during that period no energy wars, climate or nuclear disasters will occur. (Note 1): If this BAU scenario is followed, it is likely that 20% of the land area of the planet will be flooded and civilization as we know it will end. (Note 2): If the equivalent of 1–2% of the GWP was invested annually in the conversion to using renewable energy and if no new fossil or nuclear plants were built, by 2100 the transition to a free, clean, safe and inexhaustible energy economy could be completed. (Note 3): All fossil fuels are exhaustible and carbon emitting. There is no such thing as "clean coal." Continued investments in these technologies take funds away from converting to renewable energy. (Note 4): No new nuclear power plants should be built because none are safe, their waste disposal is unresolved, uranium is exhaustible and breeder reactors use weapons-grade plutonium fuel which is readily usable for building "dirty" nuclear bombs.

should build only renewable ones. I also believe that we should stop using our resources to drill for the remaining drops of oil or build longer and longer pipelines for exhaustible natural gas, but should research to find a safe way of storing nuclear wastes or to develop breeder reactors, and certainly we should not waste our resources to prepare for energy wars. Instead, we should invest in building the infrastructure for mankind's renewable energy future. Once the RFCs are mass produced, once the solar–hydrogen power plant described in this book is built, the cost of the generated electricity and hydrogen will no longer be debatable, because opinions will be replaced by facts.

5.6 Dealing with Facts

From the Sun, in less than an hour we receive all the energy we use in a year. We receive a similar amount of energy from the Moon in the form of tides. The rotation of the globe and thermal differences create ocean currents greater than all the rivers combined. Geothermal energy is nearly unlimited and in all of these cases the fuels are free. Therefore, conversion to a renewable energy economy is a completely realistic goal.

The fuel used to make solar hydrogen is free and unlimited, the raw material for hydrogen is water and the emission when burning the hydrogen (in fuel cells, internal combustion engines or power plants) is distilled water. The cost of building the solar–hydrogen plants will only be firmly known once the demonstration power plant described in this book is built. Later, this cost will drop as mass production and inventions (nanotechnology and reversible fuel cells) increase cost effectiveness.

I hope that after reading this book the reader will understand that we already have the technological know-how needed for conversion to a solar–hydrogen economy and that we can handle the cost of this transition. I believe that the time for debates is over, that the time for building has arrived, but what I believe is irrelevant! What is relevant is to stop dealing with opinions and to start making decisions based on facts.

We do not need more conferences, debates or articles! What we need is to prove that the technology for moving into the age of renewable energy exists, safe and economical! To prove that, we need not only solar roofs and solar–hydrogen demonstration plants, we also need statesmen. We need leaders who are not worried about industrial lobbies and reelection, but who care about the planet and the future of our grandchildren. We need leaders who are dedicated to lead the third industrial revolution. To do this requires vision and commitment, but so did the landing on the Moon. The United States has been the leader of the world before and she can and should now lead a global Marshall Plan to rebuild the planet.

It is debatable how much time we have or how much climate change we can live with. It is also debatable how much of our economic resources should be devoted to stabilizing and reversing mankind's growing carbon footprint. What is not debatable is that eventually we have to do it and we must not give reason to our grandchildren to ask: "Why did you not act in time?"

Appendix: Abbreviations

A

A: Analyzer (O_2, H_2O, H_2, combustibles)
AC: Accumulator
AFC: Alkaline fuel cell
ASPO: Association for the Study of Peak Oil and Gas

B

Boiler
BAU: Business as usual
BTU or btu: British thermal unit
BV: Balancing valve

C

C: Closed
CCS: Clean carbon system
CH$_3$OH: Methanol
ChV: Check valve
CIGS: Copper indium gallium selenide
CO: Condenser or compressor
CO$_2$: Carbon dioxide
CRBJT: Combined reverse Brayton Joule Thomson
CV: Control valve
CWR: Cooling water return
CWS: Cooling water supply

D

D: Deaerator
D/A: Direct acting
DARPA: Defense Advanced Research Project Agency
DMFC: Direct methanol fuel cell
DOE: U.S. Department of Energy
DR: Drain
DSG: Direct steam generation
DV: Diverting valve
DW: Distilled or demineralized water
D&P: Dryer and purifier

E

ECD: Energy conversion device
Eeg: Electric energy generation
EERE: Energy efficiency and renewable energy
Eg: Geothermal energy
EGS: Enhanced geothermal systems
EIA: Energy Information Agency
EPA: Environmental Protection Agency
Es: Insolation energy received
ET: Electrolyte tank
ETG: Expansion turbine generator
EV: Evaporator or electric vehicle

F

F: Flow
FC: Flow controller or fuel cell
FCEV: Fuel cellñbased electric vehicle
FCP: Flash compressor
FFC: Flow ratio control
FT: Flow transmitter
FW: Feedwater

G

G or G: Giga (10^9)
GDP: Gross domestic product
GHCB: Great heat conveyor belt
GHP: Geothermal heat pump
GH_2 or GH2: Hydrogen Gas
GN_2 or GN2: Nitrogen Gas
GSFC: Goddard Space Flight Center
gW or GW: Gigawatt
GWP: Gross world product

H

HE: Heat exchanger
HHV: Higher heating value
HP: High pressure
H_2G or H2G: Hydrogen gas

I

IC: Internal combustion
IEA: International Energy Agency

IPCC: International Panel on Climate Change
ISCCS: Integrated solar combined cycle system

K

k or K: Kilo
kg: Kilogram
kJ: Kilo joule
kW or KW: Kilowatt
kWh or KWH: Kilowatt hour

L

L: Level
LC: Level controller
LDV: Let down valve
LHV: Lower heating value
LH$_2$ or LH2: Liquid hydrogen
LN$_2$ or LN2: Liquid nitrogen
LNG: Liquefied natural gas
LSH: High level switch
LP: Low pressure
LT: Level transmitter

M

M: Motor
m or M: Mega
MCFC: Molten carbonate fuel cell
MTBF: Mean time between failures
mW or MW: Megawatts

N

NASA: National Aeronautical and Space Administration
NEI: Nuclear Energy Institute
Nm^3: Normal cubic meter
NREL: National Renewable Energy Laboratory
N$_2$: Nitrogen

O

O: open
O$_2$: Oxygen
OECD: Organisation for Economic Cooperation and Development
ORNL: Oak Ridge National Laboratory

P

P: Pressure
PC: Pressure controller
PCFC: Protonic ceramic fuel cell
PEM: Proton exca
PoC: Power distribution controller
PSV: Pressure safety valve
PT: Pressure transmitter
PU$_{239}$: Plutonium 239

Q

Q: Heat (total or stored)
QbR: Boiler heat input rate
QsR: Solar heat collection rate
QT1: Heat stored in T1
QT2: Heat stored in T2

R

R/A: Reverse acting
RCP: Recycle compressor
RFC: Reversible fuel cell
R/P: Resource to production ratio

S

SC: Speed controller
SEGS: Solar energy generating system
SES: Stirling Energy Systems
SM: Servo motor
SOFC: Solid oxide fuel cell
SP: Set Point
STC: Sun tracking controls

T

T: Temperature
TDC: Temperature difference controller
TuG: Turbine generator
Twb: Wet bulb temperature
T1 & T2: Oil tanks (expansion tanks not shown)

U

U_{235} **and** U_{238}**:** Uranium 235 and 238

V

V: Isolation valve
VP: Variable speed pump

W

W: Watt
Wh: Watt hour
Wp: Peak solar watt rating
WRI: World Resource Institute

Z

ZAFC: Zink air fuel cell
ZEV: Zero emission electric vehicle

x: Multiplier
ΔLC: Level difference controller
ΔPC: Pressure difference controller
ΔTC: Temperature difference controller
ΔTT: Temperature difference transmitter
$\leq$: Low Selector
$\geq$: High Selector

Index